WITHDRAWN

Carnegie Mellon

EMISSIONS OF ATMOSPHERIC TRACE COMPOUNDS

ADVANCES IN GLOBAL CHANGE RESEARCH

VOLUME 18

Editor-in-Chief

Martin Beniston, *Department of Geosciences, University of Fribourg, Perolles, Switzerland*

Editorial Advisory Board

B. Allen-Diaz, *Department ESPM-Ecosystem Sciences, University of California, Berkeley, CA, U.S.A.*

R.S. Bradley, *Department of Geosciences, University of Massachusetts, Amherst, MA, U.S.A.*

W. Cramer, *Department of Global Change and Natural Systems, Potsdam Institute for Climate Impact Research, Potsdam, Germany.*

H.F. Diaz, *Climate Diagnostics Center, Oceanic and Atmospheric Research, NOAA, Boulder, CO, U.S.A.*

S. Erkman, *Institute for Communication and Analysis of Science and Technology – ICAST, Geneva, Switzerland.*

R. García Herrera, *Facultad de Físicas, Universidad Complutense, Madrid, Spain*

M. Lal, *Centre for Atmospheric Sciences, Indian Institute of Technology, New Delhi, India.*

U. Luterbacher, *The Graduate Institute of International Studies, University of Geneva, Geneva, Switzerland.*

I. Noble, *CRC for Greenhouse Accounting and Research School of Biological Sciences, Australian National University, Canberra, Australia.*

L. Tessier, *Institut Mediterranéen d'Ecologie et Paléoécologie, Marseille, France.*

F. Toth, *International Institute for Applied Systems Analysis, Laxenburg, Austria.*

M.M. Verstraete, *Institute for Environment and Sustainability, EC Joint Research Centre, Ispra (VA), Italy.*

EMISSIONS OF ATMOSPHERIC TRACE COMPOUNDS

Edited by

Claire Granier

*Service d'Aéronomie,
Université Paris 6, Paris, France*

*CIRES/NOAA Aeronomy Laboratory,
Boulder CO, U.S.A.*

*Max-Planck-Institute for Meteorology,
Hamburg, Germany*

Paulo Artaxo

*Instituto de Física,
Universidade de São Paulo,
São Paulo, Brazil*

and

Claire E. Reeves

*School of Environmental Sciences,
University of East Anglia, Norwich, United Kingdom*

KLUWER ACADEMIC PUBLISHERS
DORDRECHT / BOSTON / LONDON

A C.I.P. Catalogue record for this book is available from the Library of Congress.

ISBN 1-4020-2166-6 (HB)
ISBN 1-4020-2167-4 (e-book)

Published by Kluwer Academic Publishers,
P.O. Box 17, 3300 AA Dordrecht, The Netherlands.

Sold and distributed in North, Central and South America
by Kluwer Academic Publishers,
101 Philip Drive, Norwell, MA 02061, U.S.A.

In all other countries, sold and distributed
by Kluwer Academic Publishers,
P.O. Box 322, 3300 AH Dordrecht, The Netherlands.

Printed on acid-free paper

All Rights Reserved
© 2004 Kluwer Academic Publishers and copyright holders
as specified on appropriate pages within.
No part of this work may be reproduced, stored in a retrieval system, or transmitted
in any form or by any means, electronic, mechanical, photocopying, microfilming, recording
or otherwise, without written permission from the Publisher, with the exception
of any material supplied specifically for the purpose of being entered
and executed on a computer system, for exclusive use by the purchaser of the work.

Printed in the Netherlands.

Table of contents

List of contributors	VII
Acknowldegments	XI
Atmospheric Composition and Surface Exchanges *Guy P. Brasseur, Will Steffen, and Claire Granier*	1

Compilation of Regional to Global Inventories of
Anthropogenic Emissions
   Carmen M. Benkovitz, Hajime Akimoto, James J. Corbett,
   J. David Mobley, Toshimasa Ohara, Jos G.J. Olivier,
   John A. van Aardenne, and Vigdis Vestreng
	17

Deriving Global Quantitative Estimates for Spatial and
Temporal Distributions of Biomass Burning Emissions
   Catherine Liousse, Meinrat O. Andreae, Paulo Artaxo,
   Paulo. Barbosa, Hélène Cachier, Jean-Marie Grégoire,
   Peter Hobbs, David Lavoué, Florent Mouillot, Joyce Penner,
   Mary Scholes, and Martin Schultz
	71

Global Organic Emissions from Vegetation
   Christine Wiedinmyer, Alex Guenther, Peter Harley,
   Nick Hewitt, Chris Geron, Paulo Artaxo,
   Rainer Steinbrecher, and Rei Rasmussen
	115

Nitrogen Emissions from Soils
   Laurens Ganzeveld, Changsheng Li, Laura Cárdenas,
   Jane Hawkins, and Grant Kirkman
	171

Global Emissions of Mineral Aerosol: Formulation and
Validation Using Satellite Imagery
   Yves Balkanski, Michael Schulz, Tanguy Claquin,
   Cyril Moulin, and Paul Ginoux
	239

Emissions from Volcanoes
   Christiane Textor, Hans-F. Graf, Claudia Timmreck,
   and Alan Robock
	269

A Satellite-based Method for Estimating Global Oceanic
DMS and its Application in a 3-D Atmospheric GCM 305
    Sauveur Belviso, Cyril Moulin, Laurent Bopp,
    Emmanuel Cosme, Elaine Chapman, and Kazushi Aranami

Sea-salt Aerosol Source Functions and Emissions 333
    Michael Schulz, Gerrit de Leeuw, and Yves Balkanski

Use of Isotopes 361
    Valerie Gros, Carl A.M. Brenninkmeijer, Patrick Jöckel,
    Jan Kaiser, Dave Lowry, Euan G. Nisbet, Phillip O'Brien,
    Thomas Röckmann, and Nicola Warwick

Determination of Emissions from Observations
of Atmospheric Compounds 427
    Claire E. Reeves, Derek. M. Cunnold,
    Richard G. Derwent, Edward Dlugokencky,
    Sandrine Edouard, Claire Granier, Richard Ménard,
    Paul Novelli, and David Parrish

Data Assimilation and Inverse Methods 477
    Richard Ménard, Sandrine Édouard, Sander Houweling,
    Gabrielle Pétron, Claire Granier, and Claire Reeves

List of acronyms 517

List of chemical species 523

Index 525

Color plates 527

Copyright acknowledgments 547

List of contributors

HAJIME AKIMOTO, Frontier Research System for Global Change, Yokohama, Japan

MEINRAT O. ANDREAE, Max-Planck-Institute for Chemistry, Mainz, Germany

KAZUSHI ARANAMI, Laboratory of Marine and Atmospheric Geochemistry, Sapporo, Japan

PAULO ARTAXO, Universidade de Sao Paulo, Sao Paulo, Brazil

YVES BALKANSKI, Laboratoire des Sciences du Climat et de l'Environnement, Gif-sur-Yvette, France

PAULO BARBOSA, Joint Research Center, Ispra, Italy

SAUVEUR BELVISO, Laboratoire des Sciences du Climat et de l'Environnement, Gif-sur-Yvette, France

CARMEN M. BENKOVITZ, Brookhaven National Laboratory, Upton, NY, USA

LAURENT BOPP, Max Planck Institute for Biogeochemistry, Iena, Germany

GUY P. BRASSEUR, Max-Planck-Institute for Meteorology, Hamburg, Germany

CARL A.M. BRENNINKMEIJER, Max-Planck-Institute for Chemistry, Mainz, Germany

HELENE CACHIER, Laboratoire des Sciences du Climat et de l'Environnement, Gif-sur-Yvette, France

LAURA CARDENAS, Institute of Grassland and Environmental Research, North Wyke, Okehampton, United Kingdom

ELAINE CHAPMAN, Pacific Northwest National Laboratory, Richland, WA, USA

TANGUY CLAQUIN, Eurobios, Cachan, France

JAMES J. CORBETT, University of Delaware, Newark, USA

EMMANUEL COSME, Laboratoire de Glaciologie et de Géophysique de l'Environnement, St-Martin-d'Hères, France

DEREK M. CUNNOLD, Georgia Institute of Technology, Atlanta, GA, USA

GERRIT DE LEEUW, TNO Physics and Electronics Laboratory, The Hague, The Netherlands

RICHARD G. DERWENT, UK Met Office, Bracknell, United Kingdom

EDWARD DLUGOKENCKY, NOAA Climate Monitoring and Diagnostics Laboratory, Boulder, CO, USA

SANDRINE EDOUARD, Meteorological Service of Canada, Dorval, Canada

LAURENS GANZEVELD, Max-Planck-Institute for Chemistry, Mainz, Germany

CHRIS GERON, U.S. Environmental Protection Agency, Research Triangle Park, NC, U.S.A

Paul GINOUX, Geophysical Fluid Dynamics Laboratory, Princeton, NJ, USA

HANS F. GRAF, Max-Planck-Institute for Meteorology, Hamburg, Germany

CLAIRE GRANIER, Service d'Aéronomie, Paris, France; CIRES/NOAA Aeronomy Laboratory, Boulder, CO, USA; Max-Planck-Institute for Meteorology, Hamburg, Germany

JEAN-MARIE GREGOIRE, Joint Research Center, Ispra, Italy

VALERIE GROS, Max-Planck-Institute for Chemistry, Mainz, Germany

ALEX GUENTHER, National Center for Atmospheric Research, Boulder, CO, USA

PETER HARLEY, National Center for Atmospheric Research, Boulder, CO, USA

JANE HAWKINS, Institute of Grassland and Environmental Research, North Wyke, Okehampton, United Kingdom

NICK HEWITT, University of Lancaster, Lancaster, United Kingdom

PETER HOBBS, University of Washington, Seattle, WA, USA

SANDER HOUWELING, Space Research Organization Netherlands, Utrecht, The Netherlands

PATRICK JÖCKEL, Max-Planck-Institute for Chemistry, Mainz, Germany

JAN KAISER, Princeton University, Princeton, NJ, USA

GRANT KIRKMAN, Max-Planck-Institute for Chemistry, Mainz, Germany

DAVID LAVOUE, Meteorological Service of Canada, Downsview, Ontario, Canada

CHANGSHENG LI, University of New Hampshire, Durham, USA

CATHERINE LIOUSSE, Laboratoire d'Aérologie, Toulouse, France

DAVID LOWRY, Royal Holloway University of London, Egham, United Kingdom

RICHARD MENARD, Meteorological Service of Canada, Dorval, Canada

J. DAVID MOBLEY, US Environmental Protection Agency, Research Triangle Park, NC, USA

FLORENT MOUILLOT, Carnegie Institution of Washington, Stanford, USA

CYRIL MOULIN, Laboratoire des Sciences du Climat et de l'Environnement, Gif-sur-Yvette, France

EUAN G. NISBET, Royal Holloway University of London, Egham, United Kingdom

PAUL NOVELLI, NOAA Climate Monitoring and Diagnostics Laboratory, Boulder, CO, USA

PHILLIP O'BRIEN, National University of Ireland, Galway, Ireland

JOS G.J. OLIVIER, Netherlands Environmental Assessment Agency, Bilthoven, The Netherlands

TOSHIMASA OHARA, Frontier Research System for Global Change, Yokohama, Japan

DAVID PARRISH, NOAA Aeronomy Laboratory, boulder, CO, USA

JOYCE PENNER, University of Michigan, Ann Arbor, MI, USA

GABRIELLE PETRON, Service d'Aéronomie, Paris, France and National Center for Atmospheric Research, Boulder, CO, U.S.A.

REI RASMUSSEN, Oregon Graduate Institute of Science & Technology, Beaverton, OR, U.S.A.

CLAIRE E. REEVES, University of East Anglia, Norwich, United Kingdom

ALAN ROBOCK, Rutgers University, New Brunswick, NJ, USA

THOMAS RÖCKMANN, Max-Planck-Institute für chemistry, Mainz, Germany

MARY SCHOLES, University of the Witwatersrand, Johannesburg, South Africa

MARTIN G. SCHULTZ, Max-Planck-Institute for Meteorology, Hamburg, Germany

MICHAEL SCHULZ, Laboratoire des Sciences du Climat et de l'Environnement, Gif-sur-Yvette, France

WILL STEFFEN, International Geopshere-Biosphere Programme Office, Stockholm, Sweden

RAINER STEINBRECHER, Fraunhofer Institut fur Atmosphärische Umweltforschung, Garmish, Germany

CHRISTIANE TEXTOR, Laboratoire des Sciences du Climat et de l'Environnement, Gif-sur Yvette, France

CLAUDIA TIMMRECK, Max-Planck-Institute for Meteorology, Hamburg, Germany

JOHN A. VAN AARDENNE, Max Planck Institute for Chemistry, Mainz, Germany

VIGDIS VESTRENG, Norwegian Meteorological Institute, Oslo, Norway

NICOLA WARWICK, Center for Atmospheric Sciences, Cambridge, United Kingdom

CHRISTINE WIEDINMYER, National Center for Atmospheric Research, Boulder, CO, USA

Acknowledgements

We would like to thank all contributing authors and reviewers for their dedication, efforts and constructive comments during the development of this book.

The idea for this book was conceived during the workshop "Emissions of Chemical Species and Aerosols into the Atmosphere" which took place in Paris in June 2001. This workshop was partly organized and funded by the European project POET (Precursors of Ozone and their Effects in the Troposphere, EVK2-1999-00011) and was endorsed by the Global Emissions Inventories Activity (GEIA) of the International Global Atmospheric Chemistry (IGAC) project of the International Geosphere-Biosphere Program (IGBP). We gratefully acknowledge the additional financial support for this workshop provided by the CNES (Centre National d'Etudes Spatiales), the Service d'Aéronomie/CNRS (Centre National de la Recherche Scientfique)/ University Paris 6 and the IPSL (Institut Pierre Simon Laplace) in Paris, France, and by the Max Planck Institut for Meteorology in Germany. We would like to thank Michèle Levasseur and Evelyne Quinsac (Service d'Aéronomie) who helped in the preparation of this workshop.

We also give special thanks to Birgit Grodtmann from the Max Planck Institute for Meteorology in Hamburg who formatted the manuscripts and finalized the camera-ready version of this book in a very efficient manner.

Atmospheric Composition and Surface Exchanges

Guy P. Brasseur, Will Steffen, and Claire Granier

1. INTRODUCTION

Atmospheric chemistry is a relatively new scientific discipline. Research in this area has intensified dramatically in the last 3-4 decades because of the concerns that growing emissions of pollutants could have severe consequences for humankind. Although the major chemical compounds present in the atmosphere were identified long ago, it is only during the second half of the 20th century that the complexity of chemical processes was highlighted and carefully investigated. Laboratory studies as well as field campaigns and modelling studies have helped elucidate the mechanisms that, for example, control the oxidizing power of the atmosphere or the formation and fate of aerosol particles. In recent years, research has emphasised the role of atmospheric chemistry in the global environment, and specifically the impact of global air pollution on climate change. The important issue of air quality, initially considered as a local or regional problem, is now regarded as a global problem. Much work has been done to assess the causes and consequences of air pollution (including health effects) in urban areas, especially since it was shown by Haagen-Smit (1952) that the high levels of ozone recorded in Los Angeles were due to the emissions of anthropogenic hydrocarbons and nitrogen oxides. Since the early 1970s, it has been known (Crutzen, 1973) that the same type of chemical processes operates even far away from highly industrialized regions, and that the entire atmosphere must be regarded as a complex chemical reactor. This is particularly true in the tropics, where the intense solar radiation and the large natural and anthropogenic sources for many reactive compounds are responsible for high photochemical activity that determines to a large extent the global chemical properties of the atmosphere.

2. ATMOSPHERIC CHEMISTRY AND THE EARTH SYSTEM

Although present in only trace quantities, reactive species in the atmosphere affect many processes in the Earth System. Water vapour (H_2O), carbon dioxide (CO_2), methane (CH_4), nitrous oxide (N_2O) and several other species contribute directly to the greenhouse effect of the atmosphere, maintaining the Earth's surface temperature about 30°C warmer than it would otherwise be. Ozone in the stratosphere filters out UV-B radiation that would be harmful for many forms of life. Atmospheric transport of particles provides critical nutrients for both terrestrial and marine ecosystems.

Emission and deposition across the land-atmosphere interface are crucial processes for the functioning of the Earth System. Many types of vegetation emit volatile organic compounds, some of which may lead to the formation of secondary organic aerosol particles. These aerosols may then act as cloud condensation nuclei, which can, in turn, affect cloud characteristics and precipitation, both of which have feedbacks on the vegetation. These effects can change the growth of and the emissions from vegetation, leading to the intriguing possibility that the vegetation and the atmosphere are linked in a feedback loop that operates through local and regional climate. Emissions of volatile organic compounds also affect the chemistry of the atmosphere, since these compounds are continually being removed via oxidation to water-soluble compounds. The primary reaction is with the hydroxyl (OH) radical, sometimes called the detergent of the atmosphere due to its importance in oxidizing a range of compounds. Thus, surface emissions from vegetation are also critical in controlling the oxidizing capacity of the atmosphere.

The ocean-atmosphere interface is equally important. Sulphur compounds, specifically dimethylsulfide (DMS) and carbonyl sulfide (COS) produced by marine biota in the surface waters, may also be involved in feedback loops that are important for the functioning of the Earth System. DMS in particular could influence climate via the formation of sulphate aerosol particles, which act as cloud condensation nuclei and thus affect the radiation balance. This in turn influences the primary production of the marine biota and hence the production of DMS, closing the feedback loop.

In the last century or so anthropogenic emissions of trace gases and aerosol particles have risen sharply and are now having a significant effect on atmospheric chemistry, with consequences for the functioning of the Earth System. For example, sulphur dioxide (SO_2), released primarily as a result of coal burning, is a major source of sulphate aerosols that interact with incoming solar radiation, and hence affect the Earth's climate. Sulphur and nitrogen emitted to the atmosphere as a by-product of energy production are converted to sulphates and nitrates in cloud water droplets, and hence increase the acidity of precipitation with adverse consequences on living organisms at the Earth's surface (land and lakes). Aerosol particles, in particular soot, are produced by combustion processes, and mineral dust, mobilized by winds over arid regions, are also affecting the radiative balance of the atmosphere, with sometimes very high radiative forcing at the regional scale. The impact of the changes in the aerosol burden and aerosol composition on cloud microphysics, and hence on the climate system, remains poorly understood, but is believed to be significant.

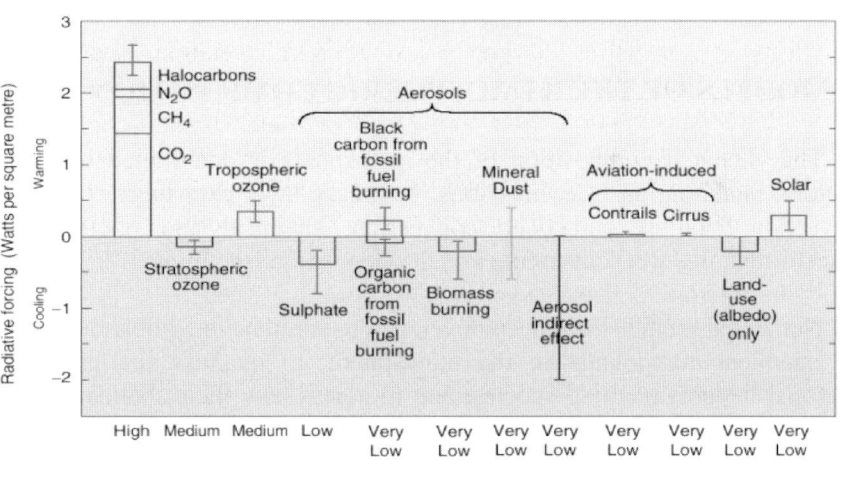

Figure 1 (see plate 1): Global, annual mean radiative forcing (Wm^{-2}) for the period from pre-industrial (1750) to present (from IPCC, 2001).

Long-lived biogenic and anthropogenic gases such as nitrous oxide, methane, or halogenated organic compounds can be transported up to the stratosphere, where they become a source of ozone-depleting radicals. Reduction in stratospheric ozone leads to enhanced fluxes of ultraviolet radiation with potential damaging effects on living organisms, including humans. Finally, the release of nitrogen oxides, carbon monoxide, and different biogenic or anthropogenic hydrocarbons affects the oxidizing power of the atmosphere, and specifically the budget of tropospheric ozone (O_3) and of the hydroxyl (OH) radical. OH is particularly efficient for oxidizing chemical compounds, including pollutants that are either injected directly in the atmosphere or produced *in situ* as a result of photochemical processes. Wet and dry deposition of gaseous species and aerosol particles at the Earth's surface affects biogeochemical cycles, both on land and in the ocean, with potential consequences for surface emissions of biogenic gases, and indirect effects on the climate system. The impact of changes in the composition of the atmosphere on climate has been extensively studied over the past decades (IPCC, 2001). Although the impact of the increase in concentrations of long-lived greenhouse gases (CO_2, CH_4, etc.) on the radiative forcing is relatively well quantified, there are still large uncertainties in the quantification of the forcing produced by changes in ozone and particle concentrations, as highlighted by Figure 1.

3. MODELS OF THE ATMOSPHERIC COMPOSITION

The chemical composition of the atmosphere is determined by several factors, including surface emissions, boundary layer exchanges, large-scale advection, shallow and deep convection, chemical and photochemical transformations, wet scavenging and dry deposition (see Figure 2).

Numerical models have been developed to simulate the global distribution of chemical compounds in the atmosphere, to quantify the global and regional budgets of these species, and to assess how their abundance could evolve in the future in response, for example, to anthropogenic forcing. These models can be viewed as mathematical tools that attempt to replicate the complex processes that occur in the atmosphere or, more generally, in the Earth system. Different types of models are being used to study the behaviour of chemical compounds in the atmosphere. Forward models derive the atmospheric concentration of chemical species for prescribed initial and boundary conditions. Surface fluxes (emissions and deposition) that are discussed in this volume, are typical examples of surface boundary conditions. Inverse modelling techniques, using similar models together with observations of atmospheric compounds, can be used to constrain surface exchanges. Deviation between predicted and observed quantities often reveals the insufficient understanding of some processes, and generates new research activities.

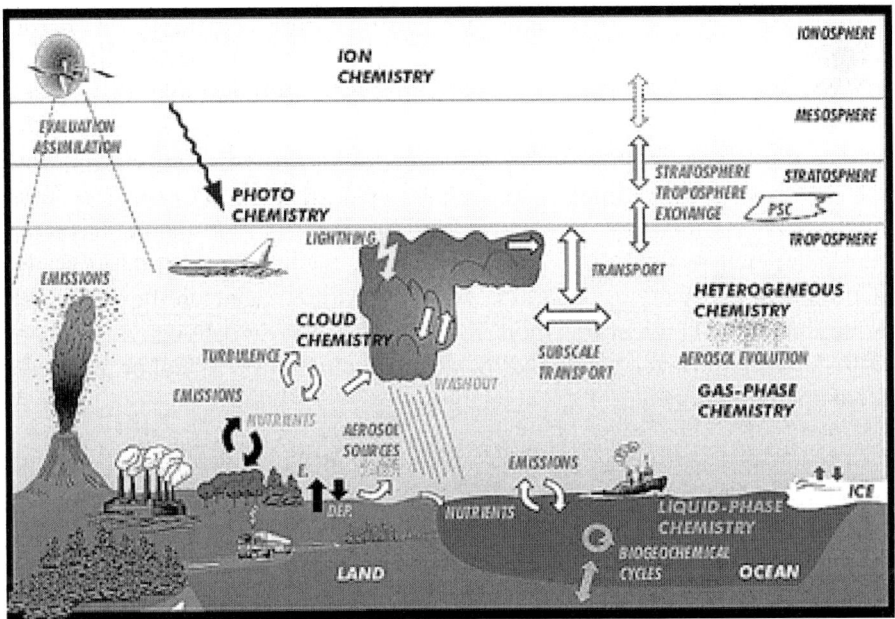

Figure 2 (see plate 2): schematic view of the main processes driving the chemical composition of the lower atmosphere. (Personal communication of M. Schultz, Max Planck Institute for Meteorology, Hamburg, Germany.)

In atmospheric models, the variation with time t of the local density n_i of a chemical species i is often expressed as the sum of the tendencies associated with different physical and chemical processes k:

$$\partial n_i / \partial t = \sum_k (\partial n_i / \partial t)_k \qquad \text{(Eq.1)}$$

The tendency associated with advective transport by the atmospheric circulation is provided by the convergence of the flux

$$(\partial n_i / \partial t)_{advection} = - \nabla.(n_i \, v) \qquad \text{(Eq.2)}$$

where v represents the wind vector. Winds are provided either from general circulation models or from meteorological analyses of observations. Transport processes that are not resolved at the smallest spatial scale reproduced explicitly by the model must be parameterized. This is the case, for example, for convective and boundary layer exchanges in most global and regional models. A simple parameterization of mixing processes is provided by the introduction of an "eddy diffusion" term

$$(\partial n_i/\partial t)_{subscale} = - \nabla [\mathcal{K} n_M \cdot \nabla (n_i/n_M)] \qquad \text{(Eq.3)}$$

where $\mathcal{K}$ is a tensor whose components are empirically prescribed and n_M is the air density. This simple approach is replaced in most models by more elaborated parameterizations, especially in the case of deep convective transport, where specific processes such as cloud entrainment and detrainment as well as updrafts and downdrafts need to be accurately represented. Chemical-transport models often drive convective transport with convective fluxes generated by a general circulation model or diagnosed from operational meteorological analyses.

Chemical reactions, which occur in the gas phase, in the aqueous or solid phase (e.g., for example in cloud droplets) or at the interface (e.g. surface reactions on aerosol particles), also modify the abundance of atmospheric compounds. The tendency associated with chemical transformations can be expressed as

$$(\partial n_i/\partial t)_{chemistry} = P_i - L_i \qquad \text{(Eq.4)}$$

where P_i and L_i represent the local chemical production and destruction rates for species i, respectively. These are derived on the basis of the chemical scheme adopted for the model, with values for the chemical and photochemical rate coefficients based on laboratory investigations. Modern three-dimensional chemical transport models include typically 100-300 reactions to describe the behaviour of 50-100 gas-phase species. State-of-the-art aerosol models estimate the size distribution and chemical composition of atmospheric particles, and their source/sink terms account for processes such as nucleation, condensation, evaporation, and coagulation.

Surface emissions E_i of gas-phase compounds or primary aerosols are prescribed either as a flux condition applied at the lower boundary of the model

$$\Phi_{i,0} = E_{,i} \qquad \text{(Eq.5)}$$

or as an additional tendency

$$(\partial n_i/\partial t)_{emission} = E_i/\Delta z \qquad (Eq.6)$$

applied only to the lower layer of the model, with Δz being the thickness of this layer.

Wet scavenging tendencies are significant only in precipitating clouds and below them; for gas-phase compounds, their values depend on the solubility of the species i. Several parameterizations have been proposed to account for this process. As wet scavenging is occurring very often in convective cells and in frontal systems, the parameterization of wet scavenging cannot be distinct from the formulation of vertical uplift by convective motions. Gravitational settling as well as wet scavenging have a substantial influence on the aerosol budget.

Finally, dry deposition at the surface can be represented in models either by a downward flux

$$\Phi_{i,0} = -n_i w_{D,i} \qquad (Eq.7)$$

specified as a lower boundary condition at the surface, or by an additional tendency term applied only in the lower layer of the model

$$(\partial n_i/\partial t)_{dry\ deposition} = -n_i w_{D,i}/\Delta z \qquad (Eq.8)$$

In these expressions, $w_{D,i}$ (≥ 0) is the deposition velocity for compound i and Δz is again the thickness of the model layer closest to the surface. The deposition velocity (or equivalently the resistance to deposition) is a function of the aerodynamical conditions near the surface, stomatal uptake by leaves of the vegetation, soil properties, etc. Detailed parameterizations used in many models are provided by Wesely (1989).

The quantification of the atmospheric budget of chemical compounds requires that all the above processes be accurately established. Laboratory studies have provided quantitative values of chemical rate constants and absorption cross sections that are needed to estimate atmospheric sources and sinks of chemical species. Field campaigns have provided detailed information for identifying important chemical and physical processes and for estimating local or regional atmospheric budgets. As highlighted in Chapter 11, ground-based monitoring of chemical compounds and space observations are used to quantify the time evolution of surface concentrations and to monitor the three-dimensional distribution of chemical compounds, respectively. Models have been developed to synthesize information provided by laboratory studies and field measurements, to quantify global budgets and to assess how these budgets have been changing and will continue to do so in response to human activities. The success of atmospheric modelling depends directly on our understanding of processes in the atmosphere and at the land/atmosphere and ocean/atmosphere interfaces. An accurate quantification of the fluxes of chemical compounds (and in certain cases of their isotopes, see Chapter 10) at the Earth's surface is key for the success of model simulations.

4. SURFACE EMISSIONS

Emission and deposition of gases and aerosols result from a multitude of natural and human-induced processes, both from biogenic and abiogenic sources. Natural emissions from land are associated with biological processes in soils and wetlands, and with the metabolism of vegetation and animals. Substantial quantities of nitrogen oxides, for example, are released to the atmosphere as the result of nitrification and denitrification processes in soils, as discussed in Chapter 5. Methane, the most abundant hydrocarbon in the atmosphere, is released as a result of microbial breakdown of organic matter in oxygen-deficient flooded areas, including lakes, swamps, bogs, boreal marshes and rice paddies, and through enteric fermentation processes in the digestive systems of animals. Emissions of methane in tropical and sub-tropical areas are regulated by flood cycles, whereas at high latitudes, they are related to water-table temperature interactions.

Approximately three-fourths of globally-emitted non-methane hydrocarbons are released by vegetation. The emission rates depend on a variety of biospheric processes and meteorological conditions. Isoprene, for example, is primarily emitted by different types of deciduous trees, while monoterpenes are released primarily by conifers. The emissions of these species are discussed in Chapter 4, together with the emissions of oxygenated volatile organic compounds, including acetone and methanol, which are also emitted by plants.

Atmospheric Composition and Surface Exchanges 9

Wildfires, agriculture waste burning, fuel-wood use and fires associated with forest and savannah clearing produce very substantial emissions, particularly in the tropics (see Chapter 3). Some of these fires, specifically those triggered by lightning, are entirely natural, while others are associated with agricultural practices or domestic activities. Biomass burning is a major source for many atmospheric compounds including nitrogen oxides, carbon monoxide and carbonaceous aerosols. Emissions from fires depend on several factors, including the temperature of the fire, the amount, nature and water content of the biomass available, soil moisture, wind velocities above the canopy, and, for each compound emitted, the emission factor. Fires occur primarily during dry periods with frequencies and intensities that are highly variable from year to year.

In deserts, large quantities of dust can be mobilized by winds and transported large distances before being deposited at the surface, as detailed in Chapter 6. This process contributes significantly to the aerosol particle load of the atmosphere and, if the deposition takes place over marine areas, to iron fertilization of the ocean. Dust particles modify the radiation field in the troposphere and hence photolytic activity, and also affect cloud microphysical properties.

Volcanic eruptions, which are discussed in Chapter 7, provide both continuously and episodically large sources of SO_2 and other compounds to the atmosphere with significant effects on the sulphate aerosol load and hence on the climate system. Lightning is also a major source of NO_x in the middle and upper troposphere.

Soluble compounds are constantly exchanged between the atmosphere and the ocean. The ocean-atmosphere exchange rate of a gas is proportional to the difference between the partial pressure of this gas in surface water and in the overlying atmosphere, to gas solubility (which is a strong function of water temperature), to atmospheric wind speed, and to the Schmidt number (ratio between kinematic viscosity of seawater and molecular diffusivity of the gas). Ocean/atmosphere exchanges are most intense when the surface wind is strong and ocean waves are high. In supersaturated areas of the ocean, the flux direction is from ocean to atmosphere, while the opposite is true in undersaturated areas. Globally, the ocean is a sink for atmospheric carbon dioxide, but the strength and even the direction of the CO_2 flux depend strongly on the temperature of the water. Warm tropical waters tend to be supersaturated in CO_2, and hence release CO_2 to the atmosphere, while cold high latitude waters tend to take CO_2 up from the atmosphere.

Many organic compounds produced by photochemical processes or by biological activity (including microbial breakdown of organic matter) can also be released from the ocean to the atmosphere. This is the case for organo-halides (CH_3Cl, CH_3Br, CH_3I), cyanides (HCN, CH_3CN), and organic sulfur compounds, especially dimethylsulfide (DMS) (see Chapter 8). DMS molecules released by the ocean are oxidized in the atmosphere and provide, during non-volcanic periods, the largest natural source of sulphate aerosols in the troposphere. Carbonyl sulfide (COS) is photochemically produced in the surface oceans and released to the atmosphere. A small oceanic emission of carbon monoxide (CO) has also been reported. Sea-salt particles are emitted in large quantities as discussed in Chapter 9, but remain mostly in the marine boundary layer of the atmosphere.

Anthropogenic emissions (Chapter 2) resulting from industrial, transport, agricultural, services and domestic activities have become large - sometimes dominant - contributors to the budget of chemical compounds in the atmosphere. Significant amounts of carbon monoxide, nitrogen oxides, sulphur dioxide and carbonaceous aerosols are released by combustion of different types of fossil fuels. Coal mining as well as the transmission and distribution of natural gas, for example, lead to large emissions of methane and other hydrocarbons. The release of nitrogen oxides (NO_x) by aircraft engines provides a small source of NO_x in the upper troposphere, which plays, however, a potentially important role because at these altitudes the chemical lifetime of NO_x particles is relatively long.

Table 1 provides an estimate of the relative importance of key sources for various atmospheric compounds. Major, significant and minor sources refer to a contribution to the total emission or *in situ* production of more than 30%, 10-30%, and less than 10%, respectively.

Table 1: Relative importance of key sources. A "?" indicates that this source has either not been identified, or quantified at the global scale.

	Biogenic and other continental sources	Biomass Burning	Fossil Fuel Burning	Ocean	Photo-chemistry
CO	significant	major	major	minor	major
CH_4	major	significant	major	minor	no
N_2O	major	significant	significant	significant	no
NOx	major	major	major	?	minor
isoprene	major	?	?	?	no
DMS	minor	?	no	major	no
SO_2	major*	minor	major	no	important
dust	major	no	no	no	no
sea-salt	no	no	no	major	no
ozone	no	no	no	no	major

*: emissions from volcanoes

5. CHEMICAL TRANSFORMATION IN THE ATMOSPHERE

Once released in the atmosphere, chemical species are transported and gradually transformed by chemical and physical processes (see, e.g., Brasseur et al., 1999 and Figure 3). A major transformation path is provided by oxidation processes of reduced species emitted at the surface, and the formation of secondary compounds. Oxidation in the atmosphere takes place through reactions primarily with the hydroxyl radical (OH) during daytime. Reactions with ozone (O_3) and with the nitrate radical (NO_3) are other possible pathways, but, in most cases, are only significant during nighttime. In the troposphere the OH radical is produced by oxidation of water vapour by the electronically excited oxygen atom (O^1D), itself produced by photolysis of ozone at wavelengths shorter than 320 nm. Once photochemically produced, OH is rapidly converted to the hydroperoxy radical (HO_2) by reaction with ozone, carbon monoxide, methane and other hydrocarbons. HO_2 is converted back to OH by reaction with ozone and nitric oxide (NO). The reaction of HO_2 with NO produces NO_2, which is photolyzed to nitric oxide (NO) and atomic oxygen (O). It leads to a net production of ozone since the atomic oxygen O reacts rapidly with O_2 to form O_3. At high concentrations of nitrogen oxides, OH is converted to nitric acid (HNO_3). This latter mechanism leads to a net loss of hydroxyl radicals when HNO_3, which is very soluble in water, is removed from the atmosphere by wet scavenging rather than being photolyzed to form OH and NO_2 (or H and NO_3). Other loss processes for odd hydrogen radicals (HO_x = OH + HO_2) include the reaction of OH with HO_2, which produces water vapour (H_2O), and of HO_2 with itself, which produces hydrogen peroxide (H_2O_2), another highly soluble compound.

This simple description of tropospheric photochemical processes shows clearly that the abundance of OH, and hence the oxidizing power of the atmosphere, depends on primary compounds such as CO, CH_4, NO_x and non-methane volatile organic compounds, in addition to ozone and water vapour. The abundance of the primary compounds (also called ozone precursors) is largely determined by the intensity of surface emissions. Thus, under natural conditions, the oxidizing power of the atmosphere is, to a large extent, controlled by biological and physical processes on land and in the ocean. Human activities, however, have altered substantially the surface emissions, with potential impacts on the level of tropospheric OH. No systematic observations of OH and of its time evolution are available to assess past changes in the oxidizing power of the atmosphere so that such studies must rely on indirect estimates (see Prinn et al. (2001), who have analyzed trends in the abundance of compounds which are known to be destroyed by OH) or on modelling studies (e.g., Krol, 1998).

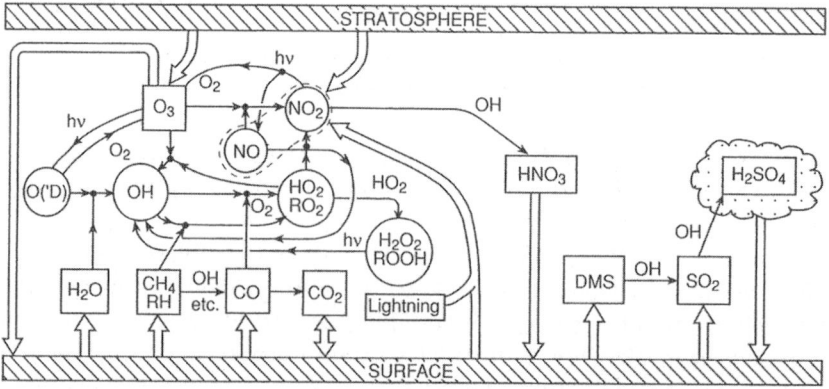

Figure 3: key chemical and photochemical processes affecting the composition of the global troposphere (adapted from Prinn, 1994).

Although no final unequivocal conclusion has yet been reached from these studies, it seems that changes in the oxidizing power of the atmosphere have been relatively limited, due to a large degree to the opposing effects of OH perturbations generated by simultaneous changes in the emissions of NO_x, CO and hydrocarbons.

Changes in surface emissions also affect the concentration of radiatively active gases and aerosol particles, and hence provide a forcing mechanism for the climate system. Emissions of methane, nitrous oxide, and halocarbons have increased substantially in the last centuries and explain the well documented changes in the atmospheric concentration of these gases, with direct consequences for the amount of terrestrial infrared radiation trapped by the Earth's atmosphere. It is intriguing to note that, over the past 400,000 years, before humankind exerted a discernible influence on the chemical composition of the global atmosphere, the variation in the atmospheric abundance of greenhouse gases such as CO_2 and CH_4 and in surface temperature were strongly correlated, as shown from the analysis of Antarctic and Arctic ice core samples (Petit et al., 1999) This important finding suggests that land-atmosphere and ocean-atmosphere exchanges have undergone large fluctuations associated with past climate variability occurring over periods of tens of thousand years.

The increasing emissions of industrially-manufactured halocarbons (CFCs) have led in the last five decades to a substantial increase (more than a factor of three) in the stratospheric abundance of reactive chlorine, with a dramatic decrease since the early 1980s in abundance of ozone in the Antarctic (10-25 km altitude) during springtime (September to November), and to a lesser extent in the Arctic (February-April). The stratospheric abundance of bromine has also increased. Ozone is destroyed by inorganic halogen species, especially in the cold air masses of the polar lower stratosphere where the presence of polar stratospheric clouds offers surface sites for heterogeneous chemical reactions to occur. As a result of international agreements in the late 1980s and 1990s, however, the production and emissions of the chlorofluorocarbons and other halocarbons has been considerably reduced and even phased-out in certain cases, so that the abundance of inorganic chlorine and bromine should decrease in the future. The rate at which the ozone layer should recover is determined by the atmospheric lifetime of anthropogenic halogen molecules and the future evolution of the temperature of the lower stratosphere, in turn influenced by future climate change.

6. TOWARDS EMISSIONS INVENTORIES AND MODELS

The determination of the distribution of chemical compounds in the atmosphere, and more generally the quantification of fluxes associated with global biogeochemical cycles in the Earth System, requires that exchange rates between the different physical components of our planet (atmosphere, ocean, land, cryosphere) be accurately determined. The construction of global emission inventories such as those established by the GEIA activity within the International Global Atmospheric Chemistry (IGAC) Project of the International Geosphere Biosphere Programme (IGBP: http://www.igbp.kva.se) has helped to address fundamental issues concerning the global atmospheric budget of several key chemical compounds. Current inventories established for many compounds at a spatial resolution of one degree in longitude and latitude (e.g., Olivier et al., 1996 and Chapter 2) are invaluable tools for the community involved in chemical and transport model development and for groups producing periodic assessments of climate change and other global change issues. Developing emission inventories requires a large number of data provided primarily by local field measurements or regional aircraft observations as well as ancillary information (including space observations and socio-economic data) that are essential for extrapolating available emission estimates to broader areas. The issues associated with the development of current emission inventories are discussed in great detail in this volume. Some chapters of the book highlight methodological progress made recently to constrain emission rates, including the development of inverse modelling techniques (Chapters 11 and 12).

It is clear that as the scientific community is developing a more integrated view of the Earth System with emphasis on coupling processes, static emission inventories used in atmospheric models will gradually be replaced by emission models, which reproduce explicitly the physical, chemical and biological processes that occur on land surfaces, in soils, in the ice and snow, and in the upper ocean. Models of such complexity will also have to account for the human dimension, which produces some interesting feedbacks in the Earth System with significant consequences for the chemical composition of the atmosphere and on other key parameters of the global environment.

7. REFERENCES

Brasseur, G.P., J.J. Orlando, and G.S. Tyndall (eds), Atmospheric Chemistry and Global Change, Oxford University Press, New York, 1999

Crutzen, P. J., A discussion of the chemistry of some minor constituents in the stratosphere and troposphere, Pure Appl. Geophys., 106, 1385, 1973.

Haagen-Smit, A.J., Chemistry and physiology of Los Angeles smog, Indus Eng. Chem., 44, 1342, 1952.

IPCC, Climate Change 2001: The Scientific Basis, Contribution of Working Group I to the Third Assessment Report of the Intergovernmental Panel on Climate Change, J.T. Houghton, Y. Ding, D.J. Griggs, M. Noguer, P.J. van der Linden, X. Dai, K. Maskell, and C.A. Johnson (eds), Cambridge University Press, U.K. and New York, NY, USA, 2001.

Krol, M., P-J. van Leeuwen, and J. Lelieveld, Global OH trend inferred from methylchloroform measurements, J. Geophys. Res., 103, 10697-10711, 1998

Olivier, J.G.J., A.F. Bouwman, C.W.M. Van de Maas, J.J.M. Berdowski, C. Veldt, J.P.J. Bloos, A.J.H. Visschedijk, P.Y.J. Zandveld, and J.L. Haverlag, Description of EDGAR version 2.0: A set of global emission inventories of greenhouse gases and ozone-depleting substances for all anthropogenic and most natural sources on a per country basis and on a 1° x 1° grid, National Institute for Publich Health and the Environment, Bilthoven, The Netherlands, 1996.

Petit, J.R. J. Jouzel, D. Raynaud, N.I. Barkov, J.M. Barnola, I. Basile, M. Bender, J. Chapellaz, J. Davis, G. Delaygue, M. Delmotte, V.M. Kotyakov, M. Legrand, V.Y. Lipenkov, C. Lorius, L. Pepin, C. Ritz, E. Saltzman, and M. Stievenard, Climate and atmospheric history of the past 420,000 years from the Vostok ice core, Antarctica, Nature, 399, 429, 1999.

Prinn, R.G. (ed), Global Atmospheric-Biospheric Chemistry, Plenum Press, New York and London, 1994.

Prinn, R.G., J. Huang, R.F. Weiss, D.M. Cunnold, P.J. Fraser, P.G. Simmonds, C. Harth, P. Salameh, S. O'Doherty, R.H.J. Wang, L. Porter, B.R. Miller, and A. McCulloch, Evidence for significant variations of atmospheric hydroxyl radicals in the last two decades, Science, 292, 1751-17793, 2001

Wesely, M.L., Parameterization of surface resistances to gaseous dry deposition in regional-scale numerical models, Atmos. Environ., 23, 1293, 1989

Compilation of Regional to Global Inventories of Anthropogenic Emissions

Carmen M. Benkovitz, Hajime Akimoto, James J. Corbett, J. David Mobley, Jos G.J. Olivier, Toshimasa Ohara, John A. van Aardenne, and Vigdis Vestreng

1. INTRODUCTION

The mathematical modelling of the transport and transformation of trace species in the atmosphere is one of the scientific tools currently used to assess atmospheric chemistry, air quality, and climatic conditions. One of the most important inputs to such models are accurate inventories of emissions of the trace species included in the study at the appropriate sectoral, spatial, temporal, and species resolution. In addition, from the management perspective detailed information on emissions is also required at high resolution for the design of policies aimed at reducing emissions of pollutants, and for the evaluation of the efficiency of such policies.

The compilation of emissions information for scientific use started in the 1970s for oxides of sulphur and nitrogen. These initial inventories calculated global emissions by large geographic areas (Várhelyi, 1985), with very little spatial and temporal resolution. In the last three decades a wealth of information has been developed on anthropogenic activities and their emissions; however, these statistics must be used wisely to provide a picture of the desired quantities.

Sections 2 to 4 present an overall description of the current methodologies used to compile regional to global inventories of anthropogenic emissions. This discussion is by no means exhaustive; rather it attempts to give the reader an overview of how emissions from anthropogenic sources were and are being estimated (Baldasano and Power, 1998). As discussed in chapter 4, these methodologies can also be applied to the estimation of emissions from biogenic sources. Section 5 describes examples of regional and global inventories that are currently used in studies of the composition of the atmosphere at different spatial scales, and Section 6 discusses the development of inventories representing anthropogenic emissions over the past decades.

2. INVENTORY DEVELOPMENT

There are two general methodologies used to estimate regional to global emissions: bottom-up and top-down. Bottom-up methodologies apply the following general equation to estimate emissions:

$$E_i = A_i (EF)_i P_{1i} P_{2i} \ldots\ldots$$

where E_i are emissions (for example, kg sulphur hr^{-1}), A_i is the activity rate for a source (or group of sources i, for example, kg of coal burned in a power plant), $(EF)_i$ is the emission factor (amount of emissions per unit activity, for example, kg sulphur emitted per kg coal burned), and P_{1i}, P_{2i} ... are parameters that apply to the specified source types and species in the inventories (for example, sulphur content of the fuel, efficiency of the control technology). Top-down methodologies, also known as inverse modelling, derive emissions estimates by inverting measurements in combination with additional information, such as the results of atmospheric transport and transformation models. The top-down methodologies are discussed in detail in chapters 11 and 12.

3. USE OF BOTTOM-UP METHODOLOGIES TO ESTIMATE EMISSIONS

The three major inputs needed to apply this methodology are the locations of the sources, the activity rates, and the emission factors. To estimate the emissions of certain trace species additional parameters are needed. For example, to estimate sulphur emissions from fossil fuel combustion the sulphur content of the fuels is needed, for certain sources with emissions control devices the efficiency of these devices is needed.

3.1. International programs

Researchers compiling national inventories for their own country usually have access to information that may not be available to others; they also have more detailed knowledge of the types of processes and operating conditions which allows the use of nation-specific emission factors or adjustment of literature values. Generally, these inventories estimate yearly emissions at the national, or for very large countries, at the level of geographic subdivisions (states, provinces, etc.). National or regional inventories may also be able to apply greater geographic resolution to sources, particularly large stationary and road and non-road mobile sources.

Researchers compiling regional or global inventories can either a) gather the national emissions estimates in the region of interest or b) estimate emissions directly using published information. To compile inventories by bringing together national estimates, the information received must be checked for transparency, consistency, completeness, comparability, and accuracy (Tarrasón and Schaug, 1999). Examples of programs that rely on national submissions to build regional or global inventories include the Co-operative Programme for Monitoring and Evaluation of the Convention on Long Range Transboundary of Air Pollutants in Europe (CLRTAP) (see section 5) and the Intergovernmental Panel for Climate Change (IPCC) (Houghton et al.,1997; Moran and Salt, 1996). Collaborations have been set up between groups concerned with gathering emissions information such as EMEP, the United Nations Economic Commission for Europe (UNECE, 1997), the United Nations Framework Convention on Climate Change (UNFCCC, 2000) and unified reporting requirements have been developed to minimize the work of the national participants and to create consistent emissions estimates for the work of these groups (Houghton et al., 1997; Moran and Salt, 1996). Step-by-step guidelines and workbooks have also been developed to help in the compilation of the required information.

In June 1985, the European Council of Ministers defined a program for gathering, coordinating, and ensuring consistency of information on the state of the environment and natural resources in the European Community. This program was called CORINE (COoRdination d'INformation Environnementale), and one of its component projects is CORINAIR (CORINe AIR emissions inventory). This program is now run by the

European Environment Agency (EEA). A prototype inventory was developed for 1985, CORINAIR85, which included a new nomenclature for source types, classification of emissions as large point sources and area sources, and especially developed software for data input and the calculation of emissions. Based on the experience gained in compiling CORINAIR85 and in collaboration with UNECE and the Organisation for Economic Cooperation and Development (OECD), the next phase of the project, CORINAIR90, has been defined (Bouscaren, 1992). In 1996 CORINAIR90 provided a complete, consistent, and transparent air pollutant emission inventory for Europe for 1990. New source sector splits have been defined (SNAP90), the list of pollutants and large point sources to be included has been expanded, and the availability of the CORINAIR software system has been extended to 30 countries in Europe. The program was taken over by the EEA in 1997, and the new source sector split was prepared in 1997 (SNAP97) which has subsequently been used in the CLRTAP/EMEP inventory programme until a new nomenclature was adopted in 2002. The joint EMEP/CORINAIR Emission Inventory Guidebook work continues, and the currently third edition is available on http://reports.eea.eu.int/technical_report_2001_3/en.

The Global Emissions Inventory Activity (GEIA), a subgroup of the International Global Atmospheric Chemistry Programme (IGAC) (Galbally, 1989), is composed of international groups of scientists attempting to establish inventories for a number of trace species, with recognized accuracy and enough spatial, temporal, and species resolution to serve as standard inventories for the international community of atmospheric scientists (Benkovitz and Graedel, 1992). In the initial compilation of the GEIA inventories of SO_2 and NO_x emissions from anthropogenic sources, a "hybrid technique" was used.

First a "default" global inventory was selected which included emissions estimates derived using a unified methodology; then national or regional emissions estimates received from cooperating researchers were studied and, if appropriate, their data were substituted for those in the default inventory (Benkovitz et al., 1996). An in-depth study of all the data was done, and differences and caveats were included in the accompanying documentation. Figure 1 presents a summary of the data used to develop the GEIA SO_2 and NO_x inventories for ca. 1985.

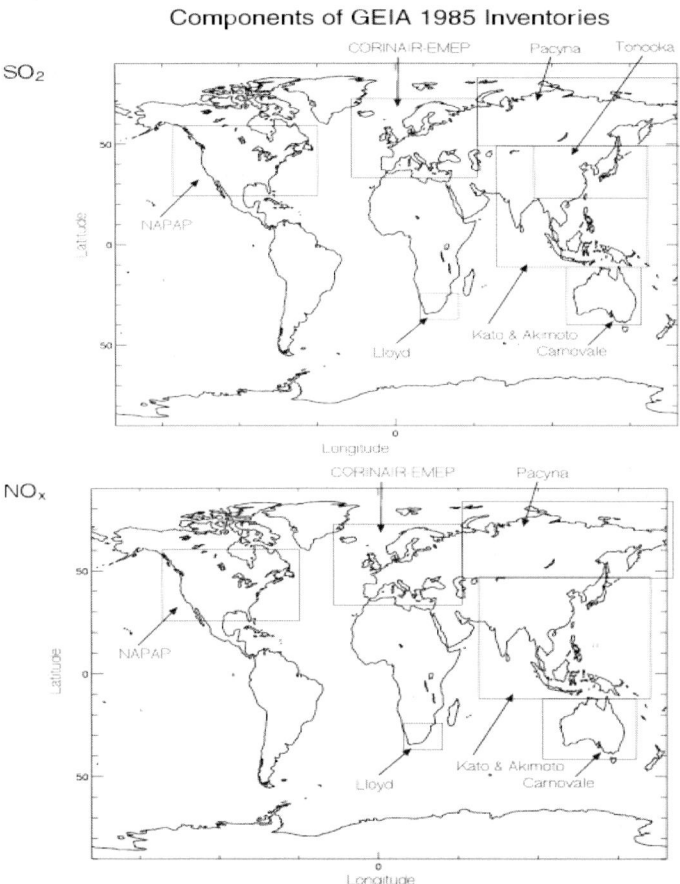

Figure 1. Regional inventories overlaid on the default global inventories of SO_2 (top panel) and NO_x (bottom panel) for the GEIA 1985 inventories. CORINAIR, EMEP described in text. NAPAP=National Acid Precipitation Assessment Program, USA. H. Akimoto, University of Tokyo, Japan. F. Carnovale, Coffey Partners International PTY Ltd., Australia. N. Kato, National Institute of Science and Technology Policy, Japan. S.M. Lloyd, South African Department of National Health and Population Development, Pretoria. J. Pacyna, Norwegian Institute for Air Research, Kjeller. Y. Tonooka, Institute for Behavioral Sciences, Tokyo. From *Benkovitz et al, 1996.*

3.2. Direct development of inventories

To directly estimate regional or global industrial emissions, published information on activity rates, emission factors, and other pertinent parameters is needed. Examples of inventories compiled using this approach include the Emission Database for Global Atmospheric Research (EDGAR) (see section 5.1) (Olivier et al., 1999a), a set of global emissions inventories of greenhouse gases and ozone-depleting substances for all anthropogenic and most biogenic sources on a per country basis, the Regional Air Pollution Information and Simulation Model (RAINS) Asia (Arndt et al., 1997), a set of SO_2 emissions data for Asia (Foell et al., 1995), and the global CO_2 emissions compiled by Marland et al. (1994, 2002). In general, the activity rates are derived from data compiled by multi-national organizations. The two major sources of statistics on energy production, use, and trade are the International Energy Agency (IEA, 2002a,b,c) and the United Nations (UN 2000, UN 2002). Other sources of activity data for specific sectors include, but are not limited to the World Energy Council (1998), the World Resources Institute (2000), the World Bank (The World Bank Group, 2000), the International Monetary Fund (International Monetary Fund Staff, 2000), the Food and Agriculture Organization of the United Nations (FAO, 2000), the US Geological Survey (USGS, 2002) and the Motor Vehicle Manufacturers Association of the United States (1998).

These compilations are usually not independent data sets because they are based on some of the same sources for national data; the published numbers are the result of slightly different questionnaires and different analysis by the respective offices and are also reported for somewhat different categories. When using any of these sources of activity rates, the data must be carefully checked for completeness and homogeneity to avoid either double counting or dropping emissions.

In general non-road mobile sources (locomotives, construction equipment, etc.) have been less well characterized, primarily because of uncertainties in the locations of activity and in the activity levels. For example, surrogates like population may not identify the location of large construction projects, and construction or logging equipment may be idle much of the time. In the case of international shipping, these uncertainties were compounded by limited information about the emissions factors assigned to marine engines.

The U.S. Environmental Protection Agency has developed and maintains a compilation of emission factors applicable to U.S. sources (U.S. Environmental Protection Agency, 1999). Canada has also developed emission factors applicable to their sources (Johnson et al., 1991). The IPCC (Houghton et al., 1997) and the United Nations Framework Convention on Climate Change (UNFCCC) (United Nations Framework Convention on Climate Change, 2000) have developed extensive guidelines, which include step by step methodologies and emission factors to be used in estimating and reporting emissions. In addition, some investigators either adapt the US or Canadian emission factors or perform literature searches for more appropriate values or for experimental data from which to develop their estimates; for examples see Kato and Akimoto (1992) and Arndt et al. (1997).

A vivid example of the impact that small differences in input data can have on emissions estimates is given by Marland et al. (1999), where the annual-by-country CO_2 emissions values in the Marland et al. inventory were compared with those in the EDGAR 2.0 inventory. Marland et al. rely mainly on statistics gathered by the UN, while EDGAR relies mainly on statistics gathered by the IEA supplemented by UN statistics when needed. After trying to account for the different sectors and emission factors included in each inventory, results presented in Figure 2 show that most of the emissions estimates cluster around the 1:1 line.

The mean of all country differences between the two inventories is about 1%, but the fractional differences tend to be larger for countries with smaller emissions, probably due to larger uncertainties in the activity rate information. However, the relative difference between the two estimates for US emissions is only 0.9%, but the absolute difference is larger than the total emissions from 147 of the 195 countries included in the analysis. For the former Soviet Union, the relative difference is about 8.1%, but the absolute difference exceeds the total emissions from all but the 14 largest emitting countries.

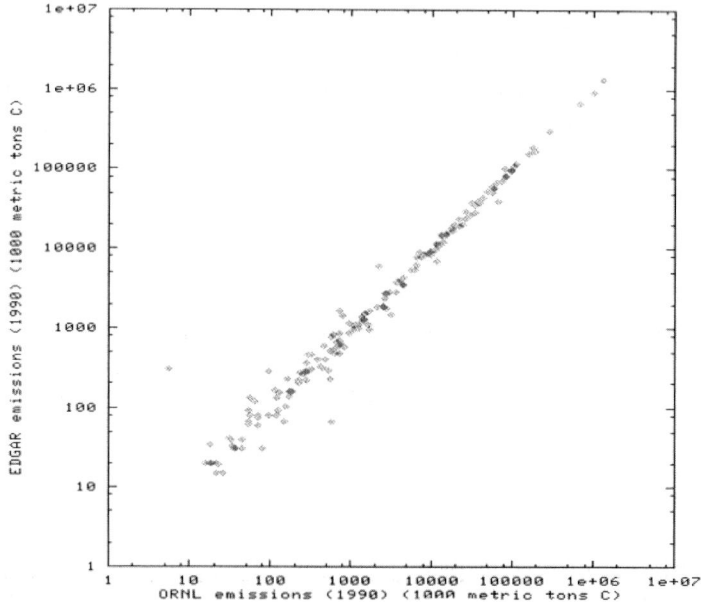

Figure 2. Correlation of countries' total emissions as calculated by EDGAR and ORNL. (from *Marland et al, 1999*).

Direct measurements of emissions can be carried out using continuous emissions monitoring (CEM) techniques; this methodology produces the most accurate emissions estimates for the sources being monitored at the temporal resolution of the measurements. In the United States, these measurements are currently in use mainly at large stationary sources, which are under strict guidelines for emissions control, such as power plants operating under state implementation plans.

4. SECTORAL, TEMPORAL, SPATIAL AND SPECIES RESOLUTION

For scientific studies and increasingly for management and policy, perspective emission inventories are needed by source sector at more detailed temporal, spatial, and species resolution. As the main driving force for the compilation of national inventories is the need to regulate sources, emissions by source sectors are usually available in these inventories.

Inventories of Anthropogenic Emissions 25

Usually, activity rate compilations are also developed by source sector. Unfortunately, the sectoral definitions are far from being uniform; they vary from the very detailed descriptions used by the US Environmental Protection Agency (source classification codes), the Canadian adaptation of this classification scheme, and CORINAIR SNAP97 to those with only a very limited number or no sectors. The EDGAR inventory and the 1990 SO_2 and NO_x inventories of anthropogenic emissions developed at the Canadian Global Interpretation Centre (CGEIC) (Van Heyst et al., 1999) have adopted sector classifications that can address the needs of regional to global studies. As an example, Table 1 presents the source classification used by the CGEIC inventories. Except for CEM measured emissions, there is a lack of information on emissions at finer temporal, spatial, and species resolutions; therefore surrogate methods are used to bring the inventory information to the desired resolutions where possible.

Main Source	Sector Division	Source Type
Power generation	Power generation	Elevated point [a]
Fuel use & combustion	Industry (including other transformation sectors)	Low level point [b]
	Residential, commercial, other	Area [c]
	Incineration	Area
Transportation	Road	Mobile
	Non-road (*e.g.* rail)	Mobile
	Air (below 1 km)	Mobile
	International shipping	Mobile
Industrial processes	Iron & steel (excludes coke ovens and blast furnaces)	Elevated point
	Non-ferro	Elevated point
	Chemicals	Low level point
	Pulp & paper	Elevated point
	Other	Low level point
Land use	Biomass burning	Area

Table 1: Sector Description for the Canadian Global Interpretation Centre (CGEIC). 1990 Global Inventory of SO_2 and NO_x Emissions from Anthropogenic Sources
[a] Stationary source with emissions discharge point at an altitude of 100 m or higher.
[b] Stationary source with emissions discharge point at an altitude of less than 100 m.
[c] A surface-level stationary source with small emissions. An aggregation of these sources in an area can have significant emissions.

The exact location (longitude, latitude) for most of the large stationary point sources such as power plants, smelters, etc. is currently available; this allows precise geographic assignment of their emissions. However, additional information needed to properly assign these emissions in the vertical, such as stack height, flow rate, temperature, etc. is rarely available. This additional information becomes critical to models where a more exact representation of plume dispersion is needed, such as the plume-in-grid (PinG) section of the community multiscale air quality modelling system (Byun and Ching, 1999). The minimum information needed to estimate the vertical release height of large stationary point source emissions is the stack height; however, with only this information accurate plume dispersion cannot be represented.

The inventories compiled for the National Acid Precipitation Program (NAPAP) were among the first to address the conversion of emissions data to the finer spatial, temporal, and species resolution required by Chemical Transport Models (CTMs, Saeger et al., 1989; Wagner et al., 1986). This work developed temporal, spatial, and species allocation factors for emissions in the United States and Canada. Temporal allocation factors were based on surrogate information such as average heating degree days, continuous traffic counts from the US Department of Transportation, operating schedules for power plants, etc. Twelve temporal categories were achieved: weekday, Saturday, and Sunday for each of the four seasons. The location of all major stationary sources was known, so the emissions were assigned to the corresponding grid cell. The spatial allocation factors for non-stationary sources were developed using the appropriate gridded surrogate information such as population, housing, land use types, etc. Speciation factors for NO_x used were those recommended in AP-42 (US Environmental Protection Agency, 1985); speciation factors for total suspended particulates (TSP), and total hydrocarbons (THC) were based on the Air Emissions Species Manual (Shareef and Bravo, 1993ab) developed for the US Environmental Protection Agency.

The EUROTRAC-2 (European Experiment on Transport and Transformation of Environmentally Relevant Trace Constituents) subproject on the Generation and Evaluation of Emission data (GENEMIS, http://www.ier.uni-stuttgart.de/genemis) has investigated and improved the methodology for the temporal, spatial and substance resolution of air pollutant emissions in Europe (Friedrich et al., 2000, 2002). The pollutants covered include anthropogenic and biogenic precursors of ozone, aerosols and acidifying substances. As methods for developing emission inventories for CO, SO_2 and NO_x are more advanced and the resultant uncertainties in their emission estimates are lower than those for VOC, PM and NH_3, work within GENEMIS largely focussed on the latter pollutants.

General aim of GENEMIS was to support the generation of validated emission data, that could be used for the development of air pollution abatement strategies in Europe. This included improvement of methods, models and emission factors for the generation of emission data. A second focus was the assessment of the accuracy and the validation of emission data and finally, the development and improvement of tools to generate emission data for atmospheric models (CTMs). GENEMIS has improved knowledge about uncertainties of emission data by carrying out or analysing results from whole city, tunnel and open road experiments and statistical analyses of uncertainties. Uncertainties are still large, this has to be taken into account when interpreting results of atmospheric models.

A number of models to set up emission inventories for street canyons, for urban areas and for regions have been developed and improved. A large number of regional and urban emission data sets have been generated and provided to groups applying atmospheric models.

Some of the surrogate data needed to give emissions inventories finer resolution are not always developed on a gridded basis; for example, population data are based on geopolitical division, transportation data are "line" sources, etc. For other data sets, such as satellite information, the "native" grid of the data may not exactly match that of the atmospheric models. The conversion of these data to gridded format (or regridding to a new grid definition) adds another source of uncertainty to the emissions data, so care must be exercised when developing and applying the gridded surrogate data. For example, in his work with the 1990 population counts Li (1996) had to take into consideration problems with multiple geographic entities (including ocean areas) within one grid cell and how this affects the distribution of both the population and ultimately the emissions.

International shipping (A modern "fleet of ships does not so much make use of the sea as exploit a highway", Joseph Conrad, The Mirror of the Sea, 1906) provides another example of using unique surrogate data to characterize the geographic domain of a sector's emissions. While ships operate along well-known shipping lanes, these "water highways" are not mapped directly and they vary by season and with the weather. In order to develop gridded emissions inventories for NO_x and SO_x from shipping, a surrogate for vessel traffic density by location was derived from atmospheric and ocean observations reported by ships participating in the worldwide weather watch. Ship observations in each gridded location are averaged and compiled by month in the Comprehensive Atmosphere Ocean Database (COADS), used by scientists and modellers (Woodruff, 1993). Using the number of shipboard observations that were averaged in each grid cell for one attribute (e.g., air temperature), the implied traffic density for commercial shipping can be obtained for each month of the year; results are presented in Figure 3.

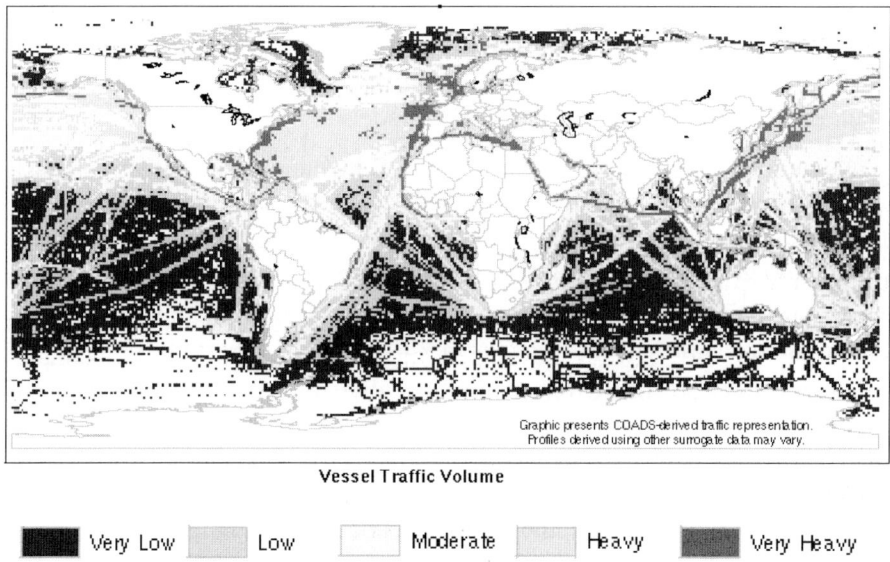

Figure 3 (see plate 3). International ship traffic for 1990 (adapted from Corbett et al., 1999 and Skjølsvik et al. 2000)

The accuracy of the resulting ship traffic profiles depends upon two fundamental assumptions: 1) a statistically significant population of ships reports to COADS; and 2) that sample of ships is representative of the international fleet overall. However, some 16% of the world cargo fleet reports to COADS, and comparison with regional inventories using more direct bottom-up traffic profiles has verified the accuracy of this approach

(Corbett, 1997; Corbett et al., 1999). Some researchers are beginning to compare the COADS-derived ship profiles with other surrogates, such as the Automated Mutual Assistance Vessel Rescue System (AMVER), to evaluate relative merits in terms of completeness, transparency, consistency, comparability and accuracy. To complete the emissions inventory, activity level for ship engine systems and emissions factors are applied as in other inventories, discussed above.

The surrogate approach has desirable features for global inventories. It directly reveals global resolution and is easily updated for different years, updated activity information, or better emissions factors. It suffers from some of the same limitations as other sectoral inventories. Chief among these is the limited resolution, which results in an increasing loss in confidence if modellers try to zoom in beyond a large region to only a few cells.

The difficulty is in finding appropriate surrogates that can be verified and applied to different years. In the case of shipping, the weather data appears to be robust over the past several decades. The data might not be a good record before 1960 because of changes in fleets or their technologies. Related to this is the challenge of reconciling port-level or regional bottom up inventories with the globally derived data. However, the agreement in many cases appears to be very good and differences are equally attributable to uncertainties in the port- or regional-scale and global-scale inventories.

5. EXAMPLES OF ANTHROPOGENIC EMISSIONS INVENTORIES

5.1. The EDGAR 3 global inventories

5.1.1. Method

The EDGAR (Emission Database for Global Atmospheric Research) project is a comprehensive task carried out jointly by the National Institute for Public Health and the Environment (RIVM) and the Netherlands Organization for Applied Scientific Research (TNO) (Olivier et al., 1996,

1999a; Olivier and Berdowski, 2001) (http://www.rivm.nl/env/int/ coredata/edgar). This set of inventories combines information on all different emission sources, and it has been used over the past few years as a reference database for many applications. The work is linked into and part of the Global Emissions Inventory Activity (GEIA) of IGBP/IGAC (Bouwman et al., 1995, 1997; Olivier et al., 1996, 1999b; Olivier and Berdowski, 2001). The initial version of the full data set, EDGAR 2.0, provides global annual anthropogenic emissions for 1990 of greenhouse gases CO_2, CH_4 and N_2O and of precursor gases CO, NO_x, NMVOC and SO_2, on a regional basis and on a $1°\times1°$ grid. Similar inventories were compiled for a number of CFCs, halons, methyl bromide, and methyl chloroform. In the follow-up project the database was extensively validated and an uncertainty analysis was carried out. Most of the applications of EDGAR 2.0 over the last few of years are in model studies, but EDGAR data are also extensively used for policy applications for which emissions data on country level were calculated with the EDGAR information system. EDGAR datasets have also been used in IPCC Assessments, both on source strengths and on spatial distribution of emissions in the development of emission scenarios (IPCC, 2001).

The database has been updated to EDGAR 3.2, which includes an update and extension from 1990 to 1995 for all gases and extended time series for direct greenhouse gases to 1970-1995, as well as the new 'Kyoto' greenhouse gases HFCs, PFCs, SF_6 (Olivier et al., 2001). Selected time profiles for the seasonality of anthropogenic sources have also been identified. For the update of version 2.0 to version 3.2, the following principles were followed:

- *Activity data*: updated by including relevant statistics for the period 1970-1995, after checking for possible changes of source categories; this implies the inclusion of the 'new' countries, e.g. for the former SU.

- *Emission factors*: only to be changed for 1990 if validation showed major discrepancies; only to be changed for 1995 compared to 1990 if there are concrete indications that major changes have occurred that cannot be neglected; the same holds for factors for 1970, in particular for direct greenhouse gases.

- *Grid maps*: only to be updated if maps available of better quality or better applicability.

The EDGAR 3.2 data set includes new types of emissions, such as unintentional coal fires at shallow coal deposits, which appeared to be considerable for China. The Gulf war oil fires in Kuwait in 1991 have been included as separate source. Recognizing the importance of emissions related to biomass burning temperate vegetation fires have been added as an emission source based on the UN/ECE forest fire statistics for 1987 to 1997 (UN/ECE-FAO, 1996; UN/ECE-FAO, 1998; see also Chapter 3). Wastewater treatment and domestic waste combustion were also added as sources contributing substantially to CH_4 emissions.

Several significant changes are obtained when compared with the previous EDGAR 2.0 inventory. For example, agricultural waste burning emissions have decreased substantially, in particular for CO, due to lower fractions assumed to be burned and a substantial decrease in the CO emission factor. NMVOC emissions from domestic waste burning and from 'miscellaneous' industrial non-combustion processes have also decreased significantly after a re-analysis of the source data. The emissions from fossil fuel fires are now taken into account, which greatly increases fuel-related emissions in China, and NO_x from international shipping has also increased substantially due to an update of the emission factors. The spatial distribution of sources allocated with the population maps has noticeably changed too due to the introduction of another base map from Li (1996) and applying urban and rural maps where appropriate (Olivier et al., 2001).

The inventory is updated regularly due to changing emission factors, and because international statistics of activity data for the most recent years tend to vary during the first few of years after initial compilation. It should also be noted that data for the former Soviet Union have become rather weak due to inconsistencies between the sum of the new countries and the 1990 data for the former SU (Olivier et al., 2001).

5.1.2. Emissions

As it is the case with all global or regional inventories, the uncertainty in the emission inventories may be substantial, due to the limited accuracy of international activity data used and in particular of emission factors used for calculating emissions on a country level (Olivier et al., 1996, 1999a, 2001). The methods used are comparable with IPCC methodologies (IPCC (1996;

2001) which are based on international information sources, and global totals comply with budgets used in atmospheric studies. At regional and global levels part of the uncertainty caused by limited precision of national activity data may be reduced by aggregation to higher spatial levels. This is, however, not true for the emission factors used to construct the emission inventories, because these are often based on literature values in which sets of emission factors were compiled and reviewed (Olivier and Peters, 2002). Therefore the dataset could contain some bias, although resulting global or regional emissions are generally in line with other estimates and with atmospheric budgets.

5.1.2.1. Methane (CH_4)

Within the framework of EDGAR 3.2 a special effort was made to compile a new global dataset for anthropogenic CH_4 emissions, aiming at providing reference data on a per country basis and at providing emission trends for the period 1970-1995. The dataset is based not only on international statistics from organizations such as IEA, FAO and UNDP, but has been supplemented with more detailed activity data where relevant for emission calculations, e.g. surface and underground mining, different ecology types of rice cultivation, etc. The CH_4 emission factors used in EDGAR 3.2 are mostly based on the IPCC 'Tier 1' default values (i.e. for the simplest standard emission calculation method), in particular for the agricultural sector (Houghton et al., 1997). The emission factors are also region-specific and time-dependent for major sources where a large degree of uniformity across time and space was not considered likely. In particular, aggregated emission factors varying in time were used for coal production, rice production, and landfills. For example, coal production, rice production, surface and underground mining, and different ecology types of rice cultivation have very different emission factors, which should be accounted for when the mix of these types changes over time. Global total anthropogenic emissions of CH_4 have been estimated at about 250 Tg in 1970 and at about 300 Tg in 1990, staying at approximately this level until 1995.

Table 2 summarizes the CH_4 emissions by region and by major source category. At a global level the largest source categories are agriculture, fossil fuels, and waste handling, contributing about 45%, 30% and almost 20% to the global anthropogenic total in 1995, respectively. Within the agricultural sector, enteric fermentation by ruminants is by far the largest source, followed by rice cultivation. Within the waste sector, the analysis by Doorn et al. (1997, 1999) concludes that landfills, domestic and industrial wastewater disposal (latrines, septic tanks, open sewers, and wastewater treatment plants) appear to contribute about the same to global methane emissions. Within the fossil fuel category coal production and gas transmission are the largest sources. The regions contributing most to the global total in 1995 are East Asia with 16% followed by South Asia, USA, Latin America and the former Soviet Union (SU), each contributing 12% to 14%. In the 25-year period 1970-1995, global emissions of CH_4 have increased by about 20%. In the 1980s the growth was about 10%, predominantly due to the strong increase in gas production and transmission in the former SU. Enteric fermentation by ruminants and wastewater disposal, particularly in less developed regions, also contributed to this growth.

The declining economy of the former SU countries in the early 1990s had a large impact on the global trend in CH_4 emissions: coal and gas production emissions dropped substantially between 1990 and 1995. It should be stressed, however, that statistics for this region are rather uncertain for this period.

Table 2: Sources and regional contribution of emissions of CH_4 in 1995 (Tg). Source: EDGAR 3.2

	Total	Canada	USA	OECD Europe	Oceania	Japan	Eastern Europe	Former SU	Latin America	Africa	Middle East	South Asia	East Asia	South East Asia
Fossil fuel	91.1	2.1	21.5	4.3	1.3	0.8	4.7	24.0	3.4	3.5	6.0	1.9	13.5	4.0
Biofuel	13.9	0.0	0.3	0.1	0.0	0.0	0.1	0.3	0.5	3.6	0.2	4.0	2.9	1.8
Indust. processes	0.8	0.0	0.1	0.1	0.0	0.1	0.0	0.1	0.0	0.0	0.0	0.0	0.2	0.0
Agriculture	134.1	1.0	7.5	7.7	4.2	0.4	2.2	7.8	21.8	14.5	2.2	28.2	23.9	12.7
Biomass burning	6.5	1.5	0.2	0.1	0.1	0.0	0.0	0.1	2.0	1.0	0.0	0.1	0.0	1.3
Waste handling	55.7	1.2	10.3	5.0	0.6	1.6	1.2	3.7	7.2	4.1	1.5	8.1	7.6	3.5
Total	301.9	5.8	39.8	17.4	6.3	2.9	8.2	36.0	35.1	26.7	9.9	42.3	48.2	23.3

Within OECD Europe, CH_4 emissions from coal production have decreased substantially as a result of the policies in Germany and the United Kingdom to reduce the amount of domestic coal production. This compensated for the increasing emissions from the waste handling sector, in particular in Asia, and caused the total global emissions of methane to stabilize during the early 1990s.

5.1.2.2. Carbon Monoxide (CO) and Nitrogen Oxides (NO_x)

In EDGAR 3.2 the anthropogenic emissions of CO and NO_x have been estimated at about 860 Tg CO and 110 Tg NO_2 in 1995. The global distributions of the emissions of NO_x and CO are presented in Figure 4.

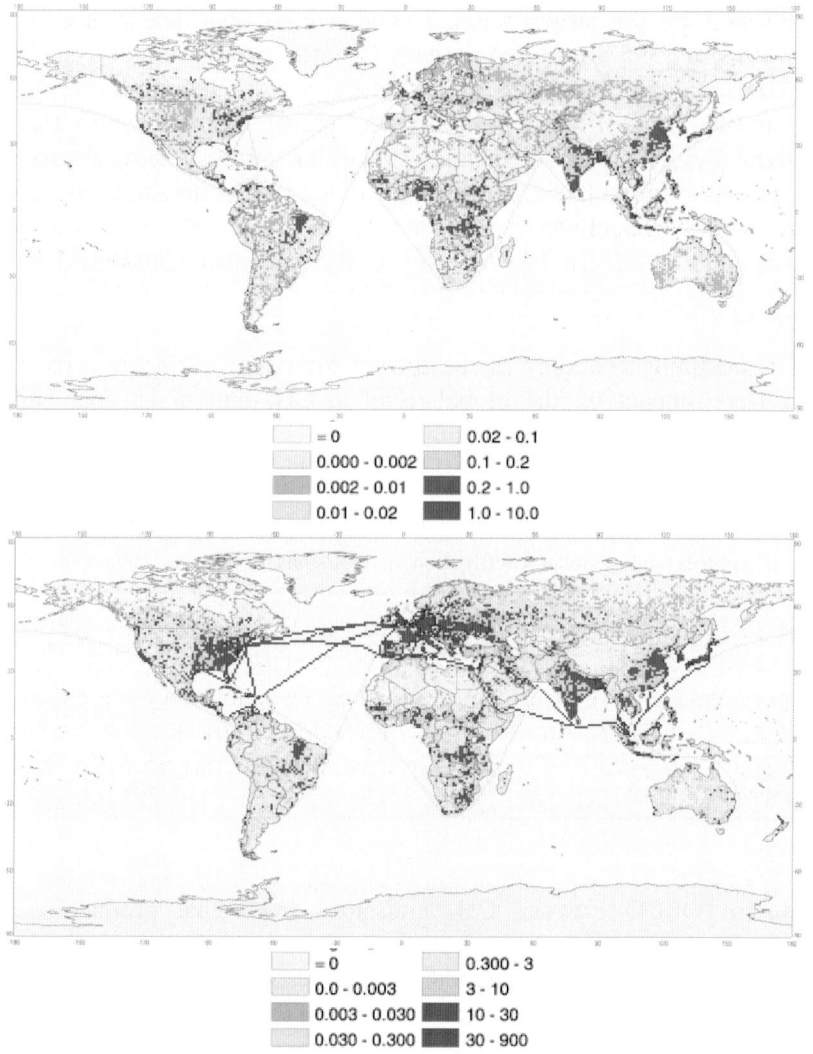

Figure 4 (see plate 4): Global distribution of CO (top, in Tg CO) and NO_x (bottom, in Gg NO_x) anthropogenic emissions in 1995. Source: EDGAR 3.2

Since 1990 total global anthropogenic emissions of these two species remained approximately constant, but the distribution of emissions across regions has changed substantially in this 5 year period: For example, in the former SU CO emissions have decreased by 45% or 35 Tg and NO_x emissions 37% or 6 Tg NO_2 due to the economic decline after the breakup of the union; in OECD Europe CO emissions decreased by 5% or 5 Tg and NO_x by 16% or 3 Tg as a result of emission control policies. The large emissions of CO in 1995 (45 Tg) are a result of the large extent of accidental vegetation fires in 1995. NO_x emissions increased significantly in Asia, by 25% to 45% (5 Tg in East Asia). Globally only the smaller NO_x sources show large changes; the largest sources (road transport, power generation and industrial combustion) remained almost constant in the early '90s, as shown in Tables 3 and 4.

Table 3: Sources and regional contribution of emissions of NO_x in 1995 (Tg NO_2). Source: EDGAR 3.2

	Total	Canada	USA	OECD Europe	Oceania	Japan	Eastern Europe	Former SU	Latin America	Africa	Middle East	South Asia	East Asia	South East Asia
Fossil fuel	83.5	1.9	19.8	13.0	1.5	2.9	2.4	9.1	5.7	3.2	4.5	4.1	12.2	3.1
Biofuel	7.7	0.0	0.5	0.1	0.0	0.0	0.0	0.2	0.5	1.5	0.1	2.3	1.5	0.9
Indust. processes	6.1	0.1	0.5	0.9	0.0	0.4	0.2	0.4	0.4	0.2	0.4	0.3	2.0	0.3
Agriculture	0.7	0.0	0.0	0.0	0.0	0.0	0.1	0.1	0.1	0.1	0.1	0.0	0.0	0.1
Biomass burning	13.0	1.8	0.2	0.1	0.7	0.0	0.0	0.1	3.3	6.0	0.0	0.1	0.1	0.7
Waste handling	0.1	0.0	0.0	0.0	0.0	0.0	0.0	0.1	0.0	0.0	0.0	0.0	0.0	0.0
Total	**111.3**	**3.8**	**21.0**	**14.2**	**2.3**	**3.4**	**2.8**	**9.9**	**10.1**	**11.1**	**5.1**	**6.7**	**15.9**	**5.1**

Over the five-year period 1990-1995 the distribution of CO emissions across sources categories has changed, but less than the regional distribution. For example, residential biofuel emissions increased by 7% or 15 Tg, road transport emissions increased by 6% or 10 Tg, and temperate forest fires increased by 25 Tg. Emissions from residential fossil fuel combustion decreased by 30% or 15 Tg and deforestation emissions decreased by 10% or 10 Tg.

Table 4: Sources and regional contribution of emissions of CO in 1995 (Tg CO). Source: EDGAR 3.2

	Total	Canada	USA	OECD Europe	Oceania	Japan	Eastern Europe	Former SU	Latin America	Africa	Middle East	South Asia	East Asia	South East Asia
Fossil fuel	278.3	6.0	76.7	36.2	4.2	6.9	8.8	26.9	27.2	11.2	16.6	5.1	42.6	10.0
Biofuel	231.6	0.4	5.6	2.1	0.6	0.1	1.2	5.8	9.6	56.3	2.7	68.4	49.2	29.6
Indust. processes	31.8	0.4	2.4	5.8	0.4	4.4	1.6	5.0	1.8	0.4	0.4	1.0	8.0	0.2
Agriculture	16.4	0.2	1.0	0.8	0.6	0.1	1.3	1.9	2.9	2.2	2.2	0.6	0.4	2.2
Biomass burning	298.9	45.7	4.9	2.8	12.8	0.1	0.0	2.3	83.3	114.4	0.4	2.2	1.4	28.5
Waste handling	3.8	0.1	1.3	0.2	0.1	0.5	0.1	0.2	0.3	0.2	0.1	0.3	0.4	0.1
Total	**860.8**	**52.7**	**91.9**	**47.9**	**18.7**	**12.0**	**13.1**	**42.2**	**125.1**	**184.8**	**22.3**	**77.6**	**101.9**	**70.7**

5.1.2.3. Non-Methane volatile organic compounds

In 1995 total global anthropogenic emissions of non-methane volatile organic compounds (NMVOCs) have been estimated at about 160 Tg (Table 5); this corresponds to an increase of about 5 Tg since 1990. The largest changes occurred in the former SU, where total emissions decreased by 40%; oil production decreasing by 200% and road transport by 70%, or 2 Tg each.. In contrast, road transport emissions in East Asia increased by 49% and in the USA by 10%, approximately 1 Tg each. Emissions from oil production in OECD Europe increased by 30% and in the Middle East by 10%, approximately 1 Tg each.

Table 5: Sources and regional contribution of emissions of NMVOC in 1995 (Tg). Source: EDGAR 3.2

	Total	Canada	USA	OECD Europe	Oceania	Japan	Eastern Europe	Former SU	Latin America	Africa	Middle East	South Asia	East Asia	South East Asia
Fossil fuel	79.0	2.4	10.7	10.1	1.4	3.0	1.4	13.5	8.4	4.8	11.5	2.0	5.5	4.5
Biofuel	28.0	0.0	0.8	0.3	0.1	0.0	0.1	0.7	1.4	6.9	0.3	8.1	5.9	3.5
Indust. processes	25.4	0.5	6.6	5.4	0.3	2.6	0.9	2.6	1.6	0.8	0.6	0.8	1.9	0.9
Agriculture	2.0	0.0	0.1	0.1	0.1	0.0	0.2	0.2	0.3	0.3	0.3	0.1	0.0	0.3
Biomass burning	22.4	6.1	0.7	0.4	0.9	0.0	0.0	0.3	4.9	7.4	0.0	0.1	0.1	1.5
Waste handling	2.7	0.1	0.7	0.5	0.0	0.3	0.1	0.3	0.2	0.1	0.1	0.1	0.2	0.1
Total	159.6	9.2	19.5	16.8	2.8	5.9	2.7	17.6	16.8	20.1	12.7	11.1	13.7	10.8

5.1.2.4. Sulfur Dioxide (SO_2)

In EDGAR 3.2 the total global anthropogenic emissions of SO_2 for 1990 have been estimated at about 155 Tg, with a slowly decreasing trend in subsequent years (Table 6). The decrease is mainly due to control measures implemented in OECD Europe and the USA and to the declining economy of the former SU countries. Decreases in power generation and industrial combustion in the former SU countries more than compensated for the high growth of SO_2 emissions in Asia, which showed an average 5-year growth rate of about 30% in the 1990-1995 timeframe.

Table 6: Sources and regional contribution of emissions of SO$_2$ in 1995 (Tg SO$_2$). Source: EDGAR 3.2

	Total	Canada	USA	OECD Europe	Oceania	Japan	Eastern Europe	Former SU	Latin America	Africa	Middle East	South Asia	East Asia	South East Asia
Fossil fuel	111.2	2.2	17.1	11.4	1.0	1.5	8.9	13.0	4.9	3.8	5.1	5.4	34.0	3.0
Biofuel	2.9	0.0	0.0	0.0	0.0	0.0	0.0	0.0	0.1	0.6	0.0	1.5	0.5	0.1
Indust. processes	25.0	0.4	1.0	5.0	0.5	0.6	1.6	3.0	4.4	1.3	0.5	0.5	5.7	0.5
Agriculture	0.2	0.0	0.0	0.0	0.0	0.0	0.0	0.0	0.0	0.0	0.0	0.0	0.0	0.0
Biomass burning	2.5	0.2	0.0	0.0	0.1	0.0	0.0	0.0	0.7	1.1	0.0	0.0	0.0	0.2
Waste handling	0.0	0.0	0.0	0.0	0.0	0.0	0.0	0.0	0.0	0.0	0.0	0.0	0.0	0.0
Total	141.9	2.8	18.1	16.5	1.7	2.2	10.5	16.0	10.1	6.9	5.7	7.5	40.2	3.8

5.2. The EMEP regional inventory

The EMEP inventories consist of emission data officially reported annually by the 49 Parties to the Convention of Long Range Transboundary Air Pollution (CLRTAP) (Vestreng and Klein, 2002, and http://www.unece.org/env/lrtap/). These inventories include emissions from 1980 to 2000, and projections for 2010 and 2020, for SO$_2$, NO$_x$, ammonia (NH$_3$), NMVOC, CO, Particulate Matter (Total Suspended Particulates, TSP; Particulate Matter with diameter < 2.5 µm, PM$_{2.5}$; Particulate Matter with diameter < 10 µm, PM$_{10}$; all by mass), selected heavy metals (HMs), and Persistent Organic Pollutants (POPs). National totals, data by sectors, and gridded data are available from WebDab, which is the web version of the UNECE/EMEP emission database, at the EMEP website: http://webdab.emep.int/ and documented in the EMEP Emission Report (Vestreng and Klein, 2002). Wherever there are spatial or temporal gaps in the inventory expert estimates of emissions are developed and included in the information available to the public. The EMEP inventory is updated during the spring each year. The gridded emission data are available in 50 x 50 km, 150 x 150 km, and 1° x 1° grid resolutions.

To assist the CLRTAP countries in their work of estimating and submitting transparent, consistent, comparable, and accurate emission estimates, an atmospheric emission inventory guidebook has been made available on the internet at http://reports.eea.eu.int/technical_ report_2001_3. It is the responsibility of the countries to assure the accuracy of the reported emissions. To serve as input to scientific studies within and outside the EMEP community the data still need to be scrutinized and in some cases replaced by or completed with expert estimates. Gaps in total national emissions are filled by data from documented sources like the GEIA inventories, the International Institute for Applied Systems Analysis (IIASA) data, or by linear interpolation. Sector emissions are completed by applying an average sector distribution developed for Europe based on reported data available in the UNECE/EMEP database. The spatial distribution is developed based on knowledge about point source and population distribution. Work is ongoing within EMEP to improve and document methods used to create expert estimates

Table 7: EMEP national total emissions for 1995 (Gg/year). Emissions officially reported by the Party to the Convention on LRTAP are displayed with no background. Values in grey shaded cells are interpolated and/or drawn from documented sources. From Vestreng and Klein, 2002 (NO_x as NO_2; NA: non available).

Area/Species	SO_2	NO_x	NH_3	NMVOC	CO	PM2.5	PM10
Albania	72	24	32	31	84	6.13	8.21
Armenia	2.5	14.9	25	23.4	173.6	4.73	6.56
Austria	53.82	182.7	74.13	275.7	1098	27.64	46.81
Belarus	275	195	142	367	1253	38.47	62.36
Belgium	245.4	324.9	97.30	250.3	1013	57.06	84.12
Bosnia and Herzegovina	480	80	31	51	280	5.67	9.85
Bulgaria	1476	266	99	173	846	37.64	93.06
Croatia	70.4	65.7	24.9	74.1	345.8	13.74	21.03
Cyprus	41	19	4	19	67	1.76	2.96
Czech Republic	1091	412	86	286	874	57.38	125.42
Denmark	149.0	261.4	112.2	152.8	688.3	12.10	27.07
Estonia	118.5	42.06	10.97	47.5	242.3	13.69	33.27
Finland	96	258	35.2	189.0	436	20.02	30.03
France	995	1709	758	1979	8880	336.00	588.00
Georgia	20.3	26.6	97	1.5	249.5	6.99	10.03
Germany	1994	1967	635	2024	6667	217.10	335.28
Greece	528	309	85	329	1316	41.87	62.44
Hungary	705.0	190.1	77.00	150.3	761.3	27.78	60.24
Iceland	23.9	28.4	3	12.0	49.4	1.62	2.03
Ireland	161.2	115.3	119.6	105.4	304.4	12.85	22.66
Italy	1322	1768	461	2368	7755	232.27	319.31
Kazakhstan	140	76	18	76	266	NA	NA
Latvia	58.98	41.76	16.82	64.04	436.5	8.57	13.21
Lithuania	94	65	38	77	286	12.68	19.83
Luxembourg	9	21	7	16	107	2.79	5.19
Netherlands	141.4	483.5	186.2	369.6	894.0	41.42	60.90
Norway	33.57	222.7	25.99	367.8	746.6	43.09	48.98

Table 7 (Cont'd). EMEP national total emissions for 1995 (Gg/year).

Poland	2376	1120	380	769	4547	126.93	313.82
Portugal	365.6	357.8	101.7	461.6	1201	36.98	50.99
Republic of Moldova	64.06	38.2	33	61.7	192	9.83	16.36
Romania	912	319	221	638	2325	92.84	185.59
Russian Federation	2969	2570	824	2857	9945	895.71	1709.1
Slovakia	239	174	39.6	159	404	22.98	40.57
Slovenia	125	67	22	44	91	6.68	13.22
Spain	1808	1355	467	1536	3569	158.46	225.58
Sweden	68.56	309.2	61	471.5	993.6	29.74	42.05
Switzerland	33.55	120	69.2	199.4	490.9	15.48	28.22
The FYR of Macedonia	105	30.4	17	19	77	9.58	27.39
Turkey	1772	800.5	321	677.3	3987	204.43	390.67
Ukraine	1639	531.0	729	811.0	2906	281.45	608.11
United Kingdom	2363	2088	318	2054	5522	132.30	237.86
Yugoslavia	462	59	90	142	207	48.88	144.33
North Africa	413	96	235	96	336	NA	NA
Remaining Asian areas	869	212	303	212	742	NA	NA
Baltic Sea	228	352	0	8	29	NA	NA
Black Sea	57	86	0	2	8	NA	NA
Mediterranean Sea	1189	1639	0	34	139	NA	NA
North Sea	454	648	0	15	59	NA	NA
Rem. N-E Atl. Ocean	901	1266	0	25	111	NA	NA
marine emissions	742	0	0	0	0	NA	NA
Volcanic emissions	2000	0	0	0	0	NA	NA
Total EMEP	32552	23407	7533	21173	74001	NA	NA

Within EMEP, the inventories are used to check the compliance of countries with different United Nations (UN) Protocols (see http://www.unece.org/env/lrtap for details) and as input to a multitude of atmospheric models used in assessment studies.

A new set of guidelines for estimating and reporting emissions data for CLRTAP were adopted in 2002 (http://www.unece.org/env/documents/2002/eb/ge1/eb.air.ge.1.2002.7.e.pdf) The major changes include the harmonization of the definition of sectors with those of the United Nations Framework Convention on Climatic Change (UNFCCC) common reporting format, and facilitating reports of information on large point sources and activity data. Table 7 lists the areas covered by the EMEP modeling domain; and the 1995 total emissions of SO_2, NO_x, NH_3, NMVOC, CO, $PM_{2.5}$, and

PM_{10} in these areas are tabulated as an example of emission data available from WebDab.

Figure 5 presents the spatial distribution of the NO_x emissions in 50 x 50 km grid resolution. A base grid representing emissions in the year 2000 is downscaled to the year 1995. The user of WebDab can however choose any presently available year: 1980-2000, 2010 and 2020.

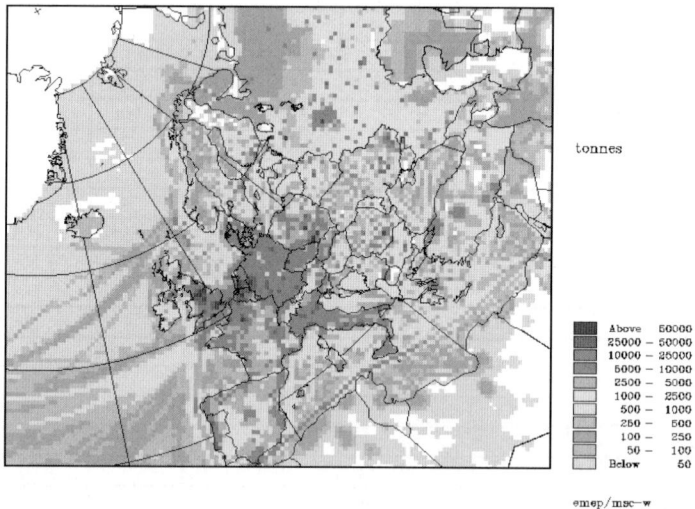

Figure 5 (see plate 5). European emissions of NO_x in 1995 at 50 km grid resolution (Mg as NO_2) (reproduced with permission of Norwegian Meteorological Institute/EMEP/MSC-W).

The comparison shown in Table 8 between the EMEP and the EDGAR3.2 emission data for NO_2 provides and excellent illustration of the point made by the Marland study of CO_2 emissions (section 3.2). While the total percentage difference between the EMEP and EDGAR3.2 data for OECD EUROPE is only 6 percent, differences up to 70 percent can be seen for individual countries. The largest differences are seen for the countries with the smallest emissions. The absolute difference in Giga Grams is larger than the emissions from 13 out of 18 countries. For the Eastern European countries, the total difference is even smaller, 3 percent, while differences for individual countries that have reported data to EMEP reach 173 percent, and almost 50 percent of the countries have emissions larger than the absolute difference.

It is interesting to note that for OECD EUROPE the EDGAR3.2 inventory in most cases underestimates emissions as compared to the countries' own estimates. For the Eastern European countries, there is a fifty-fifty split between over and under estimation while all the Former Soviet Union (SU) countries report less emissions than estimated by EDGAR. There is a large (more than 100 percent) underestimation in the reported emission data by the former SU countries relative to the EDGAR data, and the absolute difference exceeds the emissions from all the individual countries. The underestimation increases to more than 200 percent, when excluding Kazakhstan and the Russian Federation. For these two countries the EMEP estimates only consider the European part of these countries' territories, while EDGAR includes the total territory.

Table 8: Comparison between EMEP and EDGAR 3.2 emission inventories for NO2 emissions in 1995 (For the Russian Federation, EMEP figures apply for the European part of territory only).

Area	OECD EUROPE			
	NOx, EMEP (Gg)	NOx, EDGAR (Gg)	Difference (Gg)	Difference (%)
Austria	182.7	202.0	-19.3	-11
Belgium	324.9	371.4	-46.5	-14
Denmark	261.4	223.0	38.4	15
Finland	258.0	207.6	50.4	20
France	1709.0	1482.7	226.3	13
Germany	1967.0	2001.2	-34.2	-2
Greece	309.0	327.8	-18.8	-6
Iceland	28.4	8.5	19.9	70
Ireland	115.3	122.3	-7.0	-6
Italy	1768.0	1533.1	234.9	13
Luxembourg	21.0	34.0	-13.0	-62
Netherlands	483.5	450.3	33.2	7
Norway	222.7	157.9	64.8	29
Portugal	357.8	282.0	75.8	21
Spain	1355.0	1204.2	150.8	11
Sweden	309.2	234.0	75.2	24
Switzerland	120.0	124.5	-4.5	-4
United Kingdom	2088.0	2220.1	-132.1	-6
Sum	**11880.9**	**11186.6**	**694.3**	**6**

Table 8 (Cont'd) Comparison between EMEP and EDGAR 3.2 emission inventories.

EASTERN EUROPE				
Albania	24.0	13.5	10.5	44
Bosnia and Herzegovina	80.0	25.5	54.5	68
Bulgaria	266.0	195.0	71.0	27
Croatia	65.7	76.0	-10.3	-16
Czech Republic	412.0	270.7	141.3	34
Hungary	190.1	222.1	-32.0	-17
Poland	1120.0	1135.3	-15.3	-1
Romania	319.0	371.5	-52.5	-16
Slovakia	174.0	155.7	18.3	11
Slovenia	67.0	62.3	4.7	7
The FYR of Macedonia	30.4	33.6	-3.2	-11
Yugoslavia	59.0	161.3	-102.3	-173
Sum	**2807.2**	**2722.6**	**84.6**	**3**
FORMER USSR				
Armenia	14.9	18.2	-3.3	-22
Belarus	195.0	252.5	-57.5	-29
Estonia	42.1	60.6	-18.5	-44
Georgia	26.6	13.0	13.6	51
Kazakhstan	76.0	652.7	-576.7	-759
Latvia	41.8	49.6	-7.9	-19
Lithuania	65.0	81.5	-16.5	-25
Republic of Moldova	38.2	55.2	-17.0	-44
Russian Federation	2570.0	5148.1	-2578.1	-100
Ukraine	531.0	2411.9	-1880.9	-354
Sum	**3600.5**	**8743.3**	**-5142.7**	**-143**

While the EDGAR emission data rely on international activity data and emission factors from the literature, the EMEP emission data are based on the individual countries' best estimates and or measurements of activity data and emission factors. The methodology used to estimate emissions should be comparable between countries, through the use of the EMEP/CORINAIR emission inventory Guidebook. There is however currently a lack of complete transparency in the EMEP reporting of emissions, and it is difficult to validate the inventories. The new Guidelines for reporting and estimating emissions data to UNECE/EMEP (UNECE, 2002), is aiming at enhancing the completeness, transparency, consistency, comparability and accuracy of the reported data.

5.3. The United States Environmental Protection Agency (USEPA) National Inventories

The Emission Factor and Inventory Group of the United States Environmental Protection Agency (USEPA) in Research Triangle Park, NC, USA has prepared over the past years a national database of air emissions information with input from numerous US State and local air agencies, from Native American tribes, and from industry. This database contains information on stationary and mobile sources that emit so-called criteria air pollutants (Criteria air pollutants are those for which EPA has set health-based standards. Currently the following species are included as criteria pollutants: sulfur dioxide (SO_2), ozone (O_3), nitrogen oxides (NO_x), carbon monoxide (CO), lead (Pb), and particulate matter (PM)) and their precursors, as well as hazardous air pollutants (HAPs). The database includes estimates of emissions, by source, of air pollutants in each area of the country on an annual basis. The National Emission Inventory (NEI) includes emission estimates for all 50 States, the District of Columbia, Puerto Rico, and the Virgin Islands. Information and data are available at the NEI website: http://www.epa.gov/ttn/chief/trends/. Emission estimates for individual point or major sources (facilities), as well as county level estimates for area, mobile and other sources, are currently available for years 1985 through 1999 for criteria pollutants, and for years 1996 and 1999 for HAPs (see Figure 6 for resulting SO_2 emissions in 1998).

Data from the NEI are used for air dispersion modelling, regional strategy development, regulation development, air toxics risk assessment, and following trends in emissions over time. For emission inventories prior to 1999, criteria pollutant emission estimates were maintained in the National Emission Trends (NET) database and HAP emission estimates were maintained in the National Toxics Inventory (NTI) database. Beginning with 1999, criteria and HAP emissions data are being prepared in a more integrated fashion in the NEI, which combines the NET and the NTI.

Four of the six criteria pollutants are included in the NEI database: CO, NO_x, SO_2, and PM (PM_{10} and $PM_{2.5}$). The NEI includes emissions of Volatile Organic Compounds (VOC) emitted from motor vehicle fuel distribution, chemical manufacturing, and other solvent uses because VOC, along with NO_x, are ozone precursors. Emissions of NH_3 are also included in the NEI because this species is a precursor of PM.

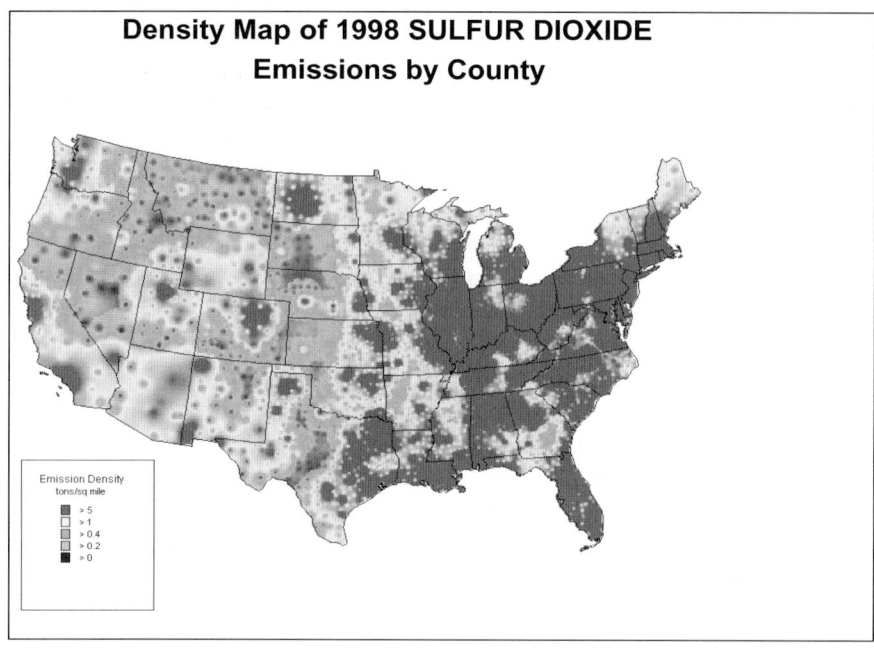

Figure 6 (see plate 6). Emissions of SO_2 in the United Stated in 1998 (short tons/mile^2). From EPA web site http://www.epa.gov/ttn/chief/trends.

The NEI database defines three classes of criteria air pollutant sources:

i) Point sources - stationary sources of emissions, such as an electric power plant, that can be identified by name and location. A "major" source emits a threshold amount (or more) of at least one criteria pollutant, and must be inventoried and reported. Many states also inventory and report stationary sources that emit amounts below the thresholds for each pollutant.

ii) Area sources - small point sources such as a home or office building, or a diffuse stationary source, such as wildfires or agricultural tilling. These sources do not individually produce sufficient emissions to qualify as point sources. Dry cleaners are one example, i.e., a single dry cleaner within an inventory area typically will not qualify as a point source, but collectively the emissions from all of the dry cleaning facilities in the inventory area may

Inventories of Anthropogenic Emissions　　　　　　　　　　　　　　　　　　　　45

be significant and therefore must be included in the inventory. The NEI includes facility data for some area sources and aggregated emission estimates at the county level for the remaining area sources.

iii) Mobile sources - any kind of vehicle or other mobile equipment with a gasoline or diesel engine: on-road vehicles, non-road 2- and 4-stroke and diesel engines, off-road vehicles, aircraft, locomotives, and commercial marine vessels. The NEI includes aggregated emission estimates at the county level for mobile sources.

The main sources of criteria pollutant emissions data for the NEI are:

- Stationary sources:
    o For electric generating units - EPA's Emission Tracking System / Continuous Emissions Monitoring Data, Department of Energy fuel use data, and EPA's Clean Air Market program.
    o For other large stationary sources - state and local environmental agencies, with data from older inventories included where no state data were submitted.

- Mobile sources:
    o For on-road mobile sources (cars, trucks, etc.) - the Federal Highway Administration's estimate of vehicle miles travelled and emission factors from EPA's MOBILE Model (US Environmental Protection Agency, 2002).
    o For non-road mobile sources (locomotives, construction equipment, etc.) - EPA's NONROAD Model (Pechan et al., 2002 and http://www.epa.gov/ttn/chief/net/index.html#1999)

- For stationary area sources - state data, EPA-developed estimates for some sources, and older inventories where no state or EPA data were submitted.

Hazardous air pollutants, also known as toxic air pollutants, are those pollutants that are known or suspected to cause serious health problems. They are defined in the US Clean Air Act (CAA) which identifies a list of 188 pollutants as HAPs. The list of HAPs includes relatively common pollutants such as benzene, formaldehyde, methanol, and asbestos, as well as numerous less common substances.

The NEI database includes emission estimates for the 188 HAPs from major stationary, area sources, and mobile sources, as defined in the CAA. As for the criteria pollutants, the HAP emission sources are grouped in three classes:

1. the major sources, which correspond to stationary (point) sources that emit or have the potential to emit 10 tons per year or more of any listed HAP or 25 tons per year or more of a combination of listed HAPs. The NEI includes data for each individual major source, including the name and location of the facility.

2. area and other stationary sources that emit or have the potential to emit less than 10 tons per year of a single HAP and less than 25 tons per year of all HAPs combined, for example, neighbourhood dry cleaners and gas stations. Although emissions from individual area sources are often relatively small, collectively their emissions can be of concern. Other stationary sources of emissions, such as wildfires and prescribed burning, are addressed through the burning policy agreed to by EPA and the Department of Agriculture. The NEI includes aggregated emission estimates at the county level for these other sources.

3. mobile sources, with the same definition as for criteria pollutants. The NEI includes aggregated emission estimates at the county level for these sources.

The estimates of these HAP emissions rely on various sources: State, local, and tribal HAP inventories, existing databases related to EPA's Maximum Achievable Control Technology (MACT) programs to reduce HAP emissions, Toxic Release Inventory (TRI) data, the mobile source methodology developed by EPA's Office of Transportation and Air Quality (OTAQ), and the use of emission factors and activity data for stationary non-point sources. More information about the NEI database and the compilation of criteria pollutant and HAP emissions inventories, and links to the database, are available on EPA's website for the Clearinghouse for Inventories and Emission Factors at www.epa.gov/ttn/chief.

5.4. Asian Inventories

Inventories of base-year 1995 or later covering East, Southeast, and/or South Asia are summarized in Table 9. Emission inventories for China, Japan and India are also included.

Table 9: Summary of research for regional emission inventory in Asia. BC refers to black carbon and OC to organic carbon.

Research group	Base year	Domain	Species	Grid size, degree	References
ACESS	2000	All Asia	SO_2, NO_x, CO, NMVOC, BC, OC, NH_3, CH_4	1	ACESS (2002)
FRSGC	1995	All Asia	SO_2, NO_x, CO, NMVOC, BC, NH_3, N_2O, CH_4	0,5	Ohara et al. (2001), Yan et al. (2002)
RAINS-ASIA	1995	All Asia	SO_2	1	IIASA (2001)
Streets et al.	1985-1997	All Asia	SO_2, NO_x	-	Streets et al. (2001, 2002)
Klimont et al.	1995	East Asia	SO_2, NO_x, NMVOC, NH_3	1	Klimont et al. (2001)
Murano et al.	1994-1996	East Asia	SO_2, NO_x, NMVOC, NH_3	0,5	Murano et al. (2002)

5.4.1 Emissions

5.4.1.1 The ACCESS inventory

Streets and co-workers at Argonne National Laboratory (ANL) prepared emission inventories for use during the ACE-Asia and TRACE-P field campaigns carried out in East Asia/Western Pacific region in spring 2002, and made the data available on the web (Ace-Asia and TPACE-P Modelling and Emission Support System: ACESS, 2002). The campaigns focused on the characterization of gaseous and aerosol species in Asian outflow to the western Pacific Ocean and the inventories includes SO_2, NO_x, CO, NMVOC, BC (black carbon), OC (organic carbon), NH_3 and CH_4. Twenty-two countries and regions in East, Southeast and South Asia are included, and $1° \times 1°$ grid maps for the base year 2000 were provided. NMVOC from biomass burning and other biogenic sources, Mg and Ca from soils, CH_4 from wetlands, and SO_2 from volcanoes were also included in the ACESS inventories.

5.4.1.2 The RAINS-Asia inventory

The Regional Air Pollution Information and Simulation (RAINS-Asia) is a project organized by the International Institute for Applied Systems Analysis (IIASA) and funded by the World Bank for the purpose of constructing a policy tool for the mitigation of acid rain in Asia. In this study emission inventories for SO_2 with base years of 1990 and 1995, and also projections from 2000 to 2030 are estimated, and the results are available in a CD-ROM (IIASA, 2001). Based on these inventories Streets et al. (2000, 2001) reported trends of country-bases emission of SO_2 and NO_x in Asia during 1985 and 1997.

5.4.1.3 The FRSGC inventory

Frontier Research System for Global Change (FRSGC) is now constructing emission inventories for the purpose of studying air quality change/climate change in Asia (Ohara et al., 2001). The inventories include country-based data for SO_2, NO_x, CO, NMVOC, and BC for 1995. Updates for 2000, as well as inventories of NO_x, NH_3, N_2O, and CH_4 from agricultural sources with consideration of regional specificity in Asia are underway (Yan et al, 2003;, Yamaji et al., 2003). Gridded data at a $0.5° \times 0.5°$ resolution will be available on the web in the near future.

5.4.2 Comparison between Asian emission inventories

5.4.2.1 Sulphur Dioxide (SO_2)

A summary of the SO_2 emissions in Asia from different inventories is presented in Table 10. Emissions from country-based inventories typically agree within 15%, and total emissions in Asia agree within 4% except for the EDGAR 3.2 values.

Table 10: Summary of Asian SO_2 emissions (Tg SO_2/year)

Inventory / Area	ACESS* (2000)	FRSGC (1995)	RAINS (1995)	Streets et al. (1995)	Klimont et al. (1995)	Murano et al. (1995,96)	EDGAR 3.2 (1995)
East Asia Total	22.6	28.3	26.5	28.6	24.6	26.3	42.3
Southeast Asia Total	3.1	3.0	3.3	3.1	-	-	3.8
Indian Subcontinent Total	7.1	5.9	6.2	6.7	-	-	7.4
Ships	1.1	-	0.8	-	-	-	*)
Asia Total	**33.9**	**37.2**	**36.8**	**38.5**	**-**	**-**	**53.5**

*) Global grid-based emissions include main shipping routes in Asia.

Table 11 compares results for China from Xue et al. (1998) and Streets et al. (2000) with the values estimated by the State Environmental Protection Administration, China (SEPA). A recent decreasing trend in SO_2 emission is discernible in SEPA's data, possibly caused by an effort to change to low sulphur coal and to convert from coal to other fuels.

Table 11: Summary of emissions in China (Tg/year)

Inventory / Component	SEPA			Xue et al.	Streets et al.		
	(1995)	(1997)	(1998)	(1995)	(1995)	(1996)	(1997)
SO_2	23.7	22.7	20.9	23.7	25.9	26.4	25.1
NO_x	-	-	-	10.7	11.2	-	12.5

EDGAR 3.2 emissions from China are estimated at 35 Tg/yr, compared to 20-25 Tg/yr from inventories by SEPA, Xue and Streets and coworkers. In EDGAR all other East Asian emissions are overestimated by more than a factor of 2-3, except emissions from Mongolia which are greatly underestimated. The region-wide estimates presented in Table 9 reflect what are considered to be the best information on sulphur content of fuels and the

results of control policies in this area. Because the EDGAR datasets are the most widely used inventories for global modelling, the significant discrepancy in East Asia should be resolved.

Table 12: Summary of emissions in India (Tg/year)

Compound	Inventory Garg et al. (1995)	Reddy & Venkataraman [1] (1996-97)	ACESS [1] (2000)	FRSGC (1995)
SO_2	4.64	4.33	5.46	4.93
NO_x	3.46	-	4.05	4.58
BC	-	0.31	0.52	0.51
OC	-	0.69 [2]	2.19	-

1) Excluding forest biomass for comparison with other results.
2) Organic Matter/OC ratio assumed as 1.3 according to Reddy and Venkatraman (2002a,b).

Table 12 compares estimates for India in regional inventories with those obtained by Indian researchers; these estimates agree within 20%. For South East Asia and South Asia agreement between EDGAR 3.2 and other studies falls within a range of 20% although EDGAR 3.2 is in the upper end of estimate.

5.4.2.2 Nitrogen Oxides (NO_x)

Estimates of NO_x emissions are shown in Table 13. Estimates are within 25% for most of the major emitting countries and for the Asian total, including EDGAR. The EDGAR 3.2 estimate is at the upper end but still within the scatter of the estimates. The estimates of Xu et al. for China agree very well with other studies. The value for India by Garg et al. (2001) is at the lower end in the 25% range.

Table 13: Summary for NO$_x$ emissions (Tg NO$_2$/year)

Area	ACESS *) (2000)	FRSGC (1995)	Streets et al. (1995)	Klimont et al. (1995)	Murano et al. (1995,96)	EDGAR 3.2 (1995)
East Asia Total	14.9	14.8	17.2	13.9	14.4	17.8
Southeast Asia Total	3.1	2.9	3.2	-	-	4.1
Indian Subcontinent Total	4.8	5.4	5.2	-	-	6.5
Asia Total	**22.7**	**23.1**	**25.6**	**-**	**-**	**28.4**

5.4.2.3 Carbon Monoxide (CO)

For CO (Table 14) substantial differences can be seen in Southeast Asia, especially between ACESS, FRSGC and EDGAR inventories. Modelling results and observational data in ACE-Asia indicate that CO emissions in China are underestimated. This underestimate is believed to be mainly in the emissions from the domestic sector where biofuels and coal are widely used. Further studies would be necessary to reduce the uncertainties in the emission factors of CO from biogenic fuels and coal, because these factors are also very important in the emissions estimates from other developing countries in Asia.

Table 14: Summary for CO emissions (Tg CO/year)

Area	ACESS *) (2000)	FRSGC (1995)	EDGAR 3.2 (1995)
East Asia Total	115.0	92.1	111.5
Southeast Asia Total	34.0	23.3	70.6
Indian Subcontinent Total	62.2	50.7	75.0
Asia Total	**211.4**	**166.1**	**257.1**

*) Excluding forest biomass burning for comparison with other results.

5.4.2.4 Other species

The agreement for NMVOC and NH_3 is remarkably good as seen in Tables 15 and 16. However, this does not mean that the accuracy of the inventories is high but rather that the values of emission factors and activity data are limited and the same values are used in most inventories. Estimates of NMVOC emissions for 1995 in the FRSGC inventory were provided by collaboration with Streets and are essentially the same as the estimates in the ACESS inventory for 2000. Table 12 includes the estimates for BC and OC It is not surprising that the discrepancies in the emissions for these species are large, and more elaboration by both top-down observation approach and bottom-up emission factor experiments are necessary.

Table 15: Summary for NMVOC emissions (Tg/year)

Inventory Country	ACESS * (2000)	FRSGC (1995)	Klimont et al. (1995)	Murano et al. (1994-96)	EDGAR 3.2 (1995)
East Asia Total	18.5	18.5	17.1	17.9	19.5
Southeast Asia Total	11.1	11.1	-	-	10.8
Indian Subcontinent Total	10.7	10.7	-	-	10.9
Asia Total	**40.2**	**40.3**	**-**	**-**	**41.1**

Table 16: Summary for NH_3 emissions (Tg NH_3/year)

Inventory Area	ACESS * (2000)	IIASA (1995)	Murano et al. (1995,96)	Bouwman et al. (1997) (1990)
East Asia Total	14.2	12.7	13.1	10.6

*) Excluding forest biomass burning for comparison with other results.

In conclusion, although SO_2 and NO_x inventories should be rather mature, large discrepancies exist between EDGAR 3.2 and other region-wide inventories, especially for China and other East Asian countries. For CO,

NMVOC, NH_3, BC, and OC verification of emission inventory data using observational data with regional representativeness is absolutely necessary.

6. PAST HISTORY OF ANTHROPOGENIC EMISSIONS

6.1 Introduction

Despite the importance of knowing the extent of past anthropogenic emissions, only a few attempts have been made to estimate time series of global historical emissions (especially before 1970), even less on a relatively detailed sectoral basis and on a high resolution grid-basis. For isolated species several inventories are available. For greenhouse gas emissions, Andres et al. (1999) estimated CO_2 emissions from fossil fuel combustion and cement production by country and 1° x 1° grid for the period 1751 to the present. Stern and Kaufmann (1995) estimated global emissions of CH_4 for the years 1860-1993. For acidifying gases several studies have estimated SO_2 emissions. Hameed and Digon (1988) estimated emissions in the period 1960-1980 for the USA, OECD Europe and the rest of the World on a 10° x 10° grid. Mylona (1996) estimated emissions for individual European countries, including Eastern European countries and SU for the period 1880-1990 on a 150x150 km grid. Lefohn et al (1999) estimated detailed emissions from 1850 to 1990 with data per country and for some sectors.

Global gridded emissions of CO_2, CO, CH_4, NMVOC, SO_2, NO_x, N_2O, and NH_3 by sector for the period 1890-1990 have been estimated by Van Aardenne et al. (2001) using a consistent and transparent methodology. The emissions have been computed using an emission factor approach. The historical activity data were taken from the HYDE database (Klein Goldewijk and Battjes, 1997) supplemented with other data and estimates developed during this work. Historical emission factors per process were based on uncontrolled emission sources included in EDGAR V2.0 (Olivier et al., 1996; 1999a). The emission database describes anthropogenic source categories such as fossil fuel production and combustion, industrial production, agricultural practices, waste handling, and land-use related activities. Although a consistent database of historical emissions is a useful tool for modeling past atmospheric changes due to human activities, users should be aware of the limitations of such a large-scale historical emission inventory. Information on activities and emission factors in the past is limited, uncertain. and sometimes non-existent, his leads in some cases to scaling back of current activity data rates using indicators and the application of global aggregated emission factors. The methodology used for estimating emissions from fossil fuel combustion will be described as an example.

6.2 Methodology for fossil fuel combustion emissions

Because emissions from fuel combustion differ between sectors, fuel types, and regions the combustion of coal, oil or gas for the power generation, industry, transport and residential energy sectors were addressed separately. For the entire period there were no resources which provided information at the required detail. Detailed energy statistics by country and for several fuel type and sectors were available from the International Energy Agency (IEA, 1994) only for the period 1970-1990. An activity data set was derived. for two other periods, 1930-1960 and 1890-1920. For the years 1930, 1940, 1950 and 1960 energy consumption information was derived based on Darmstadter (1971). The Darmstadter study provided total annual consumption statistics for three main fuel types (solids, liquids and gas) together with information on electricity and hydroelectricity consumption. The data were available for selected years only by region and for some countries without breakdown by sector. To differentiate between the four emitting sectors the following methodology was used. The amount of fossil fuel needed for power generation was determined by combining data on hydroelectricity production (Darmstadter, 1971), data on efficiency in electricity production in the past (Etemad et al., 1991) and the data on fuel mix used to produce electricity from Darmstader (1971). The amount of fuel used for non-power generation was determined by subtracting the amount and type of fuel used for power generation from the total fuel consumption of each type. To distinguish between industrial, transport and residential energy use an additional step was needed. Based on detailed IEA statistics (IEA, 1994) the ratio between industry, transport and residential energy use could be determined for the situation in the 1970's. To estimate the ratio of energy use in the industry, transport and residential sector for years prior to 1970, the sectoral split for 1970 was scaled back in time using indicators that can be associated with fuel use in the different sectors. The consumption of fuel type per sector for the years 1930, 1940, 1950 and 1960 were estimated by dividing the detailed IEA data by the 1970 indicator followed by multiplying this value with the indicator value for the other years. For industry the indicator was value-added industry (contribution of the industrial sector to Gross Domestic Product, GDP, World Bank, 1993; Madisson, 1994).

Table 17: Emission factors in rounded figures as used for calculation of fossil fuel combustion emissions in the period 1890-1970 (defined on element basis, e.g. C and N). For the period 1970-1990 country-specific emission factors were used (Olivier et al., 1999b). For a description of data sources see Van Aardenne et al. (2001).

	CO_2 kg C/GJ	CO g C/GJ	CH_4 g C/GJ	NMVOC g C/GJ	SO_2 g S/GJ	N_2O g N/GJ	NO_x g N/GJ	NH_3 gN/GJ
Power plants								
Coal	26	9	1	2	450	0.6	122	0
Oil	19	9	2	3	600	0.4	67	0
Gas	15	9	1	5	10	0.1	46	0
Domestic								
Coal	26	2100	225	200	Regional[a]	0.9	25	0
Oil	19	13	8	3		0.4	15	0
Gas	15	27	...	10	2005	0.1	15	0
Industry								
Coal	26	60	8	20	550	0.9	82	2.5
Oil	19	9	2	2	400	0.4	18	0
Gas	15	13	4	5	10	0.1	34	0
Transport								
Coal	26	64	8	20	450	0.9	82	-
Oil	19	4300	15	1300	100	0.4	183	0.1
Gas	15	...	...	10	10	0.1	...	-

[a] Domestic coal (kg S GJ^{-1}): Canada, 550; United States, 350; Latin America, 500; Africa, 300; OECD Europe, 450; eastern Europe, 450; former SU, 400; Middle East, 700; India region, 300; China region, 500; east Asia, 350; Oceania, 400; and Japan, 250.

The number of vehicles derived from the HYDE database was used to scale the energy use in the transportation sector. For the residential sector GDP per capita was used as indicator (World Bank, 1993; Madisson, 1994). For the years 1890-1920 the energy consumption per fuel (coal, oil, gas) per sector (power generation, industry, transport and residential) and per region were scaled back in time by using coal, oil and gas production data (Darmstadter, 1971; Etemad et al., 1991).

Emissions in the period 1970-1990 were calculated using country specific emissions factors (Olivier et al., 1999a). Because data were unavailable it was assumed that the 1990 EDGAR V2.0 (Olivier et al., 1999a) emission factors reflect the factors without emission controls before 1970. Pre-1970 emissions were calculated per sector and fuel type using these emission factors for regions known to have no emission controls. Because SO_2 emissions from coal depend on fuel sulphur content (for which regional data were available) regional emission factors for SO_2 from coal were used, which also represented uncontrolled emissions. Table 17 presents the emission factors used for 1890-1970.

6.3 Results and uncertainties

By using activity data and emission factors as described in Van Aardenne et al (2001) and presented for fossil combustion in Section 6.2, emissions of CO_2, CO, CH_4, NMVOC, SO_2, NO_x, N_2O, and NH_3 were calculated for the period 1890-1990 and interpolated to a 1° x 1° longitude/latitude grid. Global emissions for the eight species by source category are presented in Figure 7. As shown in the figure for several species (e.g. CO_2, SO_2) the energy sector is the most important source of emissions over the years. For other species (e.g. CH_4, N_2O and NH_3), agricultural practices are most important. For some species a shift in important emission sources is visible. For example, NO_x emissions in 1890 were mainly from agricultural lands while after the 1920's energy related emissions became important. Although these data are based on the EDGAR 2.0 methodology (Olivier 1999a), some difference between the two datasets exist. For example, the use of aggregated emission factors will lead to differences where EDGAR v2.0 applied detailed emission factors.

Inventories of Anthropogenic Emissions

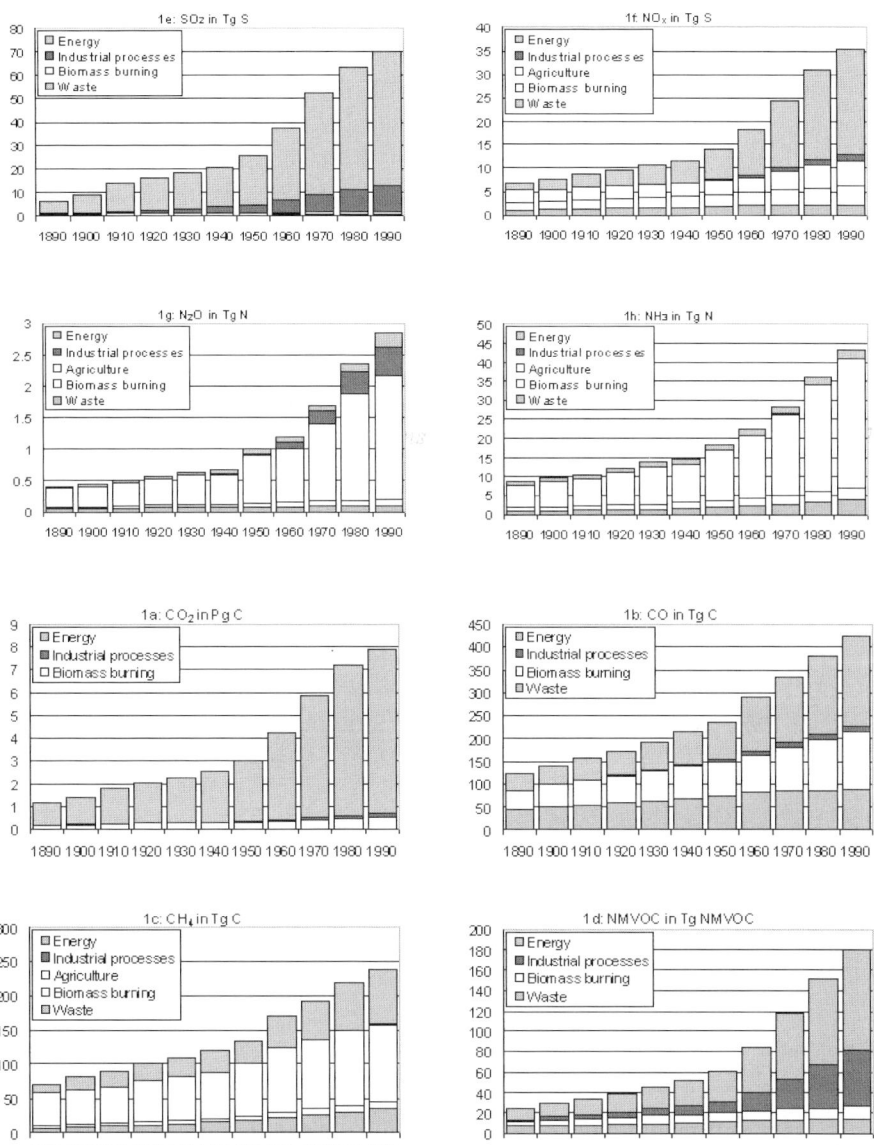

Figure 7. Estimated anthropogenic emissions in the period 1890-1990.

By comparing different emission estimates possible inaccuracies in the inventories could become apparent. For example, Figure 8 presents an overview of CO_2 emissions from fossil fuel combustion and cement production computed by Van Aardenne et al. (2001), Keeling (1994), and Marland et al. (1994). For the period 1890-1960 the global emissions are in good agreement (<5% difference). For the post-1960 years the Marland et al. estimates are somewhat higher than the Van Aardenne et al. estimates.

Possible reasons for these discrepancies are that international air transport emissions (which account for 2% of the emissions in 1990) are not included in the Van Aardenne et al. dataset and that activity data from different sources were used: Marland et al. used UN data, Van Aardenne et al used IEA data.

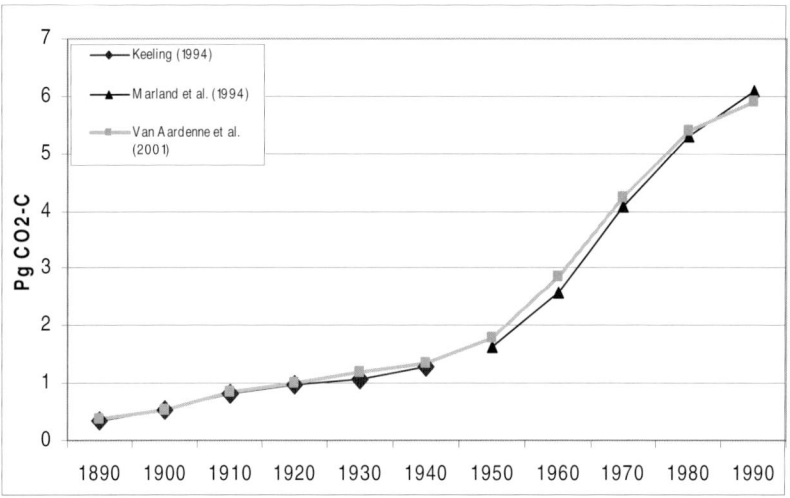

Figure 8. Comparison of fossil fuel and industrial CO_2 emissions from Keeling (1994), Marland et al. (1994) and Van Aardenne et al. (2001).

In Figure 9 global NO_x emissions for the years 1960, 1970 and 1980 are compared between Hameed and Dignon (1988) and Van Aardenne et al. (2001). The estimates of Van Aardenne et al. seem to be significant lower, which might be related to differences in the methodology. Regional estimates for the USA are compared with Gschwandtner et al. (1985). Regional emissions for the USA are in rather close agreement between the two estimates.

The results of historical emission calculations are associated with significant uncertainties. This is illustrated in Figure 9. Even for present-day emissions, estimates based on relatively more reliable activity data and emission factors, are uncertain (see section 5). Historical activity data are mostly based on studies using national or international statistics agencies. Although the quality of these data is difficult to assess, this information is probably the best available with consistent source definitions across countries. Where no activity data are available, assumptions have to be made on processes leading to the emission activity. This is of course, an important source of inaccuracy. The use of constant aggregated emission factors for the period 1890-1970 instead of representative emissions factors for processes in the past is another major source of inaccuracy. However, since prior to 1970 both verified emission factors or activity data such as IEA data are not available, assumptions on historical activities and emission factors have to be made or historical emission inventories cannot be compiled. In order to understand past atmospheric changes these types of emission inventories are necessary; however users should be aware of their large inaccuracies.

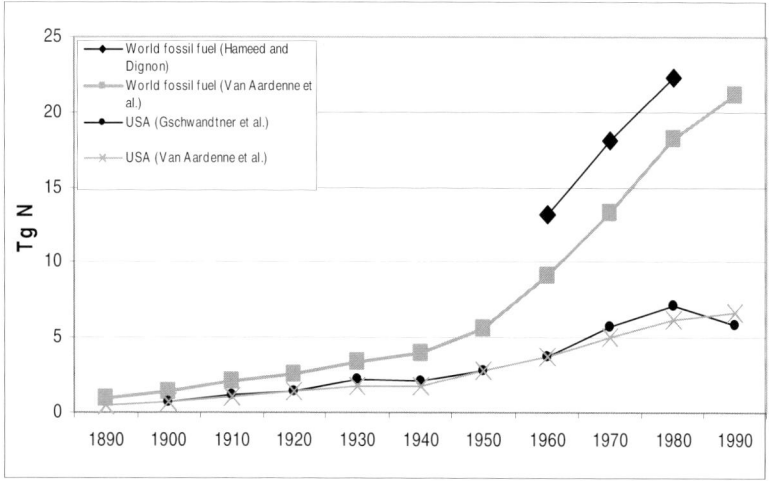

Figure 9. Comparison of global NO_x emissions (Hameed and Dignon (1988) vs. Van Aardenne et al. (2001)) and United States of America emissions (Gschwandtner et al. (1985) vs. Van Aardenne et al. (2001)).

7. FUTURE WORK

The importance of having accurate estimates of emissions to the atmosphere of trace species is becoming more and more critical in the development of our basic scientific knowledge and to better control these emissions in the mitigation of detrimental effects such as adverse effects on human health, acid rain, and climate change. As we so graphically saw in the Marland et al. (1999) article the resulting emissions estimates may be very close when comparing percentages, but can result in large absolute differences, and there is currently no way of identifying which results are more accurate. Therefore, quantitative uncertainty assessments in national and global inventories should receive more attention. Although the focus of inventory development should remain at improving the emission estimates themselves, (expert-judged) uncertainty estimates have a key function in providing a quality label needed in comparisons of different emission estimates and in setting proper priorities in inventory improvements.

The inventories described in this chapter are widely used for atmospheric chemistry studies (e.g. GEIA/IGAC) and within the environmental policy making (e.g. EMEP/IPCC). Although national projects such as EDGAR have been sponsored over the years, increased and concerted efforts in emission inventory development are necessary to improve the quality of the emission estimates and thus the understanding of the changing composition of the atmosphere and its effects on humans and nature. Moreover, high quality inventories are a prerequisite for developing cost-effective environmental policies and realistic projections of future emissions. GEIA has attempted to bring together on a voluntary basis investigators developing emissions estimates in an attempt to make the whole better than its parts. In this it has been partially successful; however, much more is needed, both in terms of acquiring additional resources for inventory compilation and in the revitalization of the cooperative efforts initiated by GEIA. Important aspects of these efforts include capacity building in developing countries to improve their emission estimates and the need for feedback from the user community on the quality of the inventories. For this a close co-operation between the atmospheric-chemistry community, the policy community, and the emission inventory community is essential.

8. SUMMARY

The mathematical modelling of the transport and transformation of trace species in the atmosphere is one of the scientific tools currently used to assess atmospheric chemistry, air quality, and climatic conditions. From the scientific but also from the management perspectives accurate inventories of emissions of the trace species at the appropriate spatial, temporal, and species resolution are required. The chapter has discussed bottom-up methodologies to estimate global and regional emissions. These methodologies are based on activity data, emission factors (amount of emissions per unit activity), and for some inventories additional parameters (such as sulphur content of fuels). To compile regional and global inventories researchers can either bring together estimates made at the national or sub-national level by national experts or directly estimate emissions based on activity rates from reports compiled by multi-national organizations such as the United Nations and the International Energy agency and on emission factors and other information available in the literature. In all cases the data used must be checked for transparency, consistency, comparability, completeness and accuracy. These emissions estimates must now be given finer spatial (usually gridded), temporal, and for some inventories species resolution. The location of major stationary sources (power plants, industrial complexes) is usually known, so the emissions can be directly assigned to the appropriate grid cell. For emissions from other activities, such as transportation, spatial resolution is obtained via the use of surrogate information, such as population distribution, land use, traffic counts, etc. which already exists in or can directly be converted to gridded form. To obtain finer temporal resolution (seasonal, daily, weekday/weekend, etc.) auxiliary information such as plant schedules, traffic counts, etc. is required. Speciation factors have been and are being developed to speciate inventories of hydrocarbons (individual species or groups of species), NO_x (NO, NO_2) and particulate matter ($PM_{2.5}$, PM_{10}; by species).

9. REFERENCES

ACESS, emission database available from http://www.cgrer.uiowa.edu/ACESS/acess_index.htm, 2002.

Andres, R. J., D. J. Fielding, G. Marland, T. A. Boden, N. Kumar, and A. T. Kearney, Carbon dioxide emissions from fossil fuel use, 1751-1950. Tellus, Serie B, 51, 759- 765, 1999.

Arndt, R.L., G.R. Carmichael, D.G. Streets, and N. Bhatti, Sulphur Dioxide Emissions and Sectoral Contributions to Sulphur Deposition in Asia, Atmos. Env., 31, 1553-1572, 1997.

Baldasano, J.M., and H. Power, Air Pollution Emissions Inventories, in Advances Series in Air Pollution. Chapter 1: Guidelines and Formulation of an Upgrade Source Emissions Model for Atmospheric Pollutants, pp. 238, Computational Mechanics Publications, Southampton, UK, 1998.

Benkovitz, C.M., M.T. Scholtz, J. Pacyna, L. Tarrasón, J. Dignon, E.V. Voldner, P.A. Spiro, J.A. Logan, and T.E. Graedel, Global Gridded Inventories of Anthropogenic Emissions of Sulfur and Nitrogen, J. Geophys. Res., 101, 29239-29253, 1996.

Benkovitz, C.M., and T.E. Graedel, The IGAC Activity for the Development of Global Emissions Inventories: Description and Initial Results., in *85th Annual Meeting of the Air and Waste Management Association, June 21-26, 1992*, pp. 2-10, Air & Waste Management Association, Pittsburgh, PA, Kansas City, MO, 1992.

Bouscaren, R., CORINAIR 1990, in First Meeting of the Task Force on Emissions Inventories, edited by J.M.P.a.H.D. G. McInnes, Norwegian Institute for Air Research, London, England, 1992.

Bouwman, A.F.,, K.W. van der Hoek, and J.G.J. Olivier. Uncertainties in the global source distribution of nitrous oxide. J. Geophys. Res., 100, 2785-2800, 1995.

Bouwman, A.F., D.S. Lee, W.A.H. Asman, F.J. Dentener, K.W. Van Der Hoek and J.G.J. Olivier. A Global High-Resolution Emission Inventory for Ammonia, Global Biogeochemical Cycles, 11, 561-587, 1997.

British Petroleum, BP Statistical Review of World Energy 2002, BP Distribution Services, Bournemouth, Dorset (UK), 2002.

Byun, D.W., and J.K.S. Ching, Science Algorithms of the EPA Models-3 Community Multiscale Air Quality (CMAQ) Modeling System, Office of Research and Development U.S Environmental Protection Agency, Research Triangle Park, NC, 1999.

Corbett, J.J. and P.S. Fischbeck, Emissions From Ships. Science, 278, 823-824, 1997.

Corbett, J.J., P.S. Fischbeck, and S.N. Pandis, Global Nitrogen and Sulfur Emissions Inventories for Oceangoing Ships, J. Geophys. Res., 104, 3457-3470, 1999.

Darmstadter, J., Energy in the World Economy. Balimore: John Hopkins Press, 876 pp, 1971.

Doorn, M.R.J., R.P. Strait, W.R. Barnard, and B. Eklund. Estimates of global greenhouse gas emissions from industrial and domestic waste water treatment. Report no. NRMRL-RTP-086. R 8/18/97. Pechan & Ass., Durham, 1997.

Doorn, M.J. and D.S. Liles. Quantification of methane emissions and discussion of nitrous oxide, and ammonia emissions from septic tanks, latrines, and stagnant open sewers in the world. EPA, Washington DC, USA. EPA report EPA-600/R-99-089, October 1999.

EMEP/CORINAIR, 2001 Atmospheric Emission Inventory Guidebook, Third Edition, 2001'. (http://reports.eea.eu.int/technical_report_2001_3/en)

Etemad, B., P. Bairoch, J. Luciani, and J.-C. Toutain, World Energy Production 1800–1985, Geneve: Libraire Droz, 272 pp., 1991.

Foell, W., M. Amann, G. Carmichael, M. Chadwick, J.P. Hettelingh, L. Hordijk, and Z. Dianwu, RAINS-ASIA: AN ASSESSMENT MODEL FOR AIR POLLUTION IN ASIA, The World Bank, Washington, DC, 1995.

Food and Agriculture Organization of the United Nations, FAOSTAT, Food and Agriculture Organization of the United Nations, 2000.

Friedrich, R., B. Wickert, U. Schwarz, and S. Reis, Improvement and Application of Methodology and Models to Calculate Multiscale High Resolution Emission Data for Germany and Europe. In: GENEMIS Annual Report 1999, pp. 119, Iernational Scientific Secretariat, GSF-Forschungszentrum für Umwelt und Gesundheit GmbH, Munich, Germany, 2000.

Friedrich, R., B. Wickert, P. Blank, S. Emeis, W. Engewald, D. Hassel, H. Hoffmann, H. Michael, A. Obermeier, K. Schäfer, T. Schmitz, A. Sedlmaier, M. Stockhause, J. Theloke, and F.-J. Weber, Development of Emission Models and Improvement of Emission Data for Germany, *J. Atmos. Chem.*, 42, 179-206, 2002.

Friedrich R and Reis S. (eds), Emissions of Air Pollutants - Measurements, Calculation, Uncertainties - Results from the EUROTRAC Subproject GENEMIS. Springer Publishers, in preparation, 2003.

Galbally, I., The International Global Atmospheric Chemistry Programme, Commission on Atmospheric Chemistry and Global Pollution of the International Association of Meteorology and Atmospheric Physics, 1989.

Garg, A., P.R. Shukla, S. Bhattacharya, and V.K. Dadhwal, Sub-region (district) and sector level SO_2 and NO_x emissions for India: assessment of inventories and mitigation flexibility, Atmospheric Environ., 35, 703-713, 2001.

Giering, R., Tangent Linear and Adjoint Biogeochemical Models, in Inverse Methods in Global Biogeochemical Cycles, edited by M.H. P. Kasibhatla, P. Rayner, N. Mahowald, R.G. Prinn, D.E. Hartley, pp. 33-48, American Geophysical Union, Washington, DC, 2000.

Gschwandtner, G., K. Gschwandtner, and K. Eldridge, Historic emissions of sulfur and nitrogen oxides in the United States from 1900 to 1980, volume I, Report EPA-600/7-85-009a. Washington: United States Environmental Protection Agency (U.S. EPA), 1985.

Hameed, S. and J. Dignon, Changes in the geographical distributions of global emissions of NO_x and SO_x from fossil fuel combustion between 1966 and 1980. Atmospheric Environment, 22, 441-449, 1988.

Houghton, J.T., M. Filho, B. Lim, K. Treanton, I. Mamaty, Y. Bonduki, D.J. Griggs, and B.A. Callander, Revised 1996 IPCC Guidelines for National Greenhouse Gas Inventories, Intergovenmental Panel For Climate Change, Paris, France, 1997.

Houweling, S., and H. Denier van der Gon, Estimation of Methane Emissions from Rice Fields Using Inverse Modeling, pp. 67-72, National Institute of Public Health and the Environment, Bilthoven, The Netherlands, 1997.

Houweling, S., T. Kaminski, F. Dentener, J. Lelieveld, and M. Heimann, Inverse Modeling of Methane Sources and Sinks Using the Adjoint of a Global Transport Model, J. Geophys. Res., 104 (D21), 26137-26160, 1999.

Houweling, S., F. Dentener, J. Lelieveld, B. Walter, and E. Dlugokencky, The Modeling of Tropospheric Methane: How Well Can Point Measurements be Reproduced by a Global Model?, J. Geophys. Res., 105 (D7), 8981-9002, 2000.

International Energy Agency (IEA), Energy statistics of OECD and non-OECD countries 1971-1992, Paris, France, 1994.

International Energy Agency (IEA). Key World Energy Statistics from the IEA, International Energy Agency, Paris, France, 2002a.

International Energy Agency (IEA), Energy Statistics of OECD countries, 1960-2000, Paris, France, 2002b.

International Energy Agency (IEA), Energy Statistics of non-OECD countries, 1971-2000, Paris, France, 2002c.

International Institute for Applied Systems Analysis (IIASA), RAINS-ASIA CD-ROM Version 7.52, 2001.

International Monetary Fund Staff, World Economic Output: Focus on Transition Economies, International Monetary Fund, Washington, DC, 2000.

IPCC, Climate Change 2001: The Scientific Basis, Contribution of Working Group I to the Third Assessment Report of the Intergovernmental Panel on Climate Change, J.T. Houghton, Y. Ding, D.J. Griggs, M. Noguer, P.J. van der Linden, X. Dai, K. Maskell, and C.A. Johnson (eds), Cambridge University Press, U.K. and New York, NY, USA, 2001.

Johnson, N.D., M.T. Scholtz, and V. Cassaday, Methods Manual for Estimating Emissions of Common Air Contaminants from Canadian Sources, ORTECH International, Mississauga, Ont., Canada, 1991.

Kato, N., and H. Akimoto, Anthropogenic Emissions of SO_2 and NO_x in Asia: Emissions Inventories (plus errata), Atmos. Environ., 26A, 2997-3017, 1992.

Keeling, C.D., Global historical CO_2 emissions. In: BODEN, T.A., ed., Trends '93: A Compendium of Data on Global Change, ORNL/CDIAC-65. Oak Ridge Tenn.: Carbon Dioxide Information Analysis Center (CDIAC), pp. 501 – 504, 1994.

Klein Goldewijk, C.G.M. and J.J. Battjes, A hundred year (1890 – 1990) database for integrated environmental assessment (HYDE version 1.1). RIVM report 422514002. Bilthoven: National Institute of Public Health and the Environment (RIVM), 1997.

Klimont, Z., J. Cofala, W. Schopp, M. Amann, D.G. Streets, Y. Ichikawa, and S. Fujita, Projection of SO_2, NO_x, NH_3 and VOC emissions in East Asia up to 2030, Water, Air, and Soil Pollution, 130, 193-198, 2001.

Lefohn, A. S., J. D. Husar, and R. B. Husar, Estimating historical anthropogenic global sulfur emission patterns for the period 1850–1990, Atmospheric Environment, 33, 3435–3444, 1999.

Li, Y.-F., Global Population Distribution Database, United Nations Environment Programme under UNEP Sub-Project FP/1205-95-12, New York, NY, 1996.

Maddison, A., Monitoring the World Economy. Paris: Organization for Economic Co-operation and Development, 1994.

Marland, G., R.J. Andres, and T.A. Boden, Global, Regional, and National CO_2 Emissions, in Trends 93: a Compedium of Data on Global Change, edited by T.A. Boden, D.P. Kaiser, R.J. Sepanski, and F.W. Stow, pp. 505-584, Carbon Dioxide Information Analysis Center, Oak Ridge, TN, 1994.

Marland, G., A. Brenkert and J. Olivier, CO_2 From Fossil Fuel Burning: A Comparison of ORNL and EDGAR Estimates of National Emissions, Environ. Sci. and Technol., 2, 265-273, 1999.

Marland, G., T.A. Boden, and R. J. Andres. Global, Regional, and National Fossil Fuel CO_2 Emissions, In Trends: A Compendium of Data on Global Change. Carbon Dioxide Information Analysis Center, Oak Ridge National Laboratory, U.S. Department of Energy, Oak Ridge, Tenn., U.S.A., 2002.

Moran, A., and J.E. Salt, International Greenhouse Gas Inventory Compilation Systems CORINAIR and IPCC, EU Directorate General XII Environment Programme, 1996.

Motor Vehicle Manufacturers Association of the United States, World Motor Vehicle Data, Motor Vehicle Manufacturers Association of the United States, Detroit, MI, 1998.

Murano, K., Y. Tonooka, and A. Kannari, Studies on the development of matrix for air pollutants emission and deposition and international cooperative field survey in East Asia, Report for global environmental research fund, Japan, (in Japanese), 2002.

Mylona, S., Sulfur dioxide emissions in Europe 1880– 1991 and their effect on sulphur concentrations and depositions, Tellus, Ser. B, 48, 662– 689, 1996.

Ohara, T., H. Akimoto, J. Kurokawa, K. Yamaji, and D.G. Streets, Development of emission inventories for anthropogenic sources in East, Southeast and South Asia, International workshop emissions of chemical species and aerosols into the atmosphere, June 19-21, Paris, France, 2001.

Olivier, J.G.J., Bouwman, A.F., Van der Maas, C.W.M., Berdowski, J.J.M., Veldt, C., Bloos, J.P.J., Visschedijk, A.J.H., Zandveld, P.Y.J. and J.L. Haverlag. Description of EDGAR Version 2.0: A set of global emission inventories of greenhouse gases and ozone depleting substances for all anthropogenic and most biogenic sources on a per country basis and on $1°x1°$ grid. RIVM, Bilthoven, December 1996, RIVM report nr. 771060 002 / TNO-MEP report nr. R96/119, 1996.

Olivier, J.G.J, A.F. Bouwman, J.J.M. Berdowski, C. Veldt, J.P.J. Bloos, A.J.H. Visschedijk, C.W.M. van der Maas, P.Y.J. Zandveld. Sectoral emission inventories of greenhouse gases for 1990 on a per country basis as well as on $1° \times 1°$, Environmental Science & Policy, 2, 241-264, 1999a.

Olivier, J.G.J, Bloos, J.P.J., Berdowski, J.J.M., Visschedijk, A.J.H. and A.F. Bouwman. A 1990 global emission inventory of anthropogenic sources of carbon monoxide on $1°x1°$ developed in the framework of EDGAR/GEIA, Chemosphere: Global Change Science, 1, 1-17, 1999b.

Olivier, J.G.J. and J. Bakker, Historical global emission trends of the Kyoto gases HFCs, PFCs and SF_6. Proceedings of *"Conference on SF_6 and the Environment: Emission Reduction Strategies"*, November 2-3, San Diego. EPA, Washington DC, USA, 2000. Conference Proceedings published at http://www.epa.gov/highgwp1/sf6/partner_resources/proceedings.html

Olivier, J. Emission inventories of national sources, IGACtivities, 22, 5-9; December, 2000.

Olivier, J.G.J., J.J.M. Berdowski, J.A.H.W. Peters, J. Bakker, A.J.H. Visschedijk and J.P.J. Bloos, Applications of EDGAR. Including a description of EDGAR V3.0: reference database with trend data for 1970-1995, RIVM report 773301 001 / NRP report 410200 052, 2001.

Olivier, J.G.J. and J.J.M. Berdowski. Global emissions sources and sinks. In: Berdowski, J., Guicherit, R. and B.J. Heij (eds.) *"The Climate System"*, pp. 33-78. A.A. Balkema Publishers/Swets & Zeitlinger Publishers, Lisse, The Netherlands, ISBN 90 5809 255 0, 2001.

Olivier, J.G.J. and J.A.H.W. Peters, Uncertainties in global, regional and national emission inventories. In: Van Ham, J., A.P.M. Baede, R. Guicherit and J.F.G.M. Williams-Jacobse (eds.):Non-CO2 greenhouse gases: scientific understanding, control options and policy aspects. Proceedings of the Third International Symposium, Maastricht, Netherlands, 21-23 January 2002, pp. 525-540. Millpress Science Publishers, Rotterdam. ISBN 90-77017-70-4, 2002.

Pechan, E.H. & Associates, Inc., Documentation for the Final 1999 National Emissions Inventory for Criteria Air Pollutants, Nonroad Sources, U.S.Environmental Protection Agency, Research Triangle Park, NC 27711, 2002.

Reddy, M.S. and C. Venkataraman, Inventory of aerosol and sulphur dioxide emissions from India: Part I - Fossil fuel combustion, Atmospheric Environ., 36, 677-697, 2002a.

Reddy, M.S. and C. Venkataraman, Inventory of aerosol and sulphur dioxide emissions from India. Part II – Biomass combustion, Atmospheric Environ., 36, 699-712, 2002b.

Saeger, M., J. Langstaff, R. Walters, L. Modica, D. Zimmerman, D. Fratt, D. Dulleba, R. Ryan, J. Demmy, W. Tax, D. Sprague, D. Mudgett, and A.S. Werner, The 1985 NAPAP Emissions Inventory (Version 2): Development of the Annual Data and Modelers' Tape, U.S. Environmental Protection Agency, Research Triangle Park, NC, 1989.

SEPA, 1998 Reports on the State of the Environment in China. State Environment Protection Administration, Beijing, China, 1998.

Shareef, G.S., and L.A. Bravo, Air Emissions Species Manual - Volume I - Volatile Organic Compounds Species Profiles, U.S. Environmental Protection Agency, Research Triangle Park, NC, 1993a.

Shareef, G.S., and L.A. Bravo, Air Emissions Species Manual - Volume II - Particulate Matter (PM) Species Profiles, U.S. Environmental Protection Agency, Research Triangle Park, NC, 1993b.

Skjølsvik, K. O., A. B. Andersen, et al. Study of Greenhouse Gas Emissions from Ships (MEPC 45/8 Report to International Maritime Organization on the outcome of the IMO Study on Greenhouse Gas Emissions from Ships). Trondheim, Norway, MARINTEK Sintef Group, Carnegie Mellon University, Center for Economic Analysis, and Det Norske Veritas, 2000.

Stern, D., and R. Kaufmann, Estimates of global anthropogenic methane emissions 1860–1993. Chemosphere, 33, 159–176, 1995.

Streets, D.G., N.Y. Tsai, H. Akimoto, and K. Oka, Sulfur dioxide emissions in Asia in the period 1985-1997, Atmospheric Environ., 34, 4413-4424, 2000.
Streets, D.G., N.Y. Tsai, H. Akimoto, and K. Oka, Trends in emissions of acidifying species in Asia, 1985-1997, Water, Air, and Soil Pollution, 130, 187-192, 2001.

Tarrasón, L., and J. Schaug, Transboundary Acid Deposition in Europe, pp. 246, Norwegian Meteorological Institute, Oslo, Norway, 1999.

United Nations Economic Commission for Europe & Food and Agriculture Organisation of the United Nations (UN/ECE-FAO), Forest Fire Statistics 1993-1995, Timber Bulletin, Vol. XLIX, No. 4, ECE/TIM/BULL/49/4, United Nations, 1996.

United Nations Economic Commission for Europe & Food and Agriculture Organisation of the United Nations (UN/ECE-FAO), Forest Fire Statistics 1995-1997, United Nations, 1998.

United Nations Economic Commission for Europe, Procedure for Estimating and Reporting Emission Data Under the Convention on Long-Range Transboundary Air Pollution, United Nations Economic Commission for Europe, Geneva, Switzerland, 1997.

United Nations Economic Commission for Europe, Guidelines for estimating and reporting emission data Under the Convention on Long-Range Transboundary Air Pollution, United Nations Economic Commission for Europe, Geneva, Switzerland, 2002, , EB.AIR/GE.1/2002/7
(http://www.unece.org/env/documents/2002/eb/ge1/eb.air.ge.1.2002.7.e.pdf), 2002.

United Nations Framework Convention on Climate Change, UNFCCC Guidelines on Reporting and Review, FCCC/CP/1999/7, Conference of the Parties, Bonn, Germany, 2000.

United Nations, 1999 Energy Statistics Yearbook, United Nations Department for Economic and Social Information and Policy Analysis, Statistics Division, New York, NY, 2002.

U.S. Environmental Protection Agency, Compilation of Air Pollutant Emission Factors, U.S. Environmental Protection Agency, Research Triangle Park, NC, 1985.

U.S. Environmental Protection Agency, Compilation of Air Pollutant Emission Factors (AP-42) Volume 1: Point and Area Sources, U.S. Environmental Protection Agency, Research Triangle Park, NC, 1999.

U.S. Environmental Protection Agency, Office of Transportation and Air Quality, Assessment and Standards Division, User's Guide to MOBILE6.0: Mobile Source Emission Factor Model, Report No.: EPA420-R-02-001, 2002.

U.S. Geological Survey, Minerals Yearbook-2000, U.S. Department of the Interior, U.S. Geological Survey, Reston, VA, 2002.

Van Aardenne, J.A., F. J. Dentener, J.G.J. Olivier, C.G.M. Klein Goldewijk and J. Lelieveld, A 1° x 1° resolution data set of historical anthropogenic trace gas emissions for the period 1890 – 1990. Global Biogeochemical Cycles, 15, 909-928, 2001.

Van Heyst, B.J., M.T. Scholtz, C.M. Benkovitz, A. Mubaraki, J.G.J. Olivier and J.M. Pacyna, 1990 Global Inventories of SO_x and NO_x on a 1° X 1° Latitude-Longitude Grid, in The Emission Inventory: Regional Strategies for the Future, Air and Waste Management Association, Raleigh, NC, 1999.

Várhelyi, G., Continental and Global Sulfur Budgets - I. Anthropogenic SO_2 Emissions, Atmos. Environ., 19, 1029-1040, 1985.

Vestreng, V. and H. Klein, Emission data reported to UNECE/EMEP: Quality assurance and trend analysis – Presentation of WebDab, EMEP/MSC-W Note 1/2002, 2002.

Wagner, J., R.A. Walters, L.J. Maiocco, and D.R. Neal, Development of the 1980 NAPAP Emissions Inventory, U.S. Environmental Protection Agency, Washington, DC, 1986.

Woodruff, S.D., et al., Comprehensive Ocean-Atmosphere Data Set (COADS) Release 1a: 1980-92. Earth System Monitor, 4(1), 1993.

World Bank, World Tables, Washington, DC, 1993.

World Bank Group, Global Economic Prospects 2001, pp. 180, World Bank, Washington, DC, 2000.

World Energy Council, Survey of Energy Resources, World Energy Council, London, UK, 1998.

World Resources Institute, World Resources 2000-2001, pp. 400, World Resources Institute, Washington, DC, 2000.

Xue, Z., Z. Yang, X. Zhao, F. Chai, and W. Wang, Distribution of SO_2 and NO_x emissions in China, 1995, Proceeding of the 6th International Conference on Atmospheric Sciences and Applications to Air Quality, Beijing, China, November 3-5, 1998.

Yamaji, K., T. Ohara, and H. Akimoto, Development of emission inventory of methane from livestock in South, Southeast and East Asia, Atmos. Enviorn. (to be published, 2003)

Yan X., T. Ohara, and H. Akimoto, Development of region-specific emission factors and estimation of methane emission from rice field in East, Southeast and south Asian countries, Global Change Biology, 9, 237-254, 2003.

Deriving Global Quantitative Estimates for Spatial and Temporal Distributions of Biomass Burning Emissions

C. Liousse, M. O. Andreae, P. Artaxo, P. Barbosa, H. Cachier, J. M. Grégoire, P. Hobbs, D. Lavoué, F. Mouillot, J. Penner, M. Scholes, and M. G. Schultz.

1. INTRODUCTION

Since the 1980's biomass burning has been recognized as a major source of global air pollution (Seiler and Crutzen, 1980; Andreae et al., 1988; Crutzen and Andreae, 1990). The majority of the emissions occur in the Tropics, due to the conjunction of anthropogenic pressure, level of development, climate, and availability of fuel. In these regions, biomass burning remains the main source for energy supply even if the contribution of fossil fuel which used to be relatively low in many countries (figure 1), has been increasing since the 1980's (for example from 1980 to 1995 fossil fuel consumption in South Africa has doubled). Because of the intensity of photochemistry and convection in tropical latitudes, biomass burning emissions in this region have an important atmospheric chemical and radiative impact. This was pointed out by numerous studies on the tropospheric ozone budget (Andreae et al., 1988; Chatfield et al., 1996; Thompson et al., 1996; Chandra et al. 2002), on the CO_2 sources and sinks (Prentice et al., 2002), and on regional and global radiation budgets (Kaufman et al., 1991; Penner et al., 1991; Cox et al. 2000; Jacobson, 2002). Recently, Wotawa and Trainer (2000) found that emissions from fires in temperate and boreal fires in the northern hemisphere may occasionally have a regional and long-range impact comparable to the emissions from fossil fuel combustion.

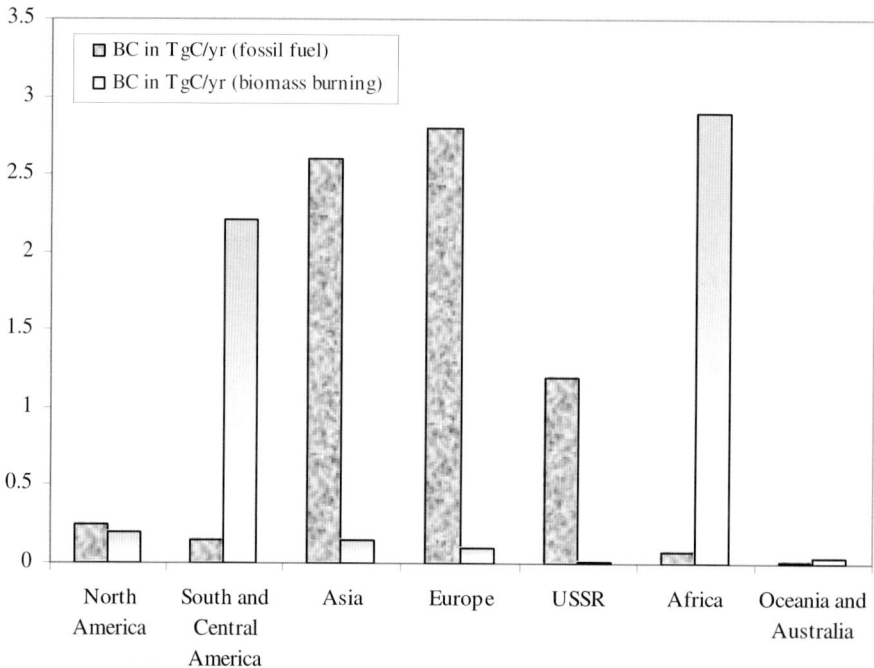

Figure 1. Comparison of the relative importance of fossil fuel and biomass burning sources for emissions of black carbon (adapted from Liousse et al., 1996ab and Cachier, 1998).

Figure 2 gives a classification of the different types of burning according to the different fuel types and purposes. Almost anywhere on the world, most fires are of anthropogenic origin. As in previous reviews (Andreae 1991, Delmas et al., 1991), we will distinguish six major classes of fuel composition and use, namely the burning of savannahs, tropical forests, extratropical forests, domestic fuels (including fuelwood and dung), and agricultural wastes, and the production and use of charcoal. Their relative importance is regionally different:

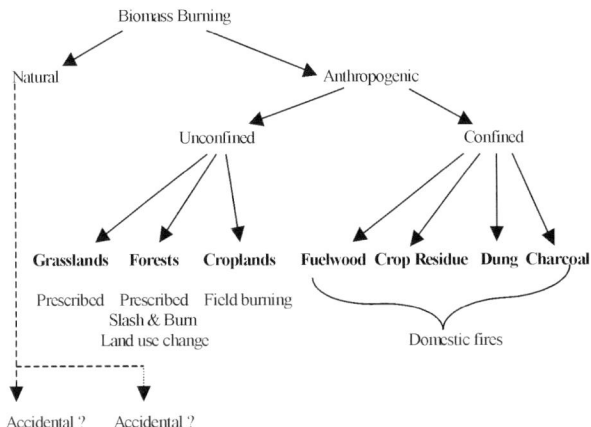

Figure 2: Different types of biomass burning

i) Savannah, tropical forest, and biofuel burning are predominant in Australia, tropical Africa, Central and South-America, and Southern Asia. Numerous international experiments (e.g. FOS-DECAFE 1991 (Lacaux et al., 1995); SAFARI 1992 and 2000 (see http://safari.gecp.virginia.edu; EXPRESSO (Delmas and Guenther, 1999); SCAR-B (Kaufman et al., 1998)) have produced considerable amounts of data, especially for Africa and South America. The measurements have shown that the different types of biomass burning have individual burning properties, different pollutant injection heights, and distinct temporal distributions. Consequently, the emission products and their spatial and temporal distributions are highly different between regions. For example, carbon monoxide (CO) emission factors (i.e. the amount of CO generated per unit mass of burnt material) from tropical forests are significantly higher than the emission factors from savannah burning. Biofuel emissions related to domestic usage (cooking and heating) using either wood, charcoal, dung or peat likely occur all year long and are related to population number. Strong seasonal fire patterns are observed for savannah, forest and agricultural fires. Indeed, such fires are ignited for land clearing, shifting cultivation, conversion of land, pest control and nutrient regeneration, and wildfire prevention, and they occur predominantly during favourable weather conditions in the tropical dry season. The interannual variability of tropical fires differs among different types of burning. Establishing past trends and estimating future emissions from savannah and tropical forest burning is difficult , because these fires depend not only on the available phytomass and on meteorological factors such as drought severity (indirectly influenced e.g. by the El Niño/Southern oscillation), but also on human practices (regional agricultural and deforestation policies, traditions, human migration, etc.). Before the beginning of the large-scale deforestation around 1960, fires in the natural tropical forests were rare (Cochrane et al., 1999; Houghton et al, 2000). In the Amazon region, deforestation followed an exponential increase (Skole et al., 1994). Between 1975 and 1980, the amount of area burnt each year had doubled (Crutzen and Andreae, 1990). Tropical forest and savannah burning in Amazonia exhibits a large interannual variability (figure 3).

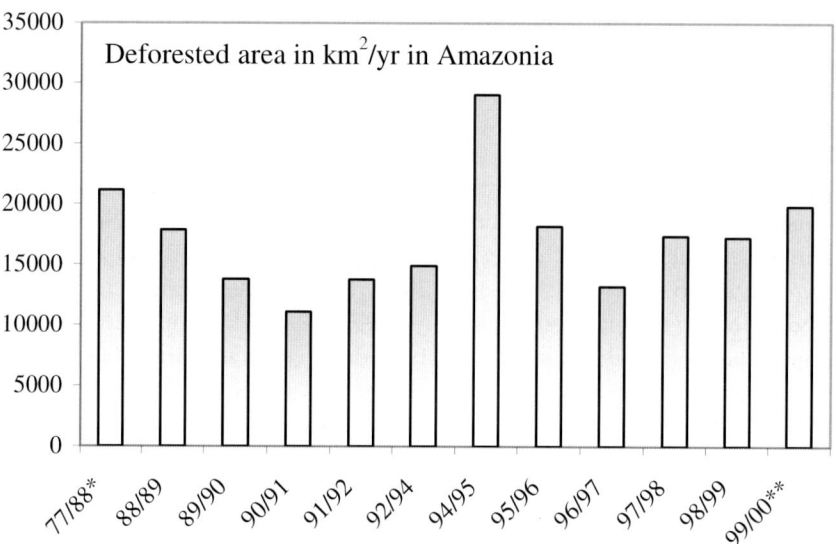

Figure 3. Evolution of deforestation in Amazonia from 1977 to 2000 (INPE data based on Landsat image analysis). (* estimate for the decade, ** estimate)

ii) In boreal ecosystems, the prevailing natural fires are mainly due to lightning strokes, and the fire danger can be more easily predicted for given fuel properties (amount and moisture). For this purpose the Fire Weather Index (FWI) which consists of 6 components that account for the effects of fuel moisture and wind on fire behaviour has been developed for forests (see http://www.fire.uni-freiburg.de/fwf/ca_fwi.htm). It uses meteorological data to estimate the temporal evolution of moisture within duff and fine fuel during the season (Flannigan et al., 1998 ; Wotton and Flannigan, 1993).

To correctly assess the impact of biomass burning on regional and global air pollution and climate, atmospheric chemists, climate modellers, and policy makers need accurate emission inventories for trace gases and aerosols. Such inventories require efforts at several different levels from a detailed catalogue of biomass burning practices to the local, regional, and global estimates of burnt fuel amounts and their resulting emissions, including the temporal and spatial distribution patterns as well as the injection height.

Different global inventories of biomass burning emissions have been developed. Hao et al. (1990) and Hao and Liu (1995) have presented emission inventories for different gaseous species for tropical areas (including savannah, forest, agricultural and domestic fires). At present, most global atmospheric chemistry models use the Hao and Liu, 1995 inventory. Since then, inventories for particulate emissions have been developed based on the work of Hao et al.. Cooke and Wilson, (1996) produced a black carbon (BC) particle emission inventory at a 1° by 1° resolution, for savannah and forest fires. This inventory is available from the GEIA database (see Chapter 1) (http://weather.engin.umich.edu/geia). The fire seasonality in their inventory has been adjusted using AVHRR satellite data (Cooke et al., 1996). Liousse et al. (1996ab), and Lavoué et al. (2000) give a comprehensive description of the different types of burning. Their global BC and organic carbon (OC) inventories account for tropical forest and savannah fires, biofuel (including charcoal, charcoal making, fuelwood, dung and agricultural fires) at a 7.5° by 4.4° resolution and for extratropical forests at a 1° by 1° resolution. Global inventories for greenhouse gases and tropospheric ozone precursors are available within the EDGAR database (versions 2.0 and 3.2) at a 1° by 1° resolution for 1990 (EDGAR 2, Olivier et al., 1996) and 1995 (EDGAR 3.2, Olivier, 2002) (See chapter 1). Recent developments include the use of different satellite data sets to estimate the seasonal and interannual variability of burning emissions (Duncan et al., 2003; Hoelzemann et al., 2003; Liousse et al., 2002; Michel et al., 2002; Schultz, 2002). Moreover, more detailed and accurate distributions of regional emissions are now appearing. At present, different estimates of biomass burning emissions and their variability have been proposed at the global and regional scale, which will need to be reconciled in the future.

The aim of this review is to present different methods currently used to build emission inventories, to compare their results and uncertainties, and to propose different ways of improving their accuracy.

2. DERIVING BIOMASS BURNING INVENTORIES WITH A BOTTOM-UP METHOD

2.1. Comparison of global budgets proposed by different existing inventories

Present biomass burning inventories differ mostly in emission factors and fuel consumption amounts. For example, figure 4 presents a comparison of different estimates of burnt biomass in Tg of dry matter.

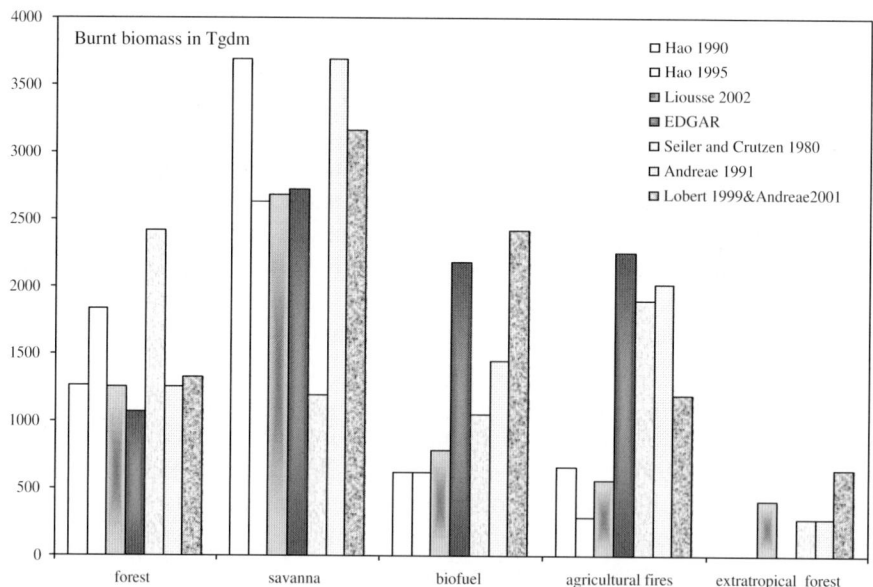

Figure 4. Comparison of burnt biomass estimates in Tgdm given by different authors: Hao et al., 1990 (Hao 1990); Hao and Liu, 1995 (Hao 1995); Liousse et al., 1996ab and Lavoué et al., 2000 (Liousse2002); Andreae, 1991; Lobert et al., 1999 and EDGAR 2.0 (http://arch.rivm.nl/env/int/coredata/edgar/intro.html).

The budgets given by Seiler and Crutzen (1980), Andreae (1991) and Lobert et al. (1999) are added for comparison. Predominance of savannah burning is a common outcome of all inventories. Hao1990 (Hao et al., 1990) and Hao1995 (Hao and Liu, 1995) give emission distributions for the Tropics (Africa, Asia and Southern America). In both studies, estimates for burnt areas are performed using a constant fire turnover time applied to a coarse resolution vegetation map. 75% of African savannahs are assumed to be consumed annually in Hao1990 whereas only 50% in Hao1995. This explains the relatively larger amount of savannah emissions in Hao1990. The global inventories of Liousse2002 (including the works of Liousse et al. (1996ab) and Lavoué et al. (2000)) and EDGAR 2.0 (Olivier et al., 1996) are mostly based on Hao1995; consequently, a general consistency may be noticed between total burnt biomass given by Hao1995, Liousse2002 and EDGAR 2.0. However, the seasonal variability of savannah burning in Liousse2002 and EDGAR 2.0 is from Hao1990.

Concerning forest burning, Liousse2002 and EDGAR 2.0 are rather equivalent to Hao1990, since they did not use the higher values for forest biomass density given by Brown and Lugo (1990), which were applied in Hao1995 (Liousse et al., 1996ab). A general agreement may be seen with budgets given by Seiler and Crutzen (1980) and Andreae (1991). With respect to biofuel, the estimates of burnt biomass in EDGAR 2.0 and Lobert et al. (1999) are about a factor of three larger than in Hao1990, Hao1995, and Liousse2002, and they exceed Crutzen (1980), and Andreae (1991) by almost a factor of two. In terms of agricultural fires, EDGAR 2.0 is closer to Crutzen (1980), and Andreae (1991), but almost a factor of two larger than Lobert et al. (1999), and a factor of three larger than Hao1990, Hao1995, and Liousse2002. It is interesting to note that in the most recent EDGAR inventory (EDGAR 3.2 (Olivier 2002)) agricultural waste burning emissions have decreased substantially due to the assumption of lower burnt fractions. However, no change has been made for biofuel sources.

Regional estimates obtained by Reddy and Venkataraman (2002) for India and recent global developments driven by Yevich and Logan (2002) indicate that the biofuel and agricultural fire inventories given by Hao1990, Hao1995 and Liousse2002 are underestimated, whereas EDGAR 2.0 appears to overestimate the emissions (Table 1). However, further investigation and a coherent integration of the patchy results is urgently needed. More details concerning biofuel emission budgets will be given in the paragraph B-2.

Table 1. Comparison of burnt biomass given by regional and global inventories for India. (Reddy02 is for Reddy and Venkataraman (2002) and YL02 for Yevich and Logan (2002))

India	Fuel use (Tgdm/yr)		
	EDGAR 2.0	Reddy 02	YL02
Biofuel Dung		121	93
Biofuel Crop Residue	534	115	87
Biofuel wood	605	300	220
Total Biofuel	**1139**	**536**	**400**
Ag/Forest burning	77	40	81
Total biomass burning	**1216**	**576**	**481**

The inventories by Cooke et al. (1996) (CK96) and by Liousse et al. (1996ab) (L96) are widely used for emissions of particles from biomass burning. They were primarily constructed for BC particles and give similar BC yearly budgets (6 and 5.6 TgC/yr respectively). This apparent global agreement masks important regional differences between these inventories, which are due to different estimates of input parameters. Recently, a new inventory has been compiled by Chin (2002) and Duncan et al. (2003) who propose a higher total flux estimate of 11 Tg BC/year.

The choice of the inventory has important implications for the analysis of observations using chemical-transport models. During the EXPRESSO experiment, carbon monoxide tracer studies have shown a high sensitivity towards the choice of the inventory applied (Cautenet et al., 1999).

Direct radiative impact calculations of carbonaceous aerosol (including black carbon (BC) and organic carbon (OC)) over Africa also show a high sensitivity to the biomass burning emission inventory. In the study presented in figure 5, two different inventories have been used for Africa: one from the use of AVHRR burnt areas for the year 1990-1991 (Barbosa et al., 1999, see paragraph 3.13) and an other one derived following Hao et al. (1990). The direct radiative forcing estimates computed with the TM3 model for the 2 cases show a high dependency on the selected inventories (Liousse et al., 2002).

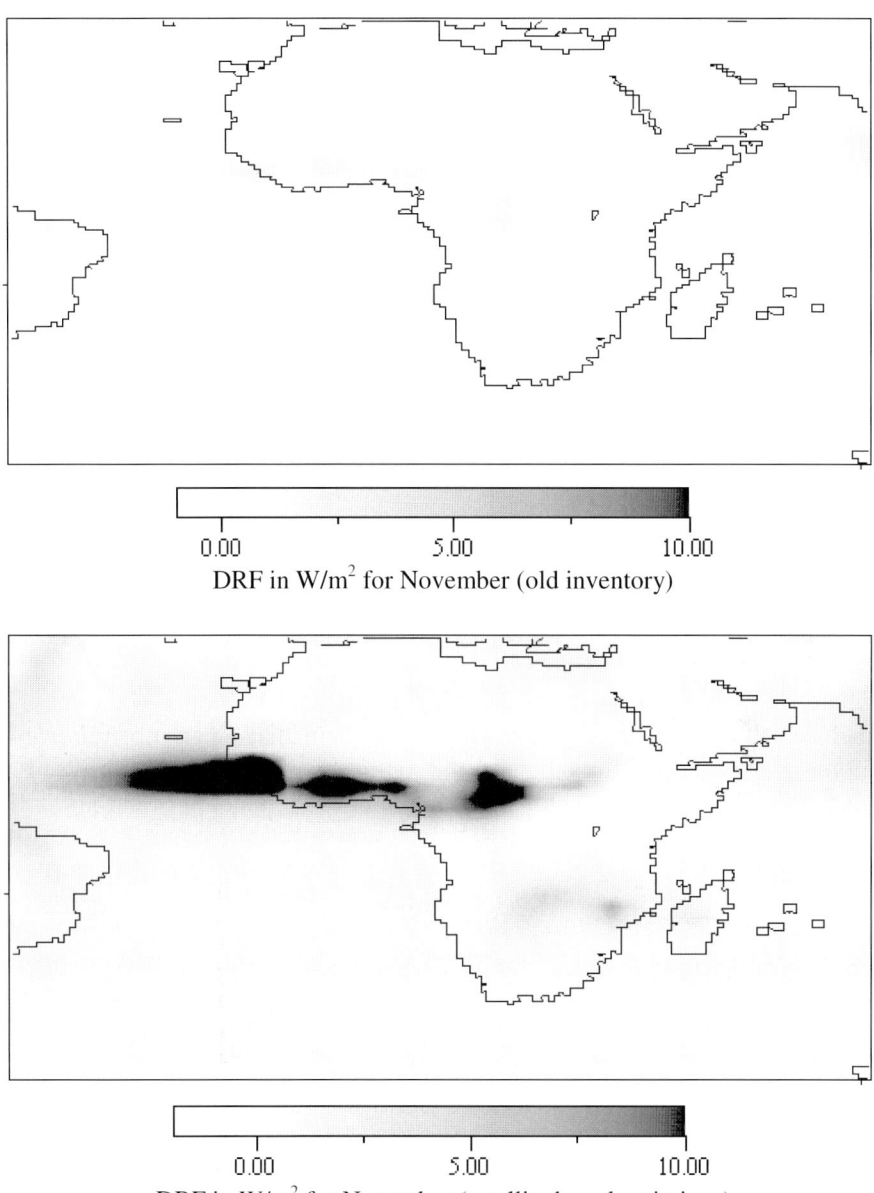

Figure 5. Radiative forcing due to carbonaceous aerosols over Africa (Liousse et al., 2002). Top figure: African emissions obtained using AVHRR burnt areas, bottom: using statistical methods derived from Hao et al. (1991)

2.2. Bottom-up method with description of main uncertainties

The commonly used method for estimating emissions from biomass burning is based on the relationship given by Seiler and Crutzen,, 1980. Gas and aerosols emissions (Q(X)) are calculated according to:

$$Q(X) = M \times EF(X) \qquad (Eq.1)$$

where M denotes the amount of burnt biomass and EF(X) is the emission factor given for the species X in gX/kg of dry matter (gX/kgdm); EF(X) is experimentally determined, but the procedure may differ for different types of burning. The method used to derive M is dependent on the ecosystem and the type of fuel.

2.2.1. Emission factor (EF)

The emission factor of gases and aerosols is obtained as:

$$EF(X) = M(X)/M(biomass) \qquad (Eq.2a)$$

where M(X) is the mass of species (X) emitted and M(biomass) the mass of fuel actually burnt. It is important to note here that EF is not relative to biomass exposed to fires but to the burnt biomass. EF is expressed in gX/kgdm. The emission factor may also refer to the amount of carbon in the dry plant M(C):

$$EF_C(X) = M(X) / M(C) \qquad (Eq.2b)$$

The two equations are related by the carbon content of the fuel [C] (EF(X) = $EF_C(X)$ x [C]). [C] is somewhat variable: typically 42-50% of the fuel dry matter is carbon. For a matter of clarity all the EF values presented in this study refer to the dry matter. The presence of two units for EF(X) (gX/kgdm and gX/kgC) may be a source of confusion and consequently lead to significant errors.

EF may be experimentally determined by:

$$EF(X) = \Delta X / (\Delta \Sigma([CO_2], [CO], [CH_4], [C_{VOC}], [C_{aerosol}],...) \times [C]biomass \qquad (Eq.3)$$

where Δ denotes the difference between concentrations in the background and in smoke conditions, and (Σ ([CO_2], [CO], [CH_4], [C_{VOC}], [$C_{aerosol}$],...) is the species concentration expressed as the amount of carbon. EF(X) is commonly referred to as ($\Delta\Sigma$ ([CO_2],[CO]) as these two species represent generally 95-99% of the carbon-containing emissions. When EF(X) corresponds to the ratio of the production of species X to the production of CO (or CO_2), it is called Emission Ratio (ER) (Andreae et Merlet, 2001).

EF(X) is highly dependent on the fire regime. In chronological order, fires begin with a drying phase, followed by a second phase including flaming, glowing and pyrolysis processes (called flaming phase), a 3rd phase with glowing and pyrolysis only (called smoldering phase) and a 4th phase with glowing and extinction. Emissions of different species are quantitatively and qualitatively governed by the relative importance of the different fire stages. The main emissions occur during phases 2 and 3. Experiments conducted in the frame of IGAC/BIBEX (http://www.mpch-mainz.mpg.de/bibex.html) over the past 12 years have allowed to attribute emissions to specific fire stages, especially over Africa and South America. The relative importance of the various stages depends on factors such as the fuel structure, fuel moisture and the ambient meteorological conditions. 7 fire regimes have been identified with typical EF values (table 2). For example, it has been found that EF for dry and wet African savannahs and for cerrado are similar for particles and driven by fuel content for sulphur and nitrogen. EF(X) is generally higher in a fire regime dominated by the smoldering phase such as tropical forest, charcoal making or dung burning, than in savannah and grassland burning, which occur primarily under flaming conditions.

Table 2 presents values given by Andreae and Merlet in 2001. This work is a compilation of various existing data sets. When specific data are missing, the values have been interpolated. It is important to note that for some types of burning, the uncertainty on EF is much lower than a few years ago, especially once they are normalised by expressing them as a function of the combustion efficiency. However, significant uncertainty still remains for some EF.

Above all factors, the uncertainty might be due to a lack of measurements, especially with respect to domestic and agricultural fires, which are important sources in developing countries, or for some specific compounds (such as oxygenated hydrocarbons). Some discrepancies may also be due to the sampling conditions: for example, aircraft and ground-based measurements do not sample the same air mass due to the rapid chemical aging of the biomass burning plume. Aircraft measurements are also likely to exhibit a bias towards flaming fires whereas ground experiments can more easily characterize smoldering emissions.

As an example, EF(BC) and EF(Total Particulate Matter, TPM) for savannah and forest burning given by Andreae et Merlet (2001) and Liousse et al. (1996ab) differ by a factor of 2. Due to the increasing interest in such particles, consistency between the different estimates is needed, and the reasons for the differences should be discussed in order to be able to recommend "best guess" values. In table 3, we compile measurements for EF(BC) from the literature describing recent experiments: SCAR-B (Kaufman et al., 1998), LBA in Brazil (Artaxo, 2002), DECAFE in West Africa (Lacaux et al., 1995, Cachier et al., 1995), EXPRESSO in Central Africa (Delmas and Guenther, 1999) and SAFARI, 1992 and 2000 (Andreae et al., 1998; Hély et al., 2003) in southern Africa. Because BC is chemically inert, EF(BC) should be independent of the sampling level (either ground-based or airplane measurements can be used).

We have restricted our analysis to data obtained with the same thermal method, because the instrumental technique used for measuring the BC concentrations is a key uncertainty parameter. This excludes the data from Yamasoe et al. (2002), who employed an optical reflectance technique and from Sinha et al. (2003) during SAFARI 2000. Due to the loss of BC observed during the thermal treatment of biomass burning samples (Reid et al., 1998), all the values obtained with the thermal method of Cachier et al. (1989) were increased by 25%. The resulting best guess EF(BC) = 0.62, Table 3) is between the estimates of Liousse et al. (1996ab) and Andreae and Merlet, (2001). For EF(TPM), the sampling height appears to be a very important parameter which points out the rapid transformations occurring in the plume. Either an increase or a decrease may be obtained when measuring TPM concentrations at ground level or on an aircraft platform. Near the source (at ground level or at the tower altitude), the vapour-particle phase partition for organics is not at equilibrium (Liousse et al., 1995, Andreae et al., 1998, Ferek et al., 1998). We estimate EF(TPM) as 7 g/kgdm, but caution that uncertainties are large, especially due to the choice of the methods of measurements which can yield either the total particulate mass by filter weighing, or the mass below 3μm diameter (PM3) (as in SAFARI, 1992) or the mass below 4μm diameter (PM4) (as in SCAR-B). Previous estimates have generally produced higher values for EF(TPM).

Table 2. Emission factors for some chemical species (extracted from Andreae et Merlet, 2001).

Species	Savanna and grassland	Tropical forest	Extratropical forest	Biofuel Burning	Charcoal making	Charcoal burning	Agricultural residues
CO_2	1613±95	1580±90	1569±131	1550±95	440	2611±241	1515±177
CO	65±20	104±20	107±37	78±31	70	200±38	92±84
CH_4	2.3±0.9	6.8±2.0	4.7±1.9	6.1±2.2	10.7	6.2±3.3	2.7
total nonmethane hydrocarbons	3.4±1.0	8.1±3.0	5.7±4.6	7.3±4.7	2.0	2.7±1.9	7.0^c
C_2H_2	0.29±0.27	0.21-0.59	0.27±0.09	0.51-0.90	0.04	0.05-0.13	0.36^c
C_2H_4	0.79±0.56	1.0-2.9	1.12±0.55	1.8±0.6	0.10	0.46±0.33	1.4^c
C_2H_6	0.32±0.16	0.5-1.9	0.60±0.15	1.2±0.6	0.10	0.53±0.48	.97^c
C_3H_4	0.022±0.014	0.013	0.04-0.06	0.024^c	--	0.06^c	0.032^c
C_3H_6	0.26±0.14	0.55	0.59±0.16	0.5-1.9	0.06	0.13-0.56	1.0^c
C_3H_8	0.09±0.03	0.15	0.25±0.11	0.2-0.8	0.04	0.07-0.30	0.52^c
isoprene	0.020±0.012	0.016	0.10	0.15-0.42	--	0.017	0.05^c
terpenes	0.015	0.15^g	0.22	0.15^g	--	0.0	0.015^c
benzene	0.23±0.11	0.39-0.41	0.49±0.08	1.9±1.0	--	0.3-1.7	0.14
PAH	0.0024	0.025^l	0.025^l	0.025^l	--	0.025^l	0.025^l
methanol	1.3^c	2.0^c	2.0±1.4	1.5^c	0.16	3.8^c	2.0^c
formaldehyde	0.26-0.44	1.4^c	2.2±0.5	0.13±0.05	--	2.6^c	1.4^c
acetaldehyde	0.50±0.39	0.65^c	0.48-0.52	0.14±0.05	--	1.2^c	0.65^c
acetone	0.25-0.62	0.62^c	0.52-0.59	0.01-0.04	0.02	1.2^c	0.63^c
2-butanone	0.26	0.43^c	0.17-0.74	0.03-0.06	--	0.83^c	0.44^c
formic acid	0.7^c	1.1^c	2.9±2.4	0.13	0.20	2.0^c	0.22
acetic acid	1.3^c	2.1^c	3.8±1.8	0.4-1.4	0.98	4.1^c	0.8

Table 2 (cont'd). Emission factors for some chemical species

Species	Savanna and grassland	Tropical forest	Extratropical forest	Biofuel Burning	Charcoal making	Charcoal burning	Agricultural residues
H_2	0.97±0.38	3.6-4.0	1.8±0.5	*1.8*	---	*4.6*[c]	*2.4*[c]
NO_x (as NO)	3.9±2.4	1.6±0.7	3.0±1.4	1.1±0.6	0.04	3.9	2.5±1.0
N_2O	0.21±0.10	*0.20*[g]	0.26±0.07	0.06	0.03	*0.20*[g]	0.07
NH_3	0.6-1.5	*1.30*[g]	1.4±0.8	*1.30*[g]	0.09	*1.30*[g]	*1.30*[g]
HCN	0.025-0.031	*0.15*[g]	*0.15*[g]	*0.15*[g]	*0.15*[g]	*0.15*[g]	*0.15*[g]
N_2	*3.1*[l]	*3.1*[l]	*3.1*[l]	*3.1*[l]	---	*3.1*[l]	*3.1*[l]
SO_2	0.35±0.16	0.57±0.23	1.0	0.27±0.30	---	*0.40*[g]	*0.40*[g]
COS	0.015±0.009	*0.04*[g]	0.030-0.036	*0.04*[g]	*0.04*[g]	*0.04*[g]	0.065±0.077
CH_3Cl	0.075±0.029	0.02-0.18	0.050-0.032	0.04-0.07	*0.01*[g]	0.012	0.24±0.14
CH_3Br	0.0021±0.0010	0.0078±0.0035	0.0032±0.0012	*0.003*[g]	*0.003*[g]	*0.003*[g]	*0.003*[g]
CH_3I	0.0005±0.0002	0.0068	0.0006	*0.001*[g]	---	*0.001*[g]	*0.001*[g]
Hg^0	0.0001	*0.0001*[g]	*0.0001*[g]	*0.0001*[g]	---	*0.0001*[g]	*0.0001*[g]
$PM_{2.5}$	5.4±1.5	9.1±1.5	13.0±7.0	7.2±2.3	---	*9*[g]	3.9
TPM	8.3±3.2	6.5-10.5	17.6±6.4	9.4±6.0	4.0	*12*[g]	13
TC	3.7±1.3	6.6±1.5	6.1-10.4	5.2±1.1	---	6.3	4.0
OC	3.4±1.4	5.2±1.5	8.6-9.7	4.0±1.2	---	4.8	3.3
BC	0.48±0.18	0.66±0.31	0.56±0.19	0.59±0.37	---	1.5	0.69±0.13
Levoglucosan	*0.28*[g]	0.42	*0.75*[g]	*0.32*[g]	---	---	*0.27*[g]
K	0.34±0.15	0.29±0.22	0.08-0.41	0.05±0.01	---	0.40	0.13-0.43

Previous particulate emission estimates for tropical forest burning were based on measurements performed over extra-tropical wild forest fires or from laboratory experiments, and a high value of 18 g/kgdm was found for EF(TPM) (Liousse et al., 1996ab). Measurements over Brazilian forests indicate that EF(TPM) values could be lower due to a smaller fire size than previously assumed (Andreae et al., 1988, Ferek et al., 1998). However, it is difficult to achieve consistency between different experiments, because the burning conditions are variable and the relative importance of the 2 main fire stages (flaming and smoldering) may be different from one fire to the next. Sampling differences between tower and ground-based measurements are also more important than for savannah fires. Consequently, only values obtained from aircraft measurements were selected in Table 3.

Table 3. Compilation of emission factors for BC and TPM particles using comparable measurement techniques, and recommendations from this study.

EF (gX/kgdm)	BC	TPM	Comments
Savanna burning			
DECAFE 1991	0.625c	7.5$_m$	Cachier et al., 1995 « ground »
SAFARI 1992	0.73c	3.1$_{pm3}$	Andreae et al. 1998 « aircraft »
	0.52c	7.7$_m$	Cachier 1998 « ground »
SCAR-B	0.5-0.65	4.8-8$_{pm4}$	Ferek et al., 1998 « aircraft »
SAFARI 2000		10$_m$	Sinha et al., 2003 « aircraft »
MEAN	**0.62**	**7**	**This paper**
Forest burning			
ABLE2A	0.8	4.5$_m$	Andreae et al., 1988 « aircraft »
SAFARI 1992	0.34c		Andreae et al. 1998 « aircraft »
SCAR-B	0.55-0.75	8.5-13.3$_{pm4}$	Ferek et al., 1998 « aircraft »
MEAN	**0.7**	**11.5***	**This paper**

(c = EF(BC) values including corrections described in paragraph 2.2.1, m = TPM mass obtained by filter weighting, pm3 and pm4 = TPM measurements for particle diameter lower than 3 and 4 µm respectively.)

Because of the difficulties of measuring high particulate emissions during the smoldering phase from aircraft, the recommended value for EF(TPM) of 11.5 g/kgdm is based on the upper part of the range proposed by the Ferek et al. (1998) measurements. Similarly, EF(BC) has been selected near the upper range of the measurements. While we have here limited our analysis to particulate emission factors, a similar discussion of the available data is needed for other compounds as well. Olivier et al. (2002) estimate uncertainties of 30 and 40% for emission factors for forest and savannah fires, respectively.

2.2.2. Burnt biomass (M)

The calculation of the amount of burnt biomass (M) relies on different relationships depending on the biomass burning sources. For tropical savannah and forest, it may be obtained following the initial relationship given by Seiler and Crutzen, (1980):

$$M = A \cdot B \cdot \alpha \cdot \beta \qquad \text{(Eq. 4)}$$

where A is the burnt area (m^2), B, the biomass density (g/m^2), α the fraction of aboveground biomass and β the burning efficiency. As detailed in paragraph 2.1, some consistency can be detected between the existing global inventories for the choice of A and B (see details in Liousse et al., 1996ab; Olivier et al., 2002), because they are all based on Hao and Liu (1995) burnt biomass estimates, with updates for Africa savannahs (following Menaut et al., 1991; Delmas et al., 1991; Lacaux et al., 1995). However, the uncertainties on the fractional area and on the fraction of the biomass burnt are at least a factor of 3. Furthermore, the biomass density and the combustion efficiency are only known to within ± 50% (e.g. for Amazonia (Houghton et al., 2001)).

Relationship (4) has been adapted for more complex ecosystems, e.g. for boreal and temperate regions (Lavoué et al., 2000). At these latitudes, each 1° by 1° cell has been considered as a mosaic of different vegetation types with different burning properties (boreal forest, temperate forest, grassland, and shrubland). Each of these vegetation types has different fuel layers (canopy, surface and ground layers), which may burn differently. Consequently, for each grid cell, the burnt biomass Mc is given by :

$$M_c = \sum_{k=1}^{m} \left[(S_k \cdot A) \cdot \sum_{i=1}^{n} (\beta_{k,i} \cdot B_{k,i}) \right] \qquad \text{(Eq. 5)}$$

where k is the number of vegetation types (m ≤ 4), i the number of fuel layers (n ≤ 3), A the area affected by fires (in m^2), S_k the percentage of the box area covered by the vegetation of type k. $B_{k,i}$ denotes the biomass load (in g/m^2), and $\beta_{k,i}$ is the combustion efficiency (in %) of the fuel layer i for the vegetation type k. For Canada and Alaska, burnt area maps have been obtained from satellite data (Stocks, 1991; Kasischke et al., 1995; Stocks et al., 1996) with documentation on interannual variability (see also section 3.1). Other parameters related to regional vegetation maps are also given with better precision. In temperate areas and the Mediterranean region, Lavoué et al. (2000) have used FAO statistics (FAO, 2001), which have the highest uncertainties for Russian ecosystems (Dixon and Krankina, 1993; Kasischke et al., 1991). The resulting uncertainties in the Lavoué et al. (2000) study are much lower than in the existing tropical savannah and forest burning inventories.

Agricultural and biofuel burning estimates are generally based on crop production, resource data and population. In developing countries, such sources are important and continue to expand, showing the need to improve the estimates which show high uncertainties and conflicting budgets. In most inventories, *Ma* which corresponds to the biomass consumed in the agricultural burning is obtained as follows:

$$Ma = P \cdot W/P \cdot Wf/W \cdot ce \qquad (Eq.6)$$

where *P* is the crop production, *W* the by-product amount (*W/P* is related to the harvest index), *Wf* is the amount of agricultural material submitted to fires, and *ce* is the burning efficiency. *Ma* may be burnt either on field as open fires or later as domestic fire fuel. In Liousse et al. (1996ab) both inventories (field and biofuel) are derived from the same relationship (Eq.6) using FAO and United Nations data. In EDGAR 2.0 and 3.2 agricultural waste burning estimates on field are based on relationship (1), whereas the data compiled by Hall et al. (1994) are used for the biofuel source. In spite of the different methodologies, some assumptions are in agreement between Liousse et al., 1996ab and EDGAR 3.2. Liousse et al. (1996ab) estimate that in developed countries 5-10% of agricultural waste is burnt in open fires, whereas EDGAR 3.2 estimates 5-20%. For developing countries, Liousse et al. (1996ab) estimate a fraction of 30-70% dependent on the crop type, whereas EDGAR 3.2 uses a fixed value of 40%. As an example, Table 4 lists the parameters used for the determination of emissions from rice straw burning (Liousse et al., 1996ab).

The same method is used to obtain other agricultural burning inventories for wheat, barley, rye, and corn. Only the distribution of sugar cane burning emissions is taken directly from FAO data. The parameters (EF, temporal distribution and injection height) to create the distribution of waste burning sources are linked to the choice of Ma, and the repartitioning of Ma between developed and developing countries and between open fires and fuel use burning. These dependencies underline the importance of decreasing the uncertainty in the determination of Ma. Regional differences also play an important role for the distribution of domestic fuel emissions. Liousse et al. (1996ab) have placed a particular emphasis on an inventory for charcoal making.

Table 4. Example of complexity in the determination of rice straw burning inventory (Liousse et al., 1996ab).

RICE	Developed countries	Developing countries
	Field burning	Field burning and domestic fires
Agricultural waste per unit of grain	1.2	1.2
Percentage of agricultural waste burnt	10	50
Combustion efficiency (%)	85	85
Nature of fires	flaming	smoldering
EF(BC) gC/kgdm	0.6	0.86
Temporal distribution harvest	yearly	
Injection Height	0-2000m	0-70m

The first studies focusing on domestic fires in developing countries were performed in Western Africa where 2-50% of fuelwood products can be attributed to charcoal making (Brocard et al., 1996). The Safari 2000 experiment has confirmed the importance of biofuel emissions (wood and charcoal) in eastern and southern Africa; the models, which have been developed for these conditions, still need to be validated for Central and Western Africa (Marufu et al., 1997). The use of charcoal is less important in

Asia, where the favourite domestic biofuels are fuelwood, dungcakes, agricultural wastes and various other waste products, which can burn (Smith et al., 1983, 2000; Joshi et al. 1992; Reddy et al. 2002; Gadi et al. 2002). More investigations with different regional focus are urgently needed to better constrain the emissions from domestic fires.

2.2.3. Temporal distribution

Finding the adequate temporal distribution of biomass burning emissions is crucial to correctly use these sources in chemistry transport models. This is particularly important for all fields burning (savannah and forest fires, and some agricultural fires). In developing countries, the domestic fire emissions are practically constant all year long except for heating purposes, which only represents a few percent of the general use. In contrast, biofuel burning in developed countries occurs predominantly in winter. The seasonal maximum of agricultural fires on fields is assumed to occur after the harvest period (see Liousse et al., 1996ab; Yevich and Logan, 2002).

Concerning tropical savannah and forest burning, the seasonality of most of the existing inventories (e.g. EDGAR 2.0, 3.2, and Liousse et al., 1996ab) are based on the Hao et al. (1990) study. Simulations using these emissions have yielded reasonable agreement with experimental and satellite data obtained in the 1990's either for Africa or Southern America. In contrast, the temporal distribution derived from Hao and Liu (1995) is not suitable for Africa. Here, the authors assumed that in Western Africa the dry season peaks in the March - June window, whereas there is good evidence that the maximum burning intensity in these regions occurs between December and January. This is of some significance, because many global chemistry transport models use the Hao and Liu (1995) inventory for their calculations (e.g. Horowitz et al., 2002).

Considerable improvement of the fire seasonality has been obtained in recent studies using satellite observations to redistribute previously computed annual emission estimates in time. Cooke et al. (1996) pioneered this approach for Africa. Other inventories with satellite-based seasonality have been developed by Generoso (2002), Heald et al. (2002), and Schultz (2002). These results are discussed in section 3.1.

2.2.4. Pollutant injection height: an often neglected parameter

The height of injection assumed for biomass burning emissions has been shown to be a critical parameter in aerosol transport simulations (Liousse et al., 1996ab). The energy generated by a fire creates a convection column, which may have different plume buoyancy depending on the burning type. Figure 6 presents a linear regression of the particle injection height versus the frontal fire intensity obtained from different experiments and observations (detailed in Lavoué et al., 2000). From this relationship it may be calculated that an injection height of 2000 meters is typical for average tropical savannah and forest fires. Significant differences appear for boreal fires, and especially for Canadian crown fires, which can inject pollutants at over 7600 meters altitude.

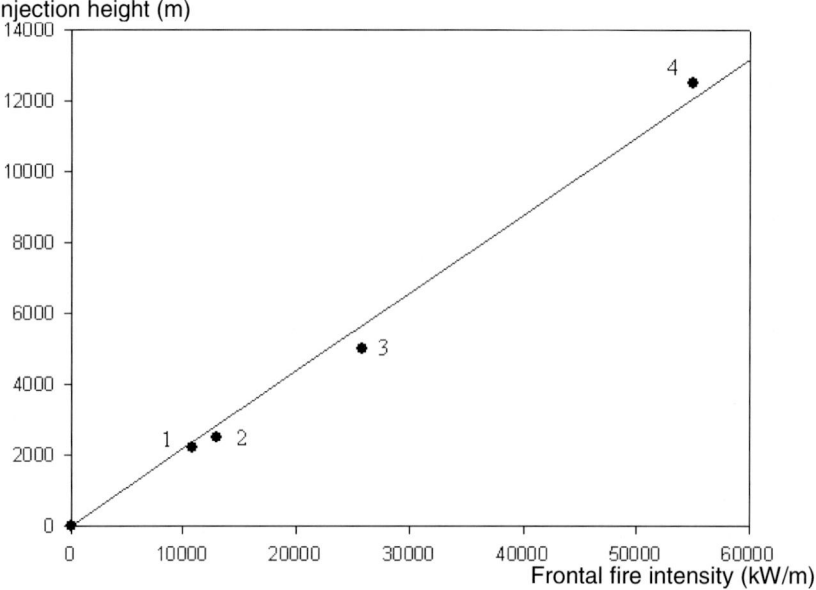

Figure 6. Relationship between frontal fire intensity and pollutant injection height (Lavoué et al. 2000). (1) Kruger National Park, South Africa, 1992, (2) International Crown Fire Modelling Experiment, Canada, 1998, (3) Bor Forest Island, Siberia, 1993, (4) Red Lake Fire 7, Canada, 1986. (© American Geophysical Union).

In future experiments, aircraft and ground-based lidar measurements may be used to experimentally determine injection heights. However, as the injection height is linked to the wind speed, fuel size, fuel moisture, and the quality of the combustion, and consequently to fire propagation in the different fuel layers, dynamic models allowing an interactive approach with meteorological conditions are now expected to be the best tool to improve the parametrizations.

2.3. Emission inventory validation by comparison between measured and modelled concentrations

2.3.1. Ground-level comparison for black carbon particles

Suitable comparisons of simulated and observed surface black carbon (BC) concentrations have been obtained above biomass burning areas. For example, figures 7a and b show BC surface concentrations obtained with the TM2z model (Heimann, 1995), for Lamto, a biomass burning site in the humid savannah of the Ivory Coast, and Amsterdam Island, a remote site in the Southern Indian ocean, which is influenced by African fires during the Southern hemisphere dry season (Cachier et al., 1994). The model, which uses the inventory of Liousse et al. (1996ab) reproduces the seasonal fire pattern in Lamto for the year 1990-1991, both in time and intensity either for black carbon and organic carbon comparisons. The agreement for Amsterdam Island is also reasonable.

2.3.2. Columnar comparison of aerosol optical depth

Modeled aerosol optical depth (AOD) values derived from the black carbon and organic carbon aerosol simulations described above are often underestimated for areas submitted to high biomass burning influence. Figure 8 shows an AOD comparison at Zambezi (Zambia) from TM2z simulations and sunphotometer measurements from the AERONET database (Holben et al., 1998). Similar results are obtained in a study with six different models, which included all the important AOD contributors (Penner et al., 2001). All models in this study use biomass emissions based on Liousse et al. (1996ab). In all models, AOD is underpredicted in a region off the west coast of Southern Africa and near Indonesia during all seasons. Another discrepancy is found in an area over the Atlantic off the coast of South America in April and July. One of the possible reasons for such a discrepancy may be due to the biomass burning emissions in tropical regions.

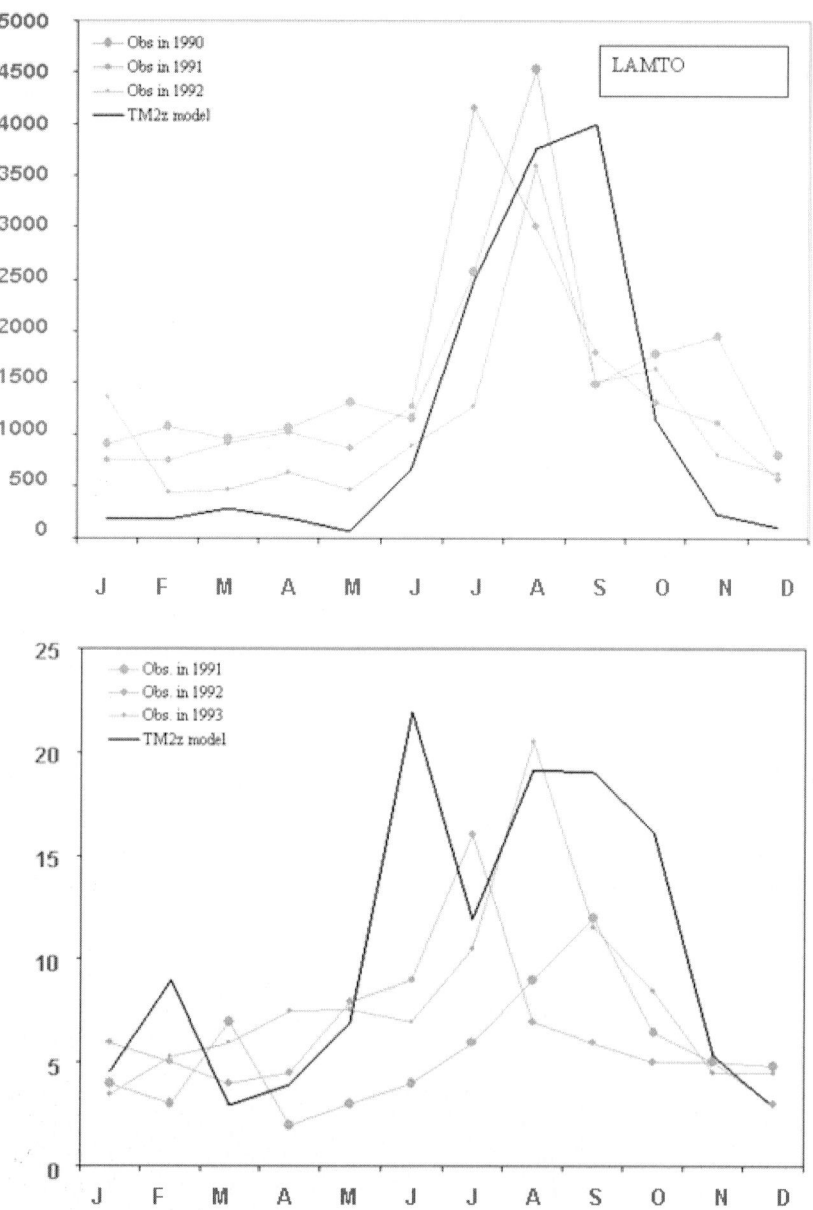

Figure 7. Comparison of modelled and observed surface black carbon concentrations in Lamto (6.21N, 5.03W) (a) and on Amsterdam Island (37.5S, 77.3E) (b).

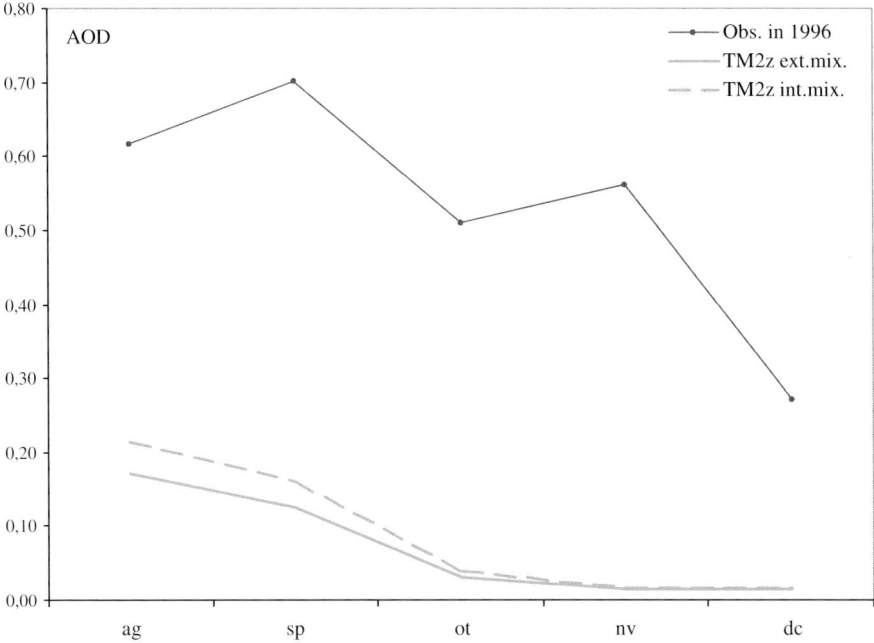

Figure 8. Comparison of modelled and observed aerosol optical depth (AOD) in Zambezi (Zambia). Modelled values are obtained from simulated surface carbonaceous aerosol concentrations and their corresponding optical properties. The BC scattering coefficient is assumed as 2, the BC absorption coefficient is set to 7, and the OC scattering coefficient is set to 5.2 (including a humidification factor of 1.3 (Kaufman et al., 1998). The radiative properties used in the sensitivity test dealing with internal mixtures are identical, except for the BC absorption coefficient, which has been doubled (Martins et al., 1998).

This model intercomparison is particularly interesting in view of other recent estimates of biomass burning emissions. Barbosa et al. (1999) estimated the average amount of burnt biomass for Africa during the 1985 - 1991 period based on satellite burn scar analyses to be in the range 704 - 2168 Tg/yr. Previous estimates (including Liousse et al., 1996ab) had proposed higher values (Hao et al., 1990; Hao and Liu, 1995). Using active fire imagery to obtain the average amount of burnt biomass, Scholes et al. (1996) obtained 178 Tg/yr for Southern Africa, almost an order of magnitude smaller than the earlier estimate of 1200 Tgdm/yr from Hao et al. (1990). If these smaller estimates of burnt biomass are correct, the AOD discrepancy would be even larger. On the other hand, more recent estimates of other biomass burning sources associated with biofuels and agriculture are considerably larger than those proposed by Liousse et al. (1996ab). Liousse et al. (1996ab) estimated approximately 1300 Tgdm/yr for the total mass of consumed biomass in biofuel and agricultural residue burning. Recent estimates from Yevich and Logan (2002) are as high as 2900 Tgdm/yr. It is expected that simulations will improve if the more recent estimates of burning based on burning scars and active fires are used concomitantly with the higher estimates of biofuel and agricultural residue burning of Yevich and Logan (2002).

3. IMPROVEMENT OF GLOBAL BIOMASS BURNING INVENTORIES

3.1. Use of satellite data

3.1.1. Why use burnt area instead of fire pixels?

As previously shown, the high interannual variability makes satellite data more appropriate than statistical data for biomass burning modelling. Indeed, biomass burning inventories require information on the amount of biomass burnt per unit area and per time unit, and information on the burning conditions. Earth Observation techniques have proved to be the only possible way to tackle this problem globally at the required temporal and spatial scales. For a long time the satellite observations were used only for the detection of fire occurrence and not for estimating the area burnt. This was due mainly to the spectral and geometric characteristics of the imagery obtained before the 1990s. The sensors were also developed for meteorological forecasts, and not for land surface applications. Most of the daytime fire monitoring studies relied on imagery provided by the Advanced Very High Resolution Radiometer (AVHRR) sensor onboard NOAA satellites (Dwyer et al., 2000). For example, daytime pixel count maps from AVHRR satellite have been used by Cooke et al.(1996). Night-time products such as those delivered by the Along Track Scanning Radiometer (ATSR) sensor (Arino et al., 1999) have also been used (e.g. Schultz, 2002). Interestingly, recent tests have shown that global simulations obtained with emissions based on daily AVHRR fire count data are negligibly different from simulations using monthly averaged fire counts (Heald et al., 2002).

Active fire counts were used mainly to improve the temporal distribution of burnt biomass estimates. Schultz (2002) and Generoso (2002) computed scaling factors of the burning intensity from ATSR fire count data and applied these to a standard inventory of biomass burning. While this approach appears to improve the seasonality of burning emissions compared to the standard inventory, it must be regarded as highly uncertain and unsuited to obtain better quantitative estimates. Tests to relate the number of fire events detected over a given region and for a given period of time to the area burnt obtained from satellite images from ATSR, AVHRR, or SPOT, clearly showed that the correlation between the number of fire events and the area burnt is highly variable as it depends on the type and conditions of the vegetation and time of the year. Moreover, the fire events detected by the polar orbiting satellite systems such as NOAA-AVHRR represent only a small portion of the fire events active during the day. There is in fact a large sampling bias in all of these products due to three main reasons: the time of the satellite overpass which usually does not correspond to the daytime maximum fire activity, the difficulty to get consistent observations during daytime and night-time (because it requires different retrieval algorithms), and the short duration of fire events, which can be less than 1 hour in some regions (e.g. in the tropical belt).

The suitable solution for chemistry modelling studies is to directly look at estimates of the area burnt. Previous examples using this methodology exist mostly for extratropical fire inventories in North America and for tropical fires in Africa and Asia (see Stocks, 1991; Stocks et al., 1996; Lavoué et al., 2000; Barbosa et al., 1999; Liousse et al., 2002; Michel et al., 2002). The first global inventory using satellite-based estimates of burnt area is the one of Hoelzemann et al. (2003). These works have become possible due to the availability of new products from two European and one American sensors: the ATSR and VEGETATION sensors onboard the ERS and SPOT platforms, respectively, and the MODIS sensor onboard the TERRA platform (launched at the end of 1999). A successor of the ATSR sensor is now also onboard the European platform ENVISAT (launched in February 2002). Two major initiatives have clearly demonstrated the feasibility to detect and globally map the major burnt areas for a specific year: the GLOBSCAR initiative launched by the European Space Agency (ESA-ESRIN; http://shark1.esrin.esa.it/ionia/ FIRE/BS/ATSR/) and the GBA2000 project, coordinated by the European Commission Joint Research Centre (Grégoire et al., 2002; http://www.gvm.sai.jrc.it/fire/gba2000_website/

index.htm). In both cases, the burnt areas have been mapped during the year 2000 and for the major land cover types of the globe. It is expected that the methodologies developed within these two projects will now be exploited on a more systematic basis for the assessment of the areas burnt globally during the period 1998-2003 (e.g. within the European GLOBCARBON project).

Special emphasis within these projects will have to be given to the detectability of agricultural fires from space. It has been shown that satellite retrievals might capture only a fraction of these fires (Michel et al., 2002), so that comparisons with ground-based statistical data are necessary in order to assess the minimum fire size that is seen by the satellite sensor. Agricultural and savannah fires with different burning properties need also to be clearly distinguished for better emission inventory constructions.

3.1.2. Example: Africa

In a recent study for African biomass burning, Liousse et al. (2002) derived black carbon emissions from the burnt area distribution given by AVHRR satellite data for 1981-1991. Following a multithreshold technique, burnt area maps have been constructed from AVHRR images (Barbosa et al., 1999). Since the percentage of pixels which are actually burnt, is somewhat uncertain, different scenarii have been assumed. Emission calculations were performed using equation (1). The scenario with the highest emissions is discussed here. The values for biomass density, combustion efficiency and emission factors have been taken from Hao et al. 1990 and Liousse et al. (1996ab). Figure 9 shows the interannual variability of BC emissions together with yearly budgets, which are in the range of 1.25 to 3.1 TgC. These estimates compare well with a previous estimate (1.8 TgC) from Liousse et al. (1996ab), based on statistical data. As already noted by Barbosa et al. (1999), the lowest value for BC emissions is found for the fire season 1987-1988, which reflects the climatological impact of El Niño on fires. The largest BC emissions are obtained at the end of the decade. No trend is observed, but a high regional interannual variability can be noticed.

This example demonstrates the possible improvement achieved by using satellite products for burnt area detection. The spatial and temporal variability of emissions are accounted for, and consistent estimates of parameter A (equation 1) can be obtained. Some uncertainty remains in terms of the burnt area estimates linked to the subpixel information. It is expected that higher resolution data from the next generation of satellites and the application of dynamic fire models will reduce the uncertainty to less than 15% in the next two years.

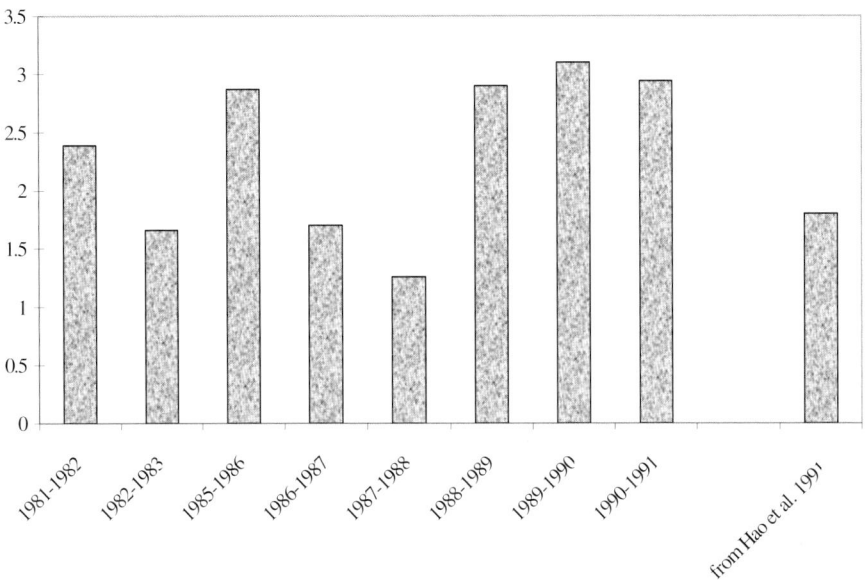

Figure 9. BC emissions in TgC/yr for African biomass burning, from burnt area distribution derived from AVHRR satellite data for 1981-1991 (Liousse et al. 2002).

The work of Barbosa et al. (1999) and Liousse et al. (2002) also documented the large interannual variability of biomass burning emissions in Africa. This implies that biomass burning inventories in Africa cannot be scaled linearly with quantities such as the population number. A complex relationship exists between climatic factors, the available phytomass, and human activities.

3.1.3. Other required parameters

While burnt area estimates are improved from the use of satellite data, the uncertainty in the assumptions of biomass density and combustion efficiency is still large and around ± 50%. The main reasons for this uncertainty are:

i) The fuel mass needs to be expressed by fuel type (dry grass, green grass, leaf litter, twigs, logs, etc.), because the combustion completeness ranges from 98% for standing dry grass to <10% for dead logs depending on the biome. Recently, studies have been performed to use global biogeochemical models to estimate the biomass content of the different carbon pools, and their combustion efficiencies (Potter et al., 2001; Hölzemann et al., 2003).

ii) Emission calculations are mostly performed at a scale where vegetation types can be mixed within a reference unit (Lavoué et al., 2000). Hence, there is a growing need to know which plant material is actually burnt in order to accurately estimate the completeness of the combustion (Cumming, 2001). In different ecosystems, fires consume different amounts of the various fuel types available. For example, crown fires, affecting the barks and twigs, represent only 20% to 30% of fires in Russia (Shvidenko and Nilsson, 1996; Conard and Ivanova, 1997), but represent more than 90% in Canada (Kasischke et al., 1991; French et al., 1999). When they do affect the crown, the completeness of the combustion is similar. Likewise, 99% of the fires in savannah regions are ground fires, affecting only the grass layer (Seiler and Crutzen, 1980; Hochberg et al., 1994). However, woody biomass represents a relatively larger amount of the phytomass susceptible to burning.

iii) Within a vegetation type, fire risk is a function of stand age, as a result of both the biomass accumulated, but also the community composition reached at different stages of succession (Mouillot et al., 2002). For example, fires in Amazonia affect the tropical forest, but most of the fires spotted by remote sensing affect early successional stages as secondary forest and grasslands, rather than the rainforest itself. On the contrary, in the boreal region, fires affect preferentially older stands according to the weibull model (Van Wagner, 1978), inducing the combustion in the stands with more biomass.

iv)The fuel composition also influences the combustion efficiency, which has an impact on the emission factors. Particularly in grasslands, fuel compaction and seasonal fuel moisture change significantly affect their combustion efficiency by a factor of 2 (Hoffa et al., 1999). In the boreal forest, fuel moisture influences the fire risk, but also drive fire spread, and the subsequent combustion efficiencies of the litter and the barks (Kasischke et al., 2000).

The concurrent development of fuel accumulation models by several groups, supported by new remote sensing inputs, and the development of much more comprehensive databases of fuel load measurements taken in several campaigns, are likely to halve the uncertainty within the next two years. Such an approach will allow to take into consideration the important inter- and intra-annual variability of fuel load submitted to fires

3.2. Regional zoom on biofuel emissions

It is now recognized that in most developing countries emissions from the preparation and consumption of domestic biomass fuels are of the same order of magnitude as emissions from other biomass burning sources. It has been estimated that the total burnt biomass has increased by 20% from 1985 to 1995 in these countries (Streets et al., 2001; Yevich and Logan, 2002). Therefore, there is a crucial need to focus on such emissions.

One of the first global emission inventories developed by Liousse et al. (1996ab) includes estimates for the burning of wood, charcoal, dung and agricultural residues. Due to the lack of information from other regions, parameters derived from western African practices had to be extrapolated globally. For example, it was assumed that half of the global fuelwood submitted to fires was used for charcoal making. Since then, it has been recognized that important differences can be found between countries and the use of charcoal is less important in Asia than in Africa. Also, agricultural residues and dung combustion, which are negligible in Africa compared to fuelwood and charcoal, are of the same intensity as fuelwood emissions in Asia (Reddy and Venkataraman, 2002; Yevich et Logan, 2002). Moreover, the Liousse et al. (1996ab) studies were based on fuel consumption given by the FAO database. More confidence may be expected from regional estimates based on field observations, which have been used by recent inventories (Joshi et al., 1992; Brocard et al., 1996; Marufu et al., 1997; Streets et al., 2001; Yevich and Logan, 2002). Table 1 summarizes biomass combustion estimates for the different classes of biofuel in India. There is a general agreement between the regional inventory and the Yevich and Logan (YL02) inventory, but further regional comparisons are needed for consistency.

Moreover, as noted above, there is a lack of emission factors for domestic and agricultural fires. In particular, the regional differences between the different ways of domestic burning in India, China or Africa need to be examined. An ongoing international intercomparison of emission factors for carbonaceous aerosol will provide tests for the consistency of the estimates made by different authors.

3.3. Other approaches

3.3.1. Development of a dynamic fire model

To reduce the uncertainties for regional emission estimates, a new concept of emission modelling for vegetation fires has been proposed, using a dynamic parameterisation for fire occurrence and intensity, including a fire propagation scheme and parameterisations for the different combustion phases. This type of model is currently under development for Canadian boreal forest wildfires (Lavoué et al., 2002).

Different models exist, which simulate the ecosystem dynamics and incorporate fire behavior and effects. Landscape-fire-succession models (LFSMs, Rupp et al., 2002) explicitly simulate the fire processes and succession in a spatial domain of 1-100 km^2 with resolutions of 10-1000 m. Other models such as the ATHAM model (Active Tracer High-Resolution Atmospheric Model) (Trentmann et al., 2002) were used to simulate the dynamical evolution of the plume from a prescribed fire during the SCAR-C experiment (Hobbs et al., 1996). ATHAM accounts for the dynamics, turbulence, transport, cloud microphysics, gas scavenging, and radiation in hot plumes, and was run in a domain of 35km x 28km x 3.75km with a minimal grid spacing of 50m x 50m x 20m. The model appears to be a valuable tool for the examination of transport processes within biomass burning plumes and might yield new insight into the interaction between transport, chemistry and microphysics in such plumes.

Coupling meteorological models, and vegetation models with emission and fire succession models on different scales raises a number of issues, which need to be resolved:

(1) The fire propagation and combustion processes are highly variable and depend on local weather, vegetation fuel characteristics, and topography. While the biomass density in forested ecosystems doesn't exhibit a large interannual variability, the aboveground biomass in grassland ecosystems is dependent on the annual net primary production, which largely depends on

climate. Years with less or more rainfall than average can produce anomalous amounts of phytomass: Gerard et al. (1999) estimate a range of +/- 0.75PgC for the tropical savannah biome. This has important implications for the capacity of a given landscape to carry large fires. This variability is currently not taken into account in usual vegetation maps, where fuel amounts are held constant. Furthermore, the high interannual variability of area burnt, as observed in Africa (Barbosa et al., 1999), strongly affects the amount of (litter) biomass available in the year following a fire, which again feeds back onto the burnt area in that year. The heterogeneity of the landscape created by patches of different biomass amount could also, in turn, affect the fire spread (Turner et al., 1989). In forested ecosystems, scars of previous disturbances remain for decades, so that the fire history must be taken into account over long time periods.

(2) The fire perimeter growth depends on the propagation of head, flank, and back fires. Back fires burn into the wind and produce much less smoke than head fires. Fire growth may be stopped by natural barriers, e.g. lakes (Richards and Bryce, 1995). Therefore, a detailed description of the landscape is necessary to model individual fires.

(3) Emission factors vary with combustion phases. Emissions may be decomposed into 2 burning phases: a flaming phase with open flames, i.e. relatively efficient burning and high BC emissions, followed by a smoldering phase with few or no open flames, with low combustion rate and low BC emissions.

(4) The injection height of the pollutants is a key parameter of the subsequent atmospheric transport. According to Manins (1985), the height is a function of both the energy released by the combustion and the length of the fire front. Convective columns carry large amounts of smoke into high altitudes during the flaming phase. After the crown fire passed, smoldering ground fires release most of the smoke in the surface boundary layer (Ferguson et al., 1998; Trentmann et al., 2002).

(5) Atmospheric concentration measurements of fire tracers indicate that smoke episodes far from the source regions last only for short time periods (Jaffrezo et al., 1998; Wotawa and Trainer, 2000). The dynamic modelling will calculate hourly or daily particulate emissions; the interest of such a temporal distribution needs to be tested in the transport model.

Ultimately, wildfire emissions will be derived from interactively coupled Earth System models, which include the necessary feedbacks between climate and the biosphere, so that they can be applied to different global warming scenarios.

3.3.2. Inverse modelling

Due to the uncertainties pertaining in regional estimates of emission inventories obtained from the bottom-up approach, different studies have used measurements of the tropospheric CO distribution together with information contained in a CTM in order to better assess the CO source distribution (top-down approach). This method, referred to as inverse modelling is discussed in details in Chapter 12. Bergamaschi et al. (2000) performed the first 3D time-independent inversion of CO sources in eight different regions using 31 CMDL stations (see Chapter 11) and the TM2 model. Table 5 summarizes the results.

Table 5. Comparison of global emission estimates with results of two inverse models (Bergamaschi et al. 2000, Petron et al. 2002)

TgCO/yr	Emission estimate	Inverse model
Bergamashi et al., 2000		
Tropical forest	139	483-633
Savannah	206	140-245
Burning at lat. > 30N	68	0-87
Total	413	623-965
Pétron et al., 2002		
Forest and savannah burning	436	606
Agricultural wastes burning and fuelwood use	384	561

The *a priori* and *a posteriori* estimates are in good agreement, except for the tropical forest. This difference may be explained by the limited number of CO data available above tropical forests. Pétron et al. (2002) adopt the time-dependent synthesis inversion technique and perform the first time-dependent inversion of regional CO sources with the IMAGES model (Müller and Brasseur, 1996). These inversion techniques do not give information on the location of sources inside large regions, they only constrains the magnitude of each emission process. The time-inversion technique proves to be a powerful tool to study the intensity and seasonality of different emission processes, provided that a sufficient number of observations exists. A key result is that the 1990 a priori estimate of the total CO flux out of Asia is significantly lower than the optimized flux constrained by the observations. Similar results were obtained from a study by Kasibhatla et al. (2002). Currently, the most severe limitation of such studies is the lack of observations, especially over tropical biomass burning areas. Further improvement is expected from the use of assimilated meteorological data and the inclusion of satellite retrieved CO columns in the inversion process. (see Chapters 11 and 12).

4. CONCLUSIONS

In this chapter different biomass burning emission inventories obtained with a bottom-up method have been discussed in light of their main differences and uncertainties. Possible improvements have been suggested.

In spite of several measurements during the last decade, uncertainties in emission factors still remain large for several species, in particular for aerosol compounds, Volatile Organic Compounds, and sulphur compounds. We have given new recommendations for aerosol emission factors based on a critical review of recent measurement results. Additional measurements are urgently needed for domestic and agricultural fires, humid forest fires, boreal and mediterranean ecosystems. Future field campaigns, such as the African Monsoon experiment (AMMA : http://www.medias.obs-mip.fr/amma/) will help to improve the situation as they place a strong focus on characterizing emissions. Intercomparisons of emission factors (e.g. the on-going activity for particles between Indian, American and European groups) are also urgently needed.

The mapping of burnt biomass is still a difficult task, but remote sensing tools were shown to be promising. Qualitative improvements are obtained from the multi-year global distribution of fire pixel counts given by ATSR, NOAA-AVHRR and SPOT Vegetation as they allow to take into account the spatial and temporal variability of fires at different scales. Quantitative improvements may be obtained from the global distribution of burnt areas given by ATSR, AVHRR, SPOT Vegetation, MODIS and future microsatellites, such as BIRD. Provided that the current problems with detection algorithms can be solved, the uncertainty of such products might be reduced to less than 15% for burnt areas. Fire emission modellers need to interact with other scientific communities in order to gain better estimates of biomass burnt per unit area, which requires better knowledge of surface properties (biomass density and combustion efficiency). This effort needs to take into account regional differences and link local measurements, remote sensing techniques and fuel models.

Different modelling methods may be used to reduce uncertainties. Improved emission estimates are expected from the development of integrated models. These models should include multi-disciplinary approaches to describe fire dynamics, the evolution of the vegetation including a detailed description of carbon pools and fire history of individual ecosystems, and meteorological parameters, which will allow to assess the impact of climate changes. This approach is on the agenda of the IGBP program (see Chapter 1), and will allow to investigate the combination of causes, effects and feedbacks relating fire and global change ecosystem succession (Lavorel et al., 2001). The combination of bottom-up with top-down methods using inverse modelling is another way to improve biomass burning emission estimates.

It is important to note that global emission estimates can only be improved by compiling more accurate data in individual regions. This is particularly true for domestic and agricultural practices, which differ significantly between regions. As outlined by Olivier (2002), a few general properties of emission inventories, e.g. flexibility, transparency (clarity of source definitions and methods), consistency and accuracy should also be a common concern for further inventory development.

5. REFERENCES

Andreae, M.O., E.V. Browell, M. Garstang, G.L. Gregory, R.C. Harriss, G.F. Hill, D.J. Jacob, M.C. Pereira, G.W. Sachse, A.W. Setzer, P.L. Silva Dias, R.W. Talbot, A.L. Torres, and S.C. Wofsy, Biomass-burning emissions and associated haze layers over Amazonia, J. Geophys. Res., 93, 1509-1527, 1988.

Andreae, M.O., Biomass burning: Its history, use and distribution and its impact on environmental quality and global climate, in Global Biomass Burning, edited by J. S. Levine, pp 1-21, MIT Press, Cambridge, Mass., 1991.

Andreae M.O., T.W. Andreae, H. Annegarn, J. Beer, H. Cachier, P. le Canut, W. Elbert, W. Maenhaut, I. Salma, F.G. Wienhold and T. Zenker, Airborne studies of aerosol emissions from savanna fires in southern Africa : 2. Aerosol chemical composition, J. Geophys. Res., 103, 32119-32128, 1998.

Andreae M.O. and P. Merlet, Emission of trace gases and aerosols from biomass burning, Global Biogeochem. Cycles, 15, 955-966, 2001.

Arino O. and J.M. Rosaz, 1997 and 1998 World Fire Atlas using ERS-2 ATSR-2 Data, in Proceedings of conference on Remote Sensing of Joint Fire Science Conference and Workshop, Boise, Idaho, p177-182, 1999.

Artaxo P., M.O. Andreae, A. Guenther, D. Rosenfeld, J.V. Martins, L.V. Rizzo, A.S. Procopio, L.V. Gatti, A.M. Cordova, P. Guyon and B. Graham, Aerosols, trace gases and climate linkages in Amazonia : what we learned so far ?, International Global Atmospheric Chemistry Conference, Hersonissos, Crete, Greece, September 18-25, 2002.

Barbosa P. M., Stroppiana D., Gregoire J.-M., Pereira J.M.C., An assessment of vegetation fire in Africa (1981-1991): Burned areas, burned biomass, and atmospheric emissions, Global Biogeochem. Cycles, 13, 993-950, 1999.

Bergamaschi P., R. Hein, C.A.M. Brenninkmeijer and P.J. Crutzen, Inverse modeling of the global CO cycle 1. Inversion of CO mixing ratio, J. Geophys. Res., 105, 1909-1927, 2000.

Brocard, D., C. Lacaux, J.P. Lacaux, G. Kouadio and V. Yoboué, Emissions from the combustion of biofuels in Western Africa, in Biomass Burning and Global Change, edited by J.S. Levine, pp. 492-508, MIT Press, Cambridge, Mass., 1996.

Brown S. and A.E. Lugo, Tropical secondary forests, J. Trop. Ecol., 6, 1-32, 1990.

Cachier, H., M. P. Brémond, and P. Buat-Ménard, Determination of atmospheric soot carbon with a simple thermal method. Tellus, 41B, 379-390, 1989.

Cachier, H., C. Liousse, A. Cachier, B. Ardouin, G. Polian, V. Kazan and A.D.A. Hansen, Black carbon aerosols at the remote site of Amsterdam Island, Paper presented at the Fifth Conference on carbonaceous aerosols, Berkeley, August 23-26, 1994.

Cachier H., C. Liousse, P. Buat-Ménard and A. Gaudichet, Particulate content of savanna fire emissions, J. Atmos. Chem., 22, 123-148, 1995.

Cachier, H., Carboneous Combustion Aerosols. Atmospheric Particles, edited by R. M. Harisson and R. Van Greiken, 296-348, 1998.

Cautenet S., D. Poulet, C. Delon, R. Delmas, J.M. Grégoire, J.M. Pereira, S. Cherchali, O. Amram, G. Flouzat, Simulation of carbon monoxide redistribution over Central Africa during biomass burning events (Experiment for Regional Sources and Sinks of Oxidants (EXPRESSO), J. Geophys. Res., 104, 30641-30658, 1999.

Chandra S., J.R. Ziemke , P.K. Bhartia and R.V Martin, Tropical tropospheric ozone : Implication for dynamics and biomass burning, J. Geophys. Res., 107, 1029-1046, 2002.

Chatfield, R. B., J. A. Vastano, H. B. Singh, and G. Sachse, A general model of how fire emissions and chemistry produce African/oceanic plumes (0, CO, PAN, smoke) in TRACE A, J. Geophys. Res., 101, 24,279, 1996.

Chin M., Black carbon aerosols : a burning question, Poster presented at the workshop « Air pollution and climate change » Hawaii, 2002.

Cochrane M.A., A. Alencar, M.D. Schulze, C.M. Souza, D.C. Nepstad, P. Lefebvre and E.A. Davidson, Positive feedbacks in the fire dynamic of closed canopy tropical forests, Science, 284, 1832-1835, 1999.

Conard, S.G. and G.A. Ivanova, Wildfire in Russian boreal forests - potential impacts of fire regime characteristics on emissions and global carbon balance estimates, Environmental Pollution, 98, 305-313, 1997

Cooke, W.C., and J.J.N. Wilson, A global black carbon aerosol model, J. Geophys. Res., 101, 19395-19409, 1996.

Cooke W.C., B. Koffi and J.M. Grégoire, Seasonality of vegetation fires in Africa from remote sensing data and application to a global chemistry model, J. Geophys. Res., 101, 21051-21065, 1996.

Cox P. M., R.A. Betts, C.D. Jones, S.A. Spall and I.J. Totterdell, Acceleration of global warming due to carbon-cycle feedbacks in a coupled climate model, Nature 408, 184-187, 2000.

Crutzen, P.J. and M.O. Andreae, Biomass burning in the Tropics : impact on atmospheric chemical and biogeochemical cycles, Science, 250, 1669-1678, 1990.

Cumming, S.G., Forest type and wildfire in the Alberta boreal mixedwood: what do fires burn?, Ecological Applications, 11, 97-110, 2001.

Delmas, R., P. Loudjani and A. Podaire, Biomass burning in Africa: an assessment of annualy burnt biomass, in Global Biomass Burning, J.S. Levine ed., MIT Press, 126-132, 1991.

Delmas R. and A. Guenther, The Experiment for regional sources and sinks of Oxidants (EXPRESSO), IGAC Newslett., 15, 1999.

Dixon, R.K. and Krankina, O.N., Forest fires in Russia: Carbon dioxide emissions to the atmosphere, Can. J. For. Res., 23, 700-705, 1993.

Duncan B.N., R.V. Martin, A.C. Staudt, R. Yevich and J.A. Logan, Interannual and seasonal variability of biomass burning emissions constrained by satellite, J. Geophys. Res., 108, doi: 10.1029/2002JD002378, 2003.

Dwyer, E., Pereira, J.M.C., Gregoire, J.M. and DaCamara, C.C. Characterisation of the spatio-temporal patterns of global fire activity using satellite imagery for the period april 1992 to March 1993. Journal of Biogeography 27: 57-69, 2000.

FAO., Global forest fire assessment, 1990-2000, Rome, 2001.

Ferek R.J., J.S. Reid, P.V. Hobbs, D.R. Blake and C. Liousse, Emission factors of hydrocarbons, halocarbons, trace gases and particles from biomass burning in Brazil, J. Geophys. Res, 103, 32107-32118, 1998.

Ferguson, S.A., D.V. Sandberg, and R. Ottmar, Modeling the effect of land-use changes on global biomass emissions, Biomass Burning and its Inter-Relationships with the Climate System, Wengen, Switzerland, September 28-October 1, 1998.

Flannigan, M., Bergeron, Y., Engelmark, O. and Wotton, B.M., Future wildfire in circumboreal forests in relation to global warming, Journal of Vegetation Science, 9, 469-476, 1998.

French, N.H.F., Kasischke, E.S., Stocks, E.S., Mudd, J.P., Martell, D.L. and Lee, B.S., Carbon release from fires in the North American Boreal forest in E. S. Kasischke and B. J. Stocks, eds. Fire, climate change and carbon cycling in the boreal forest, 1999.

Gadi R., U.C. Kulshrestha, A.K. Sarkar, S.C. Garg, and D.C. Parashar, Emissions of SO2 and NOx from biofuels in India, accepted for publication in Tellus, 2002.

Generoso S., Y. Balkanski, O. Boucher, F.M. Bréon and M. Schulz, Comparisons between aerosol data retrievals from the Polder spaceborne instrument and aerosol output from the LMdz model, , International Global Atmospheric Chemistry Conference, Hersonissos, Crete, Greece, September 18-25, 2002.

Gerard, J.C., B. Nemry, L.M. Francois, and P. Warnant, The interannual change of atmospheric CO2: contribution of subtropical ecosystems?, Geophys. Res. Lett., 26, 243-246, 1999.

Grégoire J-M., K. Tansey, and J.M.N. Silva, The GBA2000 initiative: Developing a global burned area database from SPOT-VEGETATION imagery, Int. J. Remote Sensing, in press, 2002.

Hall D.O., F. Rosillo-Calle and J. Woods, Biomass utilisation in households and industry ; energy use and development, Chemosphere, 29-5, 1099-1199, 1994.

Hao, W.M., M.H. Liu and P.J. Crutzen, Estimates of annual and regional releases of CO_2 and other trace gazes to the atmosphere from fires in the tropics, based on the FAO Statistics for the period 1975-1980, in Fire in the Tropical Biota, J.C. Goldammer ed., 440-462, 1990.

Hao, W.M. and M.-H. Liu, Spatial and temporal distribution of tropical biomass burning, Global Biogeochem. Cycles, 8, 495-503, 1995.

Heald C.L., D.J. Jacob, P.I. Palmer, M.J. Evans, J.A. Logan, G.W. Sachse, H.B. Singh, D.R. Blake and J.C. Gille, Interpreting aircraft and satellite observations of carbon monoxide and constraining Asian emission sources during the TRACE-P mission, International Global Atmospheric Chemistry Conference, Hersonissos, Crete, Greece, September 18-25, 2002.

Heimann, M., The global atmospheric model TM2, Tech. Rep. 10, Dtsch. Klimarechenzent., Modellbetreuungsgruppe, Hamburg, Germany, 1995.

Hély C., Caylor K.K, Alleaume, S. Swap, R.J. and Shugart, H.H. 2002. Release of gaseous and particulate carbonaceous compounds from biomass burning during the SAFARI 2000 dry season field campaign, J. Geophys. Res., 108, D13, 8470, doi:10.1029/2002JD002482, 2003.

Hobbs P.V., J.S. Reid, J.A. Herring, J.D. Nance, R.E. Weiss, J.L. Ross, D.A. Hegg, R.D. Ottmar and Liousse C, Particle and Trace gas measurements in the smoke from prescibed burns of forest products in the Pacific Northwest, in Biomass Burning and Global Change, edited by J.S. Levine, pp. 697-715, MIT Press, Cambridge, Mass., 1996.

Hochberg, M.E., Menaut, J.C. and Gignoux, J., The influence of tree biology and fire in the spatial structure of the Western African savannah. Journal of Ecology 82: 217-226, 1994.

Hoelzemann, J., G.P. Brasseur, C. Granier, M.G. Schultz and M. Simon, The Global Wildfire Emission Model GWEM: a new approach with global area burnt satellite data, submitted to J. Geophys. Res., 2003.

Hoffa, E.A., Ward, D.E., Hao, W.M., Susott, R.A. and Wakimoto, R.H., Seasonality of carbon emissions from biomass burning in a Zambia Savanna. J. Geophysical Research 104: 13841-13853, 1999.

Holben B.N., T.F. Eck, I. Slutsker, D. Tanré, J.P. Buis, A. Setzer, E.F. Vermote, J.A. Reagan, Y.J. Kaufman, T. Nakajima, F. Lavenu, I. Jankowiak and A. Smirnov, AERONET, A federated instrument network and data archive for aerosol characterization, Remote Sensing of Environ., 66, 1-16, 1998.

Horowitz, L.W., S. Walters, D.L. Mauzerall, L.K. Emmons, P.J. Rasch, C. Granier, X.X. Tie, J.F. Lamarque, M.G. Schultz, G.S. Tyndall, J.J. Orlando and G.P Brasseur, A global simulation of tropospheric ozone and related tracers: description and evaluation of MOZART, version 2, submitted to J. Geophys. Res., 2002.

Houghton, R.A., D.L. Skole, C.A. Nobre, J.L. Hackler, K.T. Lawrence and W.H. Chomentowski, Annual fluxes of Carbon from deforestation and regrowth in the Brazilian Amazon. Nature, 403, 301-304, 2000.

Houghton, R.A., K.T. Lawrence, J.L. Hackler, and S.T. Brown, The spatial distribution of forest biomass in the Brazilian Amazon: a comparison of estimates, Global Change Biology 7, 731-746, 2001.

Jacobson M.Z., Short-Term cooling but long-term global warming due to biomass burning particles and gases, Paper presented to AGU Fall meeting, 2002.

Jaffrezo, J.-L., C.I. Davidson, H.D. Kuhns, M.H. Bergin, R. Hillamo, W. Maenhaut, J.W. Kahl, and J.M. Harris, Biomass burning signatures in the atmosphere of central Greenland, J. Geophys. Res., 103, 31,067-31,078, 1998.

Joshi V., C.H. Sinha, M. Karuppasamy, K.K. Srivastava and B.P. Singh, Rural Energy database. Final report submitted to Non-conventional Energy sources, Government of India, Tata Energy Research Sources, 1992.

Kasibhatla P., A. Arellano, J.A. Logan, P.I. Palmer, and P. Novelli, Top-down estimate of a large source of atmospheric carbon monoxide associated with fuel combustion in Asia, Geophys. Res. Lett., 29, 2002.

Kasischke, E.S., B.J. Stocks, K. O'Neil, N.H.F. French and L.L. Bourgeau-Chavez, Direct effects of fire on the boreal forest carbon budget, 1991.

Kasischke, E.S., N.H.F. French, L.L. Bourgeau-Chavez, and N.L. Christensen, Jr., Estimating release of carbon from 1990 and 1991 forest fires in Alaska, J. Geophys. Res., 100, 2941-2951, 1995.

Kasischke, E.S., K.P. O'Neill, N.H.F. French and L.L. Bourgeau-Chavez, Controls on patterns of biomass burning in Alaskan boreal forests in E. S. Kasischke and B. J. Stocks, eds. Fire, Climate Change, and Carbon cycling in the boreal forest. Springer-Verlag, New York, 2000.

Kaufman Y.B., R.S. Fraser and R.L. Mahonet, Fossil fuel and biomass burning effect on climate ; heating or cooling ?, J. of Climate, 4 (6), 578-588, 1991.

Kaufman Y.B., P.V. Hobbs, W.J.H. Kirchhoff, P. Artaxo, L.A. Remer, B.N. Holben, M.D. King, E.M. Prins, K.M. Longo, L.F. Longo, C.A. Nobre, J.D. Spinhirne, Q. Ji, A.M. Thompson, J.F. Gleason, S.A. Christopher and S.C. Tsay, Smoke, Clouds and Radiation, J. Geophys. Res., 103, 31783-31808, 1998.

Lacaux, J.P., J.M. Brustet, R. Delmas, J.C. Menaut, L. Abbadie , B. Bonsang, H. Cachier, J.G.R. Baudet, M.O. Andreae, and G. Helas, Biomass burning in the tropical savannas of Ivory Coast: An overview of the field experiment fire of Savanna (FOS-DECAFE 91), J. Atmos. Chem., 22, 195-216, 1995.

Lavorel S., E.F. Lambin, M. Flannigan and M. Scholes, Fires in the Earth System : The need for integrated research, Global Change Newsletter, n° 48, 2001.

Lavoué D., C. Liousse, H. Cachier and J. Goldammer, Carbonaceous aerosol source inventory for temperate and boreal fires, J. Geophys. Res., 105, 26871-26890, 2000.

Lavoué, D., S.L. Gong, N,. Payne, B.J. Stocks, M. Aubé, and R. Leaitch, Dynamic modeling of Canadian forest fire emissions, International Global Atmospheric Chemistry Conference, Hersonissos, Crete, Greece, September 18-25, 2002.

Liousse, C., C. Devaux, F. Dulac, and H. Cachier, Aging of savanna biomass burning aerosols: Consequences on their optical properties, J. Atmos. Chem., 22, 1-17, 1995.

Liousse C., J.E. Penner, J.J. Walton, H. Eddleman, C. Chuang and H. Cachier, Modeling biomass burning aerosols, in Biomass Burning and Global Change, edited by J.S. Levine, pp. 492-508, MIT Press, Cambridge, Mass., 1996a.

Liousse, C., J.E. Penner, C. CHuang, J.J. Walton, H; Eddleman and H. Cachier, A global three-dimensional modeling of carbonaceous aerosols, , J. Geophys. Res., 101, 19411-19432, 1996b.

Liousse C., Chiappello I., Quesque P., Barbosa P., Grégoire J.M., Clavier F. and H. Cachier, Interannual variability of global transport and radiative impact of carbonaceous aerosol in Africa for the 1980 decade, International Global Atmospheric Chemistry Conference, Hersonissos, Crete, Greece, September 18-25, 2002.

Lobert J.M., W.C. Keene, J.A. Logan, and R. Yevich, 1999, Global chlorine emissions from biomass burning : reactive chlorine emissions inventory, J. Geophys. Res., 104, 8373-8389, 1999.

Manins, P.C., Cloud heights and stratospheric injections resulting from a thermonuclear war, Atmospheric Environment, 19, 1245-1255, 1985.

Martins, J.V., P. Artaxo, C. Liousse, J. S. Reid, P. V. Hobbs, and Y. Kaufman, Effects of Black Carbon Content, Particle Size, and Mixing on Light Absorption by Aerosol Particles from Biomass Burning in Brazil, J. Geophys. Res, 103, 32041-32050, 1998.

Marufu L., J. Ludwig, M.O. Andreae, F.X. Meixner and G. Helas, Domestic biomass burning in rural and urban Zimbabwe, Part A, Biomass Bioenergy, 12, 53-68, 1997.

Menaut, J.C., L. Abbadie, F. Lavenu, P. Loudjani and A. Podaire, Biomass burning in west African savannas, in Global Biomass Burning J. S. Levine ed., MIT Press, 133-142, 1991.

Michel C., C. Liousse, J.M. Grégoire, K. Tansey and H. Cachier, Biomass burning emission inventory in Asia with a focus during ACE-ASIA campaign,, International Global Atmospheric Chemistry Conference, Hersonissos, Crete, Greece, September 18-25, 2002.

Mishchenko, M.K., I.V. Geogdzhayev, B. Cairns, W.B. Rossow, and A.A. Lacis, Aerosol retrievals over the ocean using channel 1 and 2 AVHRR data: A sensitivity analysis and preliminary results, Applied Optics, 38, 7325-7341, 1999.

Mouillot, F., S. Rambal, and .R. Joffre, Simulating the effects of climate change on fire frequency and the dynamics of a mediterranean maquis woodland. Global Change Biology, 8, 423-437, 2002.

Müller J.-F. and Brasseur G., IMAGES : A three-dimensional chemical transport model of the global troposphere, J. Geophys. Res., 100D, 16,445-16,490, 1995.

Olivier J.G.J., A.F. Bouwmann, C.W.M. Van der Maas, J.J.M. Berdowski, C. Veldt, J.P.J. Bloos, A.J.H. Visschedijk, P.Y.J. Zandveld and J.L. Haverlag, Descritpion of EDGAR Version 2.0 : a set of global emission inventories of greenhouse gases and ozone-depleting substances for all anthropogenic and most natural sources on a per country basis and on a 1°x1° grid, Report 771060002, National Institute of Public Health and the Environment (RIVM), The Netherlands, 1996.

Olivier J.G.J., Global Inventories with Historical Emission trends; EDGAR 3.2, in « On the quality of Global Emission Inventories », Thesis Utrecht University, 57-90pp, 2002.

Penner, J.E., S.J. Ghan and J.J. Walton, The role of biomass burning in the budget and cycle of carbonaceous soot aerosols and their climate impact, in Global Biomass Burning, J. S. Levine ed., MIT Press, 387-393, 1991.

Penner, J.E., M. Andreae, H. Annegarn, L. Barrie, J. Feichter, D. Hegg, A. Jayaraman, R. Leaitch, D. Murphy, J. Nganga, and G. Pitari, Chapter 5: Aerosols, their Direct and Indirect Effects, in Climate Change, The Scientific Basis, Ed. by H.T. Houghton, Y. Ding, D.J. Griggs, M. Noguer, P.J. van der Linden, X. Dai, K. Maskell, C.A. Johnson, Report to

Intergovernmental Panel on Climate Change from the Scientific Assessment Working Group (WGI), 289-416,.Cambridge University Press, 2001.

Penner. J.E., S. Y. Zhang, M. Chin, C.C. Chuang, J. Feichter, Y. Feng, I.V. Geogdzhayev, P. Ginoux, M. Herzog, A. Higurashi, D. Koch, C. Land, U. Lohmann, M. Mishchenko, T. Nakajima, G. Pitari, B. Soden, I. Tegen, L. Stowe, A comparison of model- and satellite-derived aerosol optical depth and reflectivity, J. Atmos. Sci.,59, 441-460, 2002.

Pétron G., C. Granier, B. Khattatov, J.F. Lamarque, V. Yudin, J.F. Müller and J. Gille, Inverse modeling of carbon monoxide surface emissions using CMDL network observations, J. Geophys. Res., 107, NO. D24, 4761, doi:10.1029/2001JD001305, 2002.

Potter, C., V. Brooks Genovese, S. Klooster, M. Bobo and A. Torregrosa, Biomass burning losses of carbon estimated from ecosystem modeling and satellite data analysis for the Brazilian Amazon region, Atmospheric Environment, 35, 1773-1781, 2001.

Prentice I.C., G.D. Farquhar, M.J.R. Fasham, M.L. Goulden, M. Heimann, V.J. Jaramillo, H.S. Kheshgi, C. Le Quéré, R.J. Scholes, D.W.R. Wallace, Chapter 3 : The Carbon Cycle and Atmospheric Carbon Dioxide, Report to Intergovernmental Panel on Climate Change from the Scientific Assessment Working Group (WGI), Cambridge University Press, 2001.

Reddy M.S and C. Venkataraman, Inventory of aerosol and sulphur dioxide emissions from India. Part II-biomass combustion, Atmos. Env., 36, 699-712, 2002.

Reid J.S., P.V. Hobbs, C. Liousse, J.V. Martins, R.E. Weiss and T.F. Eck, Comparisons of techniques for measuring shortwave absorption and black carbon content of Aerosols from Biomass-Burning in Brazil, J. Geophys. Res, 103, 32031-32040, 1998.

Richards, G.D., and R.W. Bryce, A computer algorithm for simulating the spread of wildland fire perimeters for heterogeneous fuel and meteorological conditions, International Journal of Wildland Fire, 5, 73-79, 1995.

Rupp, T.S., A.M. Starfield, F.S. Chapin and P. Duffy, Modeling the impact of black spruce on the fire regime of Alaskan boreal forest, Climate Change, 55, 213-233, 2002.

Schultz M., On the use of ATSR fire count data to estimate the seasonal and interannual variability of biomass burning emissions, Atmos. Chem. Phys., 2,387-395, 2002.

Scholes, R.J., J. Kendall, and C.O. Justice, the quantity of biomass burned in southern Africa, J. Geophys. Res., 101, 23,667-23,682, 1996.

Seiler, W. and P.J. Crutzen, Estimates of gross and net fluxes of carbon between the biosphere and the atmosphere from biomass burning, Climatic Change, 2, 207-247, 1980.

Shvidenko, A. and Nilsson, S., Fire and the Carbon budget of Russian forests in E. S. Kasischke and B. J. Stocks, eds. Fire, climate change and carbon cycling in the Boreal forest, 1996.

Sinha, P., P. V. Hobbs, R. J. Yokelson, I. T. Bertschi, D. R. Blake, I. J. Simpson, S. Gao, T. W. Kirchstetter, and T. Novakov, Emissions of trace gases and particles from biomass burning in southern Africa, J. Geophys. Res., 108, D13, 8487, doi:10.1029/2002JD002325, 2003.

Skole, D.L., Chomentowski, W.H., Salas, W.A. and Nobre, A.D., Physical and human dimensions of deforestation in Amazonia. Bioscience 44: 314-322, 1994.

Smith K.R., A.L. Aggarwal and R.M. Dave, Air pollution and rural biomass fuels in developing countries: a pilot village study in India and implications for research and policy, Atmos. Env., 17, 2343-2362, 1983.

Smith K.R., R. Uma, V.V.N. Kishore, K. Lata, V. Joshi, J. Zhang, R.A. Rasmussen and M.A.K. Khalil, Greenhouse gases from small-scale combustion devices in developing countries : Part IIA, Household stoves in India. EPA-600/R-00-052. Office of research and development, US EPA, Washington, DC 20460, 2000.

Stocks, B.J., The extent and impact of forest fires in northern circumpolar countries, in Global Biomass Burning: Atmospheric, Climatic, and Biospheric Implications, edited by J.S. Levine, pp. 197-202, MIT Press, Cambridge, MA, 1991.

Stocks, B.J., D.R. Cahoon, Jr., W.R. Cofer, III, and J.S. Levine, Monitoring large-scale forest-fire behavior in northeastern Siberia using NOAA AVHRR satellite imagery, in Biomass burning and Global Change, edited by J.S. Levine, vol. 2, pp. 802-807, MIT Press, Cambridge, MA, 1996.

Streets D.G., S. Gupta, S.T. Waldhoff, M.Q. Wang, T.C. Bond and B. Yiyun, Black carbon emissions in China, Atmos. Env, 35, 4281-4296, 2001.

Thompson, A. M., R. D. Diab, G. E. Bodeker, M. Zunckel, G. J. R. Coetzee, C. B. Archer, D. P. McNamara, K E. Picketing, J. Combrink, J. Fishman, and D. Nganga, Ozone over southern Africa during SAFARI-92/TRACE A, J. Geophys. Res., 101, 23,793, 1996.

Trentmann, J., M.O. Andreae, H.-F. Graf, P.V. Hobbs, R.D. Ottmar, and T. Trautmann, Simulation of a biomass-burning plume: comparison of model results with observations, J. Geophys. Res., 107, AAC 5-1 5-15, 2002.

Turner, M.G., R.H. Gardner, V.H. Dale and R.V. O'Neill, Predicting the spread of disturbance across heterogeneous landscapes, Oïkos, 55, 121-129, 1989.

Van Wagner, C.E., Age class distribution and the forest fire cycle. Canadian Journal of Forest Research 8: 220-227, 1978.

Venkataraman C and G.U.M. Rao, Emission factors of carbon monoxide and size-resolved aerosols from biofuel combustion, Environ. Sci. and Technol. 35, 2100-2107, 2001.

Ward, D.E., A. Setzer, Y.J. Kaufman and R.A. Rasmussen R.A., Characteristics of smoke emissions from biomass fires of the Amazon region-Base-A experiment, In "Global Biomass Burning" , J. S. Levine ed., MIT Press, 394-402, 1991.

Ward D.E., R.A. Sussott, J.B. Kauffman, R.E. Babbitt, D.L. Cummings, B. Dias, B.N. Holben, Y.J. Kaufman, R.A. Rasmussen and A.W. Setzer, Smoke and fire characteristics for cerrado and deforestation burns in Brazil : BASE-B experiment, J. Geophys. Res., 97, 14601-14619, 1992.

Wotawa, G., and M. Trainer, The influence of Canadian forest fires on pollutant concentrations in the United States, Science, 288, 324-328, 2000.

Wotton, B.M. and Flanningan, M.D., Length of the fire season in changing climate. The Forestry Chronicle, 69, 187-192, 1993.

Yamasoe M.A., P. Artaxo, A.H. Miguel, A.G. Allen, Chemical composition of aerosol particles from direct emissions of vegetation fires in the Amazon basin : water-soluble species and trace elements, Atmos. Env., 34, 1641-1653, 2000.

Yevich, R. and J.A. Logan, An assessment of biofuel use and burning of agricultural waste in the developing world, submitted to Global Biogeochem. Cycles, 2002.

Global Organic Emissions from Vegetation

Christine Wiedinmyer, Alex Guenther, Peter Harley,
Nick Hewitt, Chris Geron, Paulo Artaxo, Rainer Steinbrecher,
Rei Rasmussen

1. INTRODUCTION

Organic emissions from vegetation are important inputs for global atmospheric chemistry models that simulate the processes controlling oxidant, CO, aerosol, and organic acid evolution, as well as the contribution of reactive carbon to global carbon cycles and budgets. Regional air quality policy decisions, which have large environmental and socio-economic impacts, also rely on correct natural organic emission estimates. The need for highly resolved and accurate organic emission estimates for air quality assessments in certain regions, particularly in the United States and parts of Europe, has led to recent improvements in methods for simulating emissions in these regions. Additional efforts in tropical landscapes have provided initial characterizations based on measurements whereas previous predictions were based on default assignments that treated most global landscapes in a similar manner.

In the early 1990's, a working group sponsored by the International Global Atmospheric Chemistry – Global Emission Inventory Activity (IGAC-GEIA) developed a global model of natural Volatile Organic Compound (VOC) emissions that was described by Guenther et al. (1995). At present, most global atmospheric chemistry models have either continued to use the inventory generated by this model or have incorporated the recommended model procedures. The Guenther et al. model predicts that 99% of the total global biogenic emissions are from terrestrial surfaces and nearly all of that is from vegetation. In this chapter, we summarize the considerable advances in our ability to simulate global distributions of biogenic volatile organic compound (BVOC) emissions from vegetation. Other recent reviews covering various aspects of BVOC (Kesselmeier and Staudt, 1999; Kreuzwieser et al., 1999; Harley et al., 1999; Guenther et al.,

2000; Fuentes et al., 2000) provide additional relevant information. We also include some discussion of both primary emissions of biogenic aerosol and secondary organic aerosol production from biogenic emission precursors.

2. EMISSION MEASUREMENTS

Direct and indirect measurements of VOC emissions from vegetation are essential for emission model development and evaluation. Enclosure, canopy flux and atmospheric concentration measurements have all been used for these purposes.

2.1. Enclosure measurements

The earliest studies of BVOC emissions from vegetation, conducted from the 1920s through the 1960s (e.g., Rasmussen and Went, 1964), used various types of static enclosure systems to isolate individual leaves, branches, or entire plants, and estimated emission rates from the increase in chemical concentration in the enclosure over time. These enclosure systems were first used to estimate the total BVOC emission and the relative contribution of different BVOCs, called the "fingerprint," for many plant species (e.g., Rasmussen, 1970). Later studies, using dynamic flow through enclosures, began to characterize the relationships between emissions and environmental conditions, such as temperature and light. Enclosure studies continue to be the primary source of emission factors and emission algorithms used in BVOC emission models. Current studies continue to refine and improve existing measurement techniques and extend them into new landscapes and new chemical species. With increased understanding of additional environmental controls over BVOC emissions, enclosure studies are being used to develop new algorithms that account for a more complete set of driving variables (i.e., drought stress and soil moisture, long-term temperature variations, and chlorophyll content).

By the end of 2001, there were more than 150 peer reviewed journal publications describing enclosure studies of BVOC emissions. In addition, we are aware of 12 detailed reports that describe additional BVOC enclosure measurements. The majority of these studies contain qualitative descriptions of emissions from identified vegetation species. These descriptions provide useful insights into the processes controlling emission variations but cannot be used directly in numerical emission models. Few of the studies (~18%) provide information suitable for characterizing emission factors and even fewer are useful for developing numerical emission algorithms. The reported studies have screened the emissions from over 1800 plant species. Most of these plants were examined only for isoprene (67%), and approximately 24% have been characterized for both isoprene and monoterpenes. Relatively few studies have investigated the emissions of other compounds, such as other alkenes and oxygenated VOCs, though with recent improvements in analytical techniques, these efforts are accelerating.

Efforts to compile the results from these studies into an emission rate database describing the emission characteristics of individual plant species and genera have been initiated independently by several research groups (e.g., Lancaster University, Oregon Graduate Institute of Science & Technology, National Center for Atmospheric Research (NCAR), U.S. Environmental Protection Agency (EPA), Fraunhofer Institute for Atmospheric Environmental Research (IFU), CEC-Joint Research Centre). A recent IGAC-GEIA sponsored effort to combine and extend these endeavours has led to the creation of a community database of enclosure BVOC emission measurements for individual plant species (see 'ENCLOSURE MEASUREMENTS' at URL: http://bvoc.acd.ucar.edu). This database summarizes information from all literature reports of which we are aware and identifies the plant species studied, the BVOC examined, the enclosure and analytical techniques applied, and other parameters considered. Quantitative emission rates are included when available. All data may be queried and downloaded in text, comma-delimited, or html formats by the user. This database is a work in progress and investigators are encouraged to contribute new or unpublished data to the database, following procedures summarized at URL: http://bvoc.acd.ucar.edu).

The utility of maintaining databases of this type is not always apparent and may seem out of proportion to the effort involved. However, some unexpected uses may result. For example, the Lancaster University database (URL: http://www.es.lancs.ac.uk/cnhgroup/iso-emissions.pdf), containing some 1200 entries of isoprene emission rates (including zero rates), has recently been used to test the hypothesis that competence for isoprene emission is related to the mode of phloem loading (vascular tissue that transfers the products of photosynthesis throughout plant) in plants.

Analysis of the database showed a significant association between minor vein structural type in leaves (as a proxy for phloem loading type) and capacity for isoprene synthesis plants (Kerstiens and Possell, 2001).

Data presented in the emissions database, and other databases discussed above, are input directly from the literature and are unscreened, and the validity of the measurement techniques and the accuracy of the measurements themselves are assumed to vary. Many examples exist in the database of conflicting information concerning the emission characteristics of a given plant species. Given natural variability, it is not unexpected that significant quantitative differences might occur in reported emission rates. However, in extreme cases, a given plant species is recognized as a significant isoprene emitter in one publication, and a non-emitter in another. Some of the reasons why such differences might arise are discussed in Section 4.1 for the specific case of isoprene.

Because of the uncertainties in the measurements and resulting emission rates, an important next step in this IGAC-GEIA effort will be the convening of an expert panel familiar with the enclosure measurements, their results, and associated uncertainties. The objective of this panel will be to evaluate the collection of enclosure data and to recommend emission factors for use in biogenic emissions models. The emission factor recommendations will be contained in a second database that includes emission factors for all BVOCs for individual plant species and plant functional types in an easily accessible format. This will provide a consistent, current, and easily accessible resource for biogenic emissions modellers. Another outcome of this effort will be recommended protocols for making BVOC measurements. These new protocols will remove some of the uncertainties and inconsistencies in the measurements, foster greater confidence in future data, and ensure that data are well-suited for advancing the development of regional/global databases and models. This is a community effort, and input from all interested parties is welcomed (see 'EMISSION FACTOR RECOMMENDATIONS' at URL: http://bvoc.acd.ucar.edu). All databases are freely available for use in development and evaluation of both emission and chemical models.

2.2. Above-canopy flux measurements

The earliest attempts to quantify whole canopy BVOC fluxes from above canopy measurements used indirect methods such as surface layer gradients and tracer ratios (e.g., Arnts et al., 1978, 1982). These methods relied on assumptions that could be met in ideal situations but were often violated in practice. Although these initial flux measurements were difficult and labour intensive, and associated with uncertainties of 50% or more, they were extremely important for validating the general magnitude of whole canopy fluxes and establishing that the results of enclosure methods could be extrapolated to the canopy scale with reasonable accuracy. Above-canopy flux measurement techniques have improved considerably in recent years, the most significant advancement being the capability for direct measurements of BVOC fluxes using eddy covariance (e.g., Guenther and Hills, 1998; Karl et al., 2001b). These systems not only provide more accurate flux estimates but also enable continuous measurements and the ability to look at a wider range of BVOCs (e.g., oxygenated compounds). The primary use of above-canopy flux measurements has been to evaluate and validate emission modelling procedures. However, these systems are now being used as the primary means of establishing area-averaged emission factors in regions with high species diversity, such as tropical rainforests (Rinne et al., 2002), where it is difficult to characterize the emission characteristics of all tree species present.

There have been over 40 studies of above-canopy BVOC flux measurements described in peer-reviewed journals and at least 3 summarized in reports. Databases containing individual flux measurements are available to the scientific community for model testing (e.g., LBA, URL: http://beija-flor.ornl.gov/lba; BOREAS, URL: http://boreas.gsfc.nasa.gov/html_pages/boreas_home.html; EXPRESSO, URL:http://medias.obs-mip.fr:8000/www/francais/activities/donnees/expresso/expresso.html), but they are few in number and difficult for most investigators to locate. As part of this IGAC-GEIA sponsored effort, all available flux study data has been incorporated into another online database (see 'VERTICAL FLUX MEASUREMENTS' at URL: http://bvoc.acd.ucar.edu). The creation of a central location for assessing and archiving this information should improve the utility of these data. The amount of study data available at this site is expected to increase in the future, and researchers are encouraged to include their observations in this effort.

The compiled datasets described will provide improved means for model evaluation and development. Additionally, an expert panel will evaluate the measurements described in these databases and recommend a set of consistent and reliable measurement techniques and data submission protocols for future work. It is hoped that, in this way, future investigations will be of greater utility in promoting emission model development. Comments and suggestions are encouraged as these community datasets continue to evolve.

2.3. Ambient concentration measurements

The presence of BVOC in the atmospheric boundary layer has been used to demonstrate that emissions of these compounds directly influence the chemical composition of the atmosphere. The observed ambient concentrations of BVOC are often misleading, however. Even though their emission rates are substantial, many of these compounds are very reactive and are therefore present at very low levels in the atmosphere.. Ambient concentration measurements are best used as a method for evaluating the combined predictions of emissions, chemistry, and transport in numerical models. These observations are also sometimes used as the primary means of setting emission rates. This has notably been the case for tropical forests (e.g., Zimmerman et al., 1988) where more direct emission measurements have been made only recently.

There have been over 50 studies of ambient BVOC concentration measurements described in peer-reviewed journals. Study datasets containing available concentration measurements from a variety of studies are available online as part of the database web pages mentioned above (see 'CONCENTRATION MEASUREMENTS' at URL:http://bvoc.acd.ucar.edu). As this effort increases and more researchers participate, the amount of information stored on this site for the community is expected to increase.

3. EMISSION MODELING PROCEDURES

Results from leaf-level enclosure and canopy flux studies of BVOC emissions are used to develop emission models. To accurately simulate

BVOC emissions, these models also need information about the land cover, including the density of the vegetation and the species distribution, and about the environment, such as the radiation and temperature. This section provides a discussion of the most recent advances in information on these modelling variables

3.1. Vegetation emission types

Land cover and ecological classification of landscapes are important economic and scientific topics. As a result, several databases and models have been published, some of which are very useful for characterizing vegetative sources of organic emissions. The United Nations Food and Agricultural Organization (FAO) global estimates for 1990 cover an area of 130×10^6 km^2 that include 12% cropland, 26% pasture, 33% forest, and 29% other. More recent and accurate satellite measurements provide high resolution information regarding the general vegetation types and density.

Plant functional types (PFTs) have been determined from satellite data and applied algorithms (e.g., Running et al., 1994, 1995; Zeng et al., 2000, Hansen et al., 2000). These PFT classifications distinguish broadleaf and needle leaf deciduous and evergreen trees, as well as grasses (Running et al., 1994, 1995). Defries et al. (2000) have produced a global dataset at 1 km resolution that can be useful for estimating BVOC emissions. This dataset provides a 3-tiered approach at characterizing the vegetation from Advanced Very High Resolution Radiometer (AVHRR) data: 1) percent tree cover, 2) percent tree cover for each leaf type (broadleaf and fine-leaf), and 3) dominant landcover type. Figure 1 shows the distribution of the tropical broadleaf deciduous forests for the month of July.

Ecoregions are typically defined on the basis of biogeographical surveys, the results of which are very useful in estimating biogenic emissions. One such dataset is a highly detailed description of global vegetation distribution developed by the World Wildlife Fund (WWF), which divides terrestrial landscapes into 867 unique ecoregions (see URL: http://www.worldwildlife.org/ecoregions/) (Figure 2) (Olson et al., 2001). This WWF project includes the efforts of more than 100 contributors and is ongoing. A major advantage of the WWF scheme is their accompanying detailed descriptions of the dominant plant species occurring in each ecoregion, which enables an estimation of the percent coverage by these species in each ecoregion. In some parts of the world, regional and local studies have provided more detailed ecoregion classifications or detailed vegetation inventories. Several of these efforts are described in the following sections.

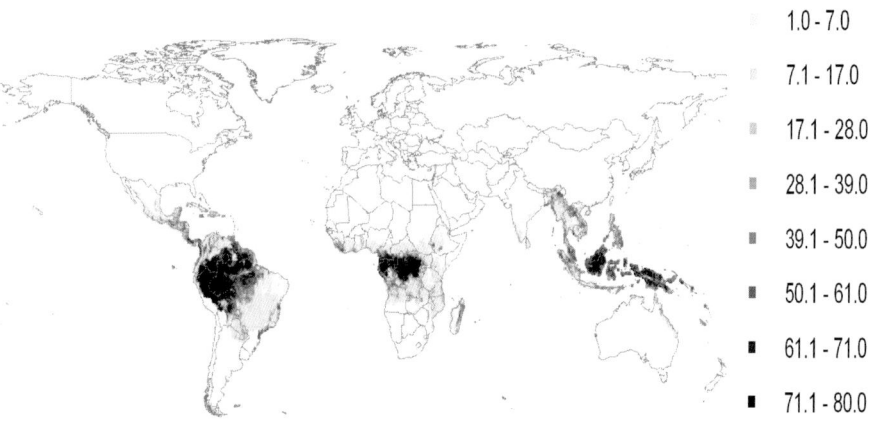

Figure 1: Global distribution of the percent of tropical broadleaf deciduous forest for the month of July. These data were derived from satellite data (Running et al., 1995) and processed for use in a land surface model as described by Bonan et al. (2002).

Land cover is rapidly changing in many parts of the world, and BVOC emissions are expected to change as a result. Therefore, it is essential to understand and track these land use changes when attempting to estimate BVOC emissions. Tropical regions have some of the highest rates of land-use change, including the conversion of tropical forests to cropland and pasture. Steiner et al. (2002) investigated the change in BVOC emissions that have resulted from land cover changes in eastern Asia. This investigation was based on present-day land cover derived from satellite data and a pre-disturbed land cover dataset based on plant functional types and climatological parameters. The authors reported that isoprene and monoterpene emissions in the studied region have been reduced by 45% and 67%, respectively, from the undisturbed to present-day land cover, primarily due to the conversion of forested regions to cropland.

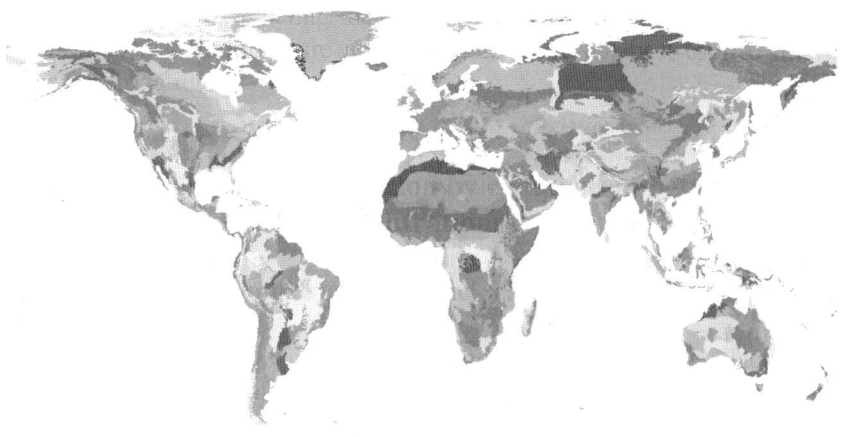

Figure 2 (see plate 7). The global distribution of ecoregions as assigned by the World Wildlife Fund ecoregion scheme. Each color represents a different ecoregion (over 850 ecoregions are assigned to the global land area) (Based on Olson et al., 2001). For more information, visit http://www.worldwildlife.org/ecoregions.

Altered land-use and succession have also influenced BVOC emissions in North America and Europe (e.g., Guenther et al., 1999b; Schaab et al., 2000). The United States Department of Agriculture (USDA) projections of future landscape change in the United States can translate into significant changes in organic compound emissions. Millions of acres of forest and agricultural land in the mid-Atlantic United States are expected to be converted to urban and suburban landscapes by the middle of the 21st century. On the other hand, increased acreage of southern pine plantations will be established in the deep south, and hardwood plantations, most of which consist of high isoprene emitting species, are anticipated for regions in southern Canada, the Lake States, the Pacific Northwest, the southern United States, and moist tropical Mexico. The same holds true in tropical landscapes where there have been increases in plantations of high VOC-emitting tree species. Reforestation programs for establishing new CO_2 sinks in the biosphere favour the planting of eucalyptus trees, which are high isoprene and monoterpene emitters. These land use and resulting BVOC emission changes will affect regional and, possibly, global atmospheric chemistry.

3.1.1. Agricultural crops

The species composition of croplands can be specified for most ecoregions using the UN-FAO database (URL: http://apps.fao.org/cgi-bin/nph-db.pl?subset=agriculture). This database provides national estimates of the relative contribution of each of over 70 crops to the total crop area in each nation. The ten globally dominant crops include grains (wheat, rice, corn, barley, sorghum), legumes (soybeans, beans), hay, cotton, and rapeseed. The next twenty-five, in order of aerial extent, are sunflowers, potatoes, sugar cane, cassava, alfalfa, oats, chickpeas, cowpeas, rye, sweet potatoes, grapes, sesame seed, sugar beets, dry peas, plantains, tobacco, pigeon peas, yams, tomatoes, lentils, cabbages, triticale, watermelon, and buckwheat. More highly spatially resolved databases are available for some ecoregions. For example, the USDA Agricultural Census (URL: http://www.nass.usda.gov/census/) provides county level crop data for the United States.

3.1.2. Tree inventories

Trees account for the largest fraction of the global total BVOC emission due to the broad spatial extent, high leaf biomass density, and high rate of emission of many important chemical species. Fortunately, the economic value of trees has led to efforts to inventory forested areas around the world and to estimate tree species composition for many of the extratropical woodlands. Aerial extents of tree plantations, excluding those grown for wood or pulp, are available in the FAO database. The ten most important in order of aerial extent on the global scale are coffee, coconuts, oil palm, olives, cocoa, rubber, apples, bananas /plantains, oranges, and mangoes. An additional 22 species are included in the database, primarily fruit and nut trees. The total global area associated with these plantations is small, but can be significant locally.

Tree inventories are available for other countries and regions, including China (Klinger et al., 2002). The Australian Bureau of Rural Sciences has developed a forest inventory for the 1.57 million km^2 of Australia that are covered by forest and woodlands (URL: http://www.affa.gov.au/nfi). North American tree inventories include the U.S. Department of Agriculture (USDA) Forest Service Forest Inventory and Analysis (FIA) database (URL: http://fia.fs.fed.us), and Canadian Ministry of Forestry Inventory (see for example URL: http://srmwww.gov.bc.ca/tib/). These periodic ground-survey-based inventories have been historically maintained for industrial and commercial applications. However, as the sponsoring agencies have recently been emphasizing ecological health and monitoring functions, these data have become even more useful for estimating coverage and density of BVOC emission sources. Table 1 lists the most abundant species and their emission characteristics for North American regions.

Table 1 (see plate 8). Dominant tree species by region in North America. Bold indicates high (> 10 µg C g^{-1} h^{-1}) VOC emission rates at standard (leaf temperature of 30°C and PAR of 1000 µmol m^{-2} s^{-1}) conditions. Red indicates species adapted to warm sunny climates, green indicates temperate adapted species, and blue denotes species found in cool, or montane, climates.

Eastern U.S.	Western U.S.	Eastern Canada	Western Canada
Pinus taeda	*Pseudotsuga menziesii*	**Populus tremuloides**	*Picea* spp
Acer rubrum	*Pinus ponderosa*	*Picea* spp	*Populus tremuloides*
Quercus alba	*Juniperus osteosperma*	*Abies* spp	*Pinus banksiana*
Liquidambar styraciflua	*Pinus contorta*	*Pinus banksiana*	*Abies* spp
Acer saccharum	*Tsuga heterophylla*	*Thuja occidentalis*	*Tsuga* spp
Quercus rubra	*Abies concolor*		
Pinus elliottii	**Picea engelmannii**	Northern Mexico	Southern Mexico
Liriodendron tulipifera	*Abies grandis*	*Pinus durangensis*	**Quercus resinoa**
Populus tremuloides	*Pinus edulis*	*Pinus arizonica*	*Pinus oocarpa*
Quercus virginiana	*Abies lasiocarpa*	*Quercus* spp	*Acacia* spp

Many of these species are becoming even more dominant in North American landscapes due to plantation forestry, and their emissions may affect future atmospheric chemistry in the western U.S. and Canada and in the southeastern U.S. These species include *Pinus taeda* and *Pinus elliottii* in the eastern U.S., *Pseudotsuga menziesii* and *Pinus ponderosa* in the western US and Canada, and *Populus tremuloides* (and numerous *Populus* hybrids) in the northern U.S. and southern Canada. Other species may be increasing in abundance due to factors such as exclusion of fire (e.g., *Acer rubrum*) or elimination of competitors by exogenous disease (e.g., Appalachian *Quercus rubra*). Similarly, selective harvesting may be decreasing the abundance of dominant tree species in some areas. Simpson et al. (1999) have compiled a national-level tree inventory for Europe using 34 types and species of trees for use in biogenic VOC emission modelling. This study reported that the Russian Federation, Sweden, and Finland have the highest forested areas, containing 41, 10, and 7%, respectively, of Europe's total forested area. The forests of these three countries are predominantly coniferous, dominated by *Picea* and *Pinus* species. *Picea abies* and *Pinus sylvestris* are dominant in the Scandinavian forests. *Quercus* and *Fagus* species contribute 4% each to the total European forests. However, in some countries, these species constitute a much larger fraction of the local forests. For example, *Quercus* species make up 35, 27 and 25% of the forests in Hungary, Albania, and the former Yugoslavia, respectively. These extensive and often coarse land use datasets can be extended to higher resolution with the results of very detailed land use surveys.

The Countryside Survey in Great Britain has produced the estimated aerial coverage of over 1000 species (e.g., Bunce et al., 1996). The most prevalent tree species by far (and 11th most important species by area overall) is *Picea sitchensis*, which dominates upland plantations, covering 4510 km^2, or 24%, of the woodland in the country. The native *Pinus sylvestris* and *Quercus spp.* are the next most prevalent tree species by area, covering 1705 and 1667 km^2, respectively (Stewart et al., 2002). Although the coarser estimates of Simpson et al. (1999) arrived at the same order of importance of these species, the estimated absolute areas covered by these species differed considerably, highlighting the need for detailed landcover data at the national scale. Similarly, the U.S. has the state-specific Gap Analysis Project (URL: http://www.gap.uidaho.edu) databases, which have spatial resolution ranging from tens of meters up to 1 km^2. Such labour-intensive survey work is clearly not feasible in most parts of the globe but may aid in providing ground-truth to the interpretation of remote sensing data in the future. The qualitative information in these databases, usually a ranking of the dominant species, can be used to estimate species composition. This approach has recently been used to produce biogenic VOC emission estimates for Africa south of the equator (Otter et al., 2003).

3.1.3. Other vegetation

There are no quantitative descriptions of species composition for many forests, particularly in the tropics. This is also the case for shrubs, mosses, grasses and other herbaceous plants in almost all ecoregions. The WWF ecoregion database provides a description of the dominant plant species for most of the given ecoregions of the globe. Within the United States and other regions where data are available, this ecoregion information can be extended with more detailed vegetation descriptions and listings (e.g., the U.S. Gap Analysis Project databases).

3.2. Foliar density

Foliage is the source of over 90% of total global BVOC emissions (Guenther et al., 1995). Emission models assume that area average emission potentials are directly correlated to foliar density. Leaf area index (LAI, m^2 crown area m^{-2} ground area) is an input to the canopy environment model that determines within-canopy radiation profiles. Since increases in LAI decrease the total light available for driving emissions of light dependent compounds such as isoprene, this factor somewhat offsets the impact of increasing foliar density. However, an accurate estimate of the total global foliar mass, and the spatial and temporal distribution, is an important requirement for BVOC emission models.

Box (1981) synthesized the results of extensive studies conducted as part of the International Biological Program and other efforts and estimated a global total foliar mass of 75 Gt and a global leaf area of 630 10^6 km^2. The Guenther et al. (1995) global emission model, which relied primarily on the Box parameterizations for foliar density, predicted a global foliar mass of 78 Gt and a global leaf area of 609 10^6 km^2. Recent advances in remote sensing estimation allow for a more direct measure of global foliar density and LAI (e.g., Preußer et al., 1999, Myneni et al., 2002) (Figure 3).

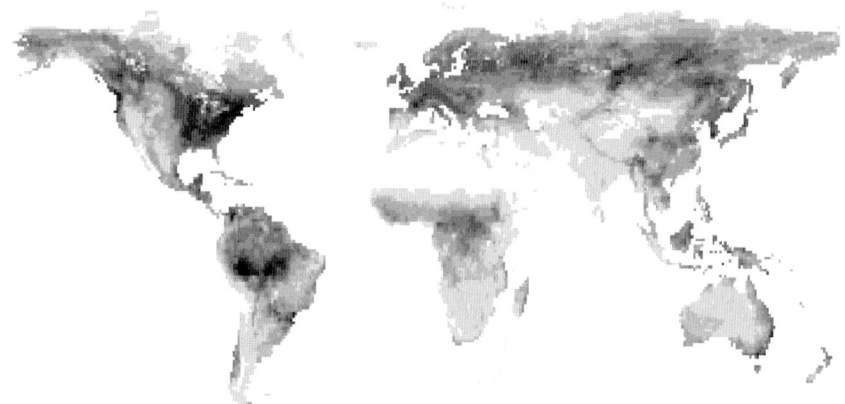

Figure 3: Example of exposed LAI for the month of July as derived by satellite data and processed for a land-surface model (Based on Bonan et al., 2002). (© American Geophysical Union).

LAI estimates obtained by MODIS (Moderate Resolution Imaging Spectroradiometer) aboard the U.S. National Aeronautics and Space Administration (NASA) Terra (EOS AM-1) satellite are provided as 1- and 8-day averages at a 1 km resolution. The differences between the monthly biomass estimates from Guenther et al. (1995) and the biomass totals from species-specific calculations for Great Britain were strongly time dependent. Summertime values (May to August) were similar, differing by an average of 18%. However differences between spring, autumn, and winter values varied from 0.3% to 100% (Stewart et al., 2002). This highlights the importance of ground-based estimates in validating satellite-derived data. In Great Britain, the peak summertime standing biomass was estimated to be 128 Mt. Of this, 35% is contributed by the grass Lolium perenne and crop Triticum aestivum (wheat) and 6% by Picea sitchensis, although the relative importance of this coniferous tree species increases significantly in wintertime (Stewart et al., 2002).

3.3. Phenology

Plant phenology (the phases of plant growth) has an influence on the emission of many biogenic VOCs. For example, flowering and bud break have been associated with elevated levels of BVOC emission. These two phenological events, however, are not thought to be major global contributors to BVOC levels, and there have been no efforts to incorporate them in regional models. Leaf phenology is of greater importance. It has been shown that isoprene emission is diminished in immature and senescing leaves, and methods for simulating this on global scales have been described (Guenther et al., 1999a). An additional method for considering seasonal effects on the BVOC emission rates is based on estimates of photosynthetic pigment content of leaves (Lehning et al., 2001).

Leaf age is also an important factor for emissions of methanol (Nemecek-Marshall et al., 1995, MacDonald and Fall, 1993) and monoterpenes (Guenther et al., 1991), and emissions of some oxygenated VOCs are greatly enhanced during leaf senescence (Fall et al., 1999, 2001). However, these effects have not yet been incorporated into emission models.

3.4. Environmental conditions

Environmental conditions, such as light and temperature, strongly influence the rate at which BVOC are emitted.

3.4.1. Relating emissions to environmental conditions

The simplest approach to varying emissions according to environmental conditions is to assume that emissions occur only during the growing season (e.g., Rassmussen, 1970). More accurate estimates must at least account for the influence of leaf temperature (for all BVOC) and photosynthetically active radiation (PAR) (for some BVOC) (Tingey et al., 1980) (PAR is the part of the spectrum that plants use and is technically defined as the flux of moles of photons in the radiant energy between 400 and 700 nm). The light and temperature algorithms suggested by Guenther et al. (1993) are still widely used for simulating these variations. They appear to be reasonable for terpenoid (e.g., isoprene, α-pinene) and methyl butenol emissions but are probably not suitable for other BVOC emissions. However, there are no widely accepted alternatives.

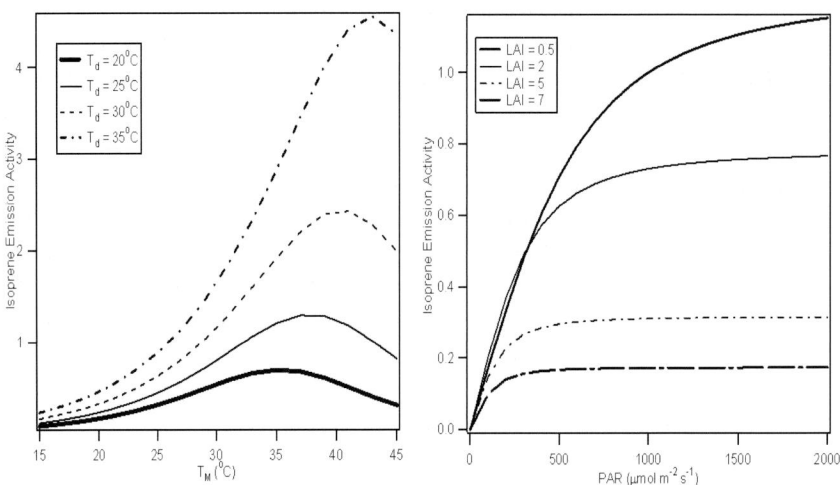

Figure 4: Estimated variations isoprene emission activity that occur as the result of changing temperature (left) and PAR (right) by the algorithms and parameters reported by Guenther et al. (1999a). T_d is the mean temperature of past 15 days, T_M is the current leaf temperature, and LAI is the leaf area index ($m^2\ m^{-2}$). (© American Geophysical Union).

Isoprene emission rates are also a function of the developmental environment, with sun leaves emitting at rates higher than shade leaves. Several recent studies (Sharkey et al., 1999; Geron et al., 2000a; Lehning et al., 2001; Pétron et al., 2000) have determined that isoprene emission capacity is strongly influenced by the temperature regimes to which a leaf is exposed over the preceding hours to days. An attempt has been made to incorporate such effects into regional models (Lehning et al., 2001; Guenther et al., 1999a) (Figure 4). A role for other environmental factors on BVOC emissions has been demonstrated (e.g., nutrients, water stress, flooding, wounding), but numerical algorithms for regional modelling have not been developed.

3.4.2. Within canopy environment

Although early modelling attempts simply used PAR and temperature measured above the canopy to drive BVOC emission models, the current approach is to divide the canopy into multiple layers, with a percentage of sun and shade leaves at each canopy depth (e.g., Guenther et al., 1995). This can result in differences in emission estimates exceeding 25% (Lamb et al., 1996) due to the non-linear relationships between emissions and temperature or light. Another consideration is the difference between the leaf and air temperatures, which can exceed several degrees (resulting in emission differences greater than 50%). BVOC emission models should take advantage of the progress being made in global climate models to correctly simulate leaf temperatures as part of a whole canopy energy balance. Also, within-canopy chemical degradation of the very reactive isoprenoids (e.g., limonene) may reduce the above-canopy isoprenoid emission by up to 30% (Forkel et al., 2001). However, secondary products (aldehydes, ketones, and carbonic acids) are formed from this chemistry within the canopy. Therefore, emission rates of these secondary compounds from the canopy may be a factor of 5 higher than estimates that omit within-canopy reactions.

3.4.3. Above canopy environmental conditions

Global environmental data necessary for predicting natural organic emissions can be obtained from various sources including ground and satellite observations, model simulation outputs or combinations of both. A variety of satellite measurements and meteorological data exists and has the potential for use in biogenic emissions models. The U.S. National Oceanic and Atmospheric Administration (NOAA) has launched several Geostationary Operational Environmental Satellites (GOES) and Polar Orbiting Environmental Satellites (POES) over the past 20 years, which provide a wealth of information about the earth's atmosphere and ground cover. GOES provide continuous information for areas as large as the western hemisphere, whereas POES provide global information on longer time scales (~ 1 to several days). Algorithms have been developed to calculate desired variables from satellite instrument output. The Advanced Very High Resolution Radiometer (AVHRR), the TIROS Operational Vertical Sounder (TOVS), and various sounding instruments aboard POES provide information about the earth surface conditions, including temperature, humidity, and land cover. The Japanese Space Agency (NASDA) also has several satellites in orbit, which provide meteorological data needed for biogenic emissions models. MODIS (Moderate Resolution Imaging Spectroradiometer), a key instrument aboard the NASA Terra (EOS AM-1) satellite launched in December 1999, is providing many global data products, including land surface temperature (Snyder et al., 1998; Wan, 1999), Fractional Photosynthetically Active Radiation (FPAR) (Myneni et al., 2002), cloud characteristics (e.g., King et al., 1997), and atmospheric water vapour (Gao and Kaufman, 1998). MODIS has finer spectral resolution and more accurate geolocation than previous remote sensing instruments. Using methods by Pinker and Laszlo (1992a, b), PAR fluxes have been estimated from satellite radiation flux estimates (Bishop and Rossow, 1991) as part of the International Satellite Cloud Climatology Project (ISCCP).

Although fine resolution meteorological data are available from processed satellite information, often the data are not sufficient for specific modelling episodes or domains. Therefore, the data can be assimilated with meteorological and climate models to provide a complete meteorological dataset for predicting BVOC emissions on both regional and global scales. Temperatures, radiation, and humidity can be estimated from the assimilation of measurements from ground level and satellite observations into models. The U.S. National Center for Environmental Prediction (NCEP) and the European Centre for Medium-Range Weather Forecasts (ECMWF) provide global meteorological data for use in global emission, climate, and photochemical models. Both supply information from available satellites, as well as assimilated datasets for use in models. For regional scales, the MM-5 model (NCAR/Penn State University) and the ERS model used by the ECMWF can be applied to estimate the temperature and radiation at the earth's surface (Richter et al., 1998). Ultimately, biogenic emissions models will be coupled with on-line land-surface, ocean, and atmospheric models that will predict simultaneously BVOC emissions and meteorological parameters (Wang and Shallcross, 2000; Levis et al., 2002). Satellite and ground-based measurements will still be necessary to constrain these models.

4. ISOPRENE EMISSIONS

Isoprene is by far the most thoroughly investigated BVOC due to its high emission rates from many plant species, predominance in many non-urban air masses, and high reactivity. We devote this section to a detailed discussion of isoprene emissions and methods of incorporating such information into regional and global models in the hope that it may serve as a model for future investigations of those important BVOC that are just beginning to receive greater attention.

4.1. Uncertainties in leaf-level isoprene emission rate measurements

As noted previously, there can be considerable variability in the reported rates of isoprene emission for a given plant species. The source of much of this variability is natural variation in the emission characteristics of individual plants; however, differences in measurement protocol and analytical technique also play a significant role. Unfortunately, the failure of many early investigators to report relevant ancillary information, such as the physiological state of the plant, leaf phenology, plant growth environment, or even PAR and leaf temperature during the measurements, sometimes makes it difficult to interpret the reported data

4.1.1. Sources of natural variability

The primary source of emission rate variability in isoprene emitting species, and the easiest to control, is the incident PAR and leaf temperature during measurement. To allow measurement comparison, the community of BVOC researchers has somewhat arbitrarily developed the useful term "emission capacity" (also variously referred to as "base emission rate", "emission factor", "standard emission rate"), representing the emission rate measured under a given "standard" PAR (generally 1000 µmol m^{-2} s^{-1}) and leaf temperature (generally 30°C, but occasionally 25°C). If PAR and leaf temperature conditions during the measurement are not reported, the resulting data, while useful for distinguishing emitting from non-emitting species, is of little value in establishing the emission capacity and, thus, estimating emissions in models. On the other hand, since all isoprene emitting species have similar PAR and temperature dependencies, to a first approximation, if measurements are made under non-standard conditions, the emission rate can be "normalized" to standard conditions using the widely accepted algorithms of Guenther et al. (1995) as long as PAR and leaf temperature are reported (though the farther from standard conditions, the less confidence in the normalization process).

Once variations in an emission rate that occur from differences in PAR and temperature conditions are factored out, we focus on remaining inconsistencies in reported emission capacities. Environmental conditions under which a plant or individual leaf develops are known to affect emission capacity. Reducing nitrogen nutrition can limit isoprene emission capacity

(Litvak et al., 1996; Harley et al., 1994). Effects of growth light environment on emission capacity, usually characterized as differences between "sun" and "shade" leaves, are well documented (Harley et al., 1996, 1997; Schuh et al., 1997; Sharkey et al., 1991, 1996), and Geron et al. (2000a) found that large discrepancies reported in the literature for certain species could be largely explained by the (often unreported) growth light environment. Leaves of high isoprene emitting species such as kudzu (*Pueraria lobata*) (Sharkey et al., 2001) and *Populus tremuloides* (Monson et al., 1994), which develop under cool temperatures (<20°C), may not produce isoprene at all until warmed above some threshold temperature; it is now well-documented that temperatures in the hours or days prior to measurement can significantly impact emission capacity (Geron et al., 2000a; Hanson and Sharkey, 2001; Pétron et al., 2000). Thus, when reporting emission results, as much information concerning prevailing environmental conditions should be included, at least in a general way (e.g., sun vs. shade leaves; top of canopy vs. understory; hot, warm, or cool period). Isoprene is not regarded as a stress-induced compound, except for possibly high temperatures (Sharkey et al., 1995), but reporting the presence of potential stresses (herbivory, drought, flooding, high oxidant levels) will certainly be important for many other BVOC.

4.1.2. Sources of experimental variability

If all isoprene emission capacity estimates could be the result of replicate measurements made on sun-adapted leaves at the top of the forest canopy using a dynamic leaf enclosure with light and temperature control and with direct sampling of the air exiting the enclosure by GC analysis, we would have a great deal more confidence in our results than we currently do. In reality, measurement techniques inevitably represent a compromise between portability, expense, power availability, canopy access, and a range of other factors. A variety of leaf enclosure techniques and analytical tools have been used to estimate isoprene emission rates from vegetation. Many of the pioneering experiments were carried out using relatively crude techniques. Static branch enclosures were used and emission rates estimated by the change in concentration within the enclosure over time. These enclosures often heated up dramatically in full sun conditions, temperature and light intensity were not often recorded, and self-shading within the enclosure was common. While valuable qualitative information was obtained in this fashion, quantitative results must be viewed with caution. In recent years, investigators have increasingly taken advantage of instruments developed for carefully controlled measurements of photosynthesis and transpiration and incorporated them into BVOC studies. These studies are performed with dynamic, flow-through instruments, often enclosing a single leaf, or part of a leaf, where leaf temperature and incident PAR are carefully measured and often carefully controlled. Using these controlled enclosure instruments, the investigator can verify that the leaves are physiologically active, and that steady-state conditions have been realized. All other things being equal, these instruments provide the most useful data for quantification of emission capacities.

In the ideal situation, samples of air inside or exiting the enclosure can be analyzed *in situ*, but logistical difficulties often require that samples be trapped by one of several means for transport to a laboratory for subsequent analysis. Air samples may be trapped into canisters of various designs, into Teflon, Tedlar, or other bags, or onto cartridges containing a variety of adsorbents. Although each of these techniques can provide very accurate analyses, presence of ozone in the storage vessel, photooxidation in transparent bags, and losses or chemical conversion on cartridge adsorbents are all potential sources of error. Storage issues may also be relevant if samples are stored for long periods prior to analysis.

A wide variety of analytical tools has been employed to identify and quantify plant emissions. Simple hand-held photoionization detectors (PID) have often been used to screen vegetation for the presence of isoprene emissions. Though hand-held PIDs provide useful qualitative information, results of these measurements should be viewed with some degree of caution. This type of instrument doesn't discriminate between isoprene and other hydrocarbons, so false positives are a possibility, particularly if the leaf tissue is damaged in the process of measurement. In addition, hand-held PIDs are not very sensitive, so false negative results are also possible. In general, quantitative analyses have been performed using gas chromatography (GC) coupled to a variety of detectors, including photoionization, flame ionization (FID), and mercuric oxide reduction gas detectors (RGD), all of which are capable of very accurate determinations of isoprene concentration if used properly and well calibrated.

One final source of potential error is inaccurate plant species identification. This is a problem particularly in the tropics, where even expert botanists have difficulty identifying plant material in the non-flowering or fruiting state. Often local experts provide only common names, which must then be subsequently matched against scientific names. However, the same common name is often applied to more than one species or even genus of plant. Whenever there is significant doubt as to the correct species identification, voucher specimens should be collected and verified by an experienced botanist. Errors in identification can lead to both false positive and false negative identification of isoprene emitting species, which may account for some of the taxonomic anomalies as discussed in the next section.

To reduce the uncertainty associated with the measurement of biogenic emissions and to increase the utility of the measurements for modelling applications, future enclosure studies should be designed to account for as many parameters as possible. A goal of these studies is to reduce uncertainties in the measurements and to produce quantitative results. A list (although not exhaustive) of important study parameters and considerations that should be determined during the study design and be recorded throughout the study to make quantitative BVOC emission capacity measurements is outlined below:

- Enclosure type
    - Nature and size
    - Static or dynamic
    - Degree of environmental control
- Sample method
    - Immediate analysis
    - Stored samples
        - Bags
        - Canisters
        - Adsorbent cartridges

- Analytical Instrumentation
    - semi-quantitative (e.g., hand-held PID)
    - quantitative
        - GC-RGD
        - GC-PID
        - GC-FID
        - GC-MS

- - Proton Transfer Reaction (PTR)-MS
  - Fast Isoprene Sensor (FIS)
- Ancillary Data
  - Plant family, genus and species
  - Location of study
  - Intact or excised leaf, branch, whole tree
  - PAR and leaf temperature during measurement
  - Growth light environment
    - sun vs. shade
    - top of canopy
    - open field vs. canopy understory
  - Growth temperature
    - Temperatures prior to study
  - Stomatal conductance
  - Rate of photosynthesis
  - Chlorophyll content
  - Stress effects on the vegetation/plant
    - Herbivory
    - Drought/ Soil moisture
    - Ambient oxidant concentrations
  - Phenological state
  - Nutrients available to plant
- Measurement methods
  - Replicate measurements
  - Inclusion of control plants

4.2. Taxonomic relationships

When assigning an isoprene emission capacity to a given plant species or regional landcovers, it is obviously preferable to have repeated measurements carried out on multiple leaves of the species in question or all important taxa of that region, ideally under carefully controlled conditions. While such data are difficult to obtain in any reasonably diverse forest ecosystem, it is clearly an impossible goal in the extremely diverse and logistically difficult tropical forests. For this reason, attempts have been made to establish taxonomic patterns of BVOC emission (Benjamin et al., 1996; Benjamin and Winer, 1998), such that, lacking data for a given species, emission capacity data from the most closely related species for which information is available might be assigned to the unknown taxa. Table 2 shows the taxonomic distribution of isoprene emitters for many important genera.

Though not without pitfalls, this strategy is a reasonable one and is currently followed in the assigning of isoprene emission capacities to regional and global land vegetation classes. Clearly, the more distantly related the two taxa, the more problematic the technique, as may be illustrated using the species-level emission data compiled in the database described in Section 2.1 (hereafter referred to as ENCLDB; see 'ENCLOSURE MEASUREMENTS' at URL: http://bvoc.acd.ucar.edu). It should be borne in mind that, despite containing isoprene emission information on over 1800 plant species, the ENCLDB contains a relatively small sample of all woody species in the world, and generalizations made here may be refuted by subsequent measurements. Generally speaking, however, different species within the same genus are highly likely to exhibit similar isoprene emission characteristics (Karlik and Winer, 2001), especially if the emission criterion is simply emitters (>5.0 µg C g^{-1} h^{-1}) vs. non-emitters (<1.0). Nevertheless, assuming all congeners behave similarly will occasionally introduce errors, particularly in large genera, such as the oaks (*Quercus*), where 11 out of 69 species examined are non-emitters, and *Acacia*, with 31 non-emitters out of 37 measured (Harley et al., 2002). In both cases, the divergence in isoprene emitting behavior is consistent with sub-generic classification schemes based on morphological characters. Within the well-characterized genus *Quercus* (≈450 total species), only those closely related Eurasian species within Section Cerris fail to emit isoprene, though interestingly, many of these species emit monoterpenes with a strong light dependency (e.g., Loreto et al., 1996). Within *Acacia* (≈800 total species), though much less thoroughly studied, those 6 species shown to emit isoprene all fall within the subgenus Aculeiferum. Less well documented examples (see ENCLDB) of emitting and non-emitting members of the same genus occur in *Albizia* (Mimosaceae), *Bauhinia* (Papilionaceae), and *Theobroma* (Sterculiaceae).

The assumption that different genera within the same plant family share isoprene emission characteristics must be adopted with a certain degree of caution. At one extreme are those large plant families in which no isoprene emitting genera have been discovered. In this group, we can include the important families Magnoliaceae, Betulaceae, Juglandaceae, Rosaceae,

Ulmaceae, Combretaceae, Aceraceae, Melastomataceae, Meliaceae, Oleaceae, and Bignoniaceae. Until additional evidence proves us wrong, the assumption that no member of these families emits isoprene is reasonable. In addition to these families, several families in the ENCLDB contain a very small percentage of emitting species, including Annonaceae (2 emitters of 19 species examined), Apocynaceae (2 of 25), Ebenaceae (2 of 13), Lauraceae (2 of 25), Rubiaceae (1 of 29), Sapindaceae (3 of 17), and Tiliaceae (2 of 15). Many of these are extremely important tropical families. Where such a small number of species within a given family appear to emit, the assumption that unmeasured species within the same family are non-emitters again seems reasonable unless they belong to an emitting genus. In addition, those few anomalous members of each family should be re-examined; if found to be non-emitters, it would clarify the situation, and if it is verified that they do in fact emit isoprene, it would be interesting to know why they are anomalous. A few families have been found, thus far, to contain a single emitting genus, and all other genera within the family may be assumed to emit no isoprene. These include Fagaceae (only the genus *Quercus* emits) and Ericaceae (only the genus *Erica* emits).

At the other extreme from these largely non-emitting families are those few plant families in which all species thus far examined have been found to emit isoprene and for which the assumption that other unmeasured species will do the same is warranted. Among these are Berberidaceae, Papaveraceae (anomalous in having some herbaceous genera), Hamamelidaceae, Platanaceae, Proteaceae, and Salicaceae.

That leaves a number of important plant families somewhere in the middle, with significant number of both emitting and non-emitting members, for which assigning an emission capacity to unmeasured genera is problematic. Among these families are Moraceae (8 emitting genera, including the important genus *Ficus*, of 17 examined), Euphorbiaceae (11 of 34), Caesalpinaceae (12 of 39), Mimosaceae (6 of 22), Papilionaceae (51 of 66), Myrtaceae (16 of 18), Anacardiaceae (6 of 15), Burseraceae (4 of 6), Rhamnaceae (8 of 12), and Poaceae (12 of 39).

For a few of the larger families in the ambiguous middle ground, relationships between isoprene emission and taxonomy extend to the subfamilial level. For example, of the 17 subfamilies of Papilionaceae that have been tested for isoprene emission, certain subfamilies (Loteae, Vicieae, Trifoileae) have thus far yielded no isoprene emitters, while others (Genisteae, Sophoreae, Milettieae) are dominated by isoprene emitting species.

It is interesting to note that the general tendency for isoprene emitters to be restricted almost exclusively to woody taxa is manifest, in this case, at the subfamily level, as each of the three non-emitting subfamilies is exclusively herbaceous. Within the Caesalpinaceae, virtually all the emitting species are found in subfamilies Cercideae and Detarieae, while none are found within Caesalpinoideae or Cassieae. The grass family, Poaceae ($\approx$10,000 species) has been only poorly sampled, but to date, all verified emitters (12 species of 55 sampled) are members of either the Tribe Bambuseae, Subfamily Bambusoideae or Tribe Arundineae, Subfamily Arundinoideae. Again, these sub-groups represent the woodiest members of the family. Table 2 summarizes these data and shows the taxonomic distribution of isoprene emitters.

Table 2. Taxonomic distribution of isoprene emitters

No genera emit isoprene	<20% of genera emit isoprene	20-80% of genera emit isoprene	>80% of genera emit isoprene	All genera emit isoprene
Aceraceae	Annonaceae	Anacardiaceae	Clusiaceae	Berberidaceae
Betulaceae	Apocynaceae	Arecaceae	Myrtaceae	Hamamelidaceae
Bignoniaceae	Ebenaceae	Burseraceae		Ochnaceae
Combretaceae	Ericaceae	Caesalpinaceae		Papaveraceae
Juglandaceae	Fagaceae	Euphorbiaceae		Platanaceae
Magnoliaceae	Lauraceae	Mimosaceae		Proteaceae
Melastomataceae	Rubiaceae	Moraceae		Salicacaeae
Meliaceae	Sapindaceae	Papilionaceae		
Rosaceae	Sapotaceae	Poaceae		
Ulmaceae	Sterculiaceae	Rhamnaceae		
	Tiliaceae			

4.3. Emissions from trees

Current estimates indicate that about 27% of the 33 Gt of global tree foliar mass emits isoprene, and approximately 70% of this isoprene-emitting tree foliar mass is broadleaf foliage. Our estimates of the isoprene emission characteristics of trees are based on forest inventory data for 5 regions that together comprise about 28% of the world's forests (e.g., Simpson et al., 1999; Guenther et al., 2000; Klinger et al., 2002). The regional percentage of trees that emit isoprene varies considerably, from about 10% for Europe, 27-30% for the U.S. and China, 47% for Canada, and 87% for Australia. Our estimates of the fraction of isoprene emitters in the southern half of Africa, containing another 10% of global forests, are based on species distributions described by Otter et al. (2003). Table 3 summarizes the fraction of isoprene emitting species for each major country/region.

Table 3. Estimated percentage of isoprene emitting trees within different regions.

	% isoprene emitters	Land Area 10^6 km^2	Forest Area 10^6 km^2	Isoprene Tree cover 10^6 km^2
Europe	9	4.8	1.6	0.14
China	27	9.6	1.3	0.35
United States	28	9.5	3	0.84
Canada	47	10	4.5	2.12
Australia	83	7.7	1.6	1.25
Africa (S. of Equator)	~30	~10	~3.5	~1
Russia	~15	16.9	7.7	~1.1
Brazil	~35	8.5	5.6	~1.9
Congo	~25	2.3	1.7	~.42
Indonesia	~30	1.8	1.1	~.33
Peru	~35	1.3	0.8	~.28
India	~30	3	0.7	~.21
Bolivia	~35	1.1	0.6	~.21
Columbia	~35	1	0.5	~.18
Venezuela	~35	0.9	0.5	~.18
Mexico	~30	1.9	0.5	~.15
Argentina	~25	2.7	0.5	~.13
CAR	~25	0.6	0.5	~.13
Rest of world	~30	~50	~4.5	~1.6
Total	**~27**	**~140**	**~40**	**~11**

Isoprene emission factors for North American tree species have been studied extensively. Isoprene emission capacities for most common species have been determined using a variety of enclosure techniques. Leaf level emission capacity for broadleaf trees expressing the isoprene synthase gene is around 100 µg C g^{-1} h^{-1} (Geron et al., 2001), but species not possessing this trait emit at less than 0.1 µg C g^{-1} h^{-1}.

The only known North American fine-leaf isoprene emitting trees are in the genus *Picea* (spruces). which have considerably lower emission rates of about 20 µg C g^{-1} h^{-1}. European spruces, on the other hand, have been found to be very low emitters of isoprene. Conversely, European and middle-eastern true firs (*Abies*) have been found to be strong isoprene emitters, although those North American species examined to date seem not to emit isoprene.

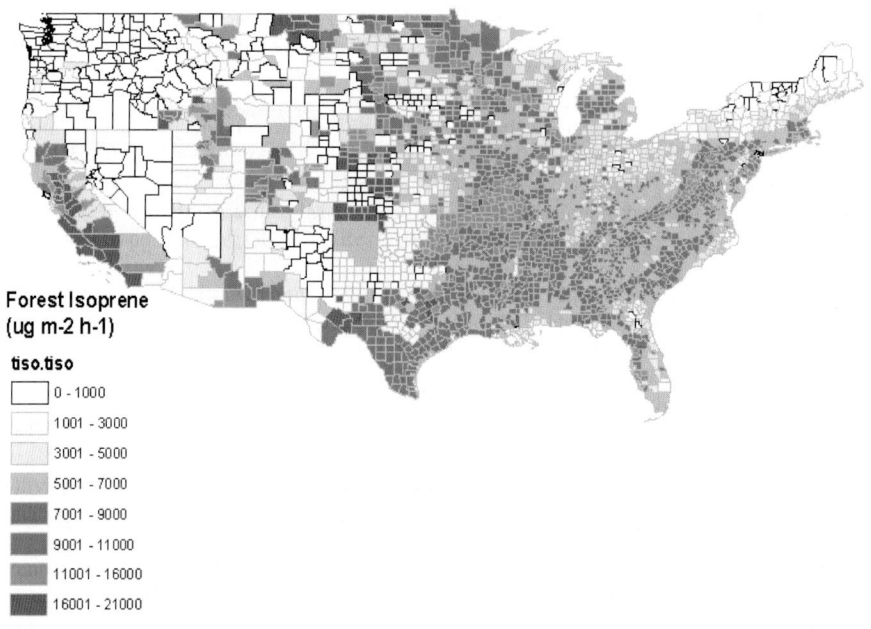

Figure 5 (Plate 9). County-level isoprene emission rates (µg m^{-2} h^{-1}) for midday summertime conditions (above-canopy 30°C and 1500 µmol m^{-2} s^{-1}). The values have been averaged over the total forest land area (not over total land area) for each county in the United States (based on Guenther et al., 2000).

Figure 5 shows the estimated county-level isoprene emission rate for the United States for midday summertime conditions (above canopy 30°C and 1500 µmol m^{-2} s^{-1}), averaged over the total forested land identified in the county. The European forest tree inventory data of Simpson et al., (1999) estimates that 9% of forest cover in Europe is composed of medium to high isoprene emitting species (>1 µg g^{-1} h^{-1} at 30°C and 1000 µmol m^{-2} s^{-1} PAR).

The isoprene emissions in Europe are dominated by *Quercus* species (2390 Gg yr^{-1}), *Populus* species (942 Gg yr^{-1}) and *Picea* species (766 Gg yr^{-1}) (Simpson et al., 1999). The aerial extent of forested areas and the species compositions of these forests can vary dramatically throughout Europe. The countries in Europe with the largest proportion of land area covered by nominally isoprene-emitting forest are Sweden and Austria (29 and 26%, respectively) while the countries with the smallest proportion are Estonia, Italy, and the Netherlands. However, a large percentage of isoprene-emitting forests is not a true indicator of isoprene emissions.

Using Great Britain as an example, despite its relatively high proportion of nominally isoprene-emitting forest area (49%), a low annual isoprene emission rate is predicted. This is due to the small proportion of the country covered by forest (8%), combined with normally low temperatures and significant cloud cover. Using 1998 meteorological data, it is estimated that the isoprene emission flux for Great Britain for 1998 was 8 kt, 40% of which was due to Sitka spruce alone (Stewart et al., 2002). This will be in contrast to those countries with large areas of forest (e.g., Scandinavian counties) and to those that experience high temperatures and PAR (e.g., southern European countries).

The Australian Bureau of Rural Sciences forest inventory estimates that 79% of Australian trees belong to one genus, *Eucalyptus*. All of the Eucalypts that have been investigated have been identified as isoprene emitters. The native Australian tree genera *Melaleuca* and *Casuarina* also have only known isoprene emitters. These two genera together dominate about 3% of the total forest and woodlands. *Acacia* is the second most prevalent Australian tree genus (8% of the total). Although this widespread genus has some species that emit isoprene, the few Australian species that have been sampled are not among them. Important tree genera that do not emit isoprene include some native Australian genera (*Callitris*, *Rhizophora*, and *Avicennia*) and some introduced plantation trees (*Pinus*), which together cover about 2% of the woodlands and forests. The remaining 8% is covered by rainforests and other native landscapes that have a mix of largely uninvestigated tree species. We estimate about 87% of Australian trees emit isoprene.

Isoprene emission characteristics have been reported for about half of the dominant tree genera in the southern extratropical portion of South America. Six genera (*Nothofagus, Drimys, Eucryphia, Araucaria, Libocedrus, Dacrydium*) have been reported as non-emitters, one as an emitter (*Jubaea*), and one (*Podocarpus*) has had reports of both. Of the other important tree genera (*Aextoxicon, Agathis, Fitzroya, Pilgerodendron, Laurelia, Quillaja, Lithrea, Weinmannia*), most are from families that tend to have few or no known isoprene emitters except for palms such as *Rhopalostylis*. Of the major tree genera planted in this area, two have high isoprene emissions (*Eucalyptus* and *Populus*), and two (*Pinus* and *Pseudotsuga*) are non-emitters. The available measurements suggest a very low isoprene emission for fine-leaf trees in this area and lower than average isoprene emission from broadleaf trees except for plantations.

Klinger et al. (2002) measured the emissions of over 500 plant species in China. Several species of *Quercus, Ficus, Salix*, and *Populus* were observed to be the high isoprene emitters. Additionally, *Cinnamomum burmani, Maackia amurensis, Syzygium jambos,* and *Caryota ochlandra* all had measured isoprene emission capacities greater than 100 µg C m^{-2} h^{-1}. Of the total forested area in the China, Pine and Oak forests made up the largest area percentages (18.9 and 15.9%, respectively), although the fir forests are substantial. Economic and bamboo forests contribute to 10.5% of the national forested areas.

4.4. Emissions from shrubs

We have estimated that global shrub foliar mass is about 19 Gt, which is about 60% of that estimated for trees. There are no shrub inventories analogous to those for trees, so both our estimated total foliar biomass and the fraction that emits isoprene are much less certain. In general, the fraction of isoprene-emitting shrubs is similar to that of trees within a region. For example, many of the dominant genera of Australian shrublands (e.g., *Banksia, Melaleuca, Hakea, Grevillea, Eucalyptus*) are isoprene emitters. Other important Australasian shrub genera (e.g., *Atriplex* and

Terminalia) have been sampled in the U.S. and Africa and found to have no isoprene emission. Isoprene emission characteristics are unknown for other shrub genera (e.g., *Owenia, Chenopodium, Arthrocnemum, Maireana*) that are important in at least some Australasian ecoregions. We estimate that over 80% of Australasian shrubs are isoprene emitters in comparison to 87% of Australian trees.

Most of the major shrubland genera for the Neotropic and Nearctic regions have been investigated for isoprene emission. Of these, all of the *Berberis* and *Condalia* species and a few of the *Acacia* are isoprene emitters. Representatives of *Alnus, Caesalpinia, Celtis, Chlorisia, Euphorbia, Larrea, Prosopis,* and *Schinus* have been investigated and none has been identified as an isoprene emitters. There are many other important shrub genera that have not been investigated. We estimate that about 22% of all shrubs are isoprene emitters and that the global total shrub emitting biomass is about 4 Gt.

4.5. Emissions from non-woody plants

The estimated global total crop foliar mass of 11 Gt of herbaceous crops is expected to make an extremely small contribution to isoprene fluxes. All of the 25 major global crop genera have been sampled and none has been found to emit isoprene. While there are a few examples of isoprene-emitting genera, including kudzu (*Pueraria lobata*) and velvet bean (*Mucuna pruriens*), they are not major crops and represent less than 0.1% of global crop foliage. The total isoprene emitting crop foliar mass is thus estimated to be <0.01% of the global total isoprene emitting foliar mass.

The global total grass foliar mass of about 18 Gt is estimated to include a similarly small fraction of isoprene emitters. The known isoprene emitting grass genera include *Phragmites* and *Arundo* and are almost exclusively in the subfamilies Arundinoideae and Bambusoideae. Assuming that less than 0.5% of all grasses emit isoprene, the contribution of grasses to the global isoprene emitting foliar mass is less than 0.1%. It should be noted, however, that the number of emitting grasses is highly uncertain and should be more thoroughly investigated. Though many bamboos are high isoprene emitters and bamboo forests may be regionally significant isoprene sources in parts of Asia and South America, for the purposes of this review they are treated as woody species.

Of the other non-woody plants, only the ferns and mosses have been associated with isoprene emission. The global fern foliar mass is probably less than 100 Tg, so that even if 30% of all ferns emit isoprene, they would constitute only 30 Tg of isoprene-emitting foliage. More significant may be the contribution from mosses. Hanson et al. (1999) found that 15 of 16 mosses sampled were found to emit detectable levels of isoprene, although only 25% of these emitted at rates exceeding 1 nmol m^{-2} (ground) s^{-1}. In moss-dominated landscapes, such as in many boreal and some temperate ecoregions, mosses may make a significant contribution to the regional isoprene flux.

Emissions of biogenic isoprene from the oceans are assumed to be small compared to those from terrestrial sources. Broadgate et al. (1997) estimated a seasonally-averaged isoprene flux of 0.07 µg isoprene m^{-2} hr^{-1} from the oceans to the atmosphere. Further studies have looked at other volatile organic compounds from the oceans (e.g., Baker et al., 2000). Although biogenic organic emissions from oceans are important, they are beyond the scope of this chapter and will not be further addressed here.

5. OTHER BVOC EMISSIONS

Many global atmospheric oxidant studies have included isoprene as the only biogenic VOC due to the limited knowledge of the emission rates and atmospheric oxidation schemes of other BVOC. There are a relatively large number of reported isoprene emission measurements because of the dominance of isoprene in most landscapes, its high reactivity and influence on atmospheric chemistry, and because many of the other BVOC are more difficult to quantify. However, BVOC other than isoprene are very important. Recent estimates for Great Britain indicate that the annual monoterpene flux exceeds that of isoprene by a factor of 10 (83 kt a^{-1} compared with 8 kt a^{-1}; Stewart et al., 2002).

Since many of these BVOCs have very different reaction rates and chemical products, the specification of explicit compound emission capacities in emissions models, rather than one lumped BVOC emission capacity as has been done in the past, is important for many chemistry and transport model investigations. Only a relatively small number (probably less than 40) of the tens of thousands of organic compounds found in plants need to be included. We have ranked these compounds into groups indicating the magnitude of their global emission (Table 4). Some recent studies that have improved our ability to make these rough estimates are summarized in this section.

Table 4. Chemical species that dominate the annual global emission of VOC from vegetation.

Annual emission (TgC yr^{-1})	Compounds
250-750	Isoprene
50-250	methanol, α-pinene
10-50	acetaldehyde, acetone, β-pinene, d-carene, ethanol, ethene, hexenal, hexenol, hexenyl-acetate
2-10	propene, formaldehyde, hexanal, butanone, sabinene, limonene, methyl butenol, butene, b-phellandrene, p-cymene, myrcene
0.4 – 2	formic acid, acetic acid, ethane, toluene, camphene, terpinolene, a-terpinolene, a-thujene, cineole, ocimene, g-terpinene, bornyl acetate, b-carophylene, camphor, piperitone, linalool, tricyclene

5.1. Mono- and Sesquiterpenes

Most efforts to characterize non-isoprene BVOC fluxes from vegetation have focused on monoterpenes. The total global monoterpene emission is dominated by contributions from a few compounds (e.g., α-pinene, β-pinene, d-limonene), and there are probably less than twenty monoterpenes that have non-negligible emissions to the global atmosphere (Table 4). An important finding since the compilation of the Guenther et al. (1995) model is the occurrence of two distinct types of monoterpene emission: direct emission from chloroplasts without storage, which is both light- and temperature-dependent, and emission of stored monoterpenes, which is independent of light. In some plant species, the total monoterpene emission is a result of both processes. The contribution of each process to the total emission is compound-specific for a plant species (Seufert et al., 1997; Schuh et al., 1997; Steinbrecher et al., 1999).

The direct emission of monoterpenes, without storage, can occur at considerably higher rates from leaves exposed to full sunlight (over 20 µg C g^{-1} h^{-1} at leaf temperatures of 30°C) (Seufert et al., 1997). The daily total emission from these plant species is decreased by the lack of emission during the night but is still much higher than from those species that emit stored monoterpenes. This type of monoterpene emission was first shown for certain species of Mediterranean oak (Steinbrecher and Hauff, 1996; Kesselmeier et al., 1997) but has since been identified as the emission behaviour for trees in southern Africa, *Colophospermum mopane* (Otter et al., 2003), in North America, *Acer saccharinum*, and in the Amazon, *Apeiba tiburbou* or *echinata* (Kuhn et al., 2002). The light and temperature dependence of these emissions is similar to that observed for isoprene.

Emission factors for stored monoterpenes are thought to be highest in conifers, although *Eucalyptus* and *Liquidambar* trees and many shrubby angiosperms feature glandular structures that store and release these compounds (Geron et al., 2000b). These genera, in addition to *Pinus* and *Abies*, are assumed to emit monoterpenes at rates of 1-5 µg C g^{-1} h^{-1} at a leaf temperature of 30°C. Emission models assume that these emissions increase exponentially with temperature (~9% increase per °C) due to the effect of temperature on terpene partial vapour pressure. Some recent studies fail to detect temperature effects, although other data indicate that short-term temperature dependence can be stronger than currently assumed in models. New information indicates that monoterpene emission capacities may be biased under some physiological conditions, and that tree and needle age, needle wetness, relative humidity, phenological state (e.g., budbreak, senescence), and stomatal control must be considered in attempts to realistically model monoterpene emissions (Kim, 2001; Schade et al., 1999). Preliminary data for pines from the southeastern U.S. (C. Geron, unpublished data) indicate high monoterpene emission rates for young needles, but much lower emissions from older needles. This suggests that existing regional and seasonal inventories of these compounds may be reasonable when canopies are dominated by young needles but are overestimated at times when older needles are dominant. In addition, leaf damage, which ruptures storage structures such as resin ducts (conifers) or glandular hairs (e.g., members of the mint family), can lead to large but short-lived monoterpene emissions.

Sesquiterpenes are also emitted from storage pools and can be important sinks for oxidants and precursors to aerosols in rural regions. Because sesquiterpenes have atmospheric lifetimes on the order of a few minutes due to rapid reaction with O_3, they are difficult to study, and little is known about their emission rates. Thus, they are not typically included explicitly in emission models. A few reports indicate that their emission may be significant from some dominant vegetation types (Helmig et al., 1999), including desert plants and many oaks, which are not known to feature storage of these compounds. Dagan et al. (1979) found that capacity for monoterpenes was highest in the basal portion of *Pinus pinaster* needles, while sesquiterpene production and storage was fairly uniform throughout the needle. Future studies to establish emission factors for sesquiterpenes must deal with the high reactivity and short lifetimes of these compounds.

5.2. Carbonyls

It is estimated that the emission of carbonyls amounts to approximately 24% of the total VOC emitted by forest ecosystems. Acetaldehyde emissions have been observed to be ~2.2 µg g^{-1} h^{-1} from *Populus* leaves, 0.4-1.5 µg g^{-1} h^{-1} from *Pinus pinea,* and 0.4-0.9 µg g^{-1} h^{-1} from *Quercus ilex* (Kesselmeier et al., 1997; Kreuzwieser et al., 1999). Acetaldehyde emissions from flooded *Populus* trees are significantly higher (up to approximately 16.8 µg g^{-1} h^{-1}, Kreuzwieser et al., 2000) and are linked to anaerobic root conditions (Fall, 1999). Emission rates of formaldehyde were in a similar range (*P. pinea* and *Q. ilex*: 0.1-0.9 µg g^{-1} h^{-1}; Kesselmeier et al., 1997). In contrast, *Q. robur* and *Carpinus betulus* showed both deposition of formaldehyde and emission of acetaldehyde and acetone (Kesselmeier, 2001). Martin et al. (1999) and Knowlton et al. (1999) found that emission rates of formaldehyde, acetaldehyde, and formic and acetic acids were significant from several western tree and shrub species, ranging from 0.1 to 5 µg C g^{-1} h^{-1} for each of these individual reactive compounds. Large differences in measured emission rates of formic and acetic acid from *Q. ilex* and *Q. pubescens* exist between laboratory and field measurements. In the field, formic and acetic acid emission rates were about a factor of ten higher than those measured in the laboratory (field, 10-50 nmol m^{-2} min^{-1};

laboratory, 1 to 2 nmol m^{-2} min^{-1}; Kesselmeier, 2001) and were of the same order of magnitude as measured total carbonyls (field: 10-50 nmol m^{-2} min^{-1}; laboratory, 1 to 2 nmol m^{-2} min^{-1}). The reason for this discrepancy remains unclear. It could be due to differences in tree age. Also, stress factors in the field could be greater than those experienced in the laboratory, and it is assumed that several stress factors such as hypoxia, drought, chilling, and wounding cause an increased production of formaldehyde in plants (see Kreuzwieser et al., 2001).

In a recent study, acetone emissions were found in other European tree species, *Picea abies* and *Pinus sylvestris* (Janson et al., 1999; Janson and de Serves, 2001). Kesselmeier et al. (1997) noted that emissions follow compensation-point behavoir. Emissions occur when ambient concentrations are below a specific threshold (the compensation point). Uptake results at ambient concentrations above the compensation point. Thus, earlier enclosure studies in which chambers were flushed with VOC-free air tended to over-predict the emission rates that would occur under ambient conditions.

5.3. 2-Methyl-3-buten-2-ol

Another terpenoid compound, 2-methyl-3-buten-2-ol (MBO), is also emitted at rates over 20 µg C g^{-1} h^{-1} at leaf temperatures of 30°C from leaves exposed to full sunlight. MBO emission has a limited taxonomic and geographic distribution and has only been observed from some species of pine (Harley et al., 1998; Baker et al., 2001). However, these trees include many major western North American pines e.g., *Pinus contorta* and *P. ponderosa* and related species, resulting in the dominance of MBO emission from at least some western North American forests. Karl et al. (2002) measured maximum daily MBO fluxes from a subalpine forest in Colorado to be 1.5 mg m^{-2} h^{-1}. In Europe, Norway spruce forests (*Picea abies*) may become significant MBO sources when infested by bark beetles (*Ips typographus*), as this compound serves as a pheromone triggering the attack behavior of the beetle predator, *Thanasimus formicarius* (Tommeras, 1985). Needle age influences MBO emission, with current year needles emitting at higher rates than 1-year-old needles, and 2 year-old needles emitted at least 50% less MBO than the younger needles (Harley et al., 1998; Schade et al., 2000). MBO emission factors were observed to be dependent upon light and temperature, including prior daytime air temperatures, in a manner similar to isoprene (Schade et al., 2000).

5.4. Methanol and other oxygenated VOCs

MacDonald and Fall (1993) found substantial emissions of methanol from a variety of North American forest and agricultural species, with rates ranging from about 1 to 17 $\mu g\ C\ g^{-1}\ h^{-1}$. Emissions were apparently highest from developing leaves and seemed to decline as leaves aged.

Warneke et al. (2002) found a methanol emission factor of ca. 4 $\mu g\ C\ g^{-1}\ h^{-1}$ from undisturbed alfalfa in Colorado and also reported measurable, but much smaller (ca. 0.1 $\mu g\ C\ g^{-1}\ h^{-1}$), emissions of hexenal, possibly from insect-related leaf wounding of the crop. Additional oxygenated VOCs (hexenylacetate, hexenal, hexenol, hexanal, butanone) are emitted at high rates when leaves are physically damaged. Other studies using on-line mass spectrometric techniques indicate that emission of acetone, methanol, pentenol/hexenol derivatives, organic acids, and other oxygenated VOCs may be important from both undisturbed and harvested and drying vegetation. Large emission pulses (hours to days) of these compounds result from cutting, frost damage, or senescence of vegetation. Table 5 summarizes recent emission factor estimates for these compounds resulting from cutting and drying of vegetation. Although the annual global flux of BVOC from disturbed vegetation is much less than that of isoprene from undisturbed vegetation, it can be significant on episodic scales in agricultural regions. Crop harvesting could be responsible for high concentrations of PAN and PPN (oxidation products of acetaldehyde and hexenol, respectively) observed in agricultural regions during crop harvesting seasons (de Gouw et al., 1999). In addition, the oxidation of carbonyls, particularly acetaldehyde and formaldehyde, by HO_2-radicals causes the generation of acetic and formic acid that significantly contribute to the atmosphere's acidity.

Table 5. Estimates (µg C g^{-1} h^{-1}) of oxygenated VOC (OVOC) released from cut and drying vegetation. The emissions are integrated over time periods of several hours to several days.

Genus & species	Common name	Compound	Emission	Reference
Medicago sativa	alfalfa	Total OVOC	175+	De Gouw et al. 2000
Trifolium repens	white clover	Total OVOC	1200	
Medicago sativa	alfalfa	methanol	90	Rinne et al. 2001
		acetaldehyde	12	
		acetone	7	
		hexenals	14	
Trifolium repens	white clover	methanol	75	De Gouw et al. 1999
		acetone	500	
		butanone	470	
Festuca rubra	red fescue	methanol	75	
		acetaldehyde	75	
		z-3-hexenal	50	
Populus tremuloides	aspen	z-3-hexenal	500	Fall et al. 1999
	alpine grasses	methanol	50	Karl et al. 2001c
		acetaldehyde	15	
		C$_4$-C$_6$ oxys	15	
Dactylis glomerata	orchard grass	hexenals	25	Karl et al. 2001a
		hexenol & hexenal	5	
		acetaldehyde	10	
		acetone	10	
Poa pratensis	blue grass	hexenals	100	
		hexenol & hexenal	30	
		acetaldehyde	25	
		acetone	20	
Mixed *P. Pratensis*, *D. glomerata*		methanol	60	Karl et al. 2001b
		butanone	25	
		pentenols	10	
	Grass			Kirstine et al. 1998
	Clover			Kirstine et al. 1998
				Fukui and Doskey, 1998
Dead & decaying plant matter		methanol	150	Warneke et al. 1999
		acetone	60	

5.5. Ethene and other hydrocarbons

Ethene is a well-known volatile plant hormone that plays a major role in plant growth and development (Abeles et al., 1992). Concentrations are known to increase when a plant is exposed to stress. Ethene has a role in triggering plant defence. Ethene production is widespread in plants and is likely to be a significant emission from most landscapes (Goldstein et al., 1996). Propene and butene are emitted at lower rates (Goldstein et al., 1996). Other VOC compounds (e.g., ethane, toluene) have been reported as emissions, but there is not yet any indication that these are significant on global scales (Guenther et al., 2000).

6. ORGANIC PARTICLE EMISSIONS FROM FORESTS

Aerosols over forested areas account for a significant mass of airborne carbonaceous species (Artaxo et al., 2002). The tropical forests, with their high biological activity, have the potential to emit large amounts of trace gases and particles to the atmosphere (Andreae and Crutzen, 1997). Aerosol sources can be divided into two types: primary and secondary. Primary aerosols are emitted directly into the atmosphere. Examples include airborne dust (see Chapter 6), sea-salt particles (see Chapter 9), or natural biogenic aerosols that consist of pollens, leave fragments, and liquids leaving the leaf surface. Secondary aerosols arise from oxidation of precursor gases, such as sulfur dioxide, nitrogen oxides, and VOCs. Two different organic aerosol size fractions exist: the fine mode component (with particle diameter (D_p) < 2 µm), which typically consists of aerosols formed by gas-to-particle conversion, and the coarse mode component (2<D_p<10 µm) known as PM_{10}, the majority of which consists of primary aerosols (Artaxo and Hansson, 1995). Although these general size and source definitions exist, there are exceptions. For example, very small (around 0.2 µm diameter) primary biogenic particles have been observed with high resolution scanning microscopy analysis. Although the majority of the aerosol mass is in the size range 5-10 µm, these small biogenic particles are very efficient as cloud condensation nuclei (CCN) and can affect cloud formation over forested areas (Roberts et al., 2001). Since particles influence important atmospheric processes such as light scattering, absorption and cloud formation (Andreae and Crutzen, 1997), it is important to gain an understanding of new particle formation from natural sources and the composition of these particles (Artaxo et al., 2001; Zhou et al., 2002).

In the wet season, when no biomass burning occurs, the average PM_{10} concentration in Amazonia is about 12μg m⁻³, with 70% of the mass in the coarse mode. (Artaxo et al., 2002; Artaxo and Hansson, 1995). The majority of this coarse mode aerosol consists of natural biogenic particles released directly by the vegetation. Only about 10% of the mass is associated with soil dust particles (Artaxo et al., 1990). In the dry season, biomass burning emissions dominate the atmospheric aerosol, with PM_{10} concentrations exceeding 600 μg m⁻³ (Artaxo et al., 1994, 2002). In both seasons, organic matter accounted for about 90% of the measured aerosol mass. During the dry season, the emission of vast amounts of smoke from forest-clearing fires at the southern perimeter of Amazonia and a predominant northward transport over the region provides an explanation for the abundance of particulate organic carbon and black carbon over large areas of Amazonia (Talbot et al., 1990; Echalar et al., 1998). The abundance of organic aerosols during the wet season is attributed to biogenic aerosol production (Kubátová et al., 2000).

6.1. Aerosol in the fine fraction originating from biogenic hydrocarbons

The amount of secondary aerosol from biogenic VOCs in the atmosphere is highly uncertain. Andreae and Crutzen (1997) have estimated that the global biogenic production of secondary organic aerosols falls in the range of 30 to 270 Tg of C per year. Griffin et al. (1999) estimated that global secondary organic aerosol formation from BVOC emissions ranges from 13-24 Tg yr⁻¹. It is estimated that gas-to-particle conversion of monoterpenes accounts for a significant fraction of the carbonaceous aerosol over forested areas. Since the 1960´s, organic aerosols have been observed to produce blue hazes in the atmosphere by way of light scattering (Went, 1960; Lodge et al., 1974). It was recently shown (O´Dowd et al., 2002) that a large number of newly formed small particles, with diameters ranging from 3 to 5 nm, consist of organic species, such as cis-pinonic acid and pinic acid, produced from the oxidation of terpenes in biogenic organic vapours released from the canopy. The formation of atmospheric particles from organic acids produced by forests in Portugal was clearly documented by

Kavouras et al. (1998) through their observations of high concentrations of pinonaldehyde and nopinone (both photo-oxidation products of α- and β-pinene). Most new particle formation over forests in temperate regions is initiated by the interaction of organic acids produced by the photo-oxidation of terpenes with other organic and inorganic species present in the atmosphere.

The formation of new particles in the ozonolysis of α- and β-pinene was investigated in laboratory studies by Hatakeyama et al. (1989) and Koch et al. (2000), who reported mass-based aerosol yields of 18.3% and 13.8%, respectively. Photochemical aerosol formation from biogenic hydrocarbons in the presence of seed aerosol was investigated using outdoor chamber experiments (Hoffmann et al., 1997). These studies revealed that the mass-based aerosol yields, ranging from a few percent up to nearly 100%, were strongly dependent on the structure of the hydrocarbon, the initial hydrocarbon-to-NO_x ratio, and the amount of seed aerosol present. Monoterpenes are oxidized in the atmosphere during the day by both OH radicals and O_3, and at night by NO_3 radicals and O_3. The reaction with O_3 leads to the formation of additional OH radicals. Carboxylic acids, such as cis-pinonic acid and trans-pinonic acid, were observed as products of the reaction of α-pinene and ozone (Hoffmann et al., 1997; Kavouras et al., 1999). Due to their low vapor pressures, these compounds almost exclusively partition into the particulate phase and contribute significantly to the water-soluble fraction (Kubátová et al., 2000).

Aerosol yields from the reactions of organic compounds with OH, O_3, and NO_3 are extremely variable and depend on the composition of the hydrocarbons as well as on the circumstances under which the oxidation reactions are taking place (Andreae and Crutzen, 1997; Kamens et al., 1999; Moldanova and Ljungstro, 2000). Yields for the daylight photo-oxidation of terpenes range from 5 to 100%, with the highest values obtained for sesquiterpenes (Yu et al., 1999; Kavouras et al., 1999; Kamens et al., 1999). For most compounds, nighttime oxidation by ozone produces even higher aerosol yields than the daytime photochemical process (Hoffmann et al., 1997). Andreae and Crutzen (1997) estimate that secondary organic aerosols are produced at a rate of 30 to 270 Tg year^{-1}, a magnitude comparable to the production of biogenic and anthropogenic sulfate aerosols (90 and 140 Tg year^{-1}, respectively).

Studies of particle hygroscopicity and growth have shown very intriguing characteristics of tropical forest aerosol (Zhou et al., 2002). In contrast to the bimodal hygroscopic behaviour (aged and processed) found in polluted continental environments, the aerosol observed in the Amazon rain forest are essentially unimodal, with diameter growth factors 1.16-1.32 from dry to 90% relative humidity. This unimodal hygroscopic behaviour has also been observed in remote marine environments; however, the measured particle growth factors are much higher in the marine atmosphere than in Amazonia. The reason for the differences in observed hygroscopic growth is that soluble inorganic compounds dominate the aerosol mass in remote marine environment, while organic species are largely responsible for water uptake in the Amazon rain forest. The observed total particle number concentrations in Amazonia were frequently between 300 and 700 particles cm^{-3}, with a mean value of around 500 particles cm^{-3}. New particle formation was only observed immediately after rain episodes, when particle numbers fell below 100 particles cm^{-3}. From this observation, it has been concluded that organic gases preferentially deposit over the surface of existing particles.

The mechanisms and the processes controlling Amazonian aerosol formation and evolution are not well understood. However, measurements made during the Large-scale Biosphere-Atmosphere Experiment in Amazonia (LBA) have enabled the development of several hypotheses: 1) Ultrafine particles are formed by nucleation aloft (not at ground level), possibly via ternary nucleation of NH_3-H_2SO_4-H_2O. These newly formed particles are ~1 nm in diameter. 2) Aitken mode particles (~ 60<D_p<100 nm) are only slightly hygroscopic. These growth factors imply that ultrafine particles grow into Aitken mode particles by condensation of low-volatility organic gases derived from the oxidation of gaseous BVOC. 3) Particles in the accumulation mode (~ 0.1<D_p<1.0 µm) are formed when Aitken-sized particles activate into cloud droplets leading to fixation of soluble gases (e.g., SO_2) in the evaporating droplets. 4) Wet deposition is an important removal mechanism for particles in the Aitken and accumulation modes (Zhou et al., 2002).

6.2. Primary biogenic aerosol in the coarse fraction

Although we have long known about the presence of biogenic aerosols

(Went, 1960), only a few studies of natural biogenic aerosols from vegetation in tropical rain forests have been undertaken (Artaxo et al., 1990, 1994, 1998; Echalar et al., 1998). Biogenic aerosols consist of many different types of particles, including pollen, spores, bacteria, algae, protozoa, fungi, fragments of leaves, and excrement and fragments of insects (Maki and Willoughby, 1978; Simoneit and Mazurek, 1982; Matthias-Maser and Jaenicke, 1995; Fish, 1972; Crozat, 1979; Beauford et al., 1977). These particles are mainly observed in the coarse size fraction (D_p>2 µm). The mechanisms of primary biogenic particle emission are still not well understood but probably include mechanical abrasion by wind, biological activity of microorganisms on plant surfaces and forest litter, and plant physiological processes (Jaenicke and Matthias-Maser, 1992). The transpiration of plants can lead to migration of Ca^{2+}, SO_4^{2-}, Cl^-, K^+, Mg^{2+} and Na^+ to the atmosphere (Nemeruyk, 1970), generating particles containing these biogenic-related elements. These elements are essential in higher plants and are released from the leaves to the atmosphere during guttation (emergence of water from vein endings in leaves at night when transpiration is not occurring) and transpiration (Nemeruyk, 1970).

Bacteria observed in forested areas were measured in the size range of 0.5 to 2.5 µm (Vali et al., 1976; Jaenicke and Mathias-Maser, 1992). The biological activity of microorganisms on leaf surfaces and forest litter can result in airborne particles. Windblown pollens certainly contribute to coarse particles in forested areas. Coarse mode natural biogenic aerosol emitted by plants play an important role on nutrient cycling in tropical ecosystems. Tropical ecosystems have a delicate nutrient balance characterized by intense internal recycling, and they depend on atmospheric input of certain mineral nutrients to achieve a nutrient balance.

It has been shown that biogenic particles may influence cloud properties (e.g. Maki and Willoughby, 1978) and can act as CCN, potentially affecting the cloud formation mechanisms and cloud dynamics. The significant light-absorbing properties of biogenic aerosol may be related to the presence of humic-like substances rather than soot carbon (Artaxo et al., 2001, 2002). In the Amazonian wet season, with no biomass burning observable, the black carbon equivalent concentration is about 150 ng m^{-3}, a relatively high value that could have regional implications for the atmospheric radiation balance (Artaxo et al., 2001; Andreae and Crutzen, 1997). Unfortunately, little information that would allow a reliable estimate of the contribution of primary biogenic particles to the organic aerosol burden in tropical regions is available. The lipid fraction of Amazonian aerosols, which has been shown to consist predominantly of microbial matter and plant waxes, constitutes 10 to 20% of the total aerosol. Given that lipids represent only a fraction of the total primary biogenic aerosol, these observations are consistent with a large primary biogenic fraction in tropical aerosols (Andreae and Crutzen, 1997). This biogenic organic aerosol may be of considerable significance for atmospheric chemistry and climate. Because both primary and secondary organic particles are effective CCN (Roberts et al., 2001; Zhou et al., 2002) and primary organic particles can be ice nuclei (Vali et al., 1976), it is likely that the optical and microphysical properties of tropical continental clouds are strongly influenced by these particles. The presence of large amounts of soluble organic matter in the aerosol would lead to a high content of dissolved organics in cloud droplets. The chemical composition in the cloud droplets can affect the cloud chemistry, formation and radiative properties. Given the high concentrations of biogenically-emitted organic material in the boundary layer, they may make a considerable contribution to the free tropospheric aerosol, even if a large fraction is removed during convective transport (Artaxo et al., 2001).

7. CONCLUSIONS

BVOC and biogenic particle emissions play an important role in regional and global chemical processes, and in the global carbon and radiation balances that affect climate. For these reasons, there has been a substantial effort to understand and estimate these emissions. In the past 10 years, significant advances have been made in our understanding of organic emissions from vegetation. Isoprene emissions and the processes that control them have been studied extensively. As better analytical techniques have developed, we have been able to observe and quantify the emissions of other gaseous and aerosol emissions from a variety of plants and landscapes. These emissions are strongly dependent on the density and type of plants that make up the terrestrial system. The ability of new remote sensing instruments to detect landcover characteristics, such as land use type and leaf area index has improved dramatically within the past few years. The combination of satellite and model capabilities has enabled us to more accurately monitor and predict the meteorological parameters that affect BVOC emissions.

Although a substantial amount of information about BVOC emissions has been gathered, there is a need to evaluate the experimental results so that they can be used for modelling investigations. A collection of data and results, such as the datasets described in Section 2, will enable a community review and use of the information and will provide a means for improving study methods and reducing uncertainty in the future. Thus, model estimates of BVOC emissions will continue to improve and contribute to a better understanding of the chemical and climatological processes of the earth.

8. ACKNOWLEDGMENTS

The authors thank Shelley Noelle Pressley of Washington State University, for her useful comments and review of the chapter, and other anonymous reviewers.

9. REFERENCES

Abeles, F.B., P.W. Morgan, M.E. Saltveit, Ethylene in Plant Biology, 2nd Edition, Academic Press, San Diego, U.S.A., 1992.

Andreae, M. O., and P. J. Crutzen, Atmospheric aerosols: biogeochemical sources and role in atmospheric chemistry, Science, 276, 1052-1058, 1997.

Arnts, R.R., R. Seila, R. Kuntz., F. Mowry, K. Knoerr, A. Dudgeon, Measurement of alpha-pinene fluxes from a Loblolly pine forest, Fourth Joint Conference on Sensing Environmental Pollutants, American Chemical Society, Washington D.C., 829-833, 1978.

Arnts, R.R., W.B. Peterson, R.L. Seila, B.W. Gay, Estimates of alpha-pinene emissions from a loblolly pine forest using an atmospheric diffusion model, Atmos. Env., 16, 2127-2137, 1982.

Artaxo, P., W. Maenhaut, H. Storms, R. Van Grieken, Aerosol characteristics and sources for the Amazon Basin during the wet season, J. Geophys. Res., 95, 16971-16985, 1990.

Artaxo, P., F. Gerab, M. A. Yamasoe, J. V. Martins, Fine Mode Aerosol Composition in Three Long Term Atmospheric Monitoring Sampling Stations in the Amazon Basin, J. Geophys. Res., 99, 22857-22868, 1994.

Artaxo, P., H-C Hansson, Size distribution of biogenic aerosol particles from the Amazon basin, Atmos. Env., 29, 393-402, 1995.

Artaxo, P., E. T. Fernandes, J. V. Martins, M. A. Yamasoe, P. V. Hobbs, W. Maenhaut, K. M. Longo, A. Castanho, Large Scale Aerosol Source Apportionment in Amazonia, J. Geophys. Res., 103, 31837-31848, 1998.

Artaxo, P., Meinrat O. Andreae, A.B. Guenther, D. Rosenfeld, LBA Atmospheric Chemistry: Unveiling the lively interactions between the biosphere and the Amazonian atmosphere, IGBP Global Change Newsletter LBA Special Issue, 12-15, 2001.

Artaxo, P., J. Vanderlei Martins, M. A. Yamasoe, A. S. Procópio, T. M. Pauliquevis, M. O. Andreae, P. Guyon, L. V. Gatti, A. M. Cordova Leal. Physical and chemical properties of aerosols in the wet and dry seasons in Rondônia, Amazonia, J. Geophys. Res., 107, LBA 49-1 - 49-14, 2002.

Baker, AR., S.M. Turner, W.J. Broadgate., A. Thompson, G.B. McFiggans, O. Vesperini,P.D. Nightingale, P.S. Liss, T.D. Jickells, Distribution and sea-air fluxes of biogenic trace gases in the eastern Atlantic Ocean, Global Biogeochemical Cycles, 14, 871-886, 2000.

Baker, B., A. Guenther, J. Greenberg, R. Fall., Canopy level fluxes of 2-methyl-3-buten-2-ol, acetone, and methanol by a portable relaxed eddy accumulation system, Environmental Science & Technology, 35, 1701-1708, 2001.

Beauford, W., J. Barber, A. R. Barringer, Release of particles containing metals from vegetation into the atmosphere, Science, 195, 571-573, 1977.

Benjamin, M.T., M. Sudol, L. Bloch, A.M. Winer, Low-emitting urban forests: a taxonomic methodology for assigning isoprene and monoterpene emission rates, Atmos. Env., 30, 1437-1452, 1996.

Benjamin, M.T. A.M. Winer, Estimating the ozone-forming potential of urban trees and shrubs, Atmos. Env., 32 53-68, 1998.

Bishop, J.K.B. W.B. Rossow, Spatial and temporal variability of global surface solar irradiance, J. Geophys. Res., 96, 16839-16858, 1991.

Bonan, G.B., S. Levis, L. Kergoat, K.W. Oleson, Landscapes as patches of plant functional types: An integrating concept for climate and ecosystem models, Global Biogeochem. Cycles, 16, doi: 10.1029/2000GB001360, 2002.

Box, E., Foliar biomass: Data base of the international biological program and other sources, Atmospheric Biogenic Hydrocarbons, Ann Arbor Science Publishers, Ann Arbor, MI, 1981.

Broadgate, W.J., P.S. Liss,S.A. Penkett, Seasonal emissions of isoprene and other reactive hydrocarbon gases from the ocean, Geophys. Res. Letters, 24, 2675-2678, 1997.

Bunce, R.G.H., C.J. Barr, R.T. Clarke, D.C. Howard, A.M.J. Lane, ITE Merlewood land classification of Great Britain, Journal of Biogeography, 23, 625-634, 1996.

Crozat, G., Sur l'émission d'un aérosol riche en potassium par la forêt tropicale, Tellus, 31, 52-57, 1979.

Dagan, C. B., J. P. Carde, et al., Etude des composes terpeniques au cours de la croissance des aiguilles du Pin maritime: comparaison de données biochimiques et ultrastructurals, Can J Bot, 57, 255-263, 1979.

Defries, R.S., M.C. Hansen, J.R.G Townshend, A.C. Janetos, T.R. Loveland, A new global 1-km dataset of percentage tree cover derived from remote sensing, Global Change Biology, 6, 247-254, 2000.

de Gouw, J. A., C. J. Howard, T.G. Custer, R. Fall, Emissions of volatile organic compounds from cut grass and clover are enhanced during the drying process, Geophys. Res. Lett., 26, 811-814, 1999.

de Gouw, J. A., C. J. Howard, T.G. Custer, B.M. Baker, R. Fall., Proton-transfer chemical-ionization mass spectrometry allows real-time analysis of volatile organic compounds released from cutting and drying of crops, Environmental Science and Technology, 34, 2640-2648, 2000.

Echalar, F., P. Artaxo, F. Gerab, M. A. Yamasoe, J. V. Martins, K. M. Longo, W. Maenhaut, B. N. Holben, Aerosol composition and variability in the Amazon basin, J. Geophys. Res., 103, 31849-31866, 1998.

Fall, R., T. Karl, A. Hansel, A. Jordom, W. Lindinger, Volatile organic compounds emitted after leaf wounding: On-line analysis by proton-transfer-reaction mass spectrometry, J. Geophys. Res., 104, 15963-15974, 1999.

Fall, R., Biogenic Emissions of volatile organic compounds from higher plants, in: Reactive Hydrocarbons in the Atmosphere, N. C. Hewitt (ed), Academic Press, London, pp. 41-96, 1999.

Fall, R., T. Karl, A. Jordan, W. Lindinger, Biogenic C5 VOCs: release from leaves after freeze-thaw wounding and occurrence in air at a high altitude observatory, Atmos. Env., 35, 3905-3916, 2001.

Fish, B.R., Electrical generation of natural aerosols from vegetation, Science, *175*, 1239-1240, 1972.

Forkel, R., R. Steinbrecher, W.R. Stockwell, Modeling of the Effect of In-Canopy Chemical Reactions on the Emission Rates of Biogenic VOC, In: Proceedings from the EUROTRAC Symposium 2000, P.M. Midgley, M. Reuther, M. Williams (Eds.), Springer Verlag Berlin, Heidelberg, GENEMIS, 725-728, 2001.

Fuentes, J.D., M. Lerdau, R. Atkinson, D. Baldocchi, J.W. Bottenheim, P. Ciccioli, B. Lamb, C. Geron, L. Gu, A. Guenther, T.D. Sharkey, W. Stockwell, Biogenic hydrocarbons in the atmospheric boundary layer: A review, Bulletin of the American Meteorological Society, 81, 1537-1575, 2000.

Fukui, Y. and P.V. Doskey, Air-surface exchange of nonmethane organic compounds at a grassland site: Seasonal variations and stressed emissions, J. Geophys. Res., 103, 13153-13168, 1998.

Gao, B.C. and Y.J. Kaufman, The MODIS Near-IR Water Vapor Algorithm. Products: MOD05_L2, MOD08_D3, MOD08_E3, MOD08_M3, ATBD Reference Number: ATBD-MOD-03 http://ltpwww.gsfc.nasa.gov/MODIS-Atmosphere/_docs/atbd_mod03.pdf, 1998.

Geron, C., A. Guenther, T. Sharkey, R. Arnts, Temporal variability in the basal isoprene emission factor, Tree Physiology, 20, 799-805, 2000a.

Geron C., R. Rasmussen, R. R. Arnts, A. Guenther, A review and synthesis of monoterpene speciation from forests in the United States, Atmos. Env., 34, 1761-1781, 2000b.

Geron C., P. Harley, A. Guenther, Isoprene emission capacity for US tree species, Atmos. Env., 35, 3341-3352, 2001.

Goldstein A.H., S.M. Fan, M.L. Goulden, J.W. Munger, S.C. Wofsy, Emissions of ethene; propene; and 1-butene by a midlatitude forest, J. Geophys. Res., 101, 9149-9157, 1996.

Griffin, R.J., D.R. Cocker III, J.H. Seinfeld, Estimate of global atmospheric organic aerosol from oxidation of biogenic hydrocarbons, Geophys. Res. Lett., 26, 2721-2724, 1999.

Guenther A., R.K. Monson, R. Fall, Isoprene and monoterpene emissions rate variability: observations with eucalyptus and emission rate algorithm development, J. Geophys. Res., 96, 10799-10808, 1991.

Guenther A., P. Zimmerman, P. Harley, R.K. Monson, R. Fall, Isoprene and monoterpene emission rate variability: model evaluations and sensitivity analyses, J. Geophys. Res., 98, 12609-12617, 1993.

Guenther A., C.N. Hewitt, D. Erickson, R. Fall, C. Geron, T. Graedel, P. Harley, L. Klinger, M. Lerdau, W.A. McKay, T. Pierce, B. Scholes, R. Steinbrecher, R. Tallamraju, J. Taylor, P. Zimmerman, A global model of natural volatile organic compound emissions, J. Geophys. Res., 100, 8873-8892, 1995.

Guenther, A. and A. Hills, Eddy covariance measurements of isoprene fluxes, J. Geophys. Res., 103, 13145-13152, 1998.

Guenther, A., B. Baugh, G. Brasseur, J. Greenberg, P. Harley, L. Klinger, D. Serca, L.Vierling, Isoprene emission estimates and uncertainties for the Central African EXPRESSO study domain, J. Geophys. Res., 104, 30625-30639, 1999a.

Guenther A., S. Archer, J. Greenberg, P. Harley, D. Helmig, L. Klinger, L. Vierling, M. Wildermuth, P. Zimmerman, S. Zitzer, Biogenic hydrocarbon emissions and landcover/climate change in a subtropical savannah, J. Phys. Chem. Earth, 24, 659-667, 1999b.

Guenther A., C. Geron, T. Pierce, B. Lamb, P. Harley, R. Fall, Natural emissions of non-methane volatile organic compounds; carbon monoxide, and oxides of nitrogen from North America, Atmos. Env., 34, 2205-2230, 2000.

Hansen, M.C., R.S. Defries, J.R.G. Townshend, R. Sohlberg, Global land cover classification at 1km spatial resolution using a classification tree approach, International Journal of Remote Sensing, 21, 1331-1364, 2000.

Hanson D.T., S. Swanson, L.E. Graham, T.D. Sharkey, Evolutionary significance of isoprene emission from mosses, American Journal of Botany, 86, 634-639, 1999.

Hanson, D. T. and T. D. Sharkey, Rate of acclimation of the capacity for isoprene emission in response to light and temperature, Plant Cell & Environment, 24, 937-946, 2001.

Harley P., M.E. Litvak, T.D. Sharkey, R.K. Monson, Isoprene Emission from Velvet Bean-Leaves – Interactions among Nitrogen Availability, Growth Photon Flux-Density, and Leaf Development, Plant Physiology, 105, 279-285, 1994.

Harley P., G. Deem, S. Flint, M. Caldwell, Effects of light; temperature and canopy position on net photosynthesis and isoprene emission from sweetgum (Liquidambar styraciflua) leaves, Tree Physiology, 16, 25-32, 1996.

Harley P., A. Guenther, P. Zimmerman, Environmental controls over isoprene emission in deciduous oak canopies, Tree Physiology, 17, 705-714, 1997.

Harley P., V. Fridd-Stroud, J. Greenberg, A. Guenther, P. Vasconcellos, Emission of 2-methyl-3-buten-2-ol by pines: A potentially large natural source of reactive carbon to the atmosphere, J. Geophys. Res., 103, 25,479-25,486, 1998.

Harley P.C., R.K. Monson, M.T. Lerdau, Ecological and evolutionary aspects of isoprene emission from plants, Oecolgia, 188, 109-123, 1999.

Harley, P., L. Otter, A. Guenther, J. Greenberg, Micrometeorological and leaf-level measurements of isoprene emissions from a southern African Savannah, J. Geophys. Res., In Press, 2002.

Hatakeyama, S., K. Izumi, T. Fukuyama, H. Akimoto, Reactions of ozone with a-pinene and b-pinene in air: yields of gaseous and particulate products, J. Geophys. Res., 94, 13013-13024, 1989.

Helmig D., L.F. Klinger, L. Vierling, C. Geron, P. Zimmerman, Biogenic Volatile Organic Compound Emissions (BVOCs) I. Identifications from Three Continental Sites in the U.S., Chemosphere, 38, 2163-2187, 1999.

Hoffmann, T., J.R. Odum, F. Bowman, D. Collins, D. Klockow, R.C. Flagan, J.H. Seinfeld, Formation of organic aerosols from the oxidation of biogenic hydrocarbons, J. Atm. Chem., 26, 189-222, 1997.

Jaenicke, R. and S. Mathias-Maser, Natural sources of atmospheric aerosol particles, in Precipitation Scavenging and Atmosphere-Surface Exchange, edited by S.E. Schwartz and W.G.N. Slinn, pp. 1617-1639, Hemisphere, Bristol, Pa, 1992.

Janson R., C. de Serves, R. Romero, Emission of isoprene and carbonyl compounds from a boreal forest and wetland in Sweden, Agricultural and Forest Meteorology, 98-99. 671-681, 1999.

Janson R. and C. de Serves, Acetone and monoterpene emissions from the boreal forest in northern Europe, Atmos. Env., 35, 4629-4637, 2001.

Kamens, R., M. Jang, C. Chien, K. Leach, Aerosol formation from the reaction of a-pinene and ozone using a gas-phase kinetics-aerosol partitioning model,. Environmental Science and Technology, 33, 1430 -1438, 1999.

Karl T., R. Fall, A. Jordon, W. Lindinger, On-line analysis of reactive VOCs from urban lawn mowing, Environmental Science and Technololgy, 35, 1926-1931, 2001a.

Karl, T., A. Guenther, A. Jordan, R. Fall, W. Lindinger, Eddy covariance measurement of biogenic oxygenated VOC emissions from hay harvesting, Atmos. Env., 35, 491-495, 2001b.

Karl, T., A. Guenther, C. Lindinger, A. Jordan, R. Fall, W. Lindinger, Eddy covariance measurement of oxygenated volatile organic compound fluxes from crop harvesting using a redesigned proton-transfer-reaction mass spectrometer, J. Geophys. Res., 106, 24157-24167, 2001c.

Karl, T., C. Spirig, P. Prevost, C. Stroud, J. Rinne, J. Greenberg, R. Fall, A. Guenther, Virtural disjunct eddy covariance measurements of organic compound fluxes from a subalpine forest using proton transfer reaction mass spectrometry, J. Atmos. Chem. Phys., 999-1033, 2002.

Karlik, J.F. and A.M. Winer, Measured isoprene emission rates of plants in California landscapes: comparison to estimates from taxonomic relationships, Atmos. Env., 35, 1123-1131, 2001.

Kavouras, I. G., N. Mihalopoulos, E. G. Stephanou, Formation of atmospheric particles from organic acids produced by the forests, Nature, 395, 683-686, 1998.

Kavouras, I.G., N. Mihalopoulos, E.G. Stephanou, Secondary organic aerosol formation vs primary organic aerosol emission: in situ evidence for the chemical coupling between monoterpene acidic photooxidation products and new particle formation over forests, Environmental Science and Technology, 33, 1028-1037, 1999.

Kerstiens, G. and M. Possell, Is competence for isoprene emission related to the mode of phloem loading?, New Phytologist, 152, 365-374, 2001.

Kesselmeier J., K. Bode, U. Hofmann, H. Müller, L. Schäfer, A. Wolf, P. Ciccioli, E. Brancaleoni, A. Cecinato, M. Frattoni, P. Foster, C. Ferrari, V. Jacob, J.L. Fugit, L. Dutaur,

V. Simon, L. Torres, Emission of short chained organic acids, aldehydes and monoterpenes from Quercus ilex L. and Pinus pinea L. in relation to physiological activities, carbon budget and emission algorithms, Atmos. Environ., 31, 119-133, 1997.

Kesselmeier J. and M. Staudt, Biogenic volatile organic compounds (VOC): An overview on emission; physiology and ecology, J. Atmos. Chem., 33, 23-88, 1999.

Kesselmeier J., Exchange of short-chain oxygenated volatile organic compounds (VOCs) between plants and the atmosphere: a compilation of field and laboratory studies, J. Atmos. Chem., 39, 219-233, 2001.

Kim, J.-C., Factors controlling natural VOC emissions in a southeastern US pine forest, Atmos. Env., 35, 3279-3292, 2001.

King, M.D., S.C. Tsay, S.E. Platnick, M. Wang, K.N. Liou,, Cloud Retrieval Algorithms for MODIS: Optical Thickness, Effective Particle Radius, and Thermodynamic Phase. Products: MOD06_L2, MOD08_D3, MOD08_E3, MOD08_M3, ATBD Reference Number: ATBD-MOD-05. http://ltpwww.gsfc.nasa.gov/MODIS-Atmosphere/_docs/atbd_mod05.pdf, 1997.

Kirstine, W., I. Galbally, Y.R. Ye, M. Hooper, Emissions of volatile organic compounds (primarily oxygenated species) from pasture, J. Geophys. Res., 103, 10605-10619, 1998.

Klinger, L.F., Li, Q.-J., Guenther, A.B., Greenberg, J.P., Baker, B., Bai, J.-H., Assessment of volatile organic compound emissions for ecosystems in China, J. Geophys. Res., 107, doi:10.1029/2001JD001076, 2002.

Knowlton, J. A., R. S. Martin, Biogenic Hydrocarbon, organic acid, and carbonyl emissions from desert shrubs, AWMA 92nd Annual Meeting, St. Louis, MO, 1999.

Koch, Stephan, R. Winterhalter, E, Uherek, A. Kollo, P. Neeb, G.K. Moortgat, Formation of new particles in the gas-phase ozonolysis of monoterpenes, Atmos. Env., 34, 4031- 4042, 2000.

Kreuzwieser, J., J.-P. Schnitzler, R. Steinbrecher, Biosynthesis of organic compounds emitted by plants, Plant Biology, 1, 149-159, 1999.

Kreuzwieser J., F. Kühnemann, A. Martis, H. Rennenberg, Urban W., Diurnal pattern of acetaldehyde emission by flooded poplar trees, Physiologia Plantarum, 108, 79-86, 2000.

Kreuzwieser J., F.J.M. Harren, L.-J. Laarhoven, I. Boamfa, S. te Lintel-Hekkert, U. Scheerer, C. Hüglin, H. Rennenberg, Acetaldehyde emission by the leaves of trees- correlation with physiological and environmental parameters, Physiologia Plantarum 113, 41-49, 2001.

Kubátová, A., R. Vermeylen, M. Claeys, J. Cafmeyer, W. Maenhaut, G. Roberts, P. Artaxo, Carbonaceous aerosol characterization in the Amazon basin, Brazil: Novel dicarboxylic acids and related compounds, Atmos. Env., 34, 5037-5051, 2000.

Kuhn, U., S. Rottenberger, T. Biesenthal, A. Wolf, G. Schebeske, P. Ciccioli, E. Brancaleoni, M. Frattoni, T.M. Tavares, J. Kesselmeier, Isoprene and Monoterpene emissions of Amazonian tree species during the wet season: direct and indirect investigations on controlling environmental functions, J. Geophys. Res., 107, LBA 38-1 – 38-13, 2002.

Lamb B., T. Pierce, D. Baldocchi, E. Allwine, S. Dilts, H. Westberg, C. Geron, A. Guenther, L. Klinger, P. Harley, P. Zimmerman, Evaluation of forest canopy models for estimating isoprene emissions, J. Geophys. Res., 101, 22787-22797, 1996.

Lehning, A., W. Zimmer, I. Zimmer, J.-P. Schnitzler,, Modeling of annual variations of oak (Quercus robur L.) isoprene synthase activity to predict isoprene emission rates, J. Geophys. Res., 106, 3157-3166, 2001.

Levis, S., C. Wiedinmyer, G. Bonan, A. Guenther, Simulating biogenic organic compound emissions in the Community Climate System Model,. Submitted to J. Geophys. Res., 2002.

Litvak M.E., F. Loreto, P. Harley, T.D. Sharkey, R.K. Monson, The response of isoprene emission rate and photosynthetic rate to photon flux and nitrogen supply in aspen and white oak trees, Plant Cell and Environment, 19, 549-559, 1996.

Lodge, J.P., P.A. Machado, J.B. Pate, D.C. Sheesley, A.F. Wartburg, Atmospheric trace chemistry in the American humid tropics, Tellus, 26, 250-259, 1974.

Loreto F; P. Ciccioli, A. Cecinato, E. Brancaleoni, M. Frattoni, D. Tricoli, Influence of environmental factors and air composition on the emission of alpha-pinene from Quercus ilex leaves, Plant Physiology, 110, 267-275, 1996.

MacDonald, R. C. and R. Fall, Detection of substantial emissions of methanol from plants to the atmosphere, Atmos. Env., 27A, 1709-1713, 1993.

Maki, L.R. and K.J. Willoughby, Bacteria as biogenic sources of freezing nuclei, J. Appl. Meteorol., 17, 1049-1053, 1978.

Martin, R. S., I. Villanueva, J.Y. Zhang, C.J. Popp, Nonmethane hydrocarbon, monocarboxylic acid, and low molecular weight aldehyde and ketone emissions from vegetation in central New Mexico, Environmental Science & Technology, 33, 2186-2192, 1999.

Matthias-Maser S. and R. Jaenicke, Examination of atmospheric bioaerosol particles with radii greater than 0.2 microns, J. Aeros. Res., 25, 1605-1613, 1995.

Moldanova and E. Ljungstro, Modeling of particles formation from NO_3 oxidation of selected monoterpenes, J. Aerosol Sci., 31, 1317 – 1333, 2000.

Monson R.K., P.C. Harley, M.E. Litvak, M. Wildermuth, A.B. Guenther, P.R. Zimmerman, R. Fall, Environmental and Developmental Controls Over the Seasonal Pattern of Isoprene Emission from Aspen Leaves, Oecologia, 99, 260-270, 1994.

Myneni, R.B., S. Hoffman, Y.Knyazikhin, J. Privette, J. Glassy, Y. Tian, Y. Wang, X. Song, Y. Zhang, G.R. Smith, A. Lotsch, M. Friedl, J.T. Morisette, P. Votava, R.R. Nemani, S.W. Running, Global products of vegetation leaf area and fraction absorbed PAR from one year of MODIS data, Remote Sensing Environment, 2002

Nemecek-Marshall M., R.C. MacDonald, J.J. Franzen, C.L. Wojociechowski, R.R. Fall, Methanol emissions from leaves, Plant Physiology, 108, pp. 1359-1368, 1995.

Nemeruyk, G.E., Migration of salts into the atmosphere during transpiration, Sov. Plant Physiol., 17, 560-566, 1970.

O'Dowd C. D., P. Aalto, K. Hmeri, M. Kulmala, T. Hoffman, Aerosol formation: Atmospheric particles from organic vapours, Nature, 416, 497- 498, 2002.

Olson, D., E. Dinerstein, E.D. Wikramanayake, N.D. Burgess, G.V.N. Powell, E.C Underwood, J.A. D'Amico, I. Itoua, H.E. Strand, J.C. Morrison, C.J. Loucks, T.F. Allnutt, T.H. Ricketts, Y. Kura, J.F. Lamoreux, W.W. Wettengel, P. Hedao, K.R. Kassem, Terrestrial ecoregions of the world: A new map of life on Earth, BioScience, 51, 933-938, 2001.

Otter, L., A. Guenther, C. Wiedinmyer, G. Fleming, P. Harley, J. Greenberg, Spatial and temporal variations in biogenic VOC emissions for Africa south of the equator, J. Geophys. Res., In Press, 2003.

Pétron, G., P. Harley, J. Greenberg, A. Guenther, Seasonal temperature variations influence isoprene emission, Geophys. Res. Lett., 28, 1707-1710, 2000.

Pinker, R.T., and I. Laszlo, Modeling surface solar irradiance for satellite solar irradiance applications on a global scale, J. Appl. Meteor., 31, 194-211, 1992a.

Pinker, R.T. I. and Laszlo, Global distribution of photosynthetically active radiation as observed from satellites, J. Climate, 5, 56-65, 1992b.

Preußer, Ch., S. Dech, P. Tungalagsaikhan, R. Steinbrecher, Development of leaf area index (LAI) distributions for Germany from NOAA/AVHRR NDVI satellite data, in: Proceedings EUROTRAC Symposium 1998, Vol 2, P. M. Borrell and P. Borrell (eds), WITpress, Southamton, 50-54, 1999.

Rasmussen R.A. and F.W. Went, Volatile Organic Material of Plant Origin in the Atmosphere. Read before the Academy, April 27, 1964.

Rasmussen, R., Isoprene: Identified as a forest-type emission to the atmosphere. Environmental Science and Technology, 4, 667-671, 1970.

Richter, K., R. Knoche, T. Schoenemeyer, G. Smiatek, R. Steinbrecher, Abschätzung der biogenen Kohlenwasserstoffemissionen in den neuen Bundesländern, Zeitschrift für Umweltchemie und Ökotoxikologie, 10, 319-324, 1998.

Rinne, H.J.I., A.B. Guenther, C. Warneke, J.A. de Gouw, S.L. Luxembouurg, Disjunct eddy covariance technique for trace gas flux measurements, Geophys. Res. Lett., 28, 3139-3142, 2001.

Rinne, H.J., A.B. Guenther, J.P. Greenberg, P.C. Harley, Isoprene and monoterpene fluxes measured above Amazonian rainforest and their dependence on light and temperature, Atmos. Env., 36, 2421-2426, 2002.

Roberts, G. C., M. O. Andreae, J. Zhou, P. Artaxo, Cloud condensation nuclei in the Amazon Basin: Marine conditions over a continent?, Geophys. Res. Lett., 28, 2807-2810, 2001.

Running, S.W., T.R. Loveland, L.L. Peerce, A vegetation classification logic based on remote sensing for using in global scale biogeochemical models, Ambio, 23, 77-81, 1994.

Running, S.W., T.R. Loveland, L.L. Pierce, R.R. Nemani, E.R. Hunt Jr., A remote sensing based vegetation classification logic for global land cover analysis, Remote Sens. Environ., 51, 39-48, 1995.

Schaab, G., R. Steinbrecher, B. Lacaze, R. Lenz, Assessment of long-term changes on potential isoprenoid emissions for a Mediterranean type ecosystem in France, J. Geophys. Res., 105, 28863-28874, 2000.

Schade, G. W., A.W. Goldstein, M.S. Lamanna, Are monoterpene emissions influenced by humidity?, Geophys. Res. Lett., 26, 2187-2190, 1999.

Schade, G. W., A. H. Goldstein, D.W. Gray, M.T. Lerdau, Canopy and leaf level 2-methyl-3-buten-2-ol fluxes from a ponderosa pine plantation, Atmos. Env., 34, 3535-3544, 2000.

Schuh G, A.C. Heiden, T. Hoffmann, J. Kahl, P. Rockel, J. Rudolph, J. Wildt, Emission of volatile organic compounds from sunflower and beech: dependence on temperature and light intensity, J. Atm. Chem., 27, 292-318, 1997.

Seufert, G., J.G. Bartzis, T. Bomboi,, P. Ciccioli,, S. Cieslik, R. Dlugi., P. Foster, C.N. Hewitt, J. Kesselmeier, D. Kotzias, R. Lenz, F. Manes, R. Perez-Pastor, R. Steinbrecher, L. Torres, R. Valentini, B. Versino, An overview of the Castelporziano experiments, Atmos. Env., 31, 5-17, 1997.

Sharkey T.D., F. Loreto, C.F. Delwiche, High carbon dioxide and sun/shade effects on isoprene emission from oak and aspen tree leaves, Plant Cell and Environment, 14, 333-338, 1991.

Sharkey T.D., E.L. Singsaas, P.J. Vanderveer, C. Geron, Responses of Isoprene and Photosynthesis to Temperature and Light in a North-Carolina Forest, Plant Physiology. 108, 60-60, 1995.

Sharkey T.D., E.L. Singsaas, P.J. Vanderveer, C. Geron, Field measurements of isoprene emission from trees in response to temperature and light, Tree Physiology. 16, 649-654, 1996.

Sharkey, T. D., E. L. Singsaas, M.T. Lerdau, C.D. Geron, Weather effects on isoprene emission capacity and applications in emission algorithms, Ecological Applications, 9, 1132-1137, 1999.

Sharkey, T.D., X. Chen, S. Yeh, Isoprene increases thermotolerance of Fosmidomycin-fed leaves, Plant Physiology, 125, 2001-2006, 2001.

Simoneit, B.R.T. and M.A. Mazurek, Organic matter of the troposphere - II. Natural background of biogenic lipid matter in aerosols over the rural western United States, Atmos. Env., 16, 2139-2159, 1982.

Simpson D., W. Winiwarter, G. Borjesson, S. Cinderby, A. Ferreiro, A. Guenther, C. Hewitt, R. Janson, M.A. Khalil, S. Owen, T.E. Pierce, H. Puxbaum, M. Shearer, U. Skiba, R. Steinbrecher, L. Tarrason, M. Oquist, Inventorying emissions from nature in Europe, J. Geophys. Res., 104, 8113-8152, 1999.

Snyder, W.C., Z. Wan, Y. Zhang, Y.Z. Feng, Classification-based emissivity for land surface temperature measurement from space, Intl. J. of Remote Sensing, 19, 2753-2774, 1998.

Steinbrecher, R. and K. Hauff, Isoprene and Monoterpene emission from Mediterranean Oaks. In: Proceedings EUROTRAC Symposium 1996, Borrell P. M., Borrell P., Kelly K., Cvitas T., Seiler W. (eds), Volume 2: Emissions, Deposition, Laboratory Work and Instrumentation. Computational Mechanics Publications, Southampton UK, 229-233, 1996.

Steinbrecher R. Hauff K., Hakola H., Rössler J., A revised parametrisation for emission modeling of isoprenoids for boreal plants, in Biogenic VOC Emission and Photochemistry in the Boreal Regionas of Europe. Th. Laurila, V. Lindfors (eds), European Communities 1999 EUR 18910 EN, ISBN 92-828-6990-3, 29-43, 1999.

Steiner, A., Luo, C., Huang, Y., Chameides, W.L., Past and present-day biogenic volatile organic compound emissions in East Asia, Atmos. Env., 36, 4895-4905, 2002.

Stewart, H.E, Hewitt, C.N., Bunce, R.G.H., Steinbrecher, R., Smiatek G., Schoenemeyer, T., A highly spatially and temporally resolved inventory for biogenic isoprene and monoterpene emissions - model description and application to Great Britain, Submitted to J. Geophys. Res., 2002.

Talbot, R.W., M.O. Andreae, H. Berresheim, P. Artaxo, M. Garstang, R.C. Harriss, K.M. Beecher, S.M. Li, Aerosol chemistry during the wet season in central Amazonia: The influence of long-range transport, J. Geophys. Res., 95, 16955-16970, 1990.

Tingey D.T., M. Manning, L.C. Grothaus, W.L. Burns, Influence of light and temperature on monoterpene emission rates from slash pine, Plant Physiology, 65, 797-801, 1980.

Tommeras, B.A., Specialization of the olfactory receptor cells in the bark beetle Ips typographus and its predator Thanasimus formicarius to bark beetle pheromones and host tree volatiles, Journal of Comparative Physiology A. 157, 335-341, 1985.

Vali, G., M. Christensen, R.W. Fresh, E.L. Galyvan, R.L. Maki, R.C. Schnell, Biogenic ice nuclei, II, bacterial sources, J. Atmos. Sci., 33, 1565-1570, 1976.

Wan, Z., "MODIS Land-Surface Temperature Algorithm Theoretical Basis Document (LST ATBD) Version 3.3, April 1999", http://modis-land.gsfc.nasa.gov/pdfs/atbd_mod11.pdf, 1999.

Wang, K.Y. and D.E. Shallcross, Modeling terrestrial biogenic isoprene fluxes and their potential impact on global chemical species using a coupled LSM-CTM model, Atmos. Env., 34, 2909-2925, 2000.

Warneke, C., T. Karl, H. Judmaier, A. Hansel A. Jordan W. Lindinger, P.J. Crutzen, Acetone, methanol, and other partially oxidized volatile organic emissions from dead plant matter by a biological processes: Significance for atmospheric HO_x chemistry, Global Biochem Cyc., 13, 9-17, 1999.

Warneke, C., S. L. Luxembourg, J.A. de Gouw., H.J.I. Rinne, A. Guenther, R. Fall, Atmospheric emissions of volatile organic compounds from undisturbed and harvested alfalfa, Submitted to J. Geophys. Res., 2002.

Went, F.W., Blue hazes in the atmosphere, Nature, 187, 641-643, 1960.

Yu, J., D.R. Cocker III, R.J. Griffin, R.J. Flagan, J.H. Seinfeld, Gas-phase ozone oxidation of monoterpenes: Gaseous and particulate products, J. Atm. Chem., 34, 207-258, 1999.

Zeng, X.B., R.E. Dickinson, A. Walker, M. Shaikh, R.S. Defries, J.G. Qi, Derivation and evaluation of global 1-km fractional vegetation cover for land modelling, J. Appl. Met., 39, 826-839, 2000.

Zimmerman P., J. Greenberg, C.E. Westberg, Measurements of atmospheric hydrocarbons and biogenic emissions in the Amazon boundary layer, J.Geophys. Res., 93, 1407-1416, 1988.

Zhou, J., E. Swietlicki, H.C. Hansson, P. Artaxo, P., Sub-micrometer aerosol particle size distribution and hygroscopic growth measured in the Amazonian rain forest during the wet season, J. Geophys. Res., 107, LBA 22-1 – 22-10, 2002.

Nitrogen Emissions from Soils

Laurens Ganzeveld, Changsheng Li, Laura Cárdenas, Jane Hawkins, Grant Kirkman

1. INTRODUCTION

The oxidized nitrogen species nitric oxide (NO), nitrogen dioxide (NO_2) and nitrous oxide (N_2O) play an important role in tropospheric and stratospheric chemistry. NO and NO_2, collectively known as NO_x, regulate the tropospheric photochemical production of ozone and the abundance of the hydroxyl radical (OH, the main oxidant of the atmosphere), whereas N_2O influences stratospheric ozone chemistry (Crutzen, 1974). Moreover, N_2O is a long-lived greenhouse gas. The reduced nitrogen species ammonia (NH_3) is involved in rain and cloud water chemistry, aerosol formation, it acidifies ecosystems (e.g. van Breemen et al., 1982) and plays a crucial role in the tropospheric sulphur cycle (e.g. Bouwman et al., 1997). Nitrogen species are also relevant for the biogeochemical cycling of terrestrial nutrients, e.g. by regulating the Net Primary Production (NPP) in a future carbon dioxide-enhanced climate (Holland et al., 1997). A large input of nitrogen via wet and dry deposition can also result in forest decline as a consequence of nitrogen saturation or acidification of the soil (Bouwman et al., 1997)

Soils are an important source of atmospheric nitrogen mainly due to a net microbial production of NO, N_2O (and N_2) in the nitrification and denitrification process, whereas NH_3 production in the soil is mainly controlled by decomposition and ammonification (Sutton, 1990). An additional contribution to the soil NO emission flux is the abiotic process of chemical decomposition of nitrite in acidic soils, the so-called chemodenitrification (e.g. van Cleemput and Baert, 1984). Nitrification, denitrification and chemodenitrification depend on biogeochemical and physical properties of the soil, e.g. microbial species, soil texture, soil water, pH, redox-potential and nutrient status. Soil emission fluxes are also tightly linked to land use management through the impact of the application of natural and synthetic fertilizers, tillage, irrigation, compaction, planting and harvesting (e.g. Frolking et al., 1998).

The focus of this chapter is to comment on soil nitrogen flux measurements, their emission controlling mechanisms, and then finally to discuss the development and use of process-based models to predict soil nitrogen emissions, including the role of canopy processes.

Of particular interest is the extent to which land cover and land use changes affect biogeochemical processes, and therefore soil nitrogen exchange in the future. Climate change and increasing anthropogenic emissions could have significant effects on global atmospheric chemistry, and thereby on emission and deposition of nitrogen. To test our understanding of these complex interactions, process-based models that explicitly simulate the dependence of nitrogen emission fluxes on environmental parameters and that can be implemented in large-scale models, e.g. global chemistry and climate models, are required. We address this in detail with a concise review of soil nitrogen exchange models, their benefits and areas for improvement.

2. GLOBAL NO_x, N_2O AND NH_3 SOIL EMISSION INVENTORIES

In order to appreciate the role which major sources of gaseous emitted soil nitrogen play in global chemistry, it is necessary to comment on global soil nitrogen emission inventories that have been compiled recently, and the complexity involved in arriving at such numbers. The regional estimation of soil nitrogen emissions is only possible through use of some form of spatial extrapolation method (Stewart et al., 1989). The simplest approach is to conduct field measurements of soil N flux randomly, and to estimate the area flux based upon the random sample. An improvement might be to sample strategically, based on sub-regional classes (Davidson and Kingerlee, 1997), and a further improvement might be to estimate regional emissions according to parameters that control trace gas fluxes such as biome, soil water or vegetation status (Matson et al., 1989). The large number of samples that is required can be prohibitive, dependent on the level of confidence deemed acceptable (Folorunso and Rolston, 1984). Since field-sampling experiments are expensive and time consuming, the number of field measurements made is limited. A compromise is therefore to use a selection of the most significant environmental predictors of soil N flux for a particular region. When these predictors are used in semi-empirical model schemes to scale(spatial extrapolation), a minimum number of strategically sampled field measurements is taken to arrive at an estimate of regional soil N flux (Kirkman et al., 2001, 2002c).

An alternative approach, the so-called top-down constraint, uses the global sink strength, calculated as the atmospheric burden/chemical lifetime plus the rate of the increase in the atmospheric burden, to infer the global emission flux. Unfortunately, this top-down constraint cannot be applied for NO_x and NH_3 due to their short lifetimes and the complexity of their global sinks (e.g. wet and dry deposition, and surface chemistry).

2.1. NO_x

Soil NO emissions control NO_x budgets in remote and rural areas, while the fossil fuel source of 20-25 Tg N yr^{-1} (Logan, 1983; Hameed and Dignon, 1988; Levy and Moxim, 1989; Delmas et al., 1997; see also Chapter 1) dominates the NO_x budget in industrialized areas. Several reviews have been published to account for estimates of the soil NO sources. Early estimates reported by Galbally and Roy (1978) were about 10 Tg N yr^{-1}. Later, Davidson (1991) compiled previously published data and derived an estimate of 20 Tg N yr^{-1}. Yienger and Levy (1995) arrived at an estimate of 5.5 TgN yr^{-1} based on an empirical model that accounts for different biomes, pulsing and the effect of the canopy uptake. Their NO soil emission flux of 11 TgN yr^{-1} (below canopy) is slightly larger compared to estimate of 9.7 TgN yr^{-1} by Potter et al. (1996), who used an ecosystem modelling approach by integrating remote sensing, climate, vegetation and soil datasets. A purely measurement-based source inventory, based on major global biomes, was derived by Davidson and Kingerlee (1997), who estimated a total global soil NO emission of 21 Tg N yr^{-1}, with major contributions from temperate and tropical cultivated land, chaparral/thorn forest and tropical savannah/woodland. The model-based estimates of the soil NO emission flux by Yienger and Levy (1995) and Potter et al. (1996), and the measurement-based estimates by Davidson (1991) and Davidson and Kingerlee (1997), disagree by a factor of 2 overall. This discrepancy suggests that the role of some important soil emission control factors such as fertilization and pulsing might not be realistically represented in the models, which is discussed at length below (Hutchinson et al., 1997).

2.2. N_2O

Tropical and agricultural soils are considered to be the most dominant sources of N_2O with a total source estimate of 6.6 Tg N yr^{-1} (IPCC, 2001), to which tropical soils contribute 4 Tg N yr^{-1}. The global source estimate, using the top-down approach, is ca. 16.4 Tg N yr^{-1} (IPCC, 2001), which is similar to the estimate given by Mosier et al. (1998) of 17.7 Tg N yr^{-1}. Estimates of the contribution by agricultural soils range from 24 % (Mosier et al., 1998) to 37% (Beauchamp, 1997), which is consistent with the IPCC (2001) estimate of 4.2 Tg N yr^{-1}. Early N_2O inventory estimates considered emissions from fertilization only but more recent inventories include contributions from biological fixation (IPCC, 1995), N leaching, runoff, NO_x and NH_3 volatilisation (Mosier et al., 1996), emissions from animal waste, N deposition and human consumption of crops and sewage treatment (Mosier et al., 1998).

2.3. NH_3

A global NH_3 emission inventory of 54 Tg N yr^{-1}, with an estimated uncertainty range of 40-70 Tg N yr^{-1}, has been provided by Bouwman et al. (1997). The emissions by natural soils are about 2.4 Tg N yr^{-1}, although there is a large range uncertainty with these estimates (1-10 Tg N yr^{-1}). Agriculture is the most important source of NH_3 and includes sources as excreta from domestic animals (Freney et al., 1983; Sommer and Hutchins, 1997; Bouwman et al., 1997; Misselbrook et al., 2000) and application of fertilizers, adding up to about 30.6 Tg N yr^{-1}. Matthews (1994) and Bouwman et al. (1995) compiled inventories of the global distribution of the use of chemical fertilizers associated with the distribution and intensity of agriculture activities.

2.4. Estimate uncertainties

A considerable amount of uncertainty prevails over global inventories of soil nitrogen fluxes. A major source of this uncertainty derives from the difficulties faced in quantifying them at relatively larger spatial and longer temporal scales compared to the representative scales of the processes that control the nitrogen emissions (Davidson and Kingerlee, 1997; Holland and Larmarque, 1997; Ludwig et al., 2001). The soil nitrogen cycling is governed by a complex interplay of soil processes at scales smaller than 1 cm, which are in turn governed by climate at the synoptic scale and land use at inter-annual scales. Complete information on soil microbial populations, species, metabolism, and environmental responses are at present not forthcoming and probably will remain so due to the shear multitude of biota found in a single gram of soil (Conrad, 1996). Soil NO, for example, is largely produced by the nitrification and denitrification process, which are influenced by soil environmental conditions (Conrad, 1996) such as soil temperature, moisture, fertility, vegetation cover, and fire. These environmental controlling factors vary spatio-temporally, often in contradiction to one another. For example, a 10 °C rise in soil temperature, which can occur within hours to days, produces a 2-5 fold increase in NO emission rates (Williams and Fehsenfeld, 1991; Valente and Thornton, 1993) whereas soil moisture, which controls soil oxygen, substrate and gas transport and thereby soil microbial processes, varies seasonally with rainfall. On the other hand, short-term changes in soil moisture can influence soil behaviour so dramatically that NO production can revert to NO consumption in a matter of hours after a rainfall event (Davidson, 1991) or it can result in large "pulses" of NO (Davidson, 1992b; Meixner et al., 1999).

Uncertainty in the classification of land use and its impact on the soil N emission fluxes, through the application of fertilizers and changes in soil properties and vegetation cover, even further complicates the compilation of large-scale inventories of soil nitrogen emissions. Land use practices, such as deforestation, which can alter soil fertility, can alter NO emissions at long temporal scales (Davidson et al., 2000). One of the specific land use practices, biomass burning, often used to eradicate cut and felled vegetation in the subtropics and tropics, results in a temporary reduction in the plant and microbial sink of soil inorganic nitrogen by removing biomass and dead litter with high C:N ratios. This results in favourable conditions for biotic and abiotic NO production, and recent observations have shown that significant amounts of NO can be produced after these fire events (e.g. Verchot et al., 1999).

An additional uncertainty in the source estimate of NO_x and NH_3 is the role of vegetation processes (e.g. Matson, 1997). Vegetative biomass, which is largely related to land use, affects NO emissions in two ways: (a) directly via dry deposition of the primary emitted species or the reaction products of fast chemical transformations occurring within the canopy and (b) indirectly by controlling the input of soil nutrients from the decomposition of litterfall, roots and organic matter. Studies by Yienger and Levy (1995) and Bouwman et al. (1997) indicate that, applying first-order estimates of the removal of NO_2 and NH_3 by dry deposition, lead to reductions of about 50% of the soil emissions. However, these first-order estimates are associated with a large uncertainty due to the complexity of the interactions between the processes involved, emissions, dry deposition, chemistry and turbulence, and the lack of observations to evaluate models that explicitly account for these canopy-interactions. More details on the role of canopy processes for soil nitrogen emissions are discussed in sections 4.2 and 5.3.

3. MEASUREMENT TECHNIQUES

The availability of accurate inventories and real time data for use in models is very much affected by the spatial and temporal variability of the fluxes. This, clearly suggested by the disagreement between different inventories, makes predictions of future fluxes very difficult. There is an urgent need for improved methods to measure these fluxes at different spatial scales in order to understand the processes and the variables that affect these fluxes. In this section we outline the main methods that are found in the literature with the aim of giving some suggestions for improvement.

The methods used to determine nitrogen species emitted by soils can be classified into three main groups:
- Laboratory scale soil incubation techniques, which operate at fine scale (from microbial scale up to a few cm soil) and are adequate for studying processes responsible for the production and removal of gases in the soil. They are perhaps the best means of evaluating the parameters that control the emissions.
- Field scale enclosure systems, which could be considered appropriate for local scale (~ 0.1m up to some meters soil) studies and allow comparative studies of the effects of soil management on the emissions.
- Micrometeorological techniques, which operate at the field and larger scales (up to some km) and can be more adequate for regional studies. These techniques depend on the homogeneity of the surface and turbulent mixing. The impact of soil management on the fluxes can also be evaluated, but large areas of land are required.

Laboratory techniques have involved the analysis of soil cores, either intact or repacked with or without the use of inhibitors. Flow-through incubation systems, where a gas or gas mixture is continuously circulated above the soil surface and emissions of gases can be analysed either by manual or automated sampling, have been used in the laboratory (Galsworthy and Burford, 1978; Wickramasinghe et al., 1978; Hutchinson and Andre, 1989; Hutchinson et al., 1993; Swerts et al., 1996; Rudolph et al., 1996a; van Dijk et al., 2001, 2002). This method has been used to measure NO fluxes (Rudolph et al., 1996a), NO production and consumption (Gödde and Conrad 1998) and N_2O fluxes (Parkin et al., 1984; Swerts et al., 1996; Scholefield et al., 1997ab).

Besides NO_x, NH_3 and N_2O, the denitrification process also produces nitrogen (N_2) Consequently, measurements of N_2 are carried out in order to study denitrification (Scholefield et al., 1997a). The N_2O/N_2 ratio has been found to be a useful parameter to study the mechanism of denitrification (Swerts et al., 1996), and the influence of several soil factors. It can have a considerable scope for minimising N_2O emission from denitrification by management of the soil control factors. For example, measurements by Jambert et al. (1997) showed that 50% of the total nitrogen emission of an agricultural soil was released as N_2, and the rest as NO, N_2O and a very small amount of NH_3. This depends on the applied fertiliser and the environmental conditions (Lockyer, 1984; Misselbrook et al., 1998). A few studies that include N_2 measurements are found in the literature, but since it is not easy to measure N_2 (Swerts et al., 1995; Scholefield et al., 1997a), inhibitors of the reduction of N_2O to N_2 are used to determine N_2 indirectly. Field and laboratory measurements have included acetylene inhibition (Yoshinary et al., 1977, Ryden et al., 1987). This can however affect other processes such as nitrification (Klemedtson et al., 1988; Hatch et al., 1990), promote extra reduction of nitrate (Swerts et al., 1995; Bollmann and Conrad 1997, 1998) and could be a source of carbon (Yeomans and Beauchamp 1982). Tiedje et al. (1984) present a review on methods to study denitrification and recent work by Del Grosso et al. (2000) has provided new insights into ways in which denitrification can be reasonably modelled.

Flow-through systems are particularly useful for N_2 analysis, and the removal of the soil atmosphere can be carried out with inert gases such as argon (Stefanson and Greenland 1970) or helium (Scholefield et al., 1997a). In the latter, this technique allows simultaneous measurement of N_2O and N_2. The technique has been greatly improved allowing continuous measurements from incubated soil cores with low detection limits. A detailed description of this system will be given in Cárdenas et al. (2003). Results of the application of nitrate and glucose on a grassland soil in the southwest of the UK showed that a total of about 30 kg N ha^{-1} were emitted as N_2O+N_2. This represents approximately 60 % of the applied N with 50.4 % emitted as N_2O and 10 % as N_2. With this system, the controlling factors of the denitrification process can be studied with the aim of defining the conditions that favour N_2 production and reduce N_2O fluxes into the atmosphere.

Chambers are used to measure fluxes in the field but require several replicates due to the large spatial variability in the soil (Velthof and Oenema 1995, Velthof et al., 1996). For NO_x measurements, the open chamber (also classed as dynamic chambers) technique is generally used with ambient air flowing continuously through the chamber, and the lid is only covered for few minutes (Sanhueza et al., 1990; Cárdenas et al., 1993; Valente and Thornton 1993; Williams and Davidson, 1993; Serca et al., 1994; Skiba et al., 1994; Harrison et al., 1995; Yamulki et al., 1995, Gut et al., 2002, Kirkman et al., 2002a). Static chambers, in which accumulation of the gas produced is allowed inside the enclosed chamber, are generally used for N_2O measurements (Sanhueza et al., 1990; Serca et al.; 1994, Skiba et al., 1994). The effect of the increasing concentration of gas inside the chamber on the diffusion gradient must be considered in the calculation of the emission (Smith and Arah, 1992), since underestimation of the fluxes is possible (Rolston et al., 1975). Skiba et al. (1992) used open chambers for N_2O measurements but trapped the gas produced on a molecular sieve for a few hours. The trap had to be desorbed later for analysis. They also used the closed chamber technique and found 95 % agreement between the two methods. However, the data frequency was low and there was a large time lag between the two methods.

There are also dynamic closed systems, where the gas is re-circulated from the gas analyser back to the chamber (Rochette et al., 1992; Yamulki and Jarvis, 1999). A comparison of a static versus a dynamic closed system for soil respiration measurements gave lower values with the former system (Rochette et al., 1992). Ambus et al. (1993) on the other hand found good agreement between closed and open chambers for N_2O measurements.

Wind tunnels have been used to measure NH_3 fluxes in the field (Lockyer, 1984), especially to identify the factors that control these fluxes (Misselbrook et al., 1998). Chambers have also been used for NH_3 measurements (McGarity and Rajaratnam 1973).

The micrometeorological methods integrate fluxes over a large scale, depending on the measurement height and turbulence, and only partly reflect the soil emission fluxes, dependent on the role of other processes involved along the transfer pathway from the soil to the measurement height. In contrast to chamber and laboratory based techniques, micro-meteorological techniques also reflect the effect of turbulent transport, advection and other production or removal processes such as dry deposition, emissions and chemical transformations, and are not affected by the artificial environments produced by chamber systems (Kirkman et al., 2002a). Among the micro-meteorological measurement methods, eddy-correlation, eddy-accumulation or flux-gradient measurement techniques are found in the literature combined with spectroscopic and GC techniques (Galle et al., 2000). Jambert et al. (1997) and Rummel et al. (2002) compared NO flux measurements by a micro-meteorological method (aerodynamic technique, the flux being inferred from the observed gradient) with a chamber technique. Jambert et al. (1997) obtained a difference of a factor of 4 when comparing the chamber with an aerodynamic technique with the chamber giving the lower values. This difference might be due to the different spatial scales reflected by the chamber and the micro-meteorological measurements, since it is generally thought that chamber measurements are quite solid.

Use of either chamber or micro-meteorological methods has some general advantages and disadvantages. The chambers are cheaper but can introduce a large error due to the already mentioned extrapolation to a larger scale. They are useful when experiments are carried out to compare different soil management regimes. Increasing the number of replicates improves the confidence of the extrapolation, but this is labour-intensive. Chambers also allow for the continuous measurement of diurnal canopy and boundary layer resistances normally not facilitated by micrometeorological methods. The micrometeorological techniques are adequate for large-scale purposes, but they are expensive and require certain characteristics for the selection of the sampling location. It is also strictly valid for non-reactive trace gases only as it has to consider chemical reactions/transformations of these trace gases during the turbulent transport between the reference height and the surface (e.g. Vilá-Guerrau de Arellano and Duynkerke, 1992).

The analytical techniques used to measure NO_x are mainly NO/O_3 chemiluminescence detection for NO determination (Williams and Davidson 1993, Skiba et al., 1994) and NO to NO_2 conversion followed by luminol chemiluminescence detection for NO_x and NO_2 determination (Williams and

Davidson, 1993; Serca et al., 1994; Rudolph et al., 1996a, Veldkamp et al., 1998). Williams and Davidson (1993) made a comparison between the two methods, where the conversion NO to NO_2 was carried out with a CrO_3 converter. They observed reasonable agreement between the two techniques, although there was a lower response from the NO/NO_2 conversion method possibly due to the presence of water vapour.

N_2O measurements are mostly done by gas chromatography (GC) electron capture detection (ECD) (Hutchinson et al., 1993; Serca et al., 1994; Skiba et al., 1994; Veldkamp et al., 1998). Photo-acoustic infrared spectrometry has also been used combined with the chamber technique (Yamulki and Jarvis, 1999) and was compared to the GC technique showing good agreement, but with slightly larger fluxes from the IR technique. Other techniques used for N_2O measurements are thermal conductivity detection (TCD) GC (Ryden et al., 1987) and tunable diode laser spectroscopy, TDLS (Wienhold et al., 1994).

Measurements of NH_3 are carried out by transferring the gas to an absorbing medium for further analysis. Continuous measurements can be performed by a continuous-flow system: examples are the trapping of NH_3 on orthophosphoric acid followed by analysis by the indophenol blue method (Lockyer, 1984; Searle, 1984), or the collection of ammonia on sulphuric acid and addition of sodium hydroxide to measure the change of pH (Lockyer, 1984). Chemical reaction and fluorimetric analysis combined with continuous flow analysis are also reported in the literature (Genfa and Dasgupta, 1989; Fuhrer et al., 1993). Diffusive samplers have been used with diffusion through a static air column, or permeation through a membrane into a static layer of air (Kirchner et al., 1999). Diffusion denuders are based on the principle that the wall of the tube acts as a sink for the gas with the coating being phosphorous or tungstic acid for ammonia (Dasgupta, 1984). Kirchner et al. (1999) carried out a comparison of ten diffusive samplers and a denuder showing an overall uncertainty between 16 to 147 %. Diffusion scrubbers have also been developed where a membrane is used to transfer the flowing gas into a collecting liquid that flows outside the membrane for further analysis (Dasgupta, 1984). Inter-comparisons of ammonia measurement methods are found in the literature (Sutton et al., 1997).

Berges and Crutzen (1996) report in-situ measurements of NH_3 by FTIR (Fourier Transform InfraRed) spectroscopy or Dräger test tubes. The latter is based on the reaction of ammonia with acid to give a change of colour. They did not get good agreement between both methods. Galle et al. (2000) found on the other hand that the FTIR method was valuable for its area integrating capabilities and high time resolution, although restricted on the detection limits for measurements of ammonia. They expect the system to be improved in terms of sensitivity and operation.

Measurements of N_2 are carried out by thermal conductivity detection (TCD) (Wickramasinghe et al., 1978; Swerts et al., 1995), but the limits of detection restrict the data to measurements immediately after the addition of fertilizer (Scholefield et al., 1997b). A helium ionisation detector has been used reaching very low detection limits (Wentworth et al., 1992).

Isotopes have not only been used to establish the sources and sinks in the atmosphere by identifying the stratospheric/tropospheric source and natural versus anthropogenic origin (Yoshida and Toyoda, 2000). They have also been used to identify gas evolution linked to added fertilizer (Rolston et al., 1975), and to study the mechanisms of production of N_2O in soils (Pérez, 2000). In addition, Leffelaar and Wessel (1988) conducted laboratory incubation studies to identify the kinetics of the sequential reactions of denitrification. Their results provide a sound basis for understanding the relative abundances of NO, N_2O and N_2 produced from denitrification, which has been incorporated in process-based models (see also section 5.3).

4. MECHANISMS

4.1. Nitrogen gas production processes in soil

This section provides a brief overview of the primary processes (biotic and abiotic) within the soil that result in the production and consumption of the gaseous forms of N. Other pathways of the nitrogen cycle, including the removal of soil N through leaching or plant uptake, and contribution to soil N through animal excretory waste products (dung and urine), electrical fixation by lightning, dry and wet deposition and application of inorganic and organic fertilizers, will not be discussed in this chapter.

Nitrogen (N) is required by all forms of life since it is an essential element of proteins and nucleic acids. Despite the relative abundance of atmospheric N_2, the triple bond between the two N atoms makes the molecule virtually inert and therefore unavailable for use by most organisms. In order for molecular N to be used for growth, it must first be "fixed" in the form of ammonium (NH_4) or nitrate (NO_3).

Only a few soil prokaryotes (mostly bacteria, e.g. rhizobia, azotobacter, azospirillium) are capable of fixing N into other forms (N-fixation). These bacteria are either free-living, or form symbiotic associations with plants or other organisms (e.g. legumes, termites) and use the enzyme nitrogenase to catalyse the conversion of dinitrogen (N_2) to NH_4 (Postgate, 1982; Sprent and Sprent, 1990).

Heterotrophic organisms require organic compounds as a source of energy and carbon (C), and are able to assimilate these compounds through the decomposition of organic matter (Dickinson and Pugh, 1974). This decomposition process is known as mineralization or ammonification and the resultant end product is NH_4^+. The energy produced from decomposition enables the decomposer organisms to assimilate NH_4^+ into proteins (immobilisation).

Nitrification is generally defined as the microbiological oxidation of ammonia (NH_3) or NH_4^+ to NO_3^- (Bremner and Blackmer, 1981) by a group of organisms known as nitrifiers, and is believed to be the major soil contributor to NO_x emissions (Anderson and Levine, 1986). The NO_3^- produced may be used as substrate by another group of organisms (denitrifiers) in an anaerobic process known as denitrification.

When soils are relatively dry, nitrification is the main contributor to N_2O emission, since the conditions favour mineralization and hence the production of NH_4^+ as an available substrate for nitrifiers. There is also a higher diffusion rate of N_2O from the soil to the atmosphere in a drier soil. However, as soils become waterlogged, the conditions favour the production of nitrous oxide from the anaerobic process of denitrification, which is the term given to the sequential reduction of N-oxides with molecular N_2 as final product. The denitrification process can therefore be considered as both producer and consumer of the two gases. Gross soil nitrification rates (200-1000 kg N ha^{-1} yr^{-1}) are usually higher than denitrification rates (< 100 kg N ha^{-1} yr^{-1}) (Li et al., 2000).

Ammonia volatilisation is the term given to the process whereby NH_3 is produced from NH_3 in solution and is returned to the atmosphere in a gaseous form. The soil gas phase concentration of NH_3 is proportional to NH_3

concentration in the aqueous phase (Glasstone, 1946; Sutton et al., 1993). The main sources of NH_3 are from the biological decomposition of organic materials, animal excretory products, plants and application of inorganic and organic fertilizers.

4.1.1. Nitrogen fixation

The mechanism of biological N-fixation is very complex, and the reactions are not completely understood. A detailed account of the current knowledge may be found in Postgate (1987). Briefly, N fixing bacteria use the enzyme nitrogenase which consists of a complex of two coproteins, one containing iron (Fe), the other containing Fe plus molybdenum (Mo). Both coproteins must be present for nitrogenase to function. Altogether 8 electrons and 16 ATP are needed to reduce N_2 to NH_3, and are indicative of the high-energy input required. In case of symbiotic organisms (e.g. rhizhobia) that live in the root nodules of plants such as legumes, the energy is obtained indirectly from the energy produced from photosynthesis in the symbiont plant. Free-living bacteria (e.g. cyano-bacteria) obtain their energy by the direct photosynthetic route and chemical reactions. The NH_3 produced is assimilated into amino acids and subsequently into proteins.

One of the major factors affecting N-fixation is that nitrogenase is susceptible of high O_2 concentrations. This does not present a problem for N fixers that live in anaerobic conditions. In contrast, a variety of ways of protecting the enzyme have been developed by N fixers that live in aerobic conditions. Some of these include the formation of protective barriers such as thickened cell walls, depletion of cellular O_2 through high respiratory activity and the rapid turnover in production of nitrogenase to counteract the inactivation caused by exposure to O_2 (Gallon, 1981).

High levels of soil N limit the rate of N-fixation, since N fixing organisms will preferentially use soil N before they start to fix atmospheric N, as it is an energetically more efficient method of N assimilation. In particular, NH_4^+ produced from N-fixation can have a repressive effect on the production of nitrogenase, as it affects the coding for the production of the enzyme. High available soil NO_3^- represses the development of root nodules and hence sites for symbiotic bacteria, although the mechanisms of inhibition are not well understood. One possibility is that available soil C becomes a limiting factor (Paul and Clark, 1989).

4.1.2. Mineralization

According to Parsons and Tinsley (1975) and Stevenson (1982), at least 90% of the N occurring in the surface layer of most soils is in organic form. Soil organic N primarily results from the products from microbial decomposition of plant and animal residues and soil microorganisms. The majority of soil N is in proteinaceous structures; however, other organic N forms include nucleic acids and constituents of microbial cell walls such as peptidoglycans and chitin. Mineralization refers to the process whereby soil organic N constituents are converted by microorganisms from organic form to inorganic or mineral form, NH_4^+ (ammonification).

During decomposition, proteins are hydrolyzed to peptides by proteinases and peptidases, and are eventually degraded to their original amino acid residues and finally to NH_4^+ or ammonia (NH_3). The accumulation of NH_4^+ in soil depends on the N requirement for microbial growth coupled with C availability. The critical C:N ratio for net mineralization to occur is <25-30 (Parnas, 1975; Harmsen and van Schreven, 1995). Mineralization is also affected by soil moisture and temperature. There is some evidence that suggests that the rate of mineralization is limited by the process of nitrification (MacDuff and White, 1985), which we discuss more extensively in section 4.1.4.

4.1.3. Immobilization

Immobilization refers to the biological uptake or assimilation of mineral forms of N (NO_3^-, NO_2^-, NH_3^+, NH_4^+) into organic compounds. Nitrate is immobilised following a sequence of enzymatic reductions via NO_2^- to NH_3^+ prior to uptake. The rate of NO_3^- assimilation or immobilisation is determined by the N requirements of the biomass and the supply of soil NH_4^+.

Ammonia or ammonium is assimilated into the biomass and incorporated into amino acids by either of two processes depending on the concentration of NH_4^+ present in the soil. At high concentrations of soil NH_4^+, the primary and reversible reaction involved is the addition of NH_4^+ to α-ketoglutarate to form glutamate, by the actions of the enzyme glutamate dehydrogenase and the coenzymes NADH or NADPH. The reversible reaction coupled with the utilization of either of the two different coenzymes by glutamate dehydrogenase facilitates a regulatory mechanism for either mineralization or immobilization of $NH4^+$. The NH_2 group of glutamate is subsequently transferred to other α-keto acids by enzyme-driven transamination reactions to produce the corresponding amino acids. Under the more usual conditions of low concentrations of soil NH_4^+, microorganisms employ two enzymes, glutamine synthetase and glutamate synthase (GS/GOGAT), that act in a sequence. Glutamine is synthesized by the addition of NH_4^+ to glutamate, following which the amine group is transferred to α-ketoglutarate to form two new glutamate molecules. Both stages require energy input from ATP and either reduced ferrodoxin or NADPH as electron donors. Microorganisms employ a suite of transaminases, each of which may be used to catalyse the transfer of the amino group from a single amino acid and produce of any one of the required amino acids.

Biological immobilization is essentially the reverse process of mineralization, and the net direction of N transformation is dependent in particular on the availability of free N in the soil. Lower free N will limit the microbial capacity of assimilating C. When C:N ratios of the substrate are higher than 30, net immobilization occurs and soil mineral N reserves are depleted. In contrast, when C:N ratios are less than 25-30, net mineralization causes an increase in soil N (Stevenson, 1986). The rate of microbe N immobilization is also dependent on the microbial C:N ratio, soil temperature, moisture, and inorganic N concentration.

It is worth noting that NH_4^+ can be abiotically immobilised by clays and organic matter in both an exchangeable and non-exchangeable form. The exchangeable form is in dynamic equilibrium with ions in the liquid phase and is thus available for microbial immobilization. The non-exchangeable form is generally considered to be unavailable for biological uptake, especially if fixed in certain clay lattices (Rosswall, 1982). Estimates of the fixed NH_4^+ that is available for biotic immobilisation have been estimated to be 30-60% (Kudeyerov, 1981).

It is estimated that at any one time, only about 0.1% of soil N is in available mineral forms i.e. NO_3^- and exchangeable NH_4^+ (Stevenson, 1986). Nitrate is also immobilised after reduction firstly to NH_4^+ and hence by the process described previously.

4.1.4. Nitrification

Nitrification is a two-stage process involving the microbial oxidative conversion of NH_4^+ to NO_2^- and NO_2^- to NO_3^- (Alexander et al., 1960). The process is principally associated with chemoautotrophic bacteria, which are obligate aerobes and derive their C sources solely from inorganic forms such as CO_2^- or carbonates, and their energy either from NH_4^+ or NO_2^- (Focht and Verstraete, 1997). The nitrifying bacteria can be categorised into two groups, which are defined according to how they derive their energy. The first group oxidises NH_4^+ to NO_2^-, with hydroxylamine (NH_2OH) as an intermediate product (Payne, 1981), and includes the genera Nitrosomonas, Nitrosolobus and Nitrosospira. The second group oxidises NO_2^- to NO_3^- and includes the genus Nitrobacter (Walker, 1978). The two stages are closely inter-linked with little accumulation of soil NO_2^- (Atlas and Bartha, 1993).

Nitrification only takes place under aerobic soil conditions, since nitrifying bacteria utilise the enzyme ammonia monooxygenase which requires molecular O_2 to oxidise NH_4^+. Both N_2O and NO are produced during nitrification. Poth and Focht (1985) have shown that N_2O is produced when O_2 is limited and NH_4^+ oxidising bacteria use NO_2^- as an electron acceptor. The mechanism for NO production is poorly understood, but possible suggestions are that NO may result from the oxidation of NH_2OH or from the reduction of NO_2^- (Firestone and Davidson, 1989).

The rate of nitrification is influenced by several factors including pH, O_2, soil water content and temperature. However, the major limiting factor is supply of NH_4^+ as a substrate. Nitrification, therefore, is highly dependent on the rate of mineralization, which is also dependent on immobilisation.

In general, the pH optimum for nitrifier growth is approximately neutral to slightly alkaline (Painter, 1970). There is some evidence to suggest that nitrifiers can make some adaptation to the prevailing soil pH (Bramley and White, 1990), and nitrification has been reported in soils with a pH as low as 4.0-5.0 (Weber and Gainey, 1962; Walker and Wickramsinghe, 1979). Although liming vastly increases the number of autotrophs (Soriano and Walker, 1973), at soil conditions at a pH higher than 7.5 the chemical equilibrium between NH_4^+ and NH_3 is shifted in favour of the latter, which has been shown to be toxic to nitrobacteria (Morril and Dawson, 1967). At high pH, NH_4^+ also inhibits the transformation of NO_2^- to NO_3^- (Paul and Clark, 1989).

Nitrifying bacteria are obligate aerobes and therefore the O_2 status of the soil has a major impact on the nitrification process. Soil moisture content and temperature exert an indirect influence on the nitrification process, since they in turn affect the availability of O_2 in the soil. High soil water contents restrict the diffusion rate of the O_2 into pore spaces, and high soil temperatures limit the solubility of O_2. Although the optimum temperature for nitrification seems to vary between soil types, in general, at temperatures below 5°C and above 40°C, the rate of nitrification is slowed down.

At present, the contribution of either oxidative or reductive mechanisms of NO and N_2O production from nitrification cannot be quantified with certainty, and therefore the NO and N_2O production rates need to be defined as fractions of predicted nitrification rates. Observations show a range in the NO production rate due to nitrification of 0.1-4%, whereas the observed range in the N_2O production rate was between 0.008 and 0.2% of the nitrification rate (Li et al., 2000).

4.1.5. Denitrification

Denitrification is the term given to a series of microbiological respiratory processes, whereby N oxides act as the terminal electron acceptor during electron transport in the absence of O_2. Oxidised forms of N, namely NO_3^- and NO_2^-, are reduced to the gaseous products NO, N_2O and N_2. The denitrification process represents one of the major mechanisms by which fixed soil N is returned to the atmosphere. Each stage of the process is catalysed by a different enzyme reductase, which transfers electrons from one N oxide intermediate to the next. A wide range of bacterial species possess the reductases necessary to reduce NO_3^- entirely to N_2; however, some species lack NO_3^- reductase and are dependent on NO_2^- whilst other lack N_2O reductase and produce N_2O as a terminal product (Knowles, 1981).

The major factors controlling denitrification are well documented and are identified as soil water content, O_2, pH, soil temperature, nutrient supply and soil antecedent conditions (Nommik, 1956; Bremner and Shaw, 1958; Firestone and Tiedje, 1979; Knowles, 1981; Aulakh et al., 1992). Although the production of N_2 from denitrification has little significance in terms of atmospheric pollution, it represents a significant loss of N from agriculture systems. The ratio of N_2O and N_2 are important when trying to balance N budgets.

The presence of O_2 has a repressive effect on all the N oxide reductases (Paul and Clark, 1989). Consequently, denitrification generally only occurs when the supply of soil O_2 for microbial respiration is limited. Such conditions arise when the soil water content exceeds certain levels (Nommik,

1956). There is an apparent linear relationship between denitrification and soil water content expressed as percentage water filled pore space (WFPS). For example, Scholefield et al. (1997b) showed that there is a 50-fold increase in denitrification with a WFPS increase from approximately 70 to 90%. Other studies on intact soil cores from grassland have shown a linear relationship between WFPS and denitrification rate itself (Colbourn, 1993).

The pH for optimum growth of denitrifiers is in the range of 6-8 (Firestone, 1982; Li et al., 2000). Denitrification is suppressed in acid soil conditions, but the mechanism for inhibition of denitrification at low soil pH is not clear. Several suggestions have been made. Low pH may reduce the availability of C (Koskinen and Keeney, 1982), which may indirectly limit the population of denitrifying organisms (Parkin et al., 1985). Firestone (1982) speculates that there may be toxicity effects due to increased NO_2^- production from NO_3^- reduction or the increased solubility of aluminium and manganese at decreasing pH. She also speculates that there may be a molybdate (Mo) deficiency (a component of nitrate reductase) at lower pH levels. A study by Bremner and Shaw (1958) showed that the addition of Mo to soil at pH 3.6 did not increase the rate of denitrification, and it also showed that moderately low pH causes the reduction of NO_3^- to NH_4^+ rather than denitrification. Despite these limitations, significant losses have been observed from acid soils (<pH 5.0) and suggest that the denitrifier population can adapt to pH levels lower than the optimum (Parkin et al., 1985, Weier and Gillam, 1986). Denitrification is suppressed at pH levels higher than 10.5 and may be due to a deficiency of soil organic matter. Several studies (Nommik, 1956; Bremner and Shaw, 1958; Scholefield et al., 1997b) have shown that the ratio of N_2O to N_2 is increased with decreased pH. For example, Scholefield et al. (1997b) showed that for a pH range of 5.1-9.4, the ratio of N_2O to N_2 decreased from 1.8 to 0.35.

Studies on the thresholds of temperature at which denitrification occurs appear to give conflicting results, particularly at the lower limit. Denitrification activity has been observed in soils at low temperatures ranging from 2 – 5 °C (Bremner and Shaw, 1958; Nommik, 1956; Stanford et al., 1975). Smid and Beauchamp (1976) suggest that denitrification could occur at temperatures as low as 0 °C, as long as there is a plentiful supply of available C. The upper temperature threshold for denitrification activity has been reported as high as 60-67 °C (Freeney et al., 1979).

Increasing NO_3^- concentrations increase total denitrification as long as C is readily available. At the same time, high NO_3^- concentrations appear to inhibit the conversion of N_2O to N_2 (Nommik, 1956; Weier et al., 1993, Scholefield et al., 1997b).

Antecedent soil conditions such as prolonged periods of dryness followed by wetting up and substrate addition have been shown to be a cause of variability in the ratio of N_2O to N_2 production (Firestone and Tiedje, 1979; Dendooven et al., 1994, Dendooven and Anderson, 1995). This is thought to be due to a retarded de-repression of the N_2O reductase compared to that of the NO_3^- reductase following the onset of anaerobiosis. It appears that N_2O reductase is less robust than NO_3^- reductase in aerobic soil conditions.

4.1.6. Chemodenitrification

Many researchers have observed that chemical decomposition of nitrite plays an important role in the emission of NO from acidic soils (Davidson, 1992 ab; Saad and Conrad 1993, Yamulki et al., 1997). Chemodenitrification is a non-enzymatic chemical reaction under aerobic conditions. It produces NO and usually occurs under acidic conditions (Chalk and Smith, 1983; Anderson and Levine, 1986). It is interesting to point out that chemodenitrification seems to be more important for the production of NO than for N_2O.

4.1.7. Ammonia volatilisation

Since the partial pressure of $NH_{3(gas)}$ in the atmosphere is generally low, NH_3 is readily volatilised out of soil solutions and from the soil. The volatilisation is a series of reversible chemical reactions occurring between solid, liquid and gaseous phases of NH_3 within the soil (Freney and Simpson, 1983).

The main source of NH_3 is from soil NH_4^+ resulting from mineralization of the organic matter and applied fertilizers such as urea. The equilibrium equation between the two N's is:

$$NH_4^+ + OH^- \leftrightarrow NH_3 + H_2O \qquad (Eq.1)$$

Soil pH determines in which way the equilibrium reaction is driven. An increase of hydroxyl ions (OH^-) in soil solution drives the reaction to the right and more NH_3 is produced. The above equation also shows that the amount of NH_4^+ in the soil has a direct effect on the production of NH_3.

Factors controlling the variability in ammonia volatilisation can be divided roughly into two, though not mutually exclusive, categories; those disturbing the equilibriums between the reactions of the different phases and those affecting the rate of removal of gaseous NH_3 to the atmosphere. For example, the chemical and physical characteristics of soils (e.g. cation exchange capacity (CEC), pH, $CaCO_3$ content), soil environmental conditions (e.g. temperature, moisture content) and agronomic practices (e.g. application of fertilizer, plant uptake) influence the equilibrium between the phases and primarily the concentration of NH_3 in solution. Climatological conditions such as windspeed and rainfall affect the rate and removal of NH_3 to the atmosphere. Details concerning these mechanisms can be found in Freney and Simpson (1983). Generally, NH_3 emissions increase with high pH, temperature, windspeed, N concentration in animal wastes and the application of artificial N fertilizers whereas a high CEC, rainfall, and a high partial pressure of atmospheric NH_3 reduce the NH_3 emissions.

In summary, the major factors controlling the soil-biogenic gaseous emissions are principally those which influence the activity of the microbial biomass, namely soil moisture, O_2 supply, temperature, pH and C and N availability.

4.2. Canopy processes

One major constraint of the interpretation of observed soil nitrogen emission fluxes is the reference height at which the nitrogen fluxes have been observed. Closed chamber techniques placed at the soil surface are considered to be representative for the "true" soil emission flux, whereas observations over the canopy do not only reflect the soil emission fluxes but also the role of within-canopy processes, e.g. the local removal by dry deposition and chemical transformations. The importance of canopy processes for the emission depends on the chemical lifetime and the dry deposition of the nitrogen species. For a relatively inert trace gas such as N_2O, canopy interactions are not relevant. All the N_2O emitted by the soil escapes the canopy, although short-term (hours) measurements will reflect the role of turbulence for N_2O exchange. In contrast, the nitrogen species NO, NO_2 and NH_3 have chemical reaction times that are comparable to the characteristic turbulent mixing times, which result in chemical transformation and subsequent surface deposition or stomatal uptake within the canopy. The difference between what is measured at the soil surface and what leaves the top of the canopy has come to be known as the canopy flux divergence.

The NO_x canopy flux divergence has been addressed in previous studies using multi-layer models for the trace gas exchanges in canopies, often in combination with field observations to constrain the models (Jacob and Wofsy, 1990; Gao et al., 1993; Duyzer et al., 1995; Joss and Graber, 1996; Walton et al., 1997). The study by Jacob and Wofsy (1990) indicates that the canopy top NO_x flux of tropical rainforest is only about 20% of the soil emission flux, whereas Gao et al. (1993) simulated a downward canopy top flux for a deciduous forest canopy, despite the presence of a significant soil source. Recently, Ganzeveld et al. (2002a) presented a study of the role of the canopy processes for the NO_x exchanges in a tropical and deciduous forest and taiga woodland (see also section 5.3).

A study by Denmead et al. (1976) showed that also for NH_3 there is a large canopy flux divergence due to the effective vegetation uptake of the NH_3. Field observations by Hooker et al. (1980) and Parton et al. (1988) have shown that ambient NH_3 can be absorbed and metabolised by the plants, and that the absorption rate is related to the NH_3 concentration in the air around the leaves (e.g. Lockyer and Whitehead, 1986), N shortage in the plants (Harper et al., 1987), leaf surface moisture (e.g. Sutton et al., 1993), and crop growing stage (e.g. Schjørring, 1991). Observed dry deposition velocities range between 0.3 to 3.4 cm s^{-1} (Hutchinson, 1972; Meyer, 1973; Cowling and Lockyer, 1981; Aneja et al., 1986; Sommer and Jensen, 1991). However, the uptake of NH_3 by vegetation is complicated due to the existence of a compensation point (e.g. Farquhar et al., 1980). It implies that for ambient concentrations smaller than the compensation point there is an additional biogenic source of NH_3 from the vegetation. Observations have also revealed that there is an NO_2 compensation point of about 0.5-1 ppbv (Johansson, 1987; Rondón et al., 1993). Lerdau et al. (2000) calculated, based on a compensation point of 1 ppbv, a vegetation emission flux of NO_2 of about 2.10^{10} molecules cm^{-2} s^{-1} which is comparable to the soil NO emission flux.

The existence of a compensation point is not only relevant for vegetation uptake but also for soil uptake. For instance, soils act predominantly as a source of NO but can under elevated ambient mixing ratios, become a NO_x

sink (Conrad, 1994; Meixner, 1994). There are reports of NO_2 reactions with humic acids and phenols on dry soils and stone surfaces (Baumgartner et al., 1992) and chemical scavenging ($NO_2 + NO_2^- \rightarrow O_3^- + NO$) in low pH, high SOM content soils (van Cleemput and Baert, 1976). Recent work also suggests that some ammonia oxidizers are able to consume NO_2 and produce NO and NO_2^- under both oxic and anoxic soil conditions (Zart and Bock, 1998; Schmidt and Bock, 1997).

Additional complications involved in the assessment of canopy N exchanges are the role of turbulent exchange and the effect of surface wetness on the vegetation uptake (Ganzeveld et al., 2002a). Studies by Fuentes et al. (1992) and Kirkman et al. (2002a) have shown an enhancement of deposition to wet foliage, but the role of the aqueous phase chemistry involved is not well understood.

5. MODELLING NO, N_2O AND NH_3 EMISSIONS

Several hypotheses gleam from the bulk of scientific work conducted on soil N emissions in the last decade. These hypotheses are the foundation upon which modern model development proceeds and they include:

- A soil N flux is the sum of all production and consumption processes that occur within the soil profile.
- There are two major soil microbial processes directly responsible for regulating NO and N_2O production/consumption in soils - including chemical, biotic and abiotic processes. The biotic processes are regulated by nitrification and denitrification and the abiotic processes are regulated by rates of chemodenitrification.
- Nitrification or denitrification is driven by the activity of nitrifiers or denitrifiers, respectively. The microbial activity is determined by the soil redox potential (or Eh), soil water potential or soil water content, and concentrations of substrates such as dissolved organic carbon (DOC), ammonium (NH_4^+), nitrogen oxides (e.g. nitrate, nitrite, NO and N_2O), and, to a lesser extent, soil temperature.
- Diffusion of gases in the soil pores can have a decisive influence on the N-involved chemical and biological processes. An increase in soil moisture generally reduces the soil air-filled porosity, and hence reduces gas diffusivity and increases the potential for microbial consumption and chemical reduction of any intermediate N gas products.

- Aerobic and anaerobic micro-sites exist simultaneously in soils. Mainly, nitrification occurs at the aerobic, and denitrification at the anaerobic micro-sites. The ratio between the aerobic and anaerobic micro-sites is determined by the soil bulk redox potential or soil water content.
- Ammonia volatilisation is mainly controlled by the chemical equilibrium between NH_4^+, $NH_{3\ (liquid)}$ and $NH_{3\ (gas)}$ in the soil, which is affected by soil temperature, pH, and soil texture.
- Canopy structure, radiation, turbulent transfer, chemical transformations, plant-physiological regulation, (e.g. leaf nitrogen concentration) and the compensation point control the total amount of soil-produced NO or NH_3 that is released to the atmospheric surface layer.

5.1. Model approaches

In the last decades, models used to predict soil NO, N_2O, or NH_3 emissions have evolved from simple empirical schemes to detailed process-oriented approaches based on the above hypotheses. An overview of the various approaches is provided below, including a more detailed discussion of two modern processes-oriented models. In addition, model approaches that account for canopy processes are discussed. We also discuss the use of process-based models in up-scaling studies and present some results of a sensitivity analysis with a process-based model to indicate the sensitivity of the soil N emissions to changes in climate and land use.

5.2. Early model schemes

An example of a "simple" empirical model is the N_2O flux model, adopted by the IPCC to calculate national inventories for agricultural N_2O emissions (IPCC, 1997). It uses a fixed N_2O release rate of 1.25% for all N applied as fertilizers, manure or fixed by leguminous crops, based on a survey of field studies by Bouwman (1996). The empirical models have simple structures and can easily be applied at different scales. However, these models cannot be used to interpret the observed temporal variability in the nitrogen emissions at the site scale due to a missing representation of the controlling mechanisms in these models.

Process-oriented models were developed to incorporate more mechanistic process descriptions, and hence to be more reliable for predictions. Mechanistic models of nitrogen trace gas fluxes must consider the ecosystem nitrogen cycle. Consequently, processes such as N mineralization, assimilation and leaching, microbially driven N transformations as well as the interaction of the nitrogen cycle with the carbon cycle have to be included (Frolking et al., 1998). The role of the canopy processes for NO_x and NH_3 surface exchanges should also be included, as discussed in the previous sections. Williams et al. (1992a) developed an algorithm to compile a regional emission inventory for the United States, that calculates soil NO emission fluxes as a function of soil temperature and land use type with specific fitting parameters based on their field measurements. A further development of this approach was made by Yienger and Levy (1995, hereafter referred to as YL95), who introduced some modifications, e.g. the role of fertilizers and precipitation. They also introduced a first-order estimate of the uptake of NO_2 by the canopy. The YL95 algorithm has subsequently been applied and modified in other studies. Ludwig et al. (2001) used the YL95 algorithm to study the contribution of the soil NO emissions to the total NO_x emissions in southwestern Germany. Further, Otter et al. (1999) and Kirkman et al. (2001) used (and modified) the YL95 pulsing scheme to study the NO emissions from tropical savannas in southern Africa. Their soil emissions models were based on work by Yang and Meixner (1997), who related laboratory studies of soil biogenic NO emission to field NO flux measurements based on a conceptual model by Galbally and Johansson (1989). This model, which does not consider any dependencies on environmental variables, has been further evaluated by comparing laboratory NO emission measurements with field measurements (Remde et al., 1993) and studying NO release rates in the laboratory (Rudolph and Conrad, 1996b), indicating that it is feasible to predict NO emission from soils with this model.

Several other models including CASA (Potter et al., 1996) and CENTURY (see section 5.4.) have adopted the so-called "hole-in-the-pipe" conceptual scheme first proposed by Firestone and Davidson (1989). This scheme describes the flow of inorganic nitrogen through an imaginary pipe as representing the rates of nitrification and denitrification in the soil, while the holes in the pipe or "leaks" describe the amount of nitrogen trace gas (NO, N_2O and N_2) that eventually escapes the soil surface (Davidson et al., 2000). In the CASA model, the primary controlling factors are gross rates of N mineralization and soil moisture (Potter et al., 1993, 1996). In the new version of the CENTURY model (Parton et al., 1996, 2001), the nitrification rate is calculated based on decomposition rate, and the nitrification-induced

N_2O flux is a fraction of the nitrification rate. The denitrification rate is a function of soil NO_3^- content, soil respiration, and soil moisture.

5.3. The DNDC model

As an example of process-oriented model, the denitrification-decomposition (DNDC) model contains detailed processes to simulate production, consumption and transport of several major N gases including N_2O, NO, N_2 and NH_3 in agro-ecosystems. The model was originally developed for predicting denitrification processes in agricultural soils (Li et al., 1992). The current version of DNDC contains a relatively complete set of functions describing N biogeochemical cycling in agro-ecosystems (Zhang et al., 2002a; Li, 2000). The principles of thermodynamics and kinetics controlling nitrification, denitrification and chemodenitrification have been included in the model to link these reactions to the environmental factors such as temperature, moisture, Eh, pH and substrate concentration gradients. The environmental factors are driven by several major ecological drivers including climate, topography, soil properties, vegetation, and anthropogenic activity. Any variation in an ecological driver will cause simultaneous changes in the environmental factors; and each change in a single environmental factor can simultaneously cause changes in several reactions, which collectively determine N transformation and transport in the ecosystem. To organize such a complex system into a computing framework, the DNDC developers adopted a concept of biogeochemical field. Biogeochemical field is an assembly of the forces regulating the biogeochemical reactions involved in a specific ecosystem (Figure 1). A biogeochemical model is a mathematical expression of the biogeochemical field and its links to the ecological drivers and elementary movement driven by the biogeochemical reactions (Li et al., 2000). Based upon this concept, DNDC was constructed with two components.

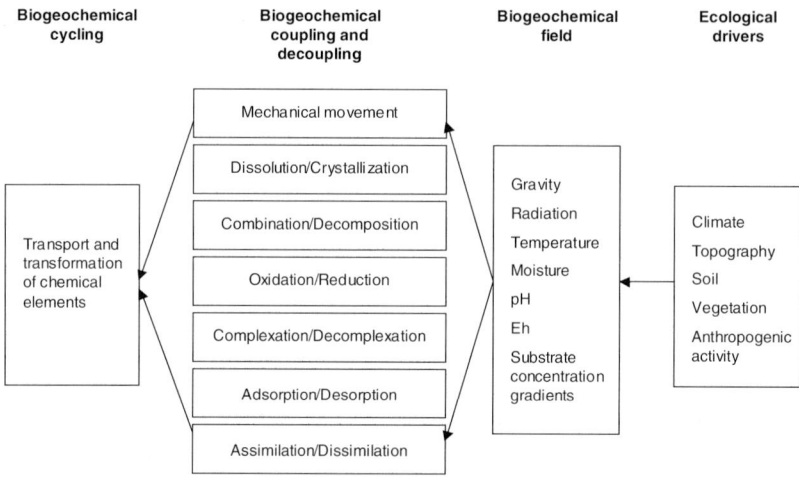

Figure 1. The DNDC model is based on the basic concepts of biogeochemical abundance, field, coupling and cycling. A biogeochemical field is an assembly of the forces regulating the concerned biochemical or geochemical reactions, which determine transport and transformation of the chemical elements in a specific ecosystem (from Li et al., 2000). (© American Geophysical Union).

The first component contains three sub-models for soil climate, plant growth and decomposition, which link the ecological drivers to soil environmental factors. The second component contains three sub-models for nitrification, denitrification and fermentation, which link the soil environmental factors to the microbiologically intermediated geochemical reactions. The soil climate sub-model predicts soil temperature, moisture and redox potential profiles based on daily meteorological data, soil texture, and plant growth. The plant sub-model calculates the daily photosynthesis, C assimilation, C allocation, N and water demands/uptake, and litter deposition. The decomposition sub-model tracks the turnover of several SOC pools including litter, microbial biomass, active and passive humus as well as CO_2 and ammonium (NH_4^+) productions. The nitrification sub-model estimates transformation rates of ammonium to nitrate driven by soil nitrifier's activity, ammonium concentration and environmental conditions. The denitrification sub-model calculates reductions of nitrate to nitrite, nitric oxide (NO), N_2O and N_2 based on the dual-substrate Michaelis-Menten equation, a simple function describing multi-nutrient-dependent growth. The fermentation sub-model predicts CH_4 production, oxidation, and transport under submerged conditions. Classical physical, chemical or biological laws in conjunction with some empirical equations generated from laboratory studies were incorporated in the sub-models to parameterize each specific reaction (Li et al., 1992; Li et al., 1994; Li, 2000). The six interacting sub-models bridge the ecological drivers (climate, soil properties, vegetation and anthropogenic activity) to N gas emissions (Figure 2). The biogeochemical framework of DNDC has been integrated with a forest physiological model, PeNT (Aber and Federer, 1992), and a nitrification model to form a forest biogeochemical model, PnET-N-DNDC (Li et al., 2000; Stange et al., 2000).

Since nitrification and denitrification can occur simultaneously in aerobic and anaerobic micro-sites, respectively, in the same soil, the model must allocate the substrates such as DOC, NH_4^+, NO_3^-, etc., to the two soil fractions with different aeration status. A simple kinetic scheme was implemented to quantify the volumetric fraction of the anaerobic micro-sites in the soil. The scheme can be described as a dynamic "anaerobic balloon" within the soil matrix. The size of the "anaerobic balloon" (i.e., volumetric fraction of the anaerobic micro-sites) is calculated based on soil redox potential. According to the Nernst equation, the soil redox potential is determined by the product of the concentrations of all the oxidizing species versus the product of the concentrations of all the reducing species in the soil liquid phase (Stumm and Morgan 1981). The common oxidizing species in soils are oxygen, NO_3^-, Mn^{4+}, Fe^{3+}, SO_4^{2-}, etc., and the common reducing species are H_2S, Fe^{2+}, Mn^{2+}, etc. Transferring an electron from a reducing species to an oxidizing species is regulated by the change in the Gibbs free energy in the system. Since oxygen possesses the lowest Gibbs free energy change to accept electrons, oxygen is usually the dominant oxidizing species in aerobic soils. Within the redox potential or Eh range from 700 to 250 mV, soil redox status is dominated by the oxygen partial pressure (pO_2). Theoretically, other oxidizing species will not be reduced until the oxygen has been depleted. The denitrifying bacteria start to use NO_3^- as electron acceptor at an Eh of 350-250 mV when oxygen is close to being depleted.

A one-dimensional soil oxygen diffusion algorithm estimates the pO_2 in the forest soil profile based on the Fick's first law. The O_2 consumption rate is the sum of soil microbial respiration and root respiration. The decomposition and plant growth sub-models predict the microbial and root respiration rates, respectively (see Figure 2).

Figure 2. Structure of the DNDC model. Six interacting sub-models bridge between the ecological drivers, soil environmental factors, and trace gas emissions (From Li, 2000).

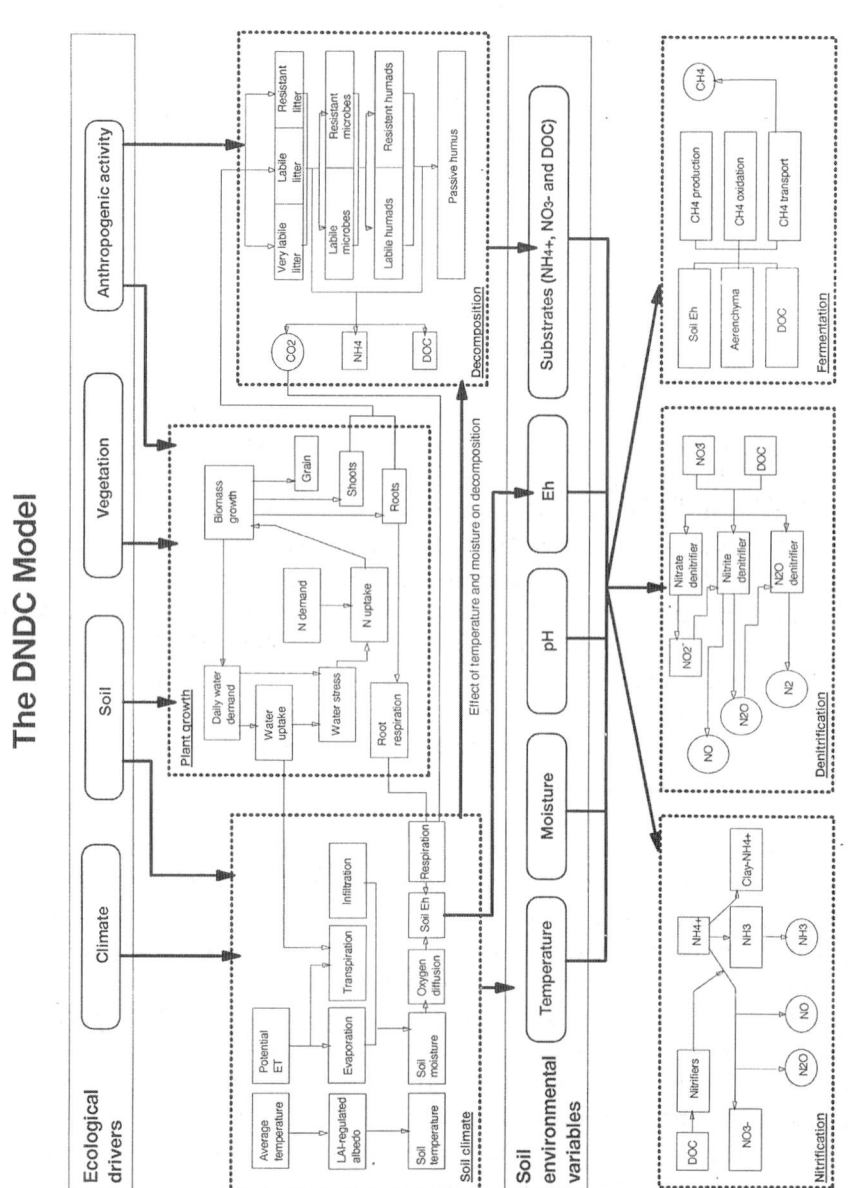

A simplified linear correlation is applied to estimate the size of the "anaerobic balloon" based on the pO_2 status. The "anaerobic balloon" regulates the nitrification and denitrification rates by distributing the substrates into the aerobic and anaerobic soil fractions. When the "anaerobic balloon" swells, (1) more substrates including DOC, NH_4^+, NO_3^-, NO and N_2O will be available in the anaerobic micro-sites for denitrification, (2) less DOC and NH_4^+ will be left in the aerobic micro-sites for nitrification, and (3) the pathway for the denitrification gas products (e.g. NO and N_2O) to escape from the anaerobic balloon will become longer, so that the probability that the N-gases are further reduced by denitrification increases. For a shrinking "anaerobic balloon", the opposite occurs. The model considers the role of dissolved organic carbon (DOC) as a key energy source for both the nitrifiers and denitrifiers, and it uses the soil-redox potential as main driver to control the nitrification and denitrification processes. This feature has made the model applicable to both dry and wet soils.

The DNDC model simultaneously simulates nitrification, denitrification and chemodenitrification at an hourly (for denitrification) or daily (for nitrification and chemodenitrification) time step. The NO or N_2O produced by the three sources forms the total gas pool. After the diffusion calculation, a fraction of the gases is emitted into the air and the remaining fraction is reallocated into the new reactions of the next time step.

The DNDC model is one of the few models that predict soil NH_3 volatilisation and emissions by calculating explicitly the $NH_{3(liquid)}$ concentration based on NH_4^+ and OH^- concentrations. The decomposition sub-model calculates the NH_4^+ concentration in the soil profile. The sub-model calculates turnover rates of soil organic matter at a daily time step (Li et al., 1992). The OH^- concentration is determined by soil pH and temperature based on Stumm and Morgan (1981). The daily-emitted fraction of the gas-phase NH_3 is a function of the soil air-filled porosity and clay content due to their effects on NH_3 gas diffusion. A maximum dry deposition velocity of 3.4 cm s^{-1} has been adopted in DNDC for calculating NH_3 absorption rate by crops. In addition, factors such as plant N status, leaf surface moisture and within-canopy NH_3 concentration profile are also included. The latter is calculated as a function of the predicted soil NH_3 flux, a constant atmospheric background NH_3 concentration (0.06 ppm, based on Ayers and Gras, 1980 and Tsunogai and Ikeuchi, 1986), and degree of closure of the canopy. The atmospheric background NH_3 concentration can be redefined based on observations.

The DNDC model has been evaluated at site scale against a number of field data sets on SOC dynamics and trace gas emissions. The evaluations conducted during the model development and comparisons have been published in several papers (Li et al., 1992, 1994; Frolking et al., 1998, Li 2000, Stange et al., 2000). A collection of researchers in northern America, Europe, Asia and Australia have independently conducted further evaluation focusing on N_2O or NO emissions (Wang et al., 1997; Smith et al., 1997; Plant et al., 1998; Smith et al., 1999 and 2002; Xiu et al., 1999; Butterbach-Bahl et al., 2001ab; Brown et al., 2002; Zhang et al., 2002a,b; Kiese et al., 2002; Cai et al., 2003). Some of the results are presented in Figure 3.

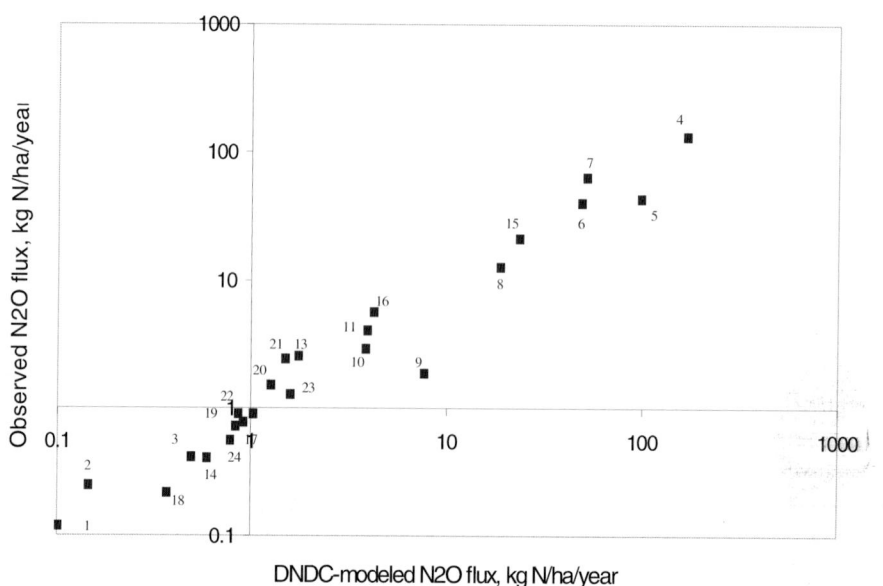

Figure 3. Comparison between observed and DNDC-modelled N_2O fluxes in grassland soils in Colorado (1, 2 and 3), organic soils in Florida (4, 5, 6, 7, 8 and 9), a pasture soil in the UK (10), fertilized and unfertilised cropland soils in Costa Rica (11 and 12), upland and paddy cropland in China (13, 14 and 15), a cropland soil in Germany (16), and cropland soils in Canada (17, 18, 19, 20, 21, 22, 23 and 24).

Geographical information system (GIS) databases were developed to enable DNDC to predict N gas emission and other biogeochemical processes at the continental scale for the U.S. and China (Li et al., 1996; Li et al., 2002b; Li et al., 2003). The up-scaling study has produced interesting results regarding the impacts of land use management on the N and C cycles of agro-ecosystems. The dominant N gas emitted from China's cropland is NH$_3$ (Figure 4), whereas U.S. croplands emit basically N$_2$O and NO (Li et al., 2002b, Figure 5).

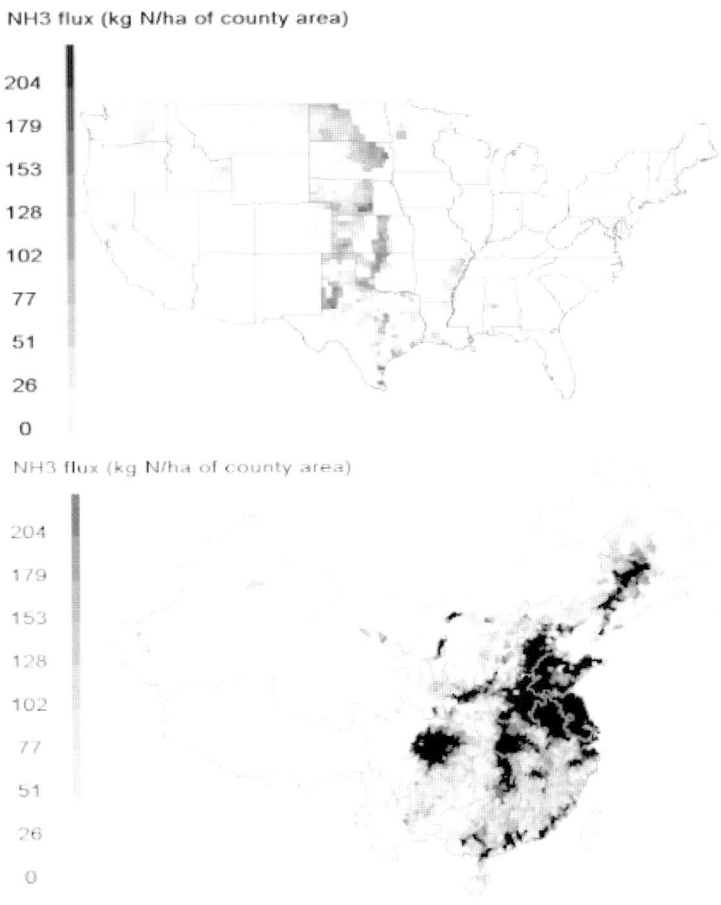

Figure 4. DNDC-simulated ammonia emissions from cropland in the U.S. and China. Chinese cropland emitted 11 Tg NH$_3$-N; and the U.S. cropland only 0.9 Tg NH$_3$-N in 1990. The difference was mainly due to the dominant types of fertilizers used as well as soil pH status in the two countries. (© American Geophysical Union).

The model simulations indicated that the difference was mainly related to the differences in the soil properties and fertilizer types used between the countries. The DNDC-predicted results revealed that the N gas emissions at regional scale were affected by both natural conditions and anthropogenic activities in a complex way. Among the many effective factors, soil C conditions dominate N gas fluxes in most of the observed and modelled cases.

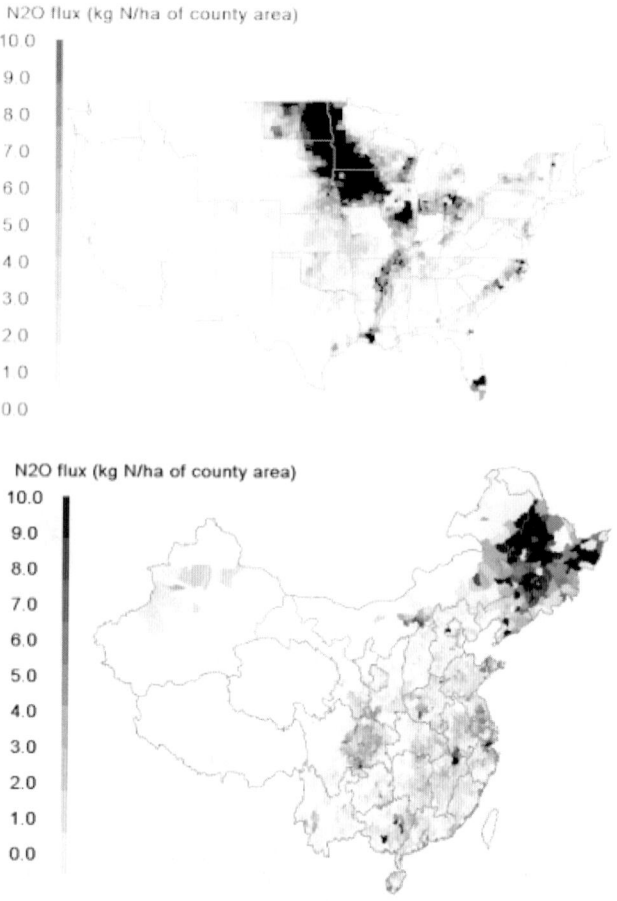

Figure 5. DNDC-simulated nitrous oxide emissions from cropland in the U.S. and China. The U.S. and Chinese croplands emitted 1.6 and 1.3 Tg N_2O-N, respectively although the fertilizer used in China was double of that in the U.S. in 1990. The lower N_2O emission rates in China were mainly related to the low contents of organic matter in the soils. (© American Geophysical Union).

5.3.1. Computational requirements

In order to indicate the requirements for applying a model like DNDC to calculate the nitrogen exchange fluxes for a larger domain, e.g. on regional or even the global scale, we have estimated the required CPU time for a one-year scenario analysis for China. This indicates if it is feasible to include a model like DNDC in mesoscale or global scale chemistry and climate models. A one-year simulation with DNDC, calculating the daily N fluxes for 30 different crop types for 3000 grid squares within China, as presented previously, requires 3 days at a 800 MHz PC, resembling about 100 CPU seconds at a parallel supercomputer with 2000 Gflop capacity. This suggests that the required computing facilities are not a limiting factor for implementation of DNDC in a chemistry and climate model like ECHAM, which has a typical grid-resolution of about 128 longitudinal and 64 latitudinal grid squares (about 2700 land grids in total, T42 resolution). Rather, it is expected that the quality of the required input data to constrain the DNDC model, the ecological drivers, especially the soil properties, shown in Table 1, is a limiting factor.

Table 1: Ecological drivers

1. Climate	- Daily max and min temperature - Daily precipitation - Atmospheric N deposition
2. Soil	- Bulk density - Texture (clay fraction) - Total organic C content - pH
3. Management	- Crop type, cultivars, and rotation - Tillage - Irrigation - Fertilization
4. Manure amendment (including crop residue)	- Weeding - Grazing

5.4. DAYCENT model

Another example of a process-oriented model is the DAYCENT plant-soil ecosystem model, which is based on the widely used CENTURY model, to simulate trace gas fluxes at a daily resolution over large timescales (centuries) (Parton et al., 1998, 2001; Kelly et al., 2000). The model simulates daily evaporation, plant production, terrestrial carbon, nitrogen transformation and NO, N_2O and N_2 gas emissions via a set of interlinking sub-models for soil organic matter (SOM), land surface (soil moisture, soil temperature and plant productivity) and trace gas fluxes. The SOM or carbon sub-model, which runs on a monthly time step is comprised of three pools, each with different turnover rates; active (0.5-1 yr.), slow (10-50 yr.) and passive (1000-5000 yr.). These soil C pools receive inputs from above- and belowground litter (each with allowable ranges of C:N) and surface microbial biomass. Above- and belowground plant residues and organic animal excreta are further partitioned into structural (below ground plant residues) and metabolic (organic animal excreta) pools as a function of the lignin to nitrogen ratio found in the plant residues. Decomposition of dead plant material supplies the SOM pool, and mineralization of SOM, N-fixation and N-deposition supply the available nutrient pools. The decomposition of both plant residues and SOM, which can be limited by anaerobic conditions (high soil water content), are assumed to be microbially mediated with an associated loss of CO_2 due to respiration. Soil physical properties, environmental variables, lignin and nitrogen concentrations of vegetation, litter and SOM all drive the flows of nutrients through these pools. The nitrogen submodel includes nitrogen inputs from symbiotic, non-symbiotic N-fixation and atmospheric deposition. Non-symbiotic N-fixation is carried out by all true free-living autotrophic or heterotrophic organisms within the plant canopy, and might include epiphytes, bryophytes, yeasts, lichens, cyanobacteria, and mosses not in direct symbioses with plants. Atmospheric (lightning & gas phase reactions) and non-symbiotic nitrogen fixation inputs are considered as wet deposition processes in the model, and are thus linearly controlled by the amount of rainfall per day. All nitrogen deposition is assumed to be wet deposited, as no explicit scheme exists for dry deposition in the model. Symbiotic N-fixation, which represents legume and rhizobial based nitrogen fixation processes occurring in the soil-plant environment, is a linear function of net primary production (NPP) and only becomes effective when there is insufficient mineral nitrogen to satisfy plant requirements.

The soil water component of the land surface submodel runs on a daily time step and is based on work by Parton et al. (1998) and Hillel (1977), which simulates the transport of water through the plant canopy, litter, and a multi-layered soil system. A portion of the rainfall is intercepted and evaporated as a function of the amount of plant biomass and/or surface litter. Non-intercepted water enters the 11-layer soil profile and is infiltrated or redistributed, depending on soil saturation conditions, and then evaporated and transpired as a function of soil water potential and root biomass demand (Parton et al., 1987). Infiltration of water is governed by the hydraulic conductivity of each soil layer and progresses downward through the soil profile with excess water allowed to flow out as base flow.

The actual soil temperature, used for decomposition, plant growth and nitrification, is the average of the maximum and minimum soil temperature, which are calculated from the maximum and minimum air temperature and the canopy biomass (e.g. Parton, 1984). The plant production submodel, which runs on a weekly time step, uses soil water content, temperature, and available nutrients to calculate plant growth, and C is allocated among leafy, woody, and root biomass based on vegetation type.

The nitrogen trace gas submodel of DAYCENT, which is described in detail in Parton et al. (2001), is based on the "hole-in-the-pipe" conceptual scheme (Firestone and Davidson, 1989). The submodel assumes that nitrification and denitrification contribute to the production of soil nitrogen trace gases. For this, the submodel has two subcomponents, "nitrify" and "denitrify". The "nitrify" component simulates nitrification, which is calculated as a function of soil NH_4^+ concentration, WFPS, soil– temperature, pH, and texture. The "denitrify" component simulates denitrification, which is calculated as a function of soil NO_3^- concentration, heterotrophic respiration (CO_2) and WFPS. The original submodel assumes that N_2O trace gas is emitted at a rate equivalent to 2 % of nitrification (from "nitrify"). The 2 % proportionality constant was derived from observed N_2O flux data (Mosier et al., 1996) and simulated model results (Frolking et al., 1998) in recent work by Parton et al. (2001). It has been widely quoted in model literature as universal for gas production (Potter et al., 1996, 1998) and has been applied in several models to describe gas emissions as either a function of mineralization as for the CASA model (Potter et al., 1996), or nitrification as for the DNDC model (Li et al., 2000), NGAS model (Parton et al., 1996), and the CENTURY model (Liu et al., 2000).

The "denitrify" component produces an entirely gaseous product of N_2O and N_2 apportioned as a function of the ratio of soil NO_3^- concentration to CO_2 emission, corrected by WFPS (Del Grosso et al., 2000). The N_2O gas produced by "nitrify" and "denitrify" are combined to produce a total soil N_2O emission.

The amount of NO emitted from the soil is none other than the product of a $NO:N_2O$ ratio and the N_2O gas emitted (produced by "nitrify" and "denitrify"). The $NO:N_2O$ ratio is calculated as a function of soil parameters (bulk density, field capacity, WFPS) that influence gas diffusivity. Studies on concurrent NO and N_2O emissions have shown that the ratio $NO:N_2O$ correlates with soil water content, with highest values for drier soils (< 60 % WFPS), decreasing as WFPS increases. (Verchot et al., 1999; Potter et al., 1996; Davidson et al., 1991, 2000). Soil diffusion was found to better represent the $NO:N_2O$ ratio than WFPS, despite substantial unexplained variability with this approach (Parton et al., 2000).

Modifications of this model have been implemented by Kirkman et al. (2002b) and include changes to the algorithms governing soil water potential, which govern the bi-directional flow of soil water under saturated and unsaturated conditions and the nitrogen trace gas sub-model. The modified trace gas model was based on the concept that NO production and consumption processes occur simultaneously in the soil, such that when NO production exceeds consumption, a net emission of NO from the soil to the atmosphere occurs (Galbally and Johansson, 1989). Further, nitrification and denitrification are both responsible for the production of NO and N_2O, such that under drier soil conditions (< 60 % WFPS) both processes are important and under reducing soil conditions (> 60% WFPS) denitrification becomes more dominant.

A significant modification to the trace gas sub-models is the shift from an approach which assumed that nitrification and denitrification produced only N_2O gas to one in which both NO and N_2O are produced by nitrification and denitrification. This facilitates the explicit simulation of NO production, doing away with the nitrification proportionality constant, which was empirically derived from N_2O data on soils that are generally too wet to produce significant amounts of NO. It also allows for the introduction of a soil NO consumption term, where the consumption of NO is proportional to the emission of N_2O (Conrad, 1996). This is achieved with a modified $NO:N_2O$ ratio as a function of WFPS.

Despite the fact that soil mineralization is essential as the first step of N "flowing" through the pipe and that nitrification depends upon the NH_4^+ supply, recent studies have shown that the concentration of NH_4^+, being a product of mineralization, is often not correlated to soil NO and N_2O emissions in the tropics (Davidson et al., 2000; Erickson et al., 2001). The proportionality constant for nitrification was replaced with a function describing the fraction of nitrogen trace gas produced as a function of the amount of NO_3^- produced by nitrification. This modification was in line with the "hole in the pipe" conceptual model and recent trace gas research findings, which suggest only a small proportion of total nitrification is responsible for NO production (Davidson et al., 2000; Erickson et al., 2001). The function was obtained from a review of studies where the measured soil nitrogen availability index, potential nitrification and soil NO and N_2O fluxes were measured (Davidson et al., 2000; Verchot et al., 1999). This index reflects the capacity of soil nitrifying bacteria to convert supplemented NH_4^+ under laboratory conditions to NO_3^- (Hart et al., 1994). After nitrification the remaining daily NO_3^- undergoes denitrification ("denitrify") producing an entirely gaseous output apportioned into N_2O and N_2 as per the original model scheme (Del Grosso et al., 2000). The difference is that the N_2O portion of "denitrify" was made to represent the production of both NO and N_2O instead of the emission of N_2O only. The NO and N_2O produced from "denitrify" and "nitrify" was then combined to represent the total nitrogen oxide trace gas production within the soil.

5.5. Canopy models

The role of canopy processes for NO_x and NH_3 surface exchange has already been discussed shortly in section 4.2. by reporting some results of site-scale studies of NO_x and NH_3 exchanges. YL95 have made a first assessment of the role of canopy processes for the global NO_x exchanges by introducing the canopy reduction factor (CRF), which is an ecosystem-dependent parameter that depends on the amount of biomass and reflects the role of canopy deposition. Bouwman et al. (1997) have used a similar approach for NH_3, using canopy absorption coefficients (Lee et al., 1997). As mentioned in section 2.4, both studies showed a reduction of the global annual soil NO and NH_3 emissions by about 50%. The DNDC model uses a dry deposition velocity (3.4 cm s^{-1}) and the plant N status, leaf surface moisture and the NH_3 concentration profile in the canopy to consider the role of NH_3 canopy deposition (see also section 5.3.).

A shortcoming of the CRF is that it only reflects the removal through dry deposition, while vertical turbulent exchange and chemical transformations are also expected to play a role. Moreover, the CRF is based on only one model study of the exchange of reactive nitrogen over the Amazon tropical rainforest during the wet season (Jacob and Wofsy, 1990). Therefore, an explicit atmosphere-biosphere trace gas exchange model which considers gas-phase chemistry, biogenic emissions, dry deposition, and turbulent exchange, has been implemented in the global chemistry and climate model ECHAM (Ganzeveld et al., 2002ab). These studies indicate that for relatively polluted locations, with the surface layer NO_x concentrations largely being controlled by the supply from anthropogenic sources, dry deposition of NO_x is the controlling process. For these locations, calculations with the traditional "big leaf" approach, which does not consider the role of canopy interactions, result in dry deposition NO_x fluxes very similar to those calculated with the multi-layer model. However, for relatively pristine sites in the sub-tropics and tropics, where soil emissions are a significant source of NO_x, there are distinct differences between the NO_x surface fluxes calculated by the multi-layer and the big leaf model. Another interesting outcome of this study is, that interpretation of the global annual NO_x budgets resulting from the multi-layer and the big leaf model including the CRF by YL95, confirms the applicability of the CRF to account for the role of within-canopy deposition for the global soil NO_x emissions.

This is indicated in Figure 6, which shows the absolute difference in the global and annual budgets of NO_x emissions, dry deposition and chemistry calculated with the multi-layer and the big leaf model. The shown difference in the NO_x emissions of about 4 Tg N yr^{-1} reflects the difference between the multi-layer soil emission flux and the "big leaf" soil-emission flux, which has already been corrected using the CRF to account for canopy deposition. This difference between the (soil- and canopy-top) emission fluxes by both approaches is comparable to the dry deposition flux of the explicit multi-layer model, which suggests that dry deposition largely controls the canopy flux divergence.

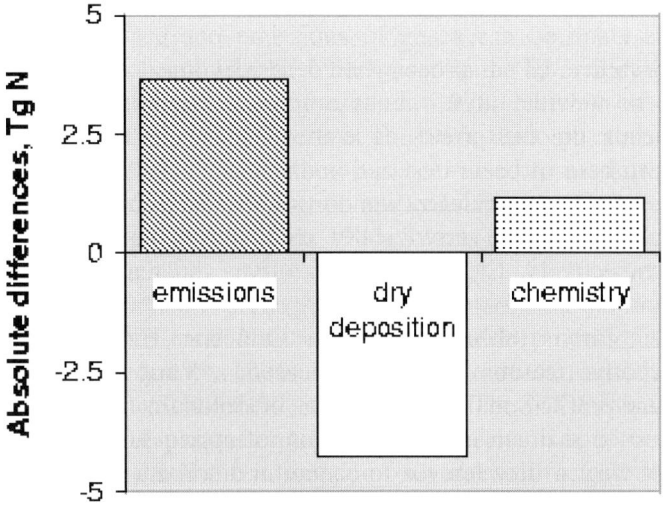

Figure 6. Absolute differences (Tg N) of the global annual tropospheric NO_x emissions, dry deposition and chemistry, comparing simulations with the big leaf and multi-layer model.

5.6. Scaling

Previous global estimates of soil N emissions have simply expanded measured or mean exchange rates by ecosystem area, and thereby ignored the distal environmental control heterogeneity (Williams et al., 1992b). Estimates have been further improved with the inclusion of various forms of biome stratification (e.g. Yienger and Levy, 1995). More elaborate process-based schemes - which include soil substrate, temperature, moisture and microbial turnover interactions at global ecosystem scales - offer the most representative approach to date (Potter et al., 1996). Spatial scaling in itself poses several problems. Since most models are developed based on the observations and empirical assumptions at the site scale, researchers are forced to ask several tough questions before they can apply these models to larger scales: (a) Are the parameters and assumptions upon which these models are based scalable in space and time? (b) Are there accurate readily available data sources for these scalable parameters? (c) Can one estimate the uncertainties involved in the up-scaling process? (d) Can a reasonable compromise be found, if the above conditions cannot be complied with?

5.6.1. Model parameters

The reliability of a process-based model used to extrapolate its predictions for a wide range of environmental conditions at large scales mainly depends on the processes embedded in the model. In general, adopting more basic process descriptions in a model allows a more flexible use of the model under different conditions. For example, nitrification and denitrification are typical biologically mediated oxidation and reduction processes, respectively. Electron transfer drives the reactions between the reductants and oxidants. Among the many environmental factors, soil redox potential (i.e., Eh) is inherently the most important one in regulating the electron exchange (Stumm and Morgan, 1981). Various models use soil moisture as a driver to simulate nitrification or denitrification rates. However, it can be questioned if soil moisture is the most appropriate driver in models, since the correlation between soil moisture and soil Eh is different for dry and wet soils. High moisture usually reduces oxygen availability in the soil, and hence decreases soil Eh. However, in submerged soils with constant soil moisture the Eh can further decrease, driven by the soil micro-organisms. In this case, soil Eh is a more basic parameter compared to soil moisture to represent the oxidation-reduction processes. Adopting Eh as a major driver will enable a model to be applicable for simulations of dry and wet soil N processes. However, the question then needs to be asked if there are readily available and accurate data sources of Eh at significant spatial scales. If not, then perhaps soil water content, which is more readily derived from cold cloud duration satellite data or can be reasonably modelled by hydrological and soil water models, is a reasonable compromise.

The spatial and temporal up-scaling also require an assessment of the relevance of processes that occur at smaller spatial and shorter timescales compared to the grid resolution and the temporal resolution, respectively, of the simulations. Due to the short lifetime and large spatial variability of the sources of NH_3 (and NO_x) it is expected to have substantial horizontal concentration gradients, which result in localized high dry deposition fluxes (Dentener and Crutzen, 1994). We recognize that a more extensive discussion about the dry deposition process is beyond the scope of the presented discussion on soil nitrogen emissions. However, the issue of sub-grid scale deposition is relevant for the NH_3 and NO_x budget related to the assumptions being made for up-scaling of emission flux. To indicate the potential role of sub-grid scale deposition, Dentener and Crutzen (1994) accounted for sub-grid scale deposition by removing directly 25% of the emitted NH_3. A study by Krüger and Tuovinen (1997), using an explicit model of the planetary boundary layer (PBL), resulted in average sub-grid deposition factors for NH_3 of about 15% in a regional-scale model.

Potter et al. (1996) simulated soil N emissions (see section 2) using monthly time intervals. This implies that the temporal variability in soil moisture occurring at shorter timescales, e.g. the role of sudden increases in soil moisture for the pulsing, which contributes according to YL95 to about 25% of the total soil NO emission flux (1.3 Tg N yr^{-1}), cannot be considered explicitly. To include the role of short-term soil moisture variability, Potter et al. (1996) introduced an index to represent the sub-timestep variability (Hutchinson et al., 1997). However, such approximations of the temporal variability obviously introduce an uncertainty in the simulated N fluxes. It demonstrates that the maximum possible temporal resolution of the simulations, as well as the uncertainty in the simulated N emission fluxes due to the unresolved temporal variability in controlling parameter, must be included when the use of a process-based model in large-scale studies is considered.

5.6.2. Available data sources

Collecting data at larger scales for up-scaling studies is always an arduous task. It is not only the quantity of required input parameters, but also the availability and quality of input parameters that can seriously limit the up-scaling. For example, a process-based model might require specific input data, which can only be quantified at the site scale with enough accuracy, which limits its application in up-scaling studies. One should then consider the use of a more empirical model using input parameters, e.g. biome types (e.g. Yienger and Levy, 1995; Kirkman et al., 2001), which have geographical distributions that are relatively well established, although there are some biomes which distribution is not well defined, e.g. savannah and woodland. Up-scaling usually requires the division of the considered domain over a number of grid cells. The applied grid resolution should not only be based on the research goal, e.g. the implementation of an emission model in a chemistry and climate model with its specific grid resolution (see section 5.3.) determining the resolution of the emission calculations, but also the resolution and quality of the databases to constrain the emission model should be considered in the selection of the grid resolution. Since it is assumed that there is a homogeneous distribution of all properties within a grid cell, the coarsest input resolution should theoretically determine the grid resolution. However, in practice the grid resolution is based on the most sensitive input parameters, which can be selected through sensitivity analyses. Priority should be given to the most sensitive factors that have been identified during the process of data collection and compilation. For example, a sensitivity study of the N_2O emissions from agricultural lands in the U.S. indicated that cropland N_2O fluxes are most sensitive to the soil organic carbon (SOC) content in simulations using DNDC. Consequently, efforts were made to collect the soil data at sub-county scale, where a county was defined as the basic grid unit to ensure each county had a range of SOC values (Li et al., 2002a).

5.6.3. Uncertainties estimations

The results of simulated nitrogen exchange fluxes by process models for larger domains should also include the estimated uncertainty in the calculated model parameters and input databases. As models improve in terms of its process representation, the quality of the input data will become more crucial for reducing uncertainties of the large-scale estimates. For the predicted large-scale N gas emissions, input-related errors can generally be attributed to two major sources: accuracy of land use and land-cover characterization and sub-grid heterogeneity of climate/soil/management conditions. The uncertainties caused by the heterogeneity within a grid cell can be quantified using resampling or randomisation tests such as a Monte-Carlo simulation. The basic idea is to build up a collection of artificial data sets of the same size as the actual data at hand, and then compute the test statistic of interest for each artificial dataset. This results in as many artificial values of the test statistic as there are artificial generated datasets. Taken together, these reference test statistics constitute an estimated null distribution, against which to compare the test statistic computed from the original data. Another approach, developed by the DNDC research group (Li et al., 2002a), calculates the gas fluxes for the combinations of the input parameters by only varying the most sensitive ones. In comparison with the randomisation method, the latter approach substantially reduces the complexity yet provides a reasonable first-order estimate of uncertainty present in up-scaling process.

5.6.4. Alternatives

The up-scaling of the soil emitted N trace gases is a complicated affair. Variations in soil atmosphere fluxes brought about by differences in climate, soil conditions and land use change result in perturbations in local and regional atmospheric concentrations at a range of spatial and temporal scales. A compromise therefore needs to be found. One such alternative is the use of a hierarchical up-scaling framework, within which estimates of regional N emissions using parameters that control soil NO emissions at the field and sub-grid scale are nested within a regional scale land use database (Matson et al., 1989). In such an approach by Kirkman et al. (2002c), an ecosystem model and satellite data were used to estimate soil NO emissions for a Brazilian region. In order to account for the inherent spatio-temporality of soil NO emissions, the scheme used the "time elapsed since primary forest deforestation" as a link between NO emissions that occur at small temporal scales and land use change (or land age), derived from LANDSAT data, at the spatial scale. Results showed that this nested hierarchical approach to up-scaling corresponded well with flux inventory-based estimates of regional soil NO emissions.

5.7. Sensitivity of soil nitrogen emission to environmental parameters

Available inventories of soil emission fluxes, presented in section 2, are useful to study, for example, the sensitivity of atmospheric chemistry to the soil emission fluxes reflecting the present-day land cover and land use. However, to what extent will land cover and land use change in the future and how this will affect biogeochemical processes, and in particular the soil nitrogen emissions? To indicate the magnitude of land cover and land use changes, estimates of the changes in land use in the past 300 hundred years show a decrease in forests and woodlands of 19 %, a 8 % decrease in grasslands and pastures and a large increase in croplands (>>100 %) (Richards, 1990). Keller et al. (1992) reported regional and global impacts of land conversion on soil nitrogen emissions, especially in the tropics whereas a study by Sanhueza et al. (1994) showed that the conversion of tropical natural grasslands to agriculture produced an increase in emissions of NO by a factor of 5.

Application of process models may help us to assess the impact of complex interactions between climate, land cover, land use, and atmospheric and biogeochemistry on nitrogen emission fluxes under present-day and also anticipated future land use and land cover conditions. To indicate the sensitivity of the soil N emissions to changes in environmental parameters as well as land management, results of a sensitivity analysis using the DNDC model are presented in Figures 7 and 8. It shows the sensitivity of the soil N fluxes for an Iowa cornfield with typical management conditions (fertilizer rate 150 kg N/ha, conservation tillage, no irrigation, no manure amendment).

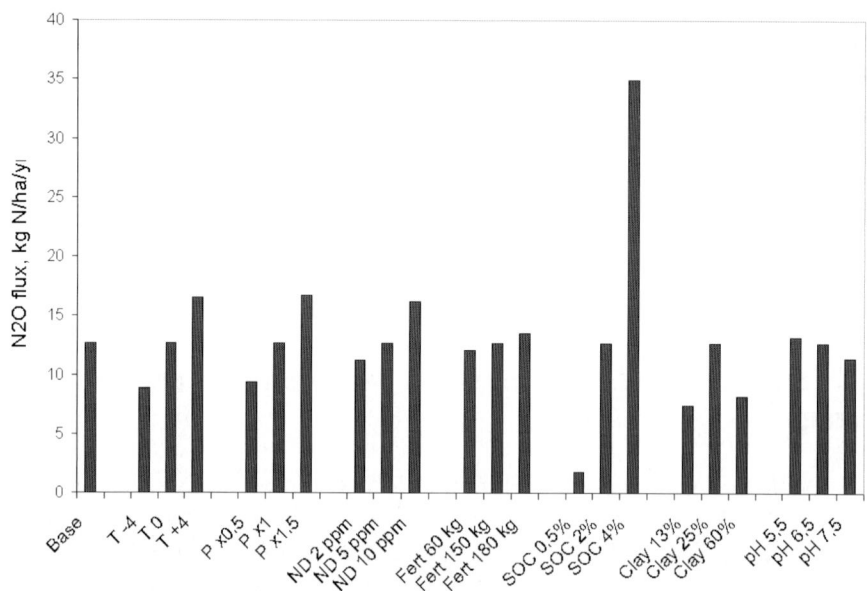

Figure 7. Sensitivity of DNDC-modeled N_2O fluxes to environmental and management factors. Soil organic carbon content is the most sensitive factor affecting N_2O emissions from the tested cropland in Iowa although climate change (e.g. changes in temperature (T), precipitation (P) and atmospheric N deposition (ND)). and other management measures (e.g. fertilizer rate (Fert.) also altered the soil N_2O fluxes.

The fluxes of N_2O, NO and NH_3 have been compared for various scenarios of changes in climate, soil properties or management measures. The simulations using DNDC indicate that the N_2O emissions from the cropland are most sensitive to the soil organic matter (SOM) content. Soil organic matter is the primary source of soil inorganic N and dissolved organic C (DOC), which both are the major drivers for nitrification and denitrification. High SOM contents often provide more inorganic N and DOC through decomposition to fuel nitrifiers or denitrifiers. The modelled relationship of N_2O with SOM is supported by many field or laboratory observations (e.g. Leffelaar and Wessel, 1988). For example, Terry et al.

(1981) observed a great amount of N_2O emitted from an organic soil in the Everglades in Florida, although there was no fertilizer used at all. SOC content, soil texture, soil pH and fertilizer rate have a significant impact on NO emissions. An increase in temperature or atmospheric N deposition also increased NO fluxes. The volatilisation and consequently the emission of NH_3 is highly sensitive to the soil pH and fertilizer rate, although changes in temperature, soil texture and SOC content also affect the NH_3 fluxes.

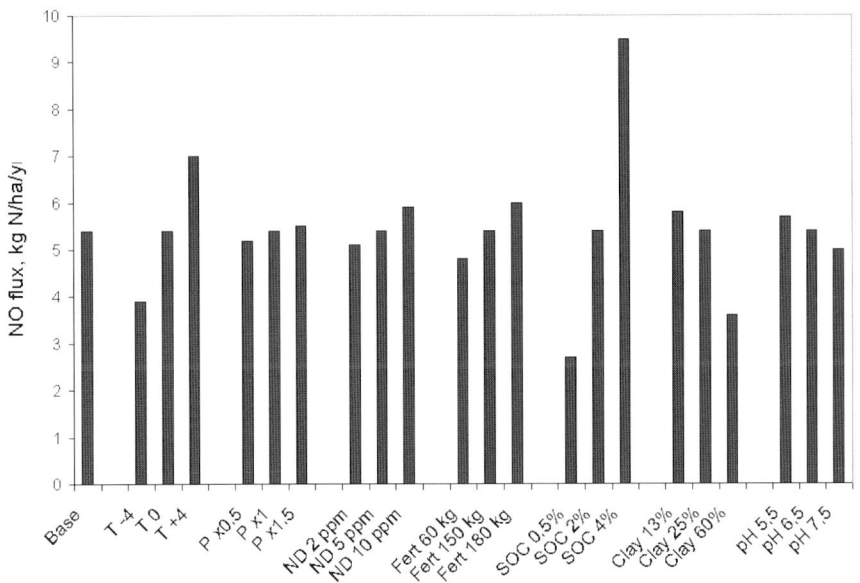

Figure 8. Sensitivity of DNDC-modeled NO fluxes to environmental and management factors. Soil organic carbon content, soil texture, soil pH and fertilizer rate had significant impacts on NO emissions. An increase in temperature (T) or atmospheric N deposition (ND) also increased NO fluxes.

Changes in crop type and tillage do not dramatically alter N emission fluxes. These sensitivities are somewhat dependent on the selected baseline conditions, but the main conclusions remain the same for a wide range of conditions. Results from such sensitivity analyses provide valuable information for up-scaling studies by indicating which parameters require the highest priority in the database compilation for use in process-based models for large-scale studies.

6. DISCUSSION

Recently it has been shown that process-oriented models of soil nitrogen emissions can be applied at the site scale as well as regional (Li et al., 2002a, Kirkman et al., 2002c) or even global scales (Potter et al., 1996). The regional and global scale studies so far have been mainly conducted constraining the models with geographical databases containing information on the major drivers. We have shown in section 5.3. that the estimated computing requirements should not impose too strong of a constraint on the model integrations using a detailed process-oriented model implemented in a chemistry and climate model like ECHAM. The advantage of implementing a process-based emission model in a chemistry and climate model will be that feedbacks between climate, atmospheric chemistry and biogeochemistry are explicitly resolved.

Probably the most limiting factor for the use of a process-based emission model in a model like ECHAM will be the characterization of the environmental parameters, such as soil and vegetation properties and land use management, at the regional and to global scale. Moreover, even the state-of-the-art process-based models inevitable contain to some extent parameterisations, thereby limiting the application to scales for which the parameterisations were supposed to be used. However, the models can be applied to identify the most sensitive parameters and, consequently, the focus of database compilation. Stewart et al. (1989) already discussed the issue of database compilation, and proposed to establish central international holding agencies to coordinate the activities required to improve the quality and accessibility of global scale databases. They also recognized the role of remote sensing for the compilation of large-scale databases. Potter et al., (1996) stated that improvements in the reliability of soil data sets are critical to progress in global trace gas model studies.

The quality of the input data should be high, so that the uncertainty range in the calculated fluxes due to the uncertainty in the applied input data does not exceed the calculated temporal and spatial variability in the soil N fluxes. However, to arrive at an estimate of the uncertainty range in the geographical databases requires highly specialized expertise concerning the use of GIS and remote sensing information. Therefore a strong collaboration with GIS and remote sensing specialists should be pursued. The modellers should indicate what input data are required to constrain the models, whereas the GIS and remote sensing specialists can indicate if these data are available, and most importantly, what the uncertainty is in these data. For example, Ganzeveld et al. (2002b) present an analysis of the uncertainty in the role of the canopy processes for NO_x surface exchanges due to uncertainties in the geographical input data. The afternoon NO_x canopy top flux calculated with a multi-layer atmosphere-biosphere trace gas exchange model for tropical rainforest with an leaf area index (LAI) of 11 (m^2 m^{-2}), which has been inferred from a high-resolution ecosystem database and satellite data, is about 50 % smaller compared to that calculated for an LAI of 7 due to a more efficient removal of NO_x by dry deposition. We recognize that an estimated LAI of 11, inferred for a few tropical forest locations, seems to be unrealistically high based on in-situ measurements of the LAI of tropical rainforest, which range from about 5 to 7 (e.g. Kruijt et al., 2000). However, the results of such analyses can be used as a feedback to experimentalists to initiate more specific measurements of some of the processes involved, e.g. NO_x dry deposition, and the remote sensing community that provides the algorithms to infer surface cover properties from satellite data.

An additional important aspect of the compilation of large-scale databases is the consistency between the databases used to constrain the calculations of different biogeochemical processes. This should prevent the calculation of an unrealistic combination of biogeochemical processes, e.g. a calculated soil N emission flux representative of an undisturbed tropical soil whereas dry deposition calculations use surface cover properties representative for a deforested area.

A major motivation for the implementation of process-based emissions models in chemistry and climate models is to analyse the interactions between biogeochemical processes that control the soil emissions, atmospheric chemistry, the climate system and land-use management for present-day as well as future conditions. This requires not only the compilation of databases describing the actual distribution of ecological and environmental drivers but also scenarios of future changes in these properties.

7. RECOMMENDATIONS

There will be a continuous demand for more field and laboratory measurements to collect good real time data to constrain the process-based models. These measurements should not only reflect the range of present-day but also anticipated future environmental conditions occurring at the large scale. Measurements representative for different scales supply complementary information, but we have to make sure that we understand the processes that control the spatial and temporal variability and can perform the up-scaling of measurements to obtain large-scale estimates.

We suggest that at smaller scales, studies of the enzymes responsible for the conversion of nitrogen species in soils should be intensified. At the field scale, improvement of measurements methods and investigation into the disagreement between methods (including inter-calibrations) should be a priority, as some of the uncertainty comes from these. Recommendations on how to improve data collection for the gaseous nitrogen species are found in the literature. For example, Veldkamp and Keller (1997) provide a list of suggestions for improving the estimates of fertilizer-induced NO emissions. Also, measurements in biomes where data are lacking should be encouraged. Short-term (pulsed) fluxes must be considered as well as the long-term fluxes. At the larger scale, recommendation and implementation of mitigation strategies should be the ultimate aim of these long-term measurement campaigns.

The assessment of the role of the canopy processes for the large-scale NO_x exchanges, and also for NH_3 exchanges, is largely limited by the availability of observations of deposition at the leaf as well as the canopy scale. Therefore, in line with a discussion by Delmas et al. (1997), we propose to pursue additional measurements of uptake (or release) of NO_x and NH_3 by different substrates such as dry and wet leaves and soils. These measurements should not only assess the fluxes for a large variety of different substrates, but also focus on the identification of the fundamental processes that control nitrogen uptake or release. This will facilitate the development of more process-based models for vegetation nitrogen exchanges. Since the role of dry deposition in controlling the canopy-flux divergence is closely connected to turbulent exchange, measurements of this process will also benefit a better understanding of the N surface exchanges. We also discussed the role of sub-grid scale dry deposition for up-scaling NH_3 and NO_x emission fluxes, indicating that this is a non-negligible sink for these species. This topic is being investigated mostly in projects focussing on local to regional scale air quality and acidification studies, which can hopefully provide improved parameterisations for implementation in regional or global scale models.

Concerning the future use of process-based soil N emission models, we will face two major challenges. One challenge is the further development of existing process models by incorporating more fundamental reactions. Currently, many so-called process models still contain many empirical functions, which reduces the capacity of the models to predict soil N emissions across climate zones, soil conditions, or land use and management regimes. Replacing the empirical functions with the fundamental mechanisms, which can be inferred from laboratory experiments or from theoretical analysis, would advance the evolution of existing models. The second challenge of the use of process-based emission models has already been discussed extensively; the development of geographical databases to constrain the process-oriented models. However, to be more specific, among all the important input parameters required by most of the existing process models, land use management and soil properties such as the soil texture and the soil organic matter are the most crucial parameters. Any effort to improve the availability as well as the quality of soil parameters at the regional to global scale will make a significant contribution to the predictions of N gas emissions at these scales.

8. CONCLUSIONS

The capacity of quantifying soil N emissions at site scale using biogeochemical models has been widely recognized through various validation and comparison studies (Frolking et al., 1998). One of the main challenges nowadays is to apply these biogeochemical models at large spatial scales, including global models. As presented here, several modelling studies (e.g. Li et al., 2002b) have provided detailed information of the spatially differentiated seasonality of soil N emissions at the regional scale. The success of these regional scale studies may imply that the major obstacle for applying biogeochemical models in global studies is shifting from model development to database construction. However, we should not wait for the entire mapping of ecological processes and process-based models to be completed. The interactions between soil emissions, canopy processes and the free troposphere are sufficiently well understood to include their representation in earth-system transport models with the understanding that there will be room for improvement in the long run.

To reduce the main uncertainties in the large-scale estimates of N gas emissions, a collaborative effort of the experimental and modelling communities is required. There will be a continuous demand for more observations to further develop the process-based models by including more fundamental process descriptions, and to evaluate the performance of the process-based emission models for a wide range of ecological and environmental conditions. On the other hand, the emission models can be used to interpret the observations by linking the observed signal to the controlling parameters, and to indicate through sensitivity analyses the parameters and processes that should be in focus of future measurement activities. Also, use of emission models at larger scales can provide additional feedback to optimise measurement strategies by identification of the ecosystems that are potentially important for the soil N emissions. Despite all the remaining uncertainties involved in the quantification of the soil N emissions at various spatial and temporal scales, the combined use of process-based models and laboratory, chamber and micrometeorological measurements seems to provide the most optimal framework to reduce the largest uncertainties in large-scale source estimates of soil nitrogen emissions.

9. ACKNOWLEDGMENTS

We would like to thank two anonymous reviewers and Phil Nightingale of the Plymouth Marine Laboratory, United Kingdom, and Franz Meixner of the Max-Planck Institute for Chemistry, Germany, for their constructive comments.

10. REFERENCES

Aber, J. D., and C. A. Federer, A generalized, lumped-parameter model of photosynthesis, evaporation and net primary production in temperate and boreal forest ecosystems, Oecologia, 92, 463-474, 1992.

Alexander, M., K. C. Marshall, and P. Hirsch, Autotrophy and heterotrophy in nitrification, Trans. Intern. Cong. Soil Sci. 7th Cong. Madison 2, 586-591, 1960.

Aneja V. P., H. H. Rogers, and E. P. Stahel, Dry deposition of ammonia at environmental concentrations on selected plant species, J. of the Air Pollution Control Association 36, 1338-1341, 1986.

Ambus, P., H. Clayton, J. R. M. Arah, K. A. Smith, and S. Christensen, Similar N_2O flux from soil measured with different chamber techniques, Atmos. Environ., 27, 121-123, 1993.

Anderson, I. C., and J. S. Levine, Relative rates of nitric oxide and nitrous oxide production by nitrifiers, denitrifiers, and nitrate respirers, Applied and Environ. Microbiol., 51, 938-945, 1986.

Atlas, R. M., and R. Bartha, Microbial Ecology, 3rd ed. The Benjamin/Cummings Publishing Company, Inc. California, USA, 1993.

Aulakh, M. S., J. W. Doran and A. R. Mosier, Soil denitrification - significance, measurement and effects of management, In: Advances in Soil Science, Stewart, B. A. (ed.), 1-57, Springer, New York, 1992.

Ayers G. P., and J. L. Gras, Ammonia gas concentrations over the Southern Ocean, Nature 284, 539-540, 1980.

Azhar, El. S., R. Verhe, M. Proot, P. Sandra, and W. Verstraete, Binding of nitrite-N on polyphenols during nitrification, Plant and Soil, 94, 369-382, 1986.

Baumgartner, M., E. Bock, and R. Conrad, Processes involved in uptake and release of nitrogen-dioxide from soil and building stones into the atmosphere, Chemosphere, 24, 1943-1960, 1992.

Beauchamp, E. G., Nitrous oxide emission from agricultural soils, Canadian J. of Soil Sci., 77, 113-123, 1997.

Berges, M. G. M. and P. Crutzen, Estimates of global N_2O emissions from cattle, pig and chicken manure, including a discussion of CH_4 emissions, J. Atmos. Chem., 24, 241-269, 1996.

Bollmann A, Conrad R, Acetylene blockage technique leads to underestimation of denitrification rates in oxic soils due to scavenging of intermediate nitric oxide, Soil Biol. Biochem., 29, 1067-1077, 1997.

Bollmann A. and R. Conrad, Influence of O_2 availability on NO and N_2O release by nitrification and denitrification in soils, Global Change Biology, 4, 387-396, 1998.

Bouwman, A. F., K. W. van der Hoek, and J. G. J. Olivier, Uncertainties in the global source distribution of nitrous oxide, J. Geophys. Res., 100, 2785-2800, 1995.

Bouwman, A. F., Direct emission of nitrous oxide from agricultural soils, Nutr. Cycl. Agroecosyst. 46, 53-70, 1996.

Bouwman, A. F. and W. A. H. Asman, Scaling of nitrogen gas fluxes from grasslands, in: Gaseous emissions from grasslands, Jarvis, S. C., and B. F. Pain (eds.), CAB international, Wallingford, UK, 311-330, 1997.

Bouwman, A. F., D. S. Lee, W. A. H. Asman, F. J. Dentener, K. W. Van Der Hoek, and J. G. J. Olivier, A global high-resolution emission inventory for ammonia, Global Biogeochem. Cycles, 11, 561-587, 1997.

Bramley, R. G. V. and R. E. White, The variability of nitrifying activity in field soils, Plant Soil, 126, 203-208, 1990.

Bremner, J. M. and K. Shaw, Denitrification in soil. II. Factors affecting denitrification, J. of Agric. Sci., 51, 40-52, 1958.

Bremner J. M. and A. M. Blackmer, Terrestrial nitrification as a source of atmospheric N_2O, In: Denitrification, Nitrification and Atmospheric N_2O. Delwiche, C. C. (ed.), 151-170, Wiley, New York, 1981.

Brown L, S. Jarvis, R. Sneath, V. Phillips, K. Goulding, and C. Li, Development and application of a mechanistic model to estimate emission of nitrous oxide from UK agriculture. Atmos. Environ., 36, 917-928, 2002.

Butterbach-Bahl K., F. Stange, H. Papen, G. Grell, and C. Li, Impact of changes in temperature and precipitation on N_2O and NO emissions from forest soils, In: Non-CO_2 Greenhouse Gases: Specific Understanding, Control and Implementation, J. van Ham et al. (eds.), 165-171, Kluwer Academic Publishers, the Netherlands, 2001.

Butterbach-Bahl, K., F. Stange, H. Papen, and C. Li, Regional inventory of nitric oxide and nitrous oxide emissions for forest soils of Southeast Germany using the biogeochemical model PnET-N-DNDC, J. Geophys. Res., 106, 34,155-34,165, 2001.

Cai, Z., Sawamoto, S., Li, C., Kang, G., Boonjawat, J., Mosier, A., and R. Wassmann, 2002, Field validation of the DNDC model for greenhouse gas emissions in East Asian cropping systems, submitted to Global Biogeochemical Cycles, 2003.

Cárdenas, L., A. Rondón, C. Johansson and E. Sanhueza, Effects of soil moisture, temperature, and inorganic nitrogen on nitric oxide emissions from acidic tropical savannah soils, J. Geophys. Res, 98, 14783-14790, 1993.

Cárdenas, L.M., J.M.B. Hawkins, D. Chadwick and D. Scholefield, A laboratory technique for the measurements of biogenic gas emissions from soils, submitted to Soil Biol. Biochem, 2003.

Chalk P.M. and C.J. Smith, Chemodenitrification. In: Gaseous Loss of Nitrogen from Plant-Soil Systems, Freney, J. R., and J.R. Simpson (eds.), Martinus Nijhoff Publishers, The Hague, 65-89, 1983.

Colbourn, P., Limits to denitrification in two pasture soils in a temperate maritime climate, Agric. Ecosyst. Env., 43, 49-68, 1993.

Colbourn, P., Denitrification and N_2O production in pasture soils. The influence of nitrogen supply and moisture, Agriculture, Ecosystems and Environment, 39, 267-278, 1997.

Conrad, R., Compensation concentration as critical variable for regulating the flux of trace gases between soil and atmosphere, Biogeochemistry, 27, 155-179, 1994.

Conrad, R., Soil micoorganisms as controllers of atmospheric trace gases (H_2, CH_4, OCS, N_2O and NO), Micro-biological Reviews, 60, 609-640, 1996.

Cowling D. W., and D. R. Lockyer, Increased growth of ryegrass exposed to ammonia, Nature 292, 337-338, 1981.

Crutzen, P. J., Estimates of possible variations in total ozone due to natural causes and human activities, Ambio, 3, 201-201, 1974.

Dasgupta, P., A diffusion scrubber for the collection of atmospheric gases, Atmos. Environ., 18, 1593-1599, 1984.

Davidson, E. A., Fluxes of nitrous oxide and nitric oxide from terrestrial ecosystems. In: Microbial production and consumption of greenhouse gases: methane, nitrogen oxides, and halomethanes, Rogers, J. E., and W. B. Whitman (eds.), 219-235. Washington: American Society for Microbiology, 1991.

Davidson E.A., Sources of nitric oxide and nitrous oxide following wetting of dry soil, Soil Soc. Sci. Am. J., 56, 95-102, 1992a.

Davidson E.A., Pulses of nitric oxide and nitrous oxide flux following wetting of dry soil: an assessment of probable sources and importance relative to annual fluxes, Ecological Bulletin, 42, 149-155, 1992b.

Davidson, E. A. and W. Kingerlee, A global inventory of nitric oxide emissions from soils, Nutr. Cycling Agroecosys., 48, 37-50, 1997.

Davidson, E. A. M. Keller, H. E. Erickson, L. V. Verchot, and E. Veldkamp, Testing a conceptual model of soil emissions of nitrous and nitric oxides, BioScience, 50, 667-680, 2000.

Del Grosso, S. J., W. J. Parton, A. R. Mosier, D. S. Ojima, A.E. Kulmala, and S. Phongpan, General model for N2O gas emissions from soils due to nitrification, Global Biogeochem. Cycles, 14, 1045-1060, 2000.

Delmas, R., D. Serça and C. Jambert, Global inventory of NOx sources, Nutr. Cycling Agroecosys., 48, 51-60, 1997.

Dendooven, L., P. Splatt, J. M. Anderson and D. Scholefield, Kinetics of the denitrification process in a soil under permanent pasture, Soil Biol. Biochem., 26, 361-370, 1994.

Dendooven, L. and J. M. Anderson, Maintenance of denitrification potential in pasture soil following anaerobic events, Soil Biol. Biochem., 27, 1251-1260, 1995.

Denmead, O. T., J. R. Freney, and J. R. Simpson, A closed ammonia cycle within a plant canopy, Soil Biol. Biochem., 8, 161-164, 1976.

Dentener, F. J., and P. J. Crutzen, A three-dimensional model of the global ammonia cycle, J. Atmos. Chem., 19, 331-369, 1994.

Dickinson, C. H. and G. J. F. Pugh, Biology of Plant Litter Decomposition, Vols. I and II, Dickinson, C. C and G. J. F. Pugh, (eds.), Academic Press, New York, 1974.

Duyzer, J., H. Weststrate, and S. Walton, Exchange of ozone and nitrogen oxides between the atmosphere and coniferous forest, Water, Air, Soil Pollut., 85, 2065-2070, 1995.

Erickson, H., M. Keller, and E. Davidson, Nitrogen oxide fluxes and nitrogen cycling during post-agricultural succession and forest fertilization in the humid tropics, Ecosystems, 4, 67-84, 2001.

Farquhar, G. D., P. M. Firth, R. Wetselaar, and B. Weir, On the gaseous exchange of ammonia between leaves and the environment: Determination of the ammonia compensation point, Plant. Physiol., 66, 710-714, 1980.

Firestone, M. K. and J. M. Tiedje, Temporal change in nitrous oxide and dinitrogen following onset of anaerobiosis, Applied Environ. Microbiol., 38, 673-679, 1979.

Firestone, M. K., Biological denitrification, Agronomy, 22, 289-326, 1982.

Firestone M. K., and E. A. Davidson, Microbiological basis of NO and N_2O production and consumption in soil, In: Exchange of trace gases between terrestrial ecosystems and the atmosphere, Andreae, M. O., and D. S. Schimel (eds.), John Wiley & Sons Ltd., Chichester, UK, 7-21, 1989.

Focht, D. D., and W. Verstraete, Biochemical ecology of nitrification and denitrification, Adv. Microbial. Ecol. 1, 135-214, 1977.

Folorunso, O. A., and D. E. Rolston, Spatial variability of field-measured denitrification gas fluxes, Soil Science Society of America Journal 48, 1214-1219, 1984.

Freeney, D. R., I. R. Fillery, and G. P. Marx, Effect of temperature on the gaseous nitrogen products of denitrification in a silt loam soil, Soil Soc. Sci. Am. J., 43, 1124-1128, 1979.

Freney, J. R., J. R. Simpson and O. T. Denmead, Volatilization of ammonia, in: Gaseous loss of nitrogen from plant-soil ecosystems, Freney, J. R., and J.R. Simpson (eds.), Developments in Plant Soil Sciences, Vol. 9, Martinus Nijhoff, The Hague, 1-32, 1983.

Freney, J.R. and J. R. Simpson (eds.), Gaseous loss of nitrogen from plant-soil systems, Kluwer Acad. Press, 1983.

Frolking, S. E., A. R. Mosier, D. S. Ojima, C. Li, W. J. Parton, C. S. Potter, E. Priesack, R. Stenger, C. Haberbosch, P. Dörsch, H. Flessa, and K. A. Smith, Comparison of N_2O emissions from soils at three temperate agricultural sites: simulations of year-round measurements by four models, Nutr. Cycl. Agroecosyst., 52, 77-105, 1998.

Fuentes, J. D., T. J. Gillespie, G. den Hartog, and H. H. Neumann, Ozone deposition onto a deciduous forest during dry and wet conditions, Agric. Forest. Meteorol., 62, 1-18, 1992.

Fuhrer, K., A. Neftel, M. Anklin and V. Maggi, Continuous measurements of hydrogen peroxide, formaldehyde, calcium and ammonium concentrations along the new grip ice core from summit, central Greenland, Atmos. Environ., 27A, 1873-1880, 1993.

Galbally, I. E., and C. R. Roy, Loss of fixed nitrogen from soils by nitric oxide exhalation, Nature, 275, 734-735, 1978.

Galbally, I.E., and C. Johansson, A model relating laboratory measurements of rates of nitric oxide production and field measurements of nitric oxides from soils, J. Geophys. Res., 94, 6473-6480, 1989.

Galle, B., L. Klemedtson, B. Bergqvist, M. Ferm, K. Törnqvist, D. W. T. Griffith, N. Jensen and F. Hansen, Measurements of ammonia emissions from spreading of manure using gradient FTIR techniques, Atmos. Environ. 34, 4907-4915, 2000.

Gallon, J. R., The oxygen sensitivity of nitrogenase: a problem for biochemists and microorganisms, Trends Biochem. Sci., 6, 19-23, 1981.

Galsworthy, A. M. and J. R. Burford, A system for measuring the rates of evolution of nitrous oxide and nitrogen from incubated soils during denitrification, J. Soil Science, 29, 5370-550, 1978.

Ganzeveld, L., J. Lelieveld, F. J. Dentener, M. C. Krol, and G.-J. Roelofs, Atmosphere-biosphere trace gas exchanges simulated with a single-column model, J. Geophys. Res. 107, 10.1029/2001 JD000684, 2002a.

Ganzeveld, L., J. Lelieveld, F. J. Dentener, M. C. Krol, A. F. Bouwman, and G.-J. Roelofs, The influence of soil-biogenic NOx emissions on the global distribution of reactive trace gases: the role of canopy processes, J. Geophys. Res., 107, 10.1029/2001 JD001289, 2002b.

Gao, W., M. L.Wesely, and P. V. Doskey, Numerical modelling of the turbulent diffusion and chemistry of NOx, O3, isoprene and other reactive trace gases in and above a forest canopy, J. Geophys. Res., 98, 18, 339-18,353, 1993.

Genfa, Z. and P. K. Dasgupta, Fluorometric measurement of aqueous ammonium ion in a flow injection system, Anal. Chem., 61, 408-412, 1989.

Glasstone, S., Textbook of Physical Chemistry. 2nd ed., 1320 pages. Van Nostrand Reinhold. New York, 1946.

Gödde, M. and R. Conrad, Simultaneous measurement of nitric oxide production and consumption in soil using a simple static incubation system, and the effect of soil water content on the contribution of nitrification, Soil Biol. Biochem., 30, 433-442, 1998.

Gut, A., S. M. Van Dijk, M. Scheibe, U. Rummel, M. Welling, C. Ammann, F. X. Meixner, G. A. Kirkman, M. O. Andreae, B. E. Lehmann, NO emission from an Amazonian rain forest soil: Continuous measurements of NO flux and soil concentration, J. Geophys. Res., 107, D20, 8057, doi:10.1029/2001JD000521, 2002

Hameed, S., and J. Dignon, Changes in the geographical distributions of global emissions of NOx and SOx from fossil-fuel combustion between 1966 and 1980, Atmos. Environ., 22, 441-449, 1988.

Harmsen, G. W. and D. A. van Schreven, Mineralization of organic nitrogen in soil. Adv. Agronomy, 7, 299-398, 1955.

Harper L. A., Sharpe R. R., Langdale G. W., and Giddens J. E, Nitrogen cycling in a wheat crop: soil-plant and aerial nitrogen transport. Agron. J. 79, 965-973, 1987.

Harrison, R. M., S. Yamulki, K. W. T. Goulding, C. P. Webster, Effect of fertilizer application on NO and N2O fluxes from agricultural fields, J. Geophys. Res. 100, 25923-25931, 1995.

Hart, S. C., J. M. Stark, E. A. Davidson, and M. K. Firestone, Nitrogen mineralization, immobilization and nitrification, in: Methods of Soil Analysis: II Microbial and Biogeochemical Properties, R. W. Weaver et al. (eds.), Soil Sci. Soc. of Am., Madison, Wis., 985-1018, 1994.

Hatch, D. J., S. C. Jarvis and L. Philips, Field measurement of nitrogen mineralisation using soil core incubation and acetylene inhibition of nitrification, Plant Soil, 124, 97-107, 1990.

Hillel, D., Computer simulation of soil-water dynamics: A compendium of recent work, Internation Development Research Center, Ottawa, Canada, 90-94, 1977.

Holland, E. A., B. H. Braswell, J.-F. Lamarque, A. Townsend, J. Sulzman, J.-F. Müller, F. Dentener, G. Brasseur, H. Levy II, J. E. Penner, and G.-J. Roelofs, Variations in the predicted spatial distribution of atmospheric nitrogen deposition and their impact on carbon uptake by terrestrial ecosystems, J. Geophys. Res., 102, 15,849-15,866, 1997.

Holland, E. A. and J.-F. Lamarque, Modeling bio-atmosphere coupling of the nitrogen cycle through NOx emissions and NOy deposition, Nutr. Cycling Agroecosys., 48, 7-24, 1997.

Hooker M. L., D. H. Sanders, G. A. Peterson, and L. A. Daigger, Gaseous N losses from winter wheat. Agron. J. 72, 789-792, 1980.

Hutchinson, G. L., Air containing nitrogen-15 ammonia: Foliar absorption by corn seedings, Science 175, 759-761, 1972.

Hutchinson, G. L. and C. E. Andre, Flow-through incubation system for monitoring aerobic soil nitric and nitrous oxide emissions, Soil Sci. Soc. of A. J., 53, 1068-1074, 1989.

Hutchinson, G. L., W. D. Guenzi and G. P. Livingston, Soil water controls on aerobic soil emission of gaseous nitrogen oxides, Soil Biol. Biochem., 25, 1-9, 1993.

Hutchinson, G. L., M. F. Vigil, J. W. Doran, and A. Kessavalou, Coarse-scale soil-atmosphere NOx exchange modelling: status and limitations, Nutr. Cycling Agroecosys., 48, 25-35, 1997.

IPCC, Climate Change 1995: The Science of Climate Change, Contribution of Working Group I to the Second Assessment Report of the Intergovernmental Panel on Climate change, Sanhueza, E., and X. Zhou, (eds.), Cambridge University Press, Cambridge, UK, 1995.

IPCC (Intergovernmental Panel on Climate Change/Organization for Economic Cooperation and Development). Guidelines for National Greenhouse Gas Inventories. OECD/ODCE, Paris, 1997.

IPCC, Climate Change 2001: The Scientific Basis. Contribution of Working Group I to the Third Assessment Report of the Intergovernmental Panel on Climate Change, Houghton, J. T., Y. Ding, D. J. Griggs, M. Noguer, P. J. van der Linden, X. Dai, K. Maskell, and C. A. Johnson (eds.), Cambridge University Press, Cambridge, United Kingdom and New York, NY, USA, 881p, 2001.

Jacob, D. J., and S. C. Wofsy, Budgets of reactive nitrogen, hydrocarbons and ozone over the Amazon forest during the wet season, J. Geophys. Res., 95, 16,737-16,754, 1990.

Jambert, C., D. Serça and R. Delmas, Quantification of N-losses as NH_3, NO, and N_2O and N_2 from fertilized maize fields in southwestern France, Nutr. Cycl. Agroecosyst., 48, 91-104, 1997.

Jansson, S. L. and F. E. Clark, Losses of nitrogen during decomposition of plant material in the presence of inorganic nitrogen, Soil Sci. Soc. Am. Proc., 16, 330-334, 1952.

Johansson, C., Pine forest: a negligible sink for atmospheric NO_x in rural Sweden, Tellus 39B, 426-438, 1987.

Joss, U., and W. K. Graber, Profiles and simulated exchange of H_2O, O_3, NO_2 between the atmosphere and the HartX Scots Pine Plantation, Theor. Appl. Climatol., 53, 157-172, 1996.

Keller, M., I. Galbally, M. Baer, E. Davidson, D. Fitzjarrald, G. Harris, C. Johansson, P. Matson, C. Nobre, E. Sanhueza and J. Stewart, Tropical land use change and trace gas emissions, Ecol. Bull., 42, 156-163, 1992.

Kelly, R. H., W. J. Parton, M. D. Hartman, L. K. Strerch, D. S. Ojima, and D. S. Schimel, Intra-annual and interannual variability of ecosystem processes in shortgrass steppe, J. Geophys. Res., 105, 20093-20100, 2000.

Kiese, R., C. Li, D. Hilbert, H. Papen, and K. Butterbach-Bahl, 2002. Regional application of PnET-N-DNDC for estimating the N2O source stength of tropical rainforests in the Wet Tropics of Australia. *Global Biogeochemical Cycles* (in review).

Kirchner, M., S. Braeutigam, M. Ferm, M. Haas, M. Hangartner, P. Hofschreuder, A. Kasper-Giebl, H. Römmelt, J. Striedner, W. Terzer, L. Thöni, H. Werner and R. Zimmerling, Field intercomparison of diffusive samplers for measuring ammonia, J. Environ. Monit., 1, 259-265, 1999.

Kirkman, G. A., W. X. Yang, and F. X. Meixner, Biogenic nitric oxide emissions up scaling: an approach for Zimbabwe, Global Biogeochem. Cycles, 15, 1005-1020, 2001

Kirkman, G. A., A. Gut, C. Ammann, L. V. Gatti, A. M. Cordova, M. A. L. Moura, M. O. Andreae, and F. X. Meixner, Surface exchange of nitric oxide, nitrogen dioxide and ozone at a cattle pasture in Rondonia, Brazil, J. Geophys. Res., in press, 2002a

Kirkman, G.A., C. Ammann, A. Gut, A., E. A. Holland, F. X. Meixner, Soil NO emissions in Rondônia, Brazil, I: A model case study, Ecological Applications, submitted, 2002b

Kirkman, G. A., C. Ammann, E. A. Holland, D. A. Roberts, and F. X. Meixner, Soil NO emissions in Rondônia, Brazil, II: A regional up-scaling, Ecological Applications, submitted, 2002c

Klemedtson, L., B. B. H. Svensson and T. Roswall, A method of selective inhibition to distinguish between nitrification and denitrification as source of nitrous oxide in soil, Biol. and Fertil. of soils, 6, 112-119, 1988.

Knowles, R., Denitrification. In: Terrestrial Nitrogen Cycles. Processes, Ecosystem Strategies and Management Impacts. Clark, F. E. and T. Roswell, (eds.), Ecol. Bull., 33, 315-329, Stockholm, 1981.

Koskinen, W. C, and D. R. Keeney, Effect of pH on the rate of gaseous products of denitrification in a silt loam soil, Soil Sci. Soc. Am. J., 46, 6, 1165-1167, 1982.

Kruijt, B., Y. Malhi, J. Lloyd, A. D. Nobre, A. C. Miranda, M. G. P. Pereira, A. Gulf, and J. Grace, Turbulence statistics above and within two Amazon rain forest canopies, Boundary Layer Meteorol., 94, 297-331, 2000.

Krüger, O., and J.-P. Tuovinen, The effect of variable sub-grid deposition factors on the results of the lagragian long-range transport model of EMEP, Atmos. Environ., 31, 4199-4209, 1997.

Kudeyerov, V. N., Mobility of fixed ammonium in soil. In: Terrestrial Nitrogen Cycles. Processes, Ecosystem Strategies and Management Impacts. Clark, F. E. and T. Rosswall (eds.), Ecol. Bull., 33, 281-290, Stockholm, 1981.

Lee, D. S., I. Köhler, E. Grobler, F. Rohrer, R. Sausen, L. Gallardo-Klenner, J. J. G. Olivier, F. J. Dentener, and A. F. Bouwman, Estimations of global NOx emissions and their uncertainties, Atmos. Environ., 31, 1735-1749, 1997.

Leffelaar P. A., and W. W. Wessel, Denitrification in a homogeneous, closed system: experiment and simulation, Soil Sci., 146, 335-349, 1988.

Lerdau, M. T., J. W. Munger, and D. J. Jacob, The NO2 flux conundrum, Science, 289, 2291-2293, 2000.

Levy II, H. and W. J. Moxim, Simulated global distribution and deposition of reactive nitrogen emitted by fossil fuel combustion, Tellus, 41, 256-271, 1989.

Li C., S. Frolking, and T.A. Frolking, A model of nitrous oxide evolution from soil driven by rainfall events: 1. Model structure and sensitivity, J. Geophys. Res., 97, 9759-9776, 1992.

Li, C. S., S. Frolking, and R. C. Harriss, Modeling carbon biogeochemistry in agricultural soils. Global Biogeochem. Cycles, 8, 237-254, 1994.

Li, C., V. Narayanan, and R. Harriss, Model estimates of nitrous oxide emissions from agricultural lands in the United States, Global Biogeochemical Cycles, 10, 297-306, 1996.

Li, C., Modeling trace gas emissions from agricultural ecosystems, Nutr. Cycl. Agroecosyst, 58, 259-276, 2000.

Li, C., J. Aber, F. Stange, K. Butterbach-Bahl, and H. Papen, A process-oriented model of N2O and NO emissions from forest soils: 1. Model development, J. Geophys. Res., 105, 4369-4384, 2000.

Li, C., Y. H. Zhuang, M. Q., Cao, P. M. Crill, Z. H. Dai, S. Frolking, B. Moore, W. Salas, W. Z. Song, X. K. Wang, Comparing a national inventory of N_2O emissions from arable lands in China developed with a process-based agro-ecosystem model to the IPCC methodology, Nutr. Cycl. Agroecosyst., 60, 159-175, 2002a.

Li, C., J Qiu, S. Frolking, X. Xiao, W. Salas, B. Moore III, S. Boles, Y. Huang, and R. Sass, Reduced methane emissions from large-scale changes in water management in China's rice paddies during 1980-2000, Geophysical Research Letters, 29, doi:10.1029/2002GL015370, 2002b.

Li, C., J Qiu, S. Frolking, X. Xiao, W. Salas, B. Moore III, S. Boles, Y. Huang, and R. Sass, Changing Water Management in China's Rice Paddies and the Decline in the Growth Rate of Atmospheric Methane 1980-2000, Geophysical Research Letters, in press, 2002c.

Li, C., Y. H. Zhuang, S. Frolking, B. Moore, J. Galloway, and D. Shimel, Soil carbon loss threatens sustainability of Chinese agriculture: Modeling soil organic carbon change in croplands of China., Ecological Applications, 13, 327-336, 2003.

Liu, S., W. A. Reiners, M. Keller, and D. S. Schimel, Simulation of nitrous oxide and nitric oxide emissions from tropical primary forests in the Costa Rican Atlantic Zone, Environ. Modelling & Software, 15, 727-743, 2000.

Lockyer, D. R., A system for the measurement in the filed of losses of ammonia through volatilisation, J. Sci. Food Agric., 35, 837-848, 1984.

Lockyer D. R., and D. C. Whitehead, The uptake of gaseous ammonia by the leaves of Italian Rygrass. J. Exp. Bot. 37, 919-927, 1986.

Logan, J. A., Nitrogen oxides in the troposphere: Global and regional budgets, J. Geophsy. Res., 88, 10,785-10,807, 1983.

Ludwig, J., F. X. Meixner, B. Vogel, and J. Förstner, Soil-air exchange of nitric oxide: An overview of processes, environmental factors, and modeling studies, Biogeochemistry, 52, 225-257, 2001.

MacDuff, J. H. and R. E. White, Net mineralisation and nitrification rates in a clay soil measured and predicted in a permanent grassland from soil temperature and moisture content, Plant Soil, 86, 151-172, 1985.

McGarity, J.W., and J.A. Rajaratnam, Apparatus for the measurement of losses of nitrogen as gas from the filed and simulated field environments, Soil Biol. Biochem., 5, 121-131, 1973.

Matson, P. A., P. M. Vitousek, and D. S. Schimel, Regional extrapolation of trace gas flux based on soils and ecosystems, in: Exchange of trace gases between terrestrial ecosystems and the atmosphere, M. O. Andreae and D. S. Schimel (eds.), 97-108, 1989.

Matson, P., NOx emission from soils and its consequences for the atmosphere and biosphere: critical caps and research directions for the future, Nutr. Cycling Agroecosys., 48, 1-6, 1997.

Matthews, R., Nitrogenous fertilizers: Global distribution of consumption and associated emissions of nitrous oxide and ammonia, Global B. Cycles, 8, 411-439, 1994.

Meixner, F. X., Surface exchange of odd nitrogen oxides, Nova Acta Leopoldina NF, 70, Nr. 288, 299-348, 1994.

Meixner, F. X., and W. Eugster, Effects of landscape patterns and topography on emissions and transport, in: Integrating Hydrology, Ecosystem Dynamics, and Biogeochemistry in Complex Landscapes, J. D. Tenhunen, and P. Kabat (eds.), Dahlem Workshop Report, 147-175, John Wiley & Sons Ltd., Chichester, 1999.

Meyer, M. W., Absorption and release of ammonia from and to the atmosphere by plants. Dissertation directed by J.H. Axley, University of Maryland, College Park, 1973.

Misselbrook, T. H., D. R. Chadwick, B. F. Pain and D. M. Headon, Dietary manipulation as a means of decreasing N losses and methane emissions and improving herbage N uptake following application of pig slurry to grassland, J. Agric. Science, 130, 183-191, 1998.

Misselbrook, T. H., T. J. Van Der Weerden, B. F. Pain, S. C. Jarvis, B. J. Chambers, K. A. Smith, V. R. Phillips and T. G. M. Demmers, Ammonia emission factors for UK agriculture, Atmos. Environ. 34, 871-880, 2000.

Morrill, L. G., and J. E. Dawson, Patterns observed for the oxidation of ammonium to nitrate by soil organisms, Soil Sci. Soc. Am. Proc., 31, 757-760, 1967.

Mosier, A. R., J. M. Duxbury, J. R. Freneny, O. Heinimeyer and K. Minami, Nitrous oxide emissions from agricultural fields: Assessment, measurement and mitigation, Plant Soil, 181, 95-108, 1996.

Mosier, A., C. Kroeze, C. Nevison, O. Oenema, S. Seitzinger and O. van Cleemput, Closing the global N_2O budget: nitrous oxide emissions through the agricultural nitrogen cycle, Nutr. Cycling Agroecosys., 52, 225-248, 1998.

Nommik, H., Investigations on denitrification in soil, Acta Agr. Scand., 6, 195-228, 1956.

Otter, L. B., W. X. Yang, M. C. Scholes, and F. X. Meixner, Nitric oxide emission from a southern African savanna, J. Geophys.. Res., 104, 18471-18485, 1999

Painter, H. A., A review of literature on inorganic nitrogen metabolism in microorganisms. Water Res., 4, 393-450, 1970.

Parkin, T. B., H. F. Kaspar, A. J. Sexstone and J. F. Tiedje, A gas-flow soil core method to measure field denitrification rates, Soil Biol. Biochem., 16, 323-330, 1984.

Parkin, T.B., A.J. Sextone, abd J.M. Tiedje, Adaptation of denitrifying populations to low soil pH, Appl. Environ. Microbiol., 49, 1053-1056, 1985.

Parnas, H., Model for decomposition of organic material by microorganisms, Soil Biol. Biochem., 7, 161-169, 1975.

Parsons, J. W. and J. Tinsley, Nitrogenous Substances. In: Soil Component, Geiseking, J. E. (ed.), New York, Springer-Verlag. 1: 263-304, 1975.

Parton W. J., Predicting soil temperatures in a shortgrass steppe, Soil Sci., 138, 93-101, 1984.

Parton W. J., D. S. Schimel, C. V. Cole, and D. S. Ojima, Analysis of factors controlling soil organic matter levels in Great Plains grasslands, Soc. America J., 51, 1173-1179, 1987.

Parton W. J., J. A. Morgan, J. M. Altenhofen, and L. A. Harper, Ammonia volatilization from spring wheat plants. Agron. J., 80, 419-425, 1988.

Parton W. J., A. R. Mosier, D. S. Ojima, D. W. Valentine, D. S. Schimel, K. Weier, and A. E. Kulmala, Generalized model for N_2 and N_2O production from nitrification and denitrification, Global Biogeochem. Cycles, 10, 401-412, 1996.

Parton W. J., M. D. Hartman, D. Ojima, and D. Schimel, DAYCENT and its land surface submodel, Global Planetary Change, 19, 35-48, 1998.

Parton W. J., E. Holland, S. Del Grosso, M. D. Hartman, R. Martin, R. Arvin, R. Mosier, D. S. Ojima, and D. S. Schimel, Generalized model for NO_x and N_2O emissions from soils, J. Geophys. Res., 106, 17,403-17,420, 2001.

Paul, E. A., and F. E. Clark, Soil Microbiology and Biochemistry, Academic Press Ltd., London, 1989.

Payne, W. J., The status of nitric oxide and nitrous oxide intermediates in denitrification, In: Denitrification, Nitrification and Atmospheric N_2O. Delwiche, C. C, (ed.), 85-103. Wiley, New York, 1981.

Pérez, T., S. E. Trumbore, S. C. Tyler, E. A. Davidson, M. Keller and P. B. de Camargo, Isotopic variability of N2O emissions from tropical forest soils, Global Biogeochem. Cycles, 14, 525-535, 2000.

Plant R. A. J., E. Veldkamp, C. Li, Modeling nitrous oxide emissions from a Costa Rican banana plantation, in: Effects of Land Use on Regional Nitrous Oxide Emissions in the Humid Tropics of Costa Rica, R. A. J. Plant (ed.), Universal Press, Veenendaal, pp. 41-50, 1998.

Postgate, J. R., The Fundamentals of Nitrogen Fixation, Cambridge University Press, Cambridge, England, 1982.

Postgate, J. R., Nitrogen Fixation, 2nd edition. Studies in Biology, no. 92, Edward Arnold, London, 1987.

Poth M. and D.D. Focht, 15-N Kinetics analysis of N2O production by Nitrosomonas europaea: an examination of nitrifier denitrification, Applied and Environ. Microbio., 49, 1134-1141, 1985.

Potter C. S., J. T. Randerson, and C. B. Field, Terrestrial ecosystem production: a process model based on global satellite and surface data. Global B. Cycles, 7, 811-841, 1993.

Potter, C. S., P. A. Matson, P. M. Vitousek, and E. Davidson, Process modeling of controls on nitrogen trace gas emissions from soils worldwide, J. Geophys. Res., 101, 1361-1377, 1996.

Potter, C. S., E. A. Davidson, S. A. Klooster, D. C. Nepstad, G. H. Negreiros, and V. Brooks, Regional application of an ecosystem production model for studies of biogeochemistry in Brazilien Amazonia, Global Change Biology, 4, 315-333, 1998.

Remde, A., J. Ludwig, F. X. Meixner, and R. Conrad, A study to explain the emission of nitroc oxide from a marsh soil, J. Atmos. Chem., 17, 249-275, 1993.

Richards, J. F., Land transformation. In: The Earth as Transformed by Human Action. Al., B.L.T.e. (ed), Cambridge Univ. Press, New York, 163-178, 1990.

Rochette, P., E. G. Gregorich and R. L. Desjardins, Comparison of static and dynamic closed chambers for measurements of soil respiration under field conditions, Can. J. Soil Sci., 72, 605-609, 1992.

Rolston, D. E., M. Fried and D. A. Goldhamer, Denitrification measured directly from nitrogen and nitrous oxide gas fluxes, Soil Sci. Soc. Am. J., 40, 259-266, 1975.

Rondon, A., C. Johansson, and L. Granat, Dry deposition of nitrogen dioxide and ozone to coniferous forest, J. Geophys. Res., 98, 5159-5172, 1993.

Rosswall T., Microbiological regulation of the biogeochemical nitrogen cycle, Plant Soil, 67, 15-34, 1982.

Rudolph, J., F. Rothfuss and R. Conrad, Flux between soil and atmosphere, vertical concentration profiles in soil and turnover of nitric oxide: I. Measurements on a model soil core, J. of Atmos. Chem., 23, 253-273, 1996a.

Rudolph, J., and R. Conrad, Flux between soil and atmosphere, vertical concentration profiles in soil and turnover of nitric oxide: II. Experiments with naturally layered soil cores, J. of Atmos. Chem., 23, 275-300, 1996b.

Rummel, U., C. Ammann, A. Gut, F. X. Meixner, and M. O. Andreae, Eddy covariance measurements of nitric oxide flux within an Amazonian Rain forest, J. Geophys. Res., in press, 2002.

Ryden, J. C., J. H. Skinner and D. J. Nixon, Soil core incubation system for the field measurement of denitrification using acetylene inhibition, Soil Biol. Biochem., 19, 753-757, 1987.

Saad, A.L.O., and R. Conrad, Temperature dependence of nitrification, denitrification and turnover of nitric oxide in different models, Biol. Fert. Soils, 15, 21-27, 1993.

Sanhueza, E., W. M. Hao, D. Scharffe, L. Donoso and P. J. Crutzen, N2O and NO emissions from soils of the northern part of the Guayana shield, Venezuela, J. Geophys. Res., 95, 22481-22488, 1990.

Sanhueza, E., L. Cárdenas, L. Donoso and M. Santana, Effect of plowing on CO_2, CO, CH_4, N_2O and NO fluxes from tropical savannah soils, J. Geophys. Res., 99, 16429-16434, 1994.

Schjørring J. K., Ammonia emission from the foliage of growing plants. In: Trace Gas Emissions by Plants. Sharkey T. D., E.A. Holland and H.A. Mooney (eds.) 267-292, Academic Press, San Diego, CA, 1991.

Schmidt, I., and E. Bock, Anaerobic ammonia oxidation with nitrogen dioxide by Nitrosomonas eutropa, Archives of Microbiology, 167, 106-111, 1997.

Scholefield, D., J. M. B. Hawkins and S. M. Jackson, Development of a helium atmosphere soil incubation technique for direct measurement of nitrous oxide and dinitrogen fluxes during denitrification, Soil Biol. Biochem., 29, 1345-1352, 1997a.

Scholefield, D., J. M. B. Hawkins and S. M. Jackson, Use of a flowing helium atmosphere incubation technique to measure the effects of denitrification controls applied to intact soil cores of a clay soil, Soil Biol. Biochem., 29, 1337-1344, 1997b.

Searle, P. L., The Berthelot or Indophenol reaction and its use in the analytical chemistry of nitrogen, Analyst, 109, 549-568, 1984.

Serca, D., R. Delmas, C. Jambert and L. Labrone, Emissions of nitrogen oxides from equatorial rain forest in central Africa: origin and regulation of NO emission from soil, Tellus, 46B, 243-254, 1994.

Skiba, U., K. J. Hargreaves, D. Fowler and K. Smith, Fluxes of nitric and nitrous oxides from agricultural soils in a cool temperate climate, Atmos. Environ., 26A, 2477-2488, 1992.

Skiba, U., D. Fowler and K. Smith, Emissions of NO and N_2O from soils, Environ. Monit. Assess., 31, 153-158, 1994.

Smid, A. E., and E. G. Beauchamp, Effects of temperature an organic matter on denitrification in soil, Can. J. Soil Sci., 56, 385-391, 1976

Smith, K. A., and J. R. M. Arah, Measurement and modelling of nitrous oxide emissions from soils, Ecol. Bull., 42, 116-123, 1992.

Smith P., J.U. Smith, D.S. Powlson et al. A comparison of the performance of nine soil organic matter models using datasets from seven long-term experiments, Geoderma 81, 153-225, 1997.

Smith, W.N., R.L. Desjardins, and E. Pattey, Testing of N_2O models and scaling up emission estimates for crop production systems in Canada. Eastern Cereal and Oilseeds Research Centre. In: Reducing nitrous oxide emissions from agroecosystems, R. Desjardins, J. Keng and K. Haugen-Kozyra, P.Ag. (eds.), International N_2O Workshop, Banff, Alberta, Canada, March 3-5, pp. 99-106, 1999.

Smith, W.N., R.L. Desjardins, B. Grant, C. Li, R. Lemke, P. Rochette, M.D., and Corre, D. Pennock, 2002, Testing the DNDC model using N2O emissions at two experimental sites in Canada. *Canada Journal of Soil Science* 82:365-374.

Sommer S. G., and E. S. Jensen, Foliar absorption of atmospheric ammonia by rygrass in the field, J. of Environ. Qual. 20, 153-156, 1991.

Sommer, S. G. and N. J. Hutchins, Components of ammonia volatilization from cattle and sheep production, in: Gaseous emissions from grasslands, Jarvis S. C. and B. F. Pain (eds.), CAB international, Wallingford, UK, 79-93, 1997.

Soriano, S., and N. Walker, The nitrifying bacteria in soils from Rothamsted classical fields and elsewhere. J. of Appl. Bacteriol., 4, 223-240, 1973.

Sprent, J. I. and P. Sprent, Nitrogen Fixing Organisms: Pure and Applied Aspects. Chapman Hall, London, 1990.

Stange, F., K. Butterbach-Bahl, H. Papen, S. Zechmeister-Boltenstern, C. Li, and J. Aber. A process oriented model of N_2O and NO emissions from forest soils: 2. Sensitivity analysis and validation, J. Geophys. Res. 105, 4385-4398, 2000.

Stanford, G. Dzienia, S. and Vander Pol, A., Effect of temperature on denitrification, Soil Soc. Am. J., 39, 867-870, 1975.

Stefanson, R. C. and D. J. Greenland, Measurement of nitrogen and nitrous oxide evolution from soil-plant systems using sealed growth chambers, Soil Science, 109, 203-206, 1970.

Stevenson, F. J., (Ed.) Nitrogen in Agricultural Soils, American Society of Agronomy, Madison, Wisconsin, USA, 1982.

Stevenson, F. J., Cycles of Soil, John Wiley & Sons, 1986.

Stewart, J. W. B., I. Aselman, A. F. Bouwman, R. L. Desjardins, B. B. Hicks, P. A. Matson, H. Rodhe, D. S. Schimel, B. H. Svensson, R. Wassmann, M. J. Whiticar, W.-X. Yang, Extrapolation of flux measurements to regional and global scale, in: Exchange of trace gases between terrestrial ecosystems and the atmosphere, M. O. Andreae and D. S. Schimel (eds.), 155-174, 1989.

Stumm W., and J. J. Morgan, Aquatic Chemistry: An Introduction Emphasizing Chemical Equilibria in Natural Waters, 2nd Edition, John Wiley & Sons, New York, pp. 418-503, 1981.

Sutton, M. A., The surface-atmosphere exchange of ammonia, Ph.D. thesis, Institute of Ecology and Resource Management, University of Edinburgh, Edinburgh, 1990.

Sutton M. A, C. E. R. Pitcairn and, D. Fowler D., The exchange of ammonia between the atmosphere and plant communities, In: Advances in Ecological Research. Vol. 24., 301-393. Academic Press Limited, 1993.

Sutton, M. A., C. Milford, U. Dragosits, R. Singles, D. Fowler, C. Ross, R. Hill, S. C. Jarvis, B. F. Pain, R. Harrison, D. Moss, J. Webb, S. Espenhahn, C. Halliwell, D. S. Lee, G. P. Wyers, J. Hill and H. M. ApSimon, Gradients of atmospheric ammonia concentrations and deposition downwind of ammonia emissions: first results of the ADEPT Burrington Moor Experiment, in: Gaseous emissions from grasslands, Jarvis, S.C., and B. F. Pain (eds.), CAB international, Wallingford, UK, 131-139, 1997.

Swerts, M., G. Uytterhoeven, R. Merckx and K. Vlassak, Semicontinuos measurement of soil atmosphere gases with gas-flow soil core method, Soil Sci. Soc. of Am. J., 59, 1336-1342, 1995.

Swerts, M., R. Merckx and K. Vlassak Influence of carbon availability on the production of NO, N_2O, N_2 and CO_2 by soil cores during anaerobic incubation, Plant Soil, 181, 145-151, 1996.

Terry, R.E., R.L. Tate III, and J.M. Duxbury, Nitrous oxide emissions from drained, cultivated organic soils of South Florida, Air Pollut. Control Assoc., 31, 1173-1176, 1981.

Tiedje J. M., A. J. Sexstone, T. B. Parkin, N. P. Revsbech, and D. R. Shelton, Anaerobic processes in soil. Plant Soil, 76, 197-212, 1984.

Tsunogai, S., and K. Ikeuchi, Ammonia in the atmosphere, Geochemical Journal 2, 157-166, 1986.
Valente, R. J. and F. C. Thornton, Emissions of NO from soils at a rural site in Central Tennessee, J. Geophys. Res., 98, 16745-16753, 1993.

van Breemen N., P. A. Burrough, E. J. Velthorst, H. F. Van Dobben, T. de Wit, T. B. Ridder, and H. F. R. Reijnders, Soil acidification from ammonium sulphate in forest canopy throughfall, Nature, 299, 548-550, 1982.

van Cleemput, O., and L. Baert, Theoretical considerations on nitrite self-decomposition reaction in soil, Soil Science Society of America Journal, 40, 322-324, 1976.

van Cleemput, O., and L. Baert, Nitrite: A key compound in N loss processes under acid conditions? Plant Soil, 76, 233-241, 1984

van Dijk, S. M., and F. X. Meixner, Production and consumption of NO in forest and pasture soils from the Amazon basin: a laboratory study, Water Air Soil Poll.: Focus, 1, 119-130, 2001.

van Dijk, S. M., A. Gut, G. A. Kirkman, B. M. Gomes, F. X. Meixner, and M. O. Andreae, Biogenic NO emissions from forest and pasture soils: relating laboratory studies to field measurements, J. Geophys. Res., 107, D20, 8058, doi:10.1029/2001JD00358 , 2002.

Veldkamp, E., and M. Keller, Nitrogen oxide emissions from banana plantations in the humid tropics, J. Geophys. Res., 102, 15889-15898, 1997.

Veldkamp, E., M. Keller and M. Nuñez, Effects of pasture management on N_2O and NO emissions from soils in the humid tropics of Costa Rica, Global Bio. Cycles, 12, 71-79, 1998.

Velthof, G. L. and O. Oenema, Nitrous oxide fluxes from grassland in the Netherlands. 1. Statistical analysis of flux-chamber measurements, Eur. J. of Soil Sci., 46, 533-540, 1995.

Velthof, G. L., S. C. Jarvis, A. Stein, A. G. Allen and O. Oenema, Spatial variability of nitrous oxide fluxes in mown and grazed grasslands on a poorly drained clay soil, Soil Bio. Biochem., 28, 1215-1225, 1996.

Verchot, L. V., E. A. Davidson, J. H. Cattanio, I. L. Ackerman, H. E. Erickson, and M. Keller, Land use change and biogeochemical controls of nitrogen oxide emissions from soils in eastern Amazonia, Global Biogeochem. Cycles, 13, 31-46, 1999.

Vilà-Guerau de Arellano, J., and P. G. Duynkerke, Influence of chemistry on the flux-gradient relationships for NO-NO3-NO2 system, Boundary Layer Meteorol., 61, 375-387, 1992.

Walker, N., On the diversity of nitrifiers in nature. In: Microbiology. Schlessinger, D. (ed.), American Society for Microbiology, Washington D.C., 346-347, 1978.

Walker, N. and K. N. Wickramasinghe, Nitrification and autotrophic nitrifying bacteria in acid tea soils. Soil Biol. and Biochem., 11, 231-236, 1979.

Walton, S., M. W. Gallagher, and J. H. Duyzer, Use of a detailed model to study the exchange of NOx and O3 above and below a deciduous canopy, Atmos. Environ., 31, 2915-2931, 1997.

Weber, D. F. and P. L. Gainey, Relative sensitivity of nitrifying organisms to hydrogen ions in soils and in solutions. Soil Science, 94, 134-145, 1962.

Weier, K. L. and Gilliam, J. W., Effect of acidity on denitrification and N_2O evolution from Atlantic coastal plain soils, Soil Sci. Am. J., 50, 1202-1205, 1986.

Weier, K. L, J. W. Doran, J. F. Power and D. T. Walters, Denitrification and the dinitrogen/nitrous oxide ration as affected by soil water, available carbon and nitrate, Soil Sci. Soc. Am. J., 57, 66-72, 1993.

Wentworth, W. E., S. V. Vasnin, S. D. Stearns and C. J. Meyer, Pulsed discharge helium ionization detector, Chromatographia, 34, 219-225, 1992.

Wickramasinghe, K. N., O. Talibudeen and J. F. Witty, A gas-flow through system for studying denitrification in soils, J. Soil Sci. 29, 527-536, 1978.

Wienhold, F. G., H. Frahm and G. W. Harris, Measurements of N2O fluxes from fertilized grassland using a fast-response Tunable Diode-Laser Spectrometer, J. Geophys. Res., 99, 16557-16567, 1994.

Williams E. J., and F. C. Fehsenfeld, Measurement of soil nitrogen oxide emissions at three North American ecosystems, J. Geophys. Res., 96, 1033-1042, 1991.

Williams E. J., A. Guenther, and F. C. Fehsenfeld, An inventory of nitric oxide emissions from soils in the United States, J. Geophys. Res., 97, 7511-7520, 1992a.

Williams E. J., G. L. Hutchinson, and F. C. Fehsenfeld, NOx and N2O emission from soil, Global Biogeochem., 6, 351-388, 1992b.

Williams, E. J. and E. A. Davidson, An intercomparison of two chamber methods for the determination of emission of nitric oxide from soil, Atmos. Environ. 27, 2107-2113, 1993.

Xiu, W.B., Hong Y.T., Chen X.H., Li C., 1999, Agricultural N_2O emissions at regional scale: A case study in Guizhou, China, Science in China, 29, 5-17, 1999.

Yamulki, S., K. W. T., Goulding, C. P. Webster and R. M. Harrison, Studies on NO and N_2O fluxes from a wheat field, Atmos. Environ., 29, 1627-1635, 1995.

Yamulki, S., R.M. Harrison, K.W.T. Goulding, and C.P Webster, N2O, NO and NO2 fluxes from a grassland: effect of soil pH, Soil Bio. Biochem., 29, 1199-1208, 1997.

Yamulki, S. and S. C. Jarvis, Automated chamber technique for gaseous flux measurements: evaluation of a photoacoustic infrared spectrometer-trace gas analyzer, J. Geophys. Res, 104, 5463-5469, 1999.

Yang, W. X., and F. X. Meixner, Laboratory studies on the release of nitric oxide from subtropical grassland soils: The effect of soil temperature and moisture, in: Gaseous nitrogen emissions from grasslands, S. C. Jarvis and B. F. Pain (eds.), CAB International, Wallingford, UK, 67-71, 1997.

Yeomans, J. C. and E. G. Beauchamp, Acetylene as a possible substrate in the denitrification process, Can. J. Soil Sci., 62,139-144, 1982.

Yienger, J. J., and H. Levy, II, Global inventory of soil-biogenic NOx emissions, J. Geophys. Res., 100, 11,447-11,464, 1995.

Yoshida, N. and S. Toyoda, Constraining the atmospheric N2O budget from intramolecular site preference in N2O isotopomers, Nature, 405, 330-334, 2000.

Yoshinary, T., R. Hynes and R. Knowles, Acetylene inhibition of nitrous oxide reduction and measurement of denitrification and nitrogen fixation in soil, Soil Biol. Biochem., 9, 177-183, 1977.

Zart, D., and E. Bock, High rate of anaerobic nitrification and denitrifcation by Nitrosomonas eutropha grown in a fermenter with complete biomass retention in the presence of gaseous NO2 and NO, Archives of Microbiology, 169, 282-286, 1998.

Zhang, Y., C. Li, X. Zhou, and B. Moore III, 2002a, A simulation model linking crop growth and soil biogeochemistry for sustainable agriculture, Ecological Modeling, 151, 75-108,2002.

Zhang, Y., C. Li, C. C. Trettin, H. Li, G. Sun, 2002b. An integrated model of soil, hydrology and vegetation for carbon dynamics in wetland ecosystems, *Global Biogeochemical Cy*cles 10.1029/2001GB001838.

Global Emissions of Mineral Aerosol: Formulation and Validation using Satellite Imagery

Yves Balkanski, Michael Schulz, Tanguy Claquin, Cyril Moulin, Paul Ginoux

1. INTRODUCTION

The most abundant aerosol components present in the atmosphere are dust and sea salt (Andreae, 1995). Li et al. (1996) show that dust dominates the aerosol light-scattering over the tropical and sub-tropical Atlantic. Satellite retrievals also illustrate the importance of dust over large regions from arid deserts to remote oceanic regions downwind of West Africa, Asia and the Persian Gulf (Husar et al., 1997; Deuzé et al., 2000; Tanré et al., 2001). Furthermore, land modification, agricultural practices and the migration of desert fringes appear to have contributed to the increase in the dust transport over the Atlantic from the 1960s to the 1980s. These perturbations to the dust cycle brought by human activity are thought to account for 15 to 50% of the atmospheric dust load (Tegen and Fung, 1995; Tegen, personnal communication, 2002).

Martin and Gordon (1988) and Falkowski et al. (1998) studied the limitation of oceanic primary production by iron-containing dust in areas of high nitrate-low chlorophyll regimes. The surface of dust also acts as a chemical reactor for key atmospheric gases that can be either adsorbed or oxidized. The species that have been examined so far to evaluate these heterogeneous pathways include HNO_3, N_2O_5, SO_2 and O_3 (Dentener et al., 1996 and references therein; Song and Carmichael, 2001). Ongoing laboratory, field and modelling studies show that O_3 and HNO_3 tropical and subtropical concentrations could be decreased by more than 10%.

Last but not least, dust may pose a public health threat since it may cause respiratory disease, e.g. development of silicosis (Patial, 1999), and by transporting spores of bacteria over long distance (Griffin et al., 2001).

The renewed interest in the atmospheric transport of mineral dust has been triggered by studies of how the direct (Tegen et al., 1997; Haywood and Boucher 2000; Harrison et al., 2001) and indirect effect of aerosols influence the Earth's radiative budget (Levin et al., 1996; Rosenfeld, 2000). Unfortunately, there is a large uncertainty in the annual flux of desert dust to the atmosphere, which is estimated to range from 128 to 5000 Mt yr^{-1} (Pye, 1987). The difficulties in modelling the atmospheric injection of mineral particles come from the episodic nature of these events, which are controlled by local soil properties and atmospheric conditions.

Dust weathering and geological processes that produce small particles which can be transported over large distances determine the actual dust source areas (Prospero et al., 2002). Information from recent satellite studies is now used to identify the regions where dust is uplifted.

This chapter has four parts: The first part describes the theory of dust generating schemes used in global models. The second part describes a global dust source scheme that is formulated on the basis of soil properties and satellite retrievals from TOMS (Total Ozone Mapping Spectrometer). The third part describes the regions where dust is preferentially uplifted. Finally, future directions to refine our knowledge of dust emissions using satellite retrievals are proposed.

2. DESCRIPTION OF DUST SUSPENSION IN THE ATMOSPHERE

There has been and still is an intense effort to characterise the generation and transport of dust through experiments using wind tunnel systems to create dust uplift (Gillette, 1978, 1999). These studies have shown that a wide range of factors influence dust generation on a micrometeorological scale. These factors include soil composition, soil moisture, surface conditions, and the wind velocity (Marticorena and Bergametti, 1995; Marticorena et al., 1997). Emission schemes based upon soil parameters observed in-situ such as local roughness height and the density of obstacles (Marticorena et al., 1999) have improved the representation of dust sources in large-scale models. But the application of such schemes to global scales is limited by the availability of data characterizing surface features in arid and semi-arid regions. Due to this lack of available data, global models have identified the location of dust sources based on the soil moisture content (Joussaume, 1990), the location of deserts from a vegetation dataset (Genthon, 1992ab; Mahowald et al., 1999), the location of sparsely vegetated area and the soil texture from vegetation and soil data sets (Tegen and Fung, 1994), or the distribution of dust storm frequencies over arid regions (Dentener et al., 1996).

Before we proceed in the description of the mobilisation of dust, we need to give the definition of a few key concepts:

Saltation layer. Layer of a few tens of centimeters depth near the surface where soil grains from 60 to 2000 μm undergo bombardment from other grains creeping along the surface. The bombardment, also called "saltation", dislodges fine particles (with a diameter of less than 60μm) that can be transported over long distances away from sources.

Horizontal flux. Total amount of dust-related material set in motion in the saltation layer per unit cross section per unit time (kg m^{-2} s^{-1}).

Vertical flux. The fraction of the horizontal flux constituted by fine particles that are transported vertically when the drag momentum exceeds gravitational settling.

Aggregates. Particles in soil are formed of aggregates that include a clay fraction and a non-clay fraction composed by order of importance of quartz, calcite, dolomite, feldspar, halite, gypsum and iron oxides.
Friction velocity. The friction velocity (m s^{-1}) is the square root of the ratio of wind shear stress to air density.

Threshold friction velocity. The threshold friction velocity is the minimum friction velocity necessary to set soil particles in motion and initiate the saltation movement.

Roughness height. The length scale (m) that characterizes the loss of wind momentum attributable to roughness elements.

Non-erodible elements. Elements such as stones that affect the ability to uplift dust from the surface are non-erodible elements.

Crusting. Crusting of the surface occurs when salt crystal formation, hydrate formation or bacterial activity occurs. Such crusts prevent any uplifting of soil grains.

2.1. Existing schemes for dust generation

Two main phenomena are at play to determine when and if dust will be transported out of the surface layer. The most important are: first, the presence of erodible material which can be the result of processes that have occurred over geological time scales; and second, the wind's shear stress on the surface that provides the energy momentum to set the dust in motion.

The mass of the particles can be computed by integrating the horizontal flux. Both wind tunnel experiments and field work have shown the dependency of this horizontal flux on the wind friction velocity u* (reviewed by Greeley and Iversen, 1985). Field measurements of the horizontal flux G over different soil types have documented the dependency of G on the friction velocity (Figure 1). For models of dust emissions, the expression of the horizontal flux of dust needs to account for the following facts:

First, the dust sets in motion over a terrain only when the friction velocity reaches a minimum threshold called threshold friction velocity. Second, the horizontal flux can be expressed as a function of the power of 3 of the friction wind velocity (Bagnold, 1941; Chepil, 1951; Gillette, 1974, 1978).

Further work (Iversen and White, 1982; Raupach, 1992, 1994; Raupach et al., 1993; Marticorena and Bergametti, 1995; Marticorena et al., 1997; Shao and Leslie, 1997; Fécan et al., 1999) have brought attention to the importance of the local properties of the terrain to assess the dust flux.

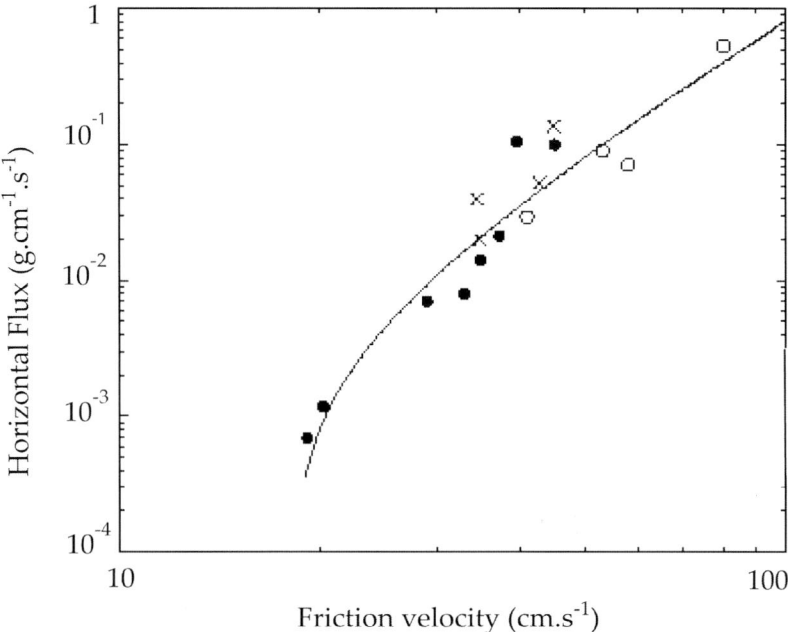

Figure 1. Dust horizontal flux, G, as a function of the threshold friction velocity, u*. The measurements made by Gillette (1974) were done for soils with different size distribution (resp. closed circles, open circles and crosses). The fitted line is a horizontal flux, $G = 1.\text{e-}6\, u^{*2}(u^*-18)$.

The properties important for the horizontal flux formulation are: the diameter of soil particles, the obstacles formed by non-erodible elements and the crusting of the soil. From these studies, the horizontal flux of dust resulting from the wind shear stress can be expressed by Equation 1:

$$G = \int_0^\infty C(D_p)\,(u^*(u,z_0))^2 \left(u^*(u,z_0) - \frac{u^*_{th}(D_p,\rho,u,q_s)}{A\gamma} \right) . dD_p \qquad \text{(Eq.1)}$$

where D_p, is the particle diameter of soil particles; $C(D_p)$ is the emission factor (kg.s^2.m^{-5}); q_s, is the soil wetness; A is a dimensionless wind attenuation factor due to non-erodible elements at the surface; z_0 is the local roughness height; ρ is the soil grain density; γ is a dimensionless factor that accounts for crusted surfaces; u^* represents the friction velocity and u^*_{th}, the threshold friction velocity. Note that this expression needs to be integrated over all the values of soil particle diameter to get a total aerosol mass flux.

Since there is very little information available to correctly estimate local roughness heights over most deserts, the assumption is often made that the friction velocity is proportional to the wind at 10-m. Hence the following simple equation which stems from the simplification of equation 1:

$$F = C. \, u^{*2} \, (u^* - u^*_{th}) \tag{Eq.2}$$

is replaced by

$$F = C'. \, u^2 \, (u - u_{th}) \tag{Eq.3}$$

where friction velocity and threshold friction velocity are replaced by u, the wind at 10-m height and the threshold velocity at 10-m height (Joussaume, 1990; Tegen and Fung, 1994; Nickovic and Dobricic, 1996; Mahowald et al., 1999). The choices of emission factors and threshold velocities in regional and global models are summarized in Table 1.

Equation 3 allows the computation of the vertical flux of dust as a function of two parameters: an emission factor C, which controls the total amount of dust injected over the time period, and a threshold velocity which controls the frequency of occurrence of dust episodes. Observational information on the frequency of storms may provide information on the threshold velocity. However, a quantitative assessment of the different formulations of dust has been long hampered by the lack of observations of the exact geographical extent of dust episodes. Remote sensing using satellite has allowed great advances in the last few years. First, several types of satellite retrievals have allowed the detection of dust over oceanic surfaces (the Meteorological Satellite (METEOSAT), the Advanced Very High Resolution Radiometer (AVHRR), the Total Ozone Mapping Spectrometer (TOMS)) and, more recently over land, TOMS, and the Lidar-In-space Technology Experiment (LITE). These observations have shown that a constant emission factor cannot correctly simulate the different dust clouds observed over desert regions as different as Asian, African, Australian Deserts or from North America (Marticorena, 1995).

Table 1. Comparison of the different dust sources that have been used for global or regional simulations.

	u^*_{thresh} threshold velocity	C Emission factor	Global emission (Mt/yr)	Remarks
Marticorena and Bergametti (1995)	$f(D_p, u^*, z_0)$	$f(D_p)$ for $D_p < 20$ μm		z_0, D_p: mapped over Sahara
Shao and Leslie (1997)	$f(D_p, z_0)$	$f(D_p)$		z_0, D_p: mapped over Australia
Nickovic and Dobricic (1996)	0.5 m s^{-1}	calibrated		
Tegen et al. (1996)	6.5 m s^{-1}	calibrated	1200 for $D_p < 20$μm	
Genthon (1992b)	0 m s^{-1}	0.7 μg s^2 m^{-5}	8000	
Mahowald et al. (1999)	5 m s^{-1}	calibrated	3000 for $D_p < 40$μm	
Dentener et al. (1996)	Statistical approach		1800 for $D_p < 20$μm; 15500 for $D_p < 60$μm	
Ginoux et al. (2001)	Based upon the work of MB95*		1500-2000	5 year simulation (see text)
Tegen et al., 2003	Based upon the work of MB95		800-1700, for $D_p < 218$μm	
Woodward (2001)	$f(D_p)$, adjusted empirically	calibrated after MB95		
This work	F(soil type)	calibrated over 12 regions (see text)	688	1 year simulation (July 1996-June 1997)

*MB95 stands for Marticorena and Bergametti (1995)

Table 1 lists the different emission schemes from detailed ones over restricted regions to global schemes. Formulations on restricted regions include works from Marticorena and Bergametti (1995) and Shao and Leslie (1997) and are based upon information on local soil properties. The unavailability of global maps for local roughness heights, soil grain size distribution and heights of non-erodible elements has led to different approaches based upon information contained in satellite retrievals. Global dust emissions have been estimated using two different methods. One is based on the physics of dust suspension and stems from equation 1, while the second accounts for the statistics of observed dust events.

3. A GLOBAL FORMULATION OF SOIL EROSION BASED UPON SOIL PROPERTIES

Here, we describe a formulation of the global dust source that uses information available at the global scale. It extends the domain over which the source was formulated in physical models from Marticorena and Bergametti (1995, hereafter referred to as MB95) and Shao and Leslie (1997). Since the availability of fine particles (<20 μm diameter) is a function of the soil type, we looked for a relationship between soil characteristics and threshold velocity.

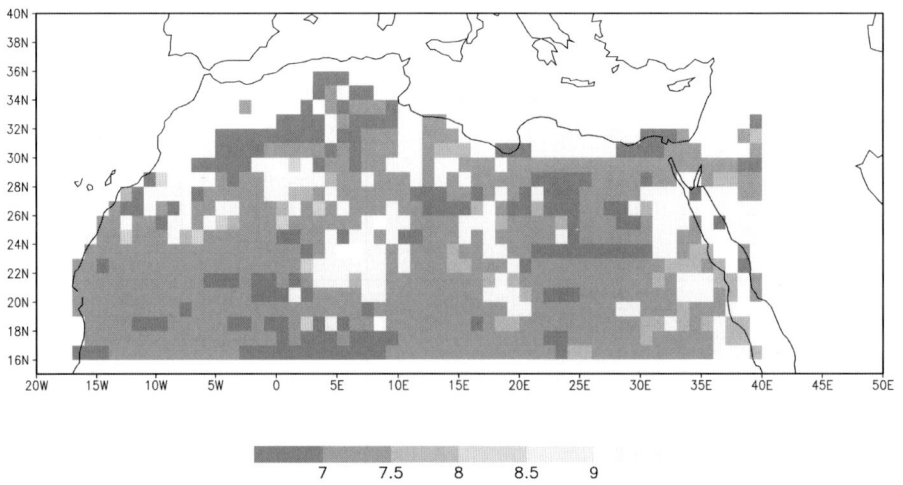

Figure 2. Threshold velocities over North Africa derived by Marticorena and Bergametti (1995). To derive these values of threshold velocities, the authors used the 10 meters wind from ECMWF (European Centre for Medium Range Forecasts) with a resolution of 1.125°x1.125°. (© American Geophysical Union).

Since we did not have available mineralogical information for all the arid and semi-arid areas of study, we used the FAO (Food Agriculture Organisation) database of soil types. The FAO Soil Map of the World (Zobler, 1986; FAO, 1995) is a complete classification of soils based upon genetic and physical characteristics (particle size, soil thickness). The global map with a resolution of 5' x 5' gives the soil type and granulometry. For this work, the map was regridded with a 1°x1° resolution. We establish a correspondence between threshold velocities derived by MB95 (shown in

Figure 2) and the different FAO soil types over the same region (not shown). As illustrated in Figure 2, most of the grid cells have associated threshold velocities of less than 14 m s^{-1}. One noticeable exception is for very rugged terrain. For smooth surfaces, the threshold velocities are never below 6.5 m s^{-1}.

Figure 3 presents all the threshold velocities as a histogram. Three different values dominate the distribution: 7.5, 12 and 20 m s^{-1}. The dominant soil over the region of Figure 2 is Yermosol for which the median threshold velocity is 7.5 m s^{-1}. This soil type is composed of fine particles. Where yermosols cover more than 90% of the 1° x 1° square, the threshold velocities lie in a very narrow range (7.2 - 7.6 m s^{-1}). In other regions where yermosols are not dominant, much higher threshold velocities are computed.

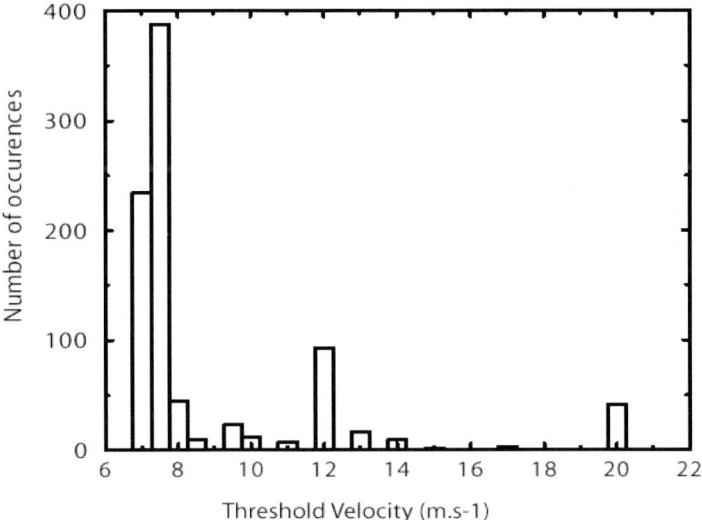

Figure 3. Histogram of the threshold velocities reproduced in Figure 2.

Lithosols have a median value for threshold velocity of 12 m s^{-1}. These areas are covered by large blocks of stones and erodible material is rare where these soils dominate. Xerosols are found in semi-arid regions such as in South Africa or over rugged terrains, as in the Atlas mountains. Erosion occurs very seldomly from these soils and they favor a scarce vegetation and contain organic matter. This soil type is the only one for which the median threshold velocity reaches 20 m s^{-1}.

Finally, dust uplift occurs seldomly in regions where slopes are present. This can be explained by the friction velocity necessary to initiate the saltation process as slopes act as an obstacle to its initiation. Flat and montainuous terrain were distinguished upon the slope criteria, the slope was used as a criteria to increase the threshold velocity. For regions with slopes between 10 and 20°, the threshold velocity was arbitrarily increased by 2 m s^{-1}. For regions with slopes greater than 20°, threshold velocities were increased by 4 m s^{-1}. Only 5% of the arid regions are affected by these criteria. The exact increment of threshold velocity as a function of slope deserves a more thorough study.

Table 2. Median threshold velocities for the dominant FAO soil types over arid and semi-arid regions.

Soils type or slope	Threshold velocity (m s^{-1})
Ferric soil present	∞ (no mobilization)
Xerosol dominant	∞ (no mobilization)
Regosol dominant	12.0
Lithosol dominant	12.0
Lithosol present	10.0
Salts flats present	10.0
Sand dunes dominant	6.5
Yermosols	7.5
Fluvisols	7.5
Slope in excess of 20°	$U_{thresh} = U_{thresh}+4$
Slope in excess of 10°	$U_{thresh} = U_{thresh}+2$

The next step is to derive the threshold velocities for all arid and semi-arid areas over the globe. Here we assume that the threshold wind velocities from MB95 can be extrapolated for all regions with similar soil characteristics. For each soil class that covers more than half of the area of a 1° x 1° square, the threshold velocity is assigned (shown in Figure 2). The median value of the threshold velocity is then derived for each FAO soil class. This median threshold velocity presented in Table 2 is then applied over the globe using the World Map of Soil from the FAO. In most areas, the threshold velocity is equal to 7.5 m s^{-1}, which is a higher value than proposed in most global simulations (see Table 1). Figure 4 illustrates the distribution of threshold velocities derived.

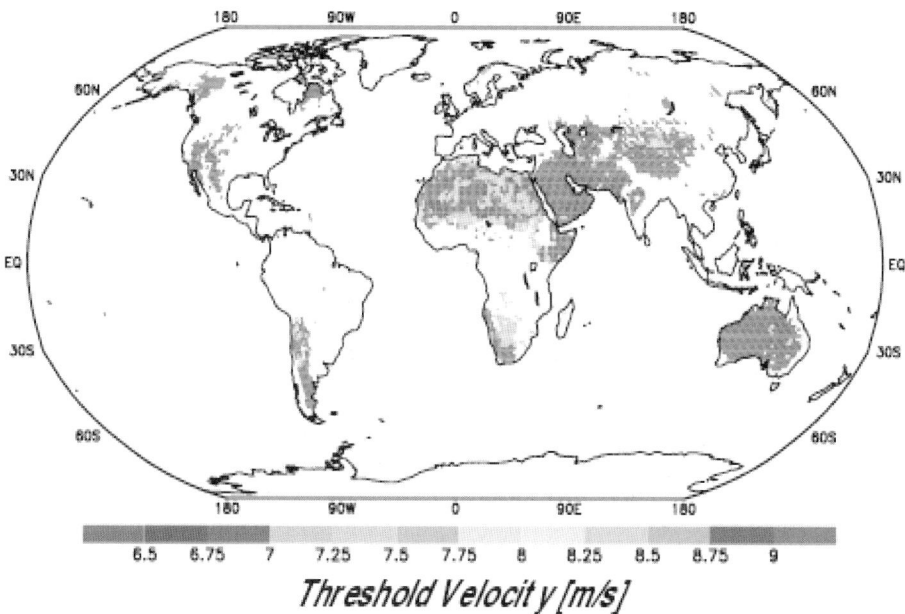

Figure 4 (Plate 10). Global map of threshold velocities over arid and semi-arid areas. Soil types were used to infer threshold velocities representative of the FAO soil types. The global FAO map was then used to affect the threshold velocities to soil types for other areas over the world.

3.1. Derivation of the vertical flux

Dust emissions mainly occur at high wind speeds: below the threshold velocity, no emission occurs whereas above this threshold, the vertical flux (like the horizontal flux) is proportional to the wind speed cubed (Shao et al., 1993). In order to estimate the vertical emission flux, the emission factor (C' from equation 3) needs to be derived.

These estimates had to be derived independently from the method used to derive threshold velocities, hence they had to be derived without using the FAO soil map. The emission factors were obtained by estimating an optical depth from the aerosol index.provided by the TOMS satellite. The semi-quantitative TOMS (Total Ozone Mapping Spectrometer) Absorbing Aerosol Index is a unique satellite archive that provides a quasi-daily global coverage of desertdust occurrence (Herman et al., 1997) and have recently been used to locate major dust sources (Prospero et al., 2002) and to monitor dust optical thickness (DOT) over land and ocean (Chiappello et al., 2002; Torres et al., 1998).

Table 3. Location of source regions and source strength derived based upon TOMS aerosol index.

REGIONS	GEOGRAPHICAL LOCATION For the comparison with TOMS	Emission Factor C ($\mu g \, s^2 \, m^{-5}$)
Sahara	Latitudes: 14°N and 30°N Longitudes: 10°W and 28°E	0.37
Sahel	Lat.: 9°N :14°N, Long.: 15°W :10°E	0.28
Takla-Makan	Lat.: 37°N :42°N, Long.: 75°E :90°E	1.80
Gobi	Lat.: 39°N :45°N, Long.: 95°E :110°E	0.15
Kyzil Kum	Lat.: 35°N : 43°N, Long.: 55°E : 63°E	0.17
Kara Kum	Lat.: 35°N : 43°N, Long.: 64°E : 68°E	0.34
Kalahari	Lat.: 19°S : 30°S, Long.: 18°E : 26°E	0.25
USA	Lat.: 33°N :40°N, Long.: 118°W :108°W	0.32
Saudi Arabia	Lat.: 15°N : 25°N, Long.: 43°E : 57°E	0.17
Thar Desert	Lat.: 22°N : 27°N, Long.: 10°W : 28°E	1.57
Australia	Lat.: 23°S : 30°S, Long.: 120°E : 142°E	0.08
Somalia	Lat.: 5°N : 10°N, Long.: 45°E : 53°E	0.04

In contrast to previous modelling efforts that derive a global strength for dust emissions (Tegen et al., 1996 or Mahowald et al., 1999), we chose twelve large regions on the basis that seasonality, altitude of dust transport and the mineral composition of dust were characteristic of a particular region. These regions and their locations are listed in Table 3. The emission factors were derived to ensure that the annual mean simulated optical depth equals the one deduced from TOMS retrievals. For this, TOMS aerosol index has to be linked to the optical thickness.

Herman et al. (1997) have recently used the data from TOMS on the NIMBUS-7 to map the distribution of absorbing aerosols. Absorbing aerosols are composed mainly of black carbon (i.e., soot) emitted primarily from biomass burning regions, and of mineral dust. Prospero et al. (2002) examined a 13-year period (1980-1992) of TOMS absorbing aerosol product for evidence of persistent dust sources. The TOMS data (Herman et al., 1997) show that on a global scale the dominant sources of mineral dust are all located in the northern hemisphere, mainly in north Africa, the Middle East, central Asia and the Indian subcontinent. TOMS also shows large seasonal changes in dust distribution patterns.

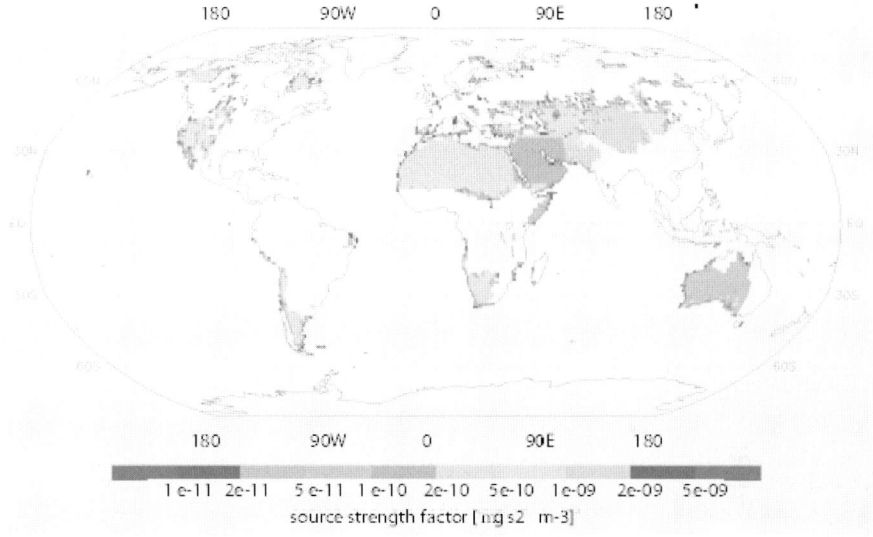

Figure 5 (Plate 11). Global distribution of source strength factors (kg m^{-5} s^2) (C' in equation 3) derived by calibrating the model derived optical depth to the optical depth corrected from the TOMS aerosol index. The comparison was made for 1990 and these coefficients are weighted by the fraction of erodible land of each model grid box.

In order to deduce an optical depth from TOMS aerosol index (AI), we assumed that dust is distributed vertically following the profiles of a one-year global dust simulation. The simulation used the TM3 model with a 3.75° x 5° horizontal resolution and 19 layers in the vertical (Heimann, 1995), with parameterisations for wet and dry deposition from Guelle et al., (2000). The meteorological fields for the simulation were extracted from the European Centre for Medium Range Forecast reanalysis for the year 1990. The vertical distribution of the dust profile was saved daily at every model grid point. To compute the aerosol depth, the relationships that Hsu et al. (1999) gave for four altitudes (0.5, 1.5, 3.0 and 6.0 km) were applied to infer an optical thickness from TOMS Aerosol Index for 1990. Less than 10% of the dust mass resides above the higher altitude (6 km), and the relationships were interpolated linearly between these altitudes. The emission factors were derived for the 12 different arid regions shown in Figure 5 and listed in Table 4 by adjusting C' to obtain the same simulated yearly mean optical depth as derived from TOMS.

Table 4. Emission factors.

TYPE	# days during which source is inhibited	Threshold precipitation amount over the period (mm)
Warm temperature/ Low clay content	1	0.82
Warm temperature/ High clay content	3	2.46
Cold temperature/ Low clay content	4	3.28
Cold temperature/ High clay content	8	6.86

The simulated dust emissions from the model were computed with the above derived threshold velocities using the ECMWF 10m winds at a 1.125°x1.125° resolution. For most regions, the emission factor lies within a factor of 3 of the constant value of 0.7 µg s^2 m^{-5} used by Tegen and Fung (1994). This emission factor is exceeded significantly for the Thar and Takla-Makan deserts, and considerably lower over Australia and Somalia. The emission factor from the African deserts with a low elevation (few hundred meters) is systematically low (between 0.25 and 0.37 µg s^2 m^{-5}). Figure 6 shows a comparison of the optical depth over the desert (near sources) derived from TOMS, and recalculated from the TM3 model. Error analysis of optical properties derived from backscattered radiances in the UV have been discussed in more details by Torres et al. (1998). Their analysis shows that the error of dust optical thickness due to uncertainty in aerosol altitude and cloud contamination could be as high as 65% and 40%, respectively. Due to the large footprint of TOMS measurements (50 km at nadir), sub-pixel cloud contamination is quite frequent and unfortunately difficult to resolve. By selecting data with low observed reflectance, it can be assumed that the data are mostly cloud free (Torres et al., 2002).

The TOMS satellite has a polar orbit with a local passing time around 11:00 am with a shift of about 30 minutes varying from year to year. When using TOMS data, it is important to recognize that the local passing time is not necessarily representative of a time dependent physical property. Statistical analysis of meteorological fields show that the most likely period of dust emission varies from source to source. The TOMS passing time does not necessarily coincide with the maximum activity. Therefore, comparison of dust emission using TOMS data will be biased towards the sources with their peak activity corresponding with TOMS passing time. For example, Ginoux et al. (2001) show that the Bodele depression which appeared as the "hot spot" of TOMS data (Prospero et al., 2002) has also the most frequent gusty winds around 11:00 am.

The seasonality of optical depth shown in Figure 6 depends upon 10m wind speed, precipitation and freezing surface temperature. We capture well this seasonality and the frequency of occurrence of the episodes for most deserts with the noticeable exception of the Sahara. Over the Sahara, the simulated emissions are stronger than the optical thickness derived from TOMS. Over the Sahel region, the period of study to calibrate the source emission strength was restricted to April to September, in order to be relatively free of the carbonaceous aerosol influence resulting from biomass burning. Individual peaks are well captured over all deserts except the Sahara, where the high background optical depth and rather smooth variations are not so well represented in the simulation.

Since to our knowledge there are no such comparisons of daily time series, it is difficult to show the improvement of this approach compared to other formulations of global dust sources.

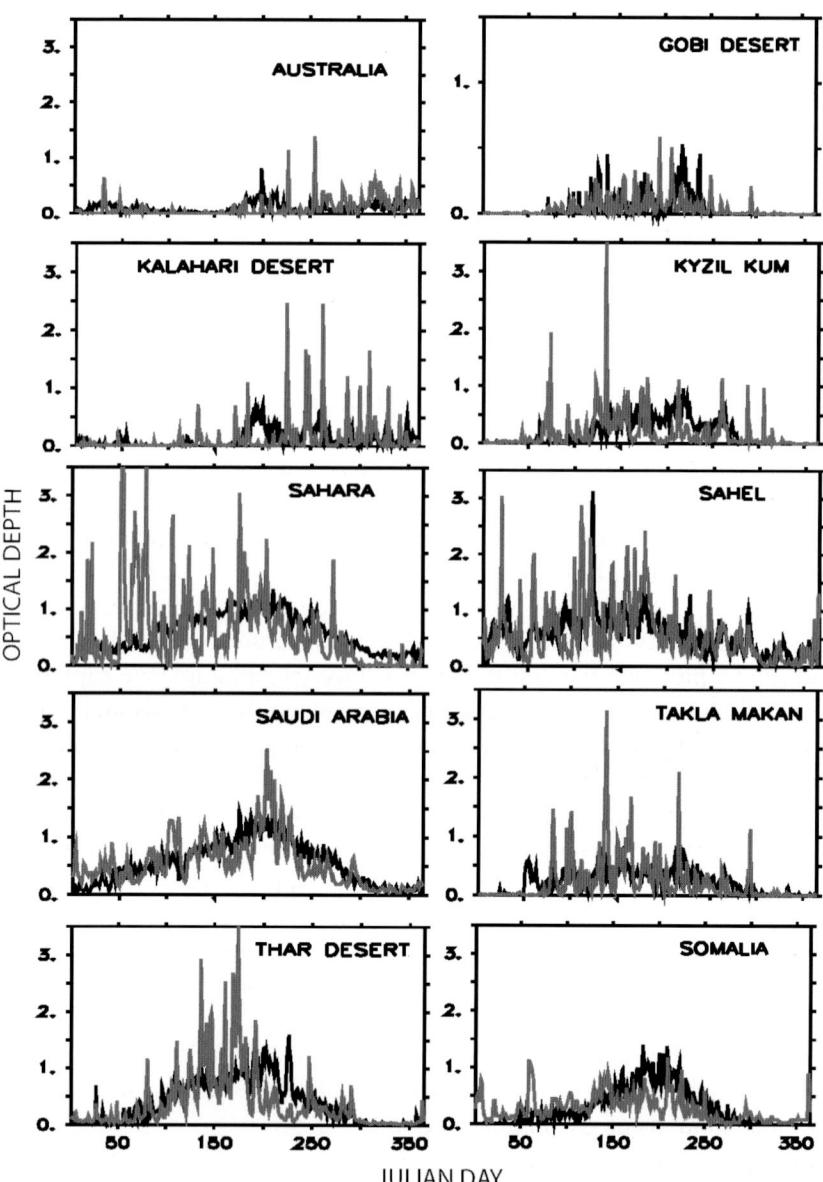

Figure 6 (Plate 12). Time series of optical depth over the areas listed in Table 3 for 10 of the 12 deserts. The black line represents TOMS-derived optical depth corrected to account for wavelength and the altitude dependence of TOMS retrieved signal. The red line is the modelled optical depth. No cloud screening was applied on the model results.

3.2. Effect of precipitation on dust emissions

Precipitation and the subsequent increase in soil moisture enhance the cohesive forces between soil grains, and will prevent emissions until the upper millimetres of soil become sufficiently dry. The two main factors that affect this enhanced adhesion of soil grain are the amounts of clay in the soil and the surface air temperature. Shao et al. (1996) discuss the dependence of the threshold velocity on soil moisture and Fécan et al. (1999) propose a relationship between threshold wind velocity and soil moisture and texture.

The retention of water by soils in arid and semi-arid regions is controlled to a large extent by its clay content. Therefore, we based the formulation of the effect of precipitation and soil freezing upon two easily accessible variables: soil clay content which controls the retention of water, and monthly mean temperature which allows the distinction between cold and warm deserts. Two deserts were selected for establishing the period during which a desert cannot emit dust after a precipitation event: the Gobi desert for its cold temperatures and the Thar desert for its occurrence of precipitation in a warm setting. Next, we had to establish a threshold number of days over which dust emission is inhibited after a precipitation event. The duration of this period was established for 4 possible cases to distinguish small and large clay content in soils, as well as cold and warm surface temperatures.

The clay content from the FAO database was considered high (resp. low) if it exceeded 10% by mass (resp. was lower than). The threshold between high and low temperature is set at 20°C. The periods with no dust emissions were calculated using the TOMS AI over the two deserts (Gobi and Thar deserts) by matching the number of occurrences of dust events over the year. The best match was found for periods of 1, 3, 4 and 8 days for respectively warm temperature/low clay, warm temperature/high clay, cold temperature/low clay, cold temperature/high clay. In addition, soil erosion is totally inhibited upon freezing. If no more than 0.82 mm/day (corresponding to 300mm/year) of precipitation occurred during these periods, then the emissions of dust were allowed to resume thereafter.

3.3. Comparison to an alternative approach

The previous section described an approach that uses global soil datasets to extend threshold wind velocities calculated using detailed datasets in north Africa. This section describes an alternative approach, based on the observation that the highest TOMS AI values are observed over regions that are topographically lower than nearby regions (e.g. Prospero et al., 2002), in addition to being dry and unvegetated. In this methodology, the threshold wind velocities are assumed to be constant everywhere, but a topographic factor is used to parameterize the relative soil erodibility. The identification of dust sources with topographical low has simplified the modeling of dust and has lead to improved results, compared to approaches when preferential sources were not identified (Ginoux et al., 2001, Tegen et al., 2003). Ginoux et al. (2001) model global dust sources using a topographical factor in the source function, wherein the source emissions are weighted by comparing the elevation of each 1 by 1 degree grid point with the surrounding hydrological basin.

Figure 7 shows the monthly mean TOMS Aerosol Index distribution for 1990 (bold lines) and the orography of the Bodele Depression, which lies south of the Tibesti and west of the Ennedi mountains. The blue area are lakes and the white spots are dry lakes or salt pans. This figure shows, that the entire basin is filled with dust all year long. This is in agreement with visibility data from meteorological stations (Mbourou et al., 1997), which show a high frequency of visibility reduction during all seasons except in autumn, when the rate decreases somewhat. Based on these elements, Ginoux et al. (2001) have defined a global distribution of dust sources by using the surface topographic features. They assume that a basin with pronounced topographic variations contains large amounts of sediments, which are accumulated essentially in the valleys and depressions, and over a relatively flat basin the amount of alluvium is homogeneously distributed. A source function is then calculated from a simple formula expressing the relative altitude of any grid point in a basin of fixed size. Only land surface with bare soil is considered as possible dust source.

Global Emissions of Mineral Aerosol   257

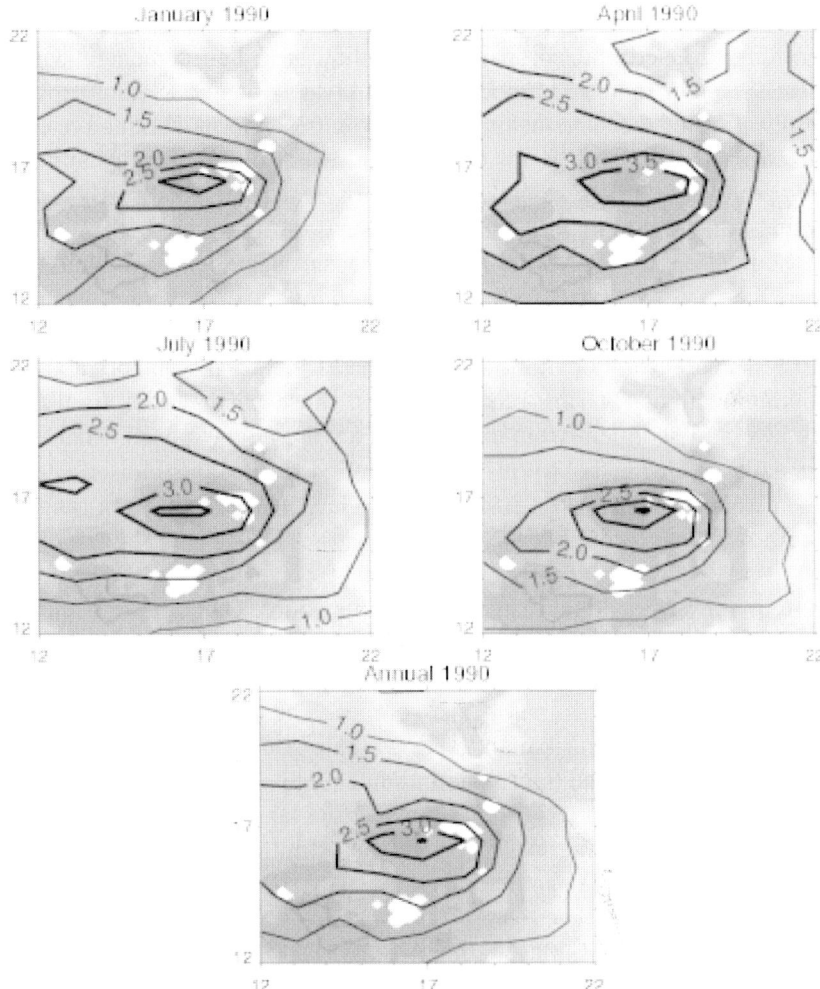

Figure 7 (Plate 13). TOMS Aerosol distribution for 1990 (bold lines) and the orography of the Bodele Depression which lies south of the Tibesti and west of the Ennedi mountains. The blue area are lakes and the white spots are dry lakes or salt pans.

A cursory examination of maps reveals that the geometry of the TOMS aerosol distributions over a specific region can often be associated with playas (undrained desert basins). There is a considerable evidence that playas can be an important source of dust (Reheis, 1997), but from the viewpoint of large-scale dust processes the playas themselves are not necessarily the dominant sources within individual regions. Playa surfaces are often hardened and compacted, and they are often cemented with salt minerals. Rather the close association of playas with TOMS "hot spots" is a manifestation of the general type of environment that is conductive to dust production: that is, arid basins that have a recent pluvial history and which today receive relatively small amounts of water either directly or from runoff. In this regard playas could serve as a convenient marker for potential dust source environments in the present or as evidence of active dust environments in the past. Another major characteristic of many sources is the presence of deep and extensive alluvial deposits, a feature that is consistent with the location of the sources in basins or juxtaposed to topographic highs. During pluvial phases, these basins were flooded and thick layers of sediment were deposited; they are now exposed. Many of the most active TOMS "hot spots" were flooded during the Pleistocene. The prime example is the Lake Chad basin, the largest source of long-range dust in north Africa and possibly the world.

Figure 8 shows the global distribution of the source function, the calculated dust emission, and the TOMS Aerosol index. Most of the maxima of the source function are collocated with the TOMS "hot spots", identified by Prospero et al. (2002) as major dust sources. The clearest examples are the Tunisian, Libyan, Mauritanian, and Malian sources in the Sahara, the Bodele depression in the Sahel, the Indian source along the Indus valley, the Taklamakan located north of the Himalaya, the lake Eyre basin in Australia, the Salton Sea in southern California, the Altiplano and Patagonia in the Andes, and the Namibian source in southwest Africa.

By using such emissions, Ginoux et al. (2001) are able to reproduce observed global scale dust distribution patterns (Figure 8). The total emission is estimated between 1500 and 2000 Tg/yr in a five-year simulation. This result is to be compared with emissions of 800 to 1700 Tg/yr obtained in Tegen et al. (2003).

Global Emissions of Mineral Aerosol 259

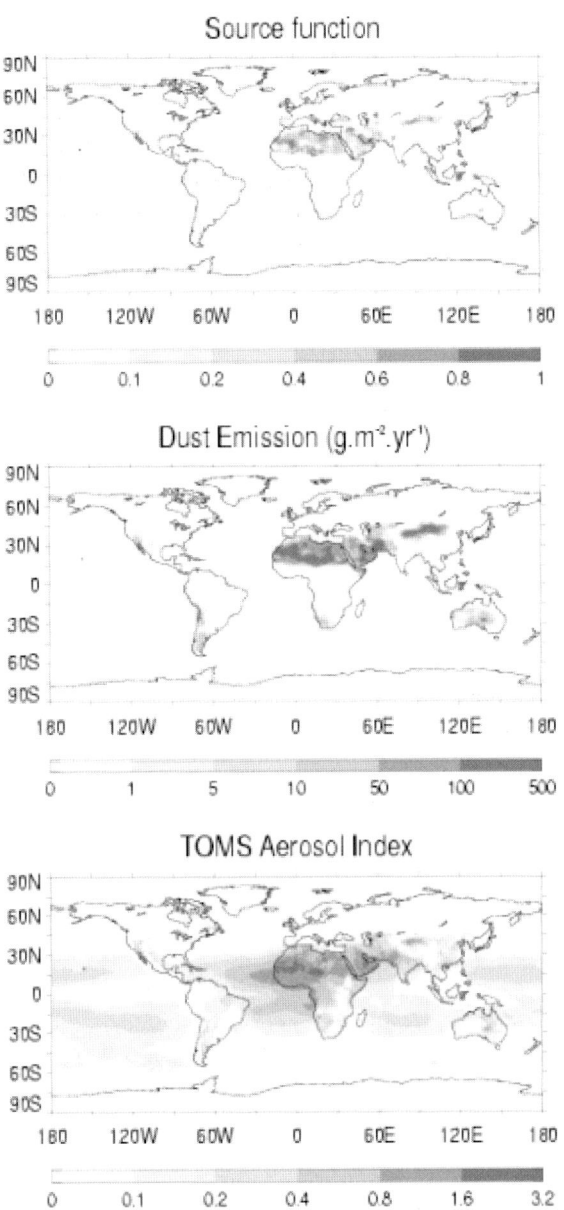

Figure 8. Comparison of model results with in-situ and satellite data indicates that the combination of observed meteorology to drive transport model and a dust sources located in topographic lows with bare soil are key factors to reproduce successfully dust features.

4. PERSPECTIVES TO INFER GLOBAL EMISSIONS OF MINERAL AEROSOL FROM SATELLITE RETRIEVALS

Currently, two semi-quantitative indices, the TOMS AI and the METEOSAT IDDI (Infrared Difference Dust Index) which makes use of the thermal infrared channel of the geostationary METEOSAT sensor, are the only satellite data currently available on the large scale and for long time periods that provide information on the dust source location and strength. These two indices do not have the same sources of error, and they can thus be viewed as "complementary". So far, no systematic daily comparison of these two indices has been performed.

As revealed from the different studies cited and the work presented in this paper, the TOMS AI archive constitutes an easy-to-use satellite data set for developing/validating schemes of the emission of mineral dust at the global scale. This product however has severe drawbacks in terms of quantitative validation of the dust emissions, since it remains uncertain how well it can be directly linked to the aerosol optical thickness. On the other hand, recent studies suggest that optical thickness can be retrieved with a reasonable accuracy from the AI value (Chiapello and Moulin, 2002; Torres et al., 2002). The METEOSAT IDDI (Infrared Difference Dust Index, Legrand et al., 2001), is also problematic to use for quantitative studies. This index is based on the observation in a cloud-free environment of a significant decrease of the brightness temperature, and thus of the IR numeric counts during daytime over arid regions, when dust concentration increases in the atmosphere. This apparent cooling of the surface is due to the attenuation of the infrared radiance, emitted by the hot arid soil, by the colder atmospheric dust layer.

The launch of a new generation of more accurate aerosol oriented sensors such as POLDER (POLarization and Directionality of the Earth's Reflectance) and MODIS (Moderate Resolution Imaging Spectroradiometer) at the end of the 90's did not yet improve our knowledge on the dust emissions. POLDER was launched in 1996 onboard the Japanese platform ADEOS (ADvanced Earth Orbiting Satellite) and provided data from November 1996 to June 1997. One of its most interesting features for aerosol studies over the continents was to measure the polarisation of the light scattered by the Earth-atmosphere system. Because most of the terrestrial surfaces have a low polarization efficiency and because aerosols strongly polarize the incident solar light, POLDER provided the first high resolution global monitoring of the aerosol optical thickness over the continents (Tanré et al., 2001). This monitoring unfortunately turned out to not be possible for mineral dust because these large (>1μm) and non-spherical particles have a very low polarization efficiency. In a similar manner, the current MODIS algorithm fails in retrieving aerosol optical thickness over deserts because of the too large brightness of the surface in the near infrared. Arid surfaces have, however, some particular optical properties that could be used to improve mineral dust detection over the deserts from MODIS, POLDER or similar multi-spectral and multi-angular sensors. Desert reflectances are relatively low and constant in the blue part of the solar spectrum, so that mineral dust optical thickness could be retrieved directly from blue band measurements using accurate radiative transfer computations. Such a method is currently not used for the operational processing of any sensor, but it has already been tested using POLDER data (Colzy and Bréon, personal communication). The land/sea continuity in retrieved aerosol optical thickness obtained for some summertime dust events suggests that realistic optical thickness can be retrieved over deserts, when the viewing geometry is favourable. The main difficulty with this approach is to create a database of clear sky desert reflectances in the blue wavelength domain to be used within the inversion procedure. This may be an interesting option to fill the gap in aerosol optical thickness over the desert and to fulfil the need to monitor dust emissions.

Dust optical thickness could also be a useful product of the forthcoming Lidar in space measurements. CALIPSO-CENA (Cloud-Aerosol Lidar and Infrared Pathfinder Satellite Observations) on the Aqua platform will provide vertical profiles of the dust layers over the desert coincident with MODIS and PARASOL (Polarization and Anisotropy of Reflectances for Atmospheric Sciences coupled with Observations from a Lidar). It would thus be most valuable to combine these different data to get the best possible estimates of the dust horizontal and vertical distribution over potential source regions.

It must be emphasized that the aerosol characterisation over the sea surface is much easier and thus much more complete than what can be achieved over bright surfaces such as deserts. POLDER and MODIS, as well as other "ocean colour" oriented sensors such as SEAWIFS (Moulin et al., 2001), can provide relatively accurate information on the particle size distribution and single scattering albedo in addition to the optical thickness. Provided that the source region is not too distant from the coast and that the dust is transported directly towards the ocean, all these information could be used to validate and improve the size distribution of the mineral dust chosen by models in source regions. Another source of information could be provided by the channels around 1 and 2 µm from the MODIS (and from the forthcoming MSG), these channels are specific to mineral dust and sea salt and could bring further information on aerosol over the ocean. Most of the aerosols have small sizes compared to these wavelengths and are therefore not detected. For mineral dust, only the largest particles will scatter the incident solar light.

5. CONCLUSIONS

Although considerable progress has been made in the last decades in our understanding of the physics of dust emissions, attempts at establishing a global source of dust have been hampered by the lack of information on local soil properties such as soil grain size distribution and local roughness height.

Topographical lows have been shown to act as active dust sources, although their relative importance compared to other surfaces has not clearly been established.

In this paper we propose a global formulation of dust sources based upon a source scheme where threshold velocities and emission factors are computed over arid and semi-arid regions. Threshold velocities were derived by associating a value which was calculated with detailed data in North Africa to each dominant FAO soil type and extrapolated globally. The source strengths of 12 individual arid areas were deduced on the basis of the retrieval of the aerosol optical depth from TOMS data.

Satellite data have opened the way for validation of dust emission schemes and of atmospheric dust distribution. Where most needed, i.e., in the vicinity of source regions, a reliable quantitative estimate of the aerosol burden remains evasive. Nonetheless, the conjunction of multi-spectral captors and active sensors have created new directions of research for a quantification of the dust emissions.

6. ACKNOWLEDGMENTS

We would like to acknowledge the contribution from Nathalie Mahowald and the comments of another anonymous reviewer. This work was supported by the European Projects Syndicate ENV4CT970483 and Phoenics EVK2-2001-00002.

7. REFERENCES

Andreae, M. O., "Climatic effects of changing atmospheric aerosol levels," in World Survey of Climatology, 16, Future climates of the world, A. Henderson-Sellers, Ed., Elsevier, Amsterdam, 347-398, 1995.

Bagnold, R.A., The Physics of Blown Sand and Desert Dunes, Mehuen, London, 265 pp., 1941.

Chepil, W.S., Properties of soil which influence wind erosion, 4: State of dry aggregate structure, Soil Sci., 72, 387-401, 1951.

Chiapello I. and C. Moulin, TOMS and METEOSAT satellite records of the variability of Saharan dust transport over the Atlantic during the last two decades (1979-1997), Geophys. Res. Lett., 29, 17-20, 2002.

Dentener, F. J., G. R. Carmichael, Y. Zhang, J. Lelieveld, and P. Crutzen, Role of mineral aerosol as a reactive surface in the global troposphere, J. Geophys. Res., 101, 22,869-22,889, 1996.

Deuzé J.L., P. Goloub, M. Herman, A. Marchand, G. Perry, S. Susana, and D. Tanré, Estimate of the aerosol properties over the oceans with POLDER, J. Geophys. Res., 105, 15329-15346, 2000.

Falkowski, P. G., R. T. Barber, and V. Smetacek, Biogeochemical controls and feedbacks on Ocean primary production, Science, 281, 200-206, 1998.

Fécan F., B. Marticorena and G. Bergametti, Parametrization of the increase in aeolian erosion threshold wind velocity due to soil moisture for arid and semi-arid areas, Ann. Geophysicae, 19, 149-157, 1999.

Food and Agriculture Organization of the United Nations (FAO-UNESCO), Digital Soil Map of the World and Derived Properties. Rome, Italy,1995.

Genthon, C., Simulations of the long range transport of desert dust and sea-salt in a general circulation model, in Precipitation, Scavenging and Atmospheric Surface Exchange, ed. S. E. Schwartz and W. G. N. Slinn, pp. 1783-1794, Taylor and Francis, Philadelphia, PA, 1992a.

Genthon,C. Simulations of desert dust and sea-salt aerosols in Antarctica with a general circulation model of the atmosphere, Tellus, Ser. B, 44, 371-389, 1992b.

Gillette, D. A., On the production of wind erosion aerosol having potential for long range transport, J. Rech. Atmos., 8, 734-744, 1974.

Gillette, D. A., A wind-tunnel simulation of the erosion of soil: Effect of soil texture, sandblasting, wind speed, and soil consolidation on dust production, Atm. Environ., 12, 1735-1743, 1978.

Gillette, D. A., A qualitative geophysical explantion for "hot-spot" dust emitting source regions, Contr. Atmos. Phys., 72, 67-77, 1999.

Ginoux, P., M. Chin, I. Tegen, J. M. Prospero, B. Holben, O. Dubovik, and S-J Lin, Sources and distributions of dust aerosols simulated with the GOCART model, J. Geophys. Res., 106, 20,255-20,273, 2001.

Greeley R. and J.D. Iversen, Wind as a Geological Process on Earth, Venus, Mars, and Titan, Cambridge University Press, Cambridge, 333 pp., 1985.

Griffin, D. W., V. H. garrison, J. R. Herman, and E. A. Shinn, African desert dust in the Caribebbean atmosphere: Microbiology and public health, Aerobiologia, 17, 203-213, 2001.

Guelle, W., Y. Balkanski, M. Schulz, B. Marticorena, G. Bergametti, C. Moulin, R. Arimoto, and K. D. Perry, Modelling the atmospheric distribution of mineral aerosol: Comparison with ground measurements and satellite observations for yearly and synoptic time scales over the North Atlantic, J. Geophys. Res, 105, 1997-2005, 2000.

Harrison, S. P., K. E. Kohfeld, C. Roelandt, and T. Claquin, The role of dust in climate changes today, at the last glacial maximum an in the future, Earth Sci. Rev., 54, 43-80, 2001.

Haywood, J., and O. Boucher, Estimates of the direct and indirect radiative forcing dir to tropospheric aerosols, Rev. Geophys., 38, 513-543, 2000.

Heimann, M., The global atmospheric model TM2 (model description and user manual), Technical report, Deutsches Klimarechenzentrum, Hamburg, Technical Report No 10, 47 pp., 1995.

Herman, J. R., P. K. Bhartia, O. Torres, C. Hsu, C. Seftor, and E. Celarier, Global distribution of UV-absorbing aerosols from Nimbus 7/TOMS data, J. Geophys. Res., 102, 16,911-16,922, 1997.

Hsu, N. C., J. R. Hermann, O. Torres, B. N. Holben, D. Tanre, T. F. Eck, A. Smirnov, B. Chatenet, and F. Lavenu, Comparisons of the TOMS aerosol index with Sun-photometer aerosol optical depth: Results and applictations, J. Geophys. Res., 104, 6269-6279, 1999.

Husar, R. B., J. M. Prospero, and L. L. Stowe, Characterization of tropospheric aerosols over the oceans with the NOAA Advanced Very High Resolution Radiometer optical thickness operational product, J. Geophys. Res., 102, 16,889-16,909, 1997.

Iversen, J.D. and J.R. White, Saltation threshold on Earth, Mars and Venus, Sedimentology, 29, 111-119, 1982.

Joussaume, S., Three dimensional simulations of the atmospheric cycle of desert dust particles using a general circulation model, J. Geophys. Res., 95, 1909-1941, 1990.

Legrand M., A. Plana-Fattori, C. N'doumé, Satellite detection of dust using the IR imagery of Meteosat, 1. Infrared difference dust index, J. Geophys. Res., 106, 18251-18273, 2001.

Levin, Z., E. Ganor, and V. Gladstein, The effects of desert particles coated with sulfate on rain formation in the eastern Mediterranean, J. Appl. Metorol., 35, 1511-1523, 1996.

Li X., H.B. Maring, D. Savoie, K. Voss and J.M. Prospero, Dominance of mineral dust in aerosol light scattering in the North Atlantic trade winds, Nature, 380, 416-419, 1996.

Mahowald, N., K. Kohfeld, M. Hansson, Y. Balkanski, S. P. Harrison, I. C. Prentice, M. Schulz, and H. Rodhe, Dust sources and deposition during the last glacial maximum and current climate: A comparison of model results with paleodata from ice cores and marine sediments, J. Geophys. Res., 104, 15,895-15,916, 1999.

Marticorena, B. and G. Bergametti, Modeling the atmospheric dust cycle: 1. Design of a soil-derived dust emission scheme, J. Geophys. Res., 100, 16415-16430, 1995.

Marticorena, B., Modeling the production of desert dust in arid and semi-arid regions: development and validation of a computational code for large-scale transport (in French), Ph.D thesis Paris University, 1995.

Marticorena, B., G. Bergametti, D. A. Gillette, and J. Belnap, Factors controlling threshold friction velocity in semi-arid and arid areas of the United States, J. Geophys. Res., 102, 23,277-23,287, 1997.

Marticorena, B., G. Bergametti, and M. Legrand, Comparison of emission models used for large scale simulation of the mineral dust cycle, Contr. Atmos. Phys., 72, 151-160, 1999.

Martin, J.H. and R.M. Gordon, Northeast Pacific iron distributions in relation to phytoplankton productivity, Deep-Sea Res., 35, 177-196, 1988.

Mbourou, G.N., J.J. Bertrand and S.E. Nicholson, The diurnal and seasonal cycles of wind-borne dust over Africa north of the Equator. J. Appl. Meteor. 36: 265-273, 1997.

Nickovic, S., and Dobricic, S., A model for long-range transport of desert dust, Mon. Wea. Rev., 124, 2537-2544, 1996.

Patial R., Mountain desert silicosis, J. Assoc. Phyis. India, 47, 503-504, 1999.

Prospero, J. M., P. Ginoux, O. Torres, S. Nicholson, and T. Gill, Environmental characterization of global sources of amospheric soil dust identified with the NIMBUS-7 TOMS Absorbing Aerosol Product, Rev. Geophys., 40, 1-31 2002.

Pye, K., Aeolian Dust and Dust Deposits, Academic Press, San Diego, California,1987.

Raupach, M.R., Drag and drag partition on rough surfaces, Bondary Layer Meteorology, 60, 375-395, 1992.

Raupach, M.R., D.A. Gillette and J.F. Leys, The effect of roughness elements on wind erosion threshold, J. Geophys. Res., 98, 3023-3029, 1993.

Raupach, M.R., Simplified expression for vegetation roughness length and zero-plane displacement as function of canopy height and area index, Bondary Layer Meteorol., 71, 211-216, 1994.

Reheis, M. C.. Dust deposition downwind of Owens (dry) Lake, 1991-1994: Preliminary findings, J. Geophys. Res., 102, 25999 – 26008, 1997.

Rosenfeld, D., Suppression of rain and snow by urban and industrial air pollution, Science, 287, 1793-1796, 2000.

Shao Y., M.R. Raupach, and P.A. Findlater, Effect of saltation bombardment on the entrainment of dust by wind, J. Geophys. Res., 12719-21726, 1993.

Shao Y., M.R. Raupach, and J.F. Leys, A model for predicting eolian sand drift and dust entrainment on scales from paddock to region, Aust. J. Soi. Res., 34, 309-342, 1996.

Shao Y. and L. Leslie, Wind erosion prediction over the Australian continent, J. Geophys. Res., 102, 30091-30105, 1997.

Song C. H., and G.R. Carmichael, A three-dimensional modeling investigation of the evolution processes of dust and sea-salt particles in east Asia, J. Geophys. Res., 106, 18,131-18,154 2001.

Tanré D., F.M. Bréon, J.L. Deuzé, M. Herman, P. Goloub, F. Nadal, and A. Marchand, Global observation of anthropogenic aerosols from satellite, Geophys. Res. Lett., 28, 4555-4558, 2001.

Tegen, I., and I. Fung, Modeling of mineral dust in the atmosphere: Sources, transport, and optical thickness, J. Geophys. Res., 99, 22,8970-22,914, 1994.

Tegen, I., and I. Fung, Contribution to the atmospheric mineral aerosol load from land surface modification, J. Geophys. Res., 100, 18,707-18,726, 1995.

Tegen, I., and A. A. Lacis, Modeling of particle size distribution and its influence on the radiative properties of mineral dust aerosol, J. Geophys. Res., 101, 19,237-19244, 1996.

Tegen, I., A.A. Lacis and I. Fung, The influence on climate forcing of mineral aerosols from disturbed soils, Nature, 380, 419-422, 1996.

Tegen, I., P. Hollrig, M. Chin, I. Fung, D. Jacob, and J. Penner, Contribution of different aerosol species to the global aerosol extinction optical thickness; estimates from model results, J. Geophys. Res., 102, 23,895-23,916, 1997.

Tegen, I., S.P. Harrison, K. Kohfeld, C. Prentice, M. Heimann, The impact of vegetation and preferential source areas on global dust aerosols: Results from a model study, J. Geophys. Res, in press, 2003.

Torres, O., P. K. Bhartia, J.R. Herman, Z. Ahmad, and J. Gleason, Derivation of aerosol properties from a backscattered measurement of ultraviolet radiation : Theoretical basis, J. Geophys. Res, 108, 17,099-17,110, 1998.

Torres, O., P.K. Bhartia, J.R. Herman, A. Sinyuk, P. Ginoux, B. Holben, A Long-Term Record of Aerosol Optical Depth from TOMS observations and comparison to AERONET measurements, J. Atmos. Sci., 59, 398-413, 2002.

Woodward, S. ,Modeling the atmospheric life cycle and radiative impact of mineral dust in the Hadley Centre climate model, J. Geophys. Res., 106, pp. 18,155-18166 , 2001.

Zobler, L., A world file for global climate modelling. Technical Support, NASA-TM-87802, 1986.

Emissions from volcanoes

Christiane Textor, Hans-F. Graf, Claudia Timmreck, and Alan Robock

1. GLOBAL VOLCANISM

Around 380 volcanoes were active during the last century, with around 50 volcanoes active per year (Andres and Kasgnoc, 1998). Volcanic activity is not randomly distributed over the Earth, but is linked to the active zones of plate tectonics, as shown in figure 1. More than 2/3 of the world's volcanoes are located in the northern hemisphere, and in tropical regions. The emission of volcanic gases depends on the thermodynamic conditions (pressure, temperature) and on the magma type (i.e., its chemical composition, which in turn depends on the tectonic environment).

Globally, most of the magma mass erupted is of basaltic composition. Basaltic volcanoes erupt most frequently along mid-oceanic ridges in deep ocean water, where magma rises along the oceanic plate boundaries. On very rare locations, these volcanoes erupt into the atmosphere (called subaerial eruptions), such as in Iceland and the Azores, where the eruption rate is so large that volcanic islands have been formed along the mid-Atlantic ridge. Their basaltic magmas are primitive melts from the Earth's mantle. Basaltic magma is rich in magnesium and iron, and poor in silicate. In general, this magma type is characterized by low viscosity and low gas content, and eruptions are mostly effusive. They consist of a high portion of CO_2 and sulphur in the gas fraction. Long lasting basaltic lava streams can cover large areas (e.g., the Deccan traps in India, Yellowstone in North America, and the Laki fissure in Iceland). Basaltic magmas are also typical for intra-plate volcanoes that evolve over very hot regions in the Earth's mantle at around 100 km depth. The island chain of Hawaii is an example of this volcano type. Basaltic magma contributes only a minor fraction to the volcanic sulphur emissions into the atmosphere and only in rare cases reaches the stratosphere.

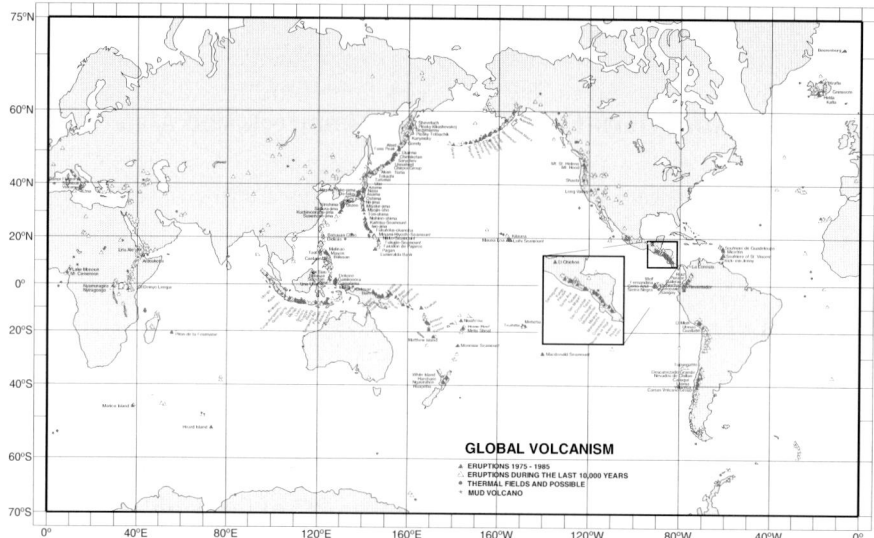

Figure 1 (see plate 14): Active volcanoes from 1975-1985 (solid triangles) and sites with volcanic activity during the last 10,000 years (open triangles). Taken from Graf et al. (1997), based on McClelland et al. (1989), adapted by permission of Prentice Hall.

Felsic magma stems from differentiation processes (i.e., chemical alteration) in the magma chamber, or from melting of earth crust material. This highly differentiated magma is rich in silicate and alkali. It contains a higher content of dissolved gases, especially water. It is highly viscous and eruptions are generally more explosive. In the petrological nomenclature, magma of this type is called rhyolite, dacite, or phonolite, depending on the specific chemical composition. Magmas of intermediate silicate content are andesitic magmas, and are typical for volcanoes at convergent plate boundaries. Subduction zone volcanoes develop in regions where the continental plate overrides the oceanic plate, such as in the Andes. When oceanic plates converge, island arc volcanoes evolve (e.g., Indonesia). Felsic and andesitic volcanoes erupt less frequently than basaltic volcanoes. They can release high amounts of magma and energy on short time scales, however, sometimes injecting ashes and gases directly into the stratosphere. In addition, many permanently emit gases during non-explosive phases. Subduction zone volcanoes contribute the largest part to the total global volcanic sulphur emission.

2. CHEMICAL SPECIES EMITTED BY VOLCANOES

The composition of volcanic gases at the volcanic vent is in general controlled by the equilibrium between a hydrous fluid (exsolved gas) at the top and the silicate melt in the magma chamber below (Symonds et al., 1994). It varies widely between volcanoes depending on the magma type, and is also dependent on the individual volcano's state of activity. An overview is given in table 1.

Table 1: Characteristic composition of volcanic gases at the vent (e.g., Symonds et al., 1988, Cadle, 1980; Symonds et al., 1994; and Chin and Davis, 1993, see also table 2).

Species	H_2O	CO_2	SO_2	H_2S	COS	CS_2	HCl	HBr	HF
%/vol	50-90	1-40	1-25	1-10	10^{-4}-10^{-2}	10^{-4}-10^{-2}	1-10	?	$< 10^{-3}$
Tg/year	?	75	1.5-50	1-2.8	0.006-0.1	0.007-0.096	0.4-11	0.0078-0.1	0.06-6

Water vapour (H_2O) is the most prevalent volcanic gas, contributing between 50 and 90% by volume, however the contribution to the global H_2O inventory is negligible in comparison to the atmospheric concentration. The second important volcanic gas is carbon dioxide (CO_2), which ranges from 1 to 40% by volume. Volcanic emissions contribute less than 1% to the total global CO_2 emission (Cadle, 1980; Gerlach, 1991). Anthropogenic annual CO_2 emissions are by a factor of 100 higher than total natural degassing of the Earth (Schmincke, 1993).

Sulphur gases contribute typically 2 to 35%/vol of volcanic gas emissions. They are the most relevant species concerning the climatic impact of volcanic events. The dominant sulphur component is sulphur dioxide (SO_2), with yearly emissions ranging from 1.5 to 50 Tg SO_2. These data are discussed in more detail below in section 5. The second important S-species is hydrogen sulphide (H_2S), which is only observed by direct sampling. It is often co-emitted and subsequently converted to SO_2 in the atmosphere (e.g., Bluth et al., 1995). The H_2S fraction increases with increasing pressure (i.e., depth of the magma chamber) and with decreasing temperature and oxygen concentration of the magma (Gerlach et al., 1986). Stoiber et al. (1987) have estimated that the amount of H_2S and SO_4^{2-} is commonly less than 1%/vol, although it may reach 10% in some cases. Berresheim and Jaeschke (1983) estimated total sulphur emissions (H_2S plus SO_4^{2-}) to be about 4.2 Tg S per year, but these figures are highly variable and uncertain. In the case of the El Chichón eruption, possibly most of the ~3.5 Tg S was emitted as H_2S. Considering all emitted sulphur as SO_2 will not lead to a big error, however, since H_2S oxidizes to SO_2 within a few days. Other sulphur species add minor portions. Carbonyl sulphide (COS) and its precursor carbon disulphide (CS_2) contribute a small fraction of 10^{-4} to 10^{-2} %/vol. However, COS has a residence time of several years in the atmosphere (e.g., Kjellström et al., 1998). Because of its high stability it is an important source for the sulphate aerosol layer in the stratosphere. Volcanoes contribute less than 1% to the total global atmospheric COS emission (Cadle, 1980; Andres and Kasgnoc, 1998). The data suggest that volcanoes contribute insignificant amounts of COS and CS_2 to the atmosphere, but the exact quantity has not yet been determined.

The main halogen component of volcanic emissions is hydrogen chloride (HCl), with 1-10 %/vol (Symonds et al., 1988). The volcanic contribution of 0.4-11 Tg HCl per year (Symonds et al., 1988; Cadle, 1980) to the total chlorine budget is approximately equal to the anthropogenic emissions, but the emissions from oceans are orders of magnitudes higher. HCl is highly soluble and is therefore rapidly washed out of the atmosphere. Hence, small eruptions and silent degassing will not be of importance for atmospheric composition. Hydrogen bromide (HBr) contributes a small proportion in the range of 10^{-6} parts per volume. Quantification of global annual emissions needs to be improved, in particular because of its enormous ozone depletion potential, which is higher than that of chlorine (e.g., McElroy et al., 1992). Estimates for HBr range from 0.0078 Tg to 0.1 Tg per year (Cadle 1980, Bureau et al., 2000). Its fate in the atmosphere is similar to that of HCl. Only under very specific conditions can its lifetime be significant, as discussed in section 4.

Very little information is available on hydrogen fluoride (HF). Its fraction in volcanic gas emissions is less than 1 ppm; the annual global emission is 0.06-6 Tg (Symonds et al., 1988). HF is not of importance in general, but during specific events (e.g., Laki in 1783, Mt. Hudson in 1990) HF emissions may be extreme and lead to severe environmental contamination with hazards to plants and livestock.

Emissions from Volcanoes 273

A variety of other gaseous components is erupted by volcanoes. Volcanic emissions can be detrimental for the local environment. However, this paper focuses on impacts on the atmosphere on larger scales in time and space.

3. OBSERVATIONAL TECHNIQUES

The determination of gas and particle concentrations in a plume is extremely difficult because of the cloud's opaqueness and the inherent risks of directly observing and sampling the volcanic cloud. Volcanic emissions can be studied remotely by airborne and ground-based instruments and by satellite observations.

Among the most abundant volcanic gases, SO_2 is the species that has been mainly measured remotely, because its concentration in a plume exceeds atmospheric background concentration. Ultraviolet (Correlation Spectrometer, COSPEC) and infrared (Fourier Transform Infrared Spectrometer, FTIR) spectroscopy have been employed (e.g., Symonds et al., 1988; Andres and Rose, 1995). These techniques are most important for the detection of injections into the troposphere.

In recent years, the launches of new satellites and new developments in remote-sensing techniques have expanded the capability to monitor volcanoes from space (Rose et al., 2000). Satellite observations (e.g., by TOMS (Total Ozone Mapping Spectrometer), AVHRR (Advanced Very High Resolution Radiometer), GOME (Global Ozone Monitoring Experiment), SCIAMACHY (SCanning Imaging Absorption SpectroMeter for Atmospheric CHartographY)) for SO_2 and ash particles are only useful for strong sources (Bluth et al., 1993). SO_2 is the only volcanic gas so far monitored operationally by satellite. Since the first TOMS data in 1979, which could only measure the presence of SO_2 from larger eruptions, improved instruments and retrieval algorithms today can detect SO_2 gas from smaller eruptions and from the passive degassing of some volcanoes if the detection limit of 5-20 kt SO_2 is exceeded. Satellites have even detected SO_2 from several eruptions not known from ground observations.

SO_2 emission rates cannot be measured directly, but have to be inferred from a multitude of other data. The errors vary with the volcanic plume mass and area, and they depend on the quality of the estimation of the plume's trajectory and dilution. Since these instruments only measure the SO_2

abundance in a vertical (satellites) and/or horizontal/slant (COSPEC and FTIR) path, the flux must be estimated using information on the wind field. This is not always available with the necessary accuracy. In addition, the results are disturbed by meteorological clouds and affected by the high optical depth of the plume and the reflectivity of the underlying surface. For example, the error of the total SO_2 mass retrieved from TOMS data ranges from 15 to 30% (Krueger et al., 1995). In addition, most of the techniques mentioned above work only during the day.

The petrologic method is employed to estimate magmatic gas content by comparing pre- and post-eruptive volatile concentrations trapped in ash deposits. One compares the gas content of matrix glass inclusions (representative of the primitive magma) to that in the degassed magma. This technique works better for sulphur for volcanoes erupting along spreading ridges, such as the Icelandic volcanoes, where the erupted material comes directly from primitive magma, as compared to volcanoes along subduction zones, such as El Chichón or Pinatubo, where there is a relatively high concentration in the atmospheric plumes but little sulphur in the inclusions, which come from reprocessed material. Palais and Sigurdsson (1989) compared petrological data to ice core estimates of volatile emissions for several major eruptions. They found that the atmospheric yield of sulphur, chlorine and fluorine is not only dependent on total erupted mass, but largely determined by the composition of the erupting magmas. The "excess sulphur" found in satellite data after Pinatubo was explained by a separate SO_2-rich fluid phase which coexists with the magma (Gerlach et al., 1996). This means that gases can dissolve from un-erupted magma, and form a separate fluid in the magma chamber which is emitted co-eruptive, or by fuming (non-steady degassing during phases of higher activity, but not highly explosive) and fumarole (steady degassing) activity.

The enormous differences in the observational data are due to uncertainties in each individual measuring technique, but they also result from the fact that the plume is investigated at different distances from the crater and during different states of volcanic activity or of the eruption. Changes in sulphur emissions at (and between) single sources over orders of magnitude can take place depending on the state of activity (e.g., Augustine or Etna) and magma type. Only a few of the about 560 potential sources have ever been measured, and only a handful of these were observed more than episodically. For important regions (e.g., Kamchatka) there are no published data at all. In some cases (e.g., Etna), the emission at the single volcano can reach mean values of highly polluted industrial areas (Langmann and Graf, 1997; Graf et al., 1998).

Seasonal and annual variations of stratospheric aerosols have been observed by satellite instruments (Hitchman et al., 1994), by balloon-borne particle counters (Hofmann et al., 1983), and by lidars (Menzies and Tratt, 1995). Each of these techniques, however, has problems. The Stratospheric Aerosol and Gas Experiment (SAGE) and Stratospheric Aerosol Measurement (SAM) projects (McCormick et al., 1979; Mauldin et al., 1985; McCormick, 1987; Thomason, 1991; Veiga, 1993) have provided more than 20 years of three-dimensional data of stratospheric aerosol spectral extinctions, the longest such record. Hitchman et al. (1994) and Stevermer et al. (2000) used these data to study the zonal mean aerosol climatology. However, a significant level of uncertainty exists in aerosol characterization during the period of SAGE observations from 1978 to present. For example, because of the "saturation" effect, SAGE II fails to measure aerosols in a large part of the most important equatorial region for almost a year after the Mt. Pinatubo eruption. The eruption of El Chichón in 1982 was not covered by SAGE observations because the SAGE I instrument failed in 1981, and SAGE II was only launched in 1984. Therefore, to produce a data set of stratospheric aerosols, SAGE observations would have to be enhanced and blended with other available satellite and ground-based measurements. For example, Antuña et al. (2002) used lidar data to fill gaps in SAGE II observations after the Mt. Pinatubo eruption. After Pinatubo, 29 balloon flights carrying aerosol counting instruments were made at Laramie, Wyoming (41.3°N, 105.6°W) by Deshler et al. (1992, 1993), but there were no observations from other latitudes. Lidars give excellent vertical distribution of aerosols, but until recently they could only operate under clear sky conditions and during the night. The distribution of lidar observatories is uneven, with none between 19°N and 23°S, with the exception of the one in Bandung, Indonesia (6.9°S, 107.6°E), which is plagued by bad weather. There are at least three lidar networks worldwide, one in Asia (Uchino et al., 1992) and two in Europe (Fiocco et al., 1996; Bösenberg et al., 1998), but none of these operates using standardized instruments and processing software. A data assimilation system, incorporating all these different observations in the context of a model of stratospheric dynamics is needed to produce a uniform aerosol data set for use in climate studies.

4. PROCESSES IN THE ERUPTION COLUMN

Little is known about the processes in volcanic plumes that determine the amount and specification of emissions into the atmosphere due to the difficulties with observations described in the previous section. Many eruptions are accompanied by a water cloud. Hydrometeors and ash particles are able to scavenge volcanic emissions (e.g, Turco et al., 1983). Emissions are altered by chemical reactions if the plume is not too opaque to prevent photochemical activity (Textor et al., 2002). For explosive eruptions, the time for the rise from the crater to the stratosphere is less than 10 minutes. Hence, the residence time in the plume is comparatively much too short for chemical transformations of the volcanic gases to take place before the reaching the stratosphere.

Sulphur species (SO_2 and H_2S) are only slightly soluble in liquid water; hence, they are only slightly removed by cloud and rain drops at lower heights. HCl, on the other hand, is highly soluble in liquid water. Volcanic gases can also be scavenged by frozen hydrometeors via direct gas incorporation during diffusional growth of ice as proposed by Textor et al. (2002, 2003). The mechanism of gas trapping in ice within a volcanic plume is supported by observational evidence of eruptions, where a lot of water was present in the plume due to interaction with sea water. For example, in one observation extremely high amounts of ice were accompanied by unusually low SO_2 concentrations (Rose et al., 1995).

The scavenging of volcanic gases in a plume has been investigated through numerical simulations by Tabazadeh et al. (1993), who used a one-dimensional plume model (Wilson, 1976; Woods, 1988, 1993) without ice microphysics, and by Textor et al. (2002, 2003), who worked with a more complex model (Oberhuber et al., 1998; Herzog et al., 1998) including ice microphysics and ash aggregation. The results of the two studies are in general accordance, however, the latter authors showed that scavenging efficiency is determined by the amount of condensed water or ice, which in turn depends on the volcanic conditions (composition of the magma, strength of the eruption) and on the meteorological conditions (stability of the atmosphere, wind shear) in the ambient atmosphere (Graf et al., 1999). These simulations showed that a large portion of gaseous SO_2 and H_2S reaches the umbrella region of the plume. The scavenging efficiency for HCl is much higher, but it might not be completely removed from the gas phase under dry conditions (Textor et al., 2002, 2003). Furthermore, the simulations showed that hydrometeors containing dissolved volcanic gases can reach the stratosphere where sublimation is possible, releasing them into the stratosphere.

5. VOLCANIC SULPHUR EMISSIONS INTO THE TROPOSPHERE

The total amount of volcanic tropospheric sulphur emissions per year was estimated by several authors (table 2). There is a large range in estimates, from 0.75 Tg (Kellogg et al., 1972) to 25 Tg (Lambert et al., 1988) sulphur per year. Lambert et al. (1988) have employed the atmospheric concentration of the polonium isotope ^{210}Po to estimate the volcanic output of SO_2. The other authors used emission data from monitored volcanoes to extrapolate to global volcanic emissions.

The contribution of silent degassing to volcanic sulphur input into the troposphere is still under discussion, the numbers vary between 1% (Andres and Kasgnoc, 1998) and 95% (Berresheim and Jaeschke, 1983). This great uncertainty is due to the poor database, the suspicious consideration of silent degassing, and to different extrapolation techniques to include non-monitored volcanoes. Graf et al. (1997) estimated 14 ± 6 Tg S per year, which is in the upper level among other recent estimates. Halmer et al. (2002) have performed a comprehensive literature survey on volcanic activity of the last century. Based on volcanological parameters and on the volcanic SO_2 index (VSI) (see section 8), the information about 50 monitored volcanoes was extrapolated to include 310 unmonitored active subaerial volcanoes.

Volcanoes contribute about 36% to the tropospheric sulphur burden (Graf et al., 1997). The proportion of H_2S in sulphur emissions is still under discussion. However, H_2S is quickly oxidized to SO_2 in about two days in the troposphere (e.g., Seinfeld and Pandis, 1998). SO_2 in the atmosphere is transferred to sulphate by chemical reactions - promoted by the presence of water drops - within some days in the troposphere. Although a variety of SO_2 oxidation rates have been determined, an agreement on the correct rate has still not been found (Bluth et al., 1992; Facchini et al., 1992; Rani et al., 1992; Fung et al., 1991; Grgi et al., 1991; Hansen et al., 1991; Joos and

Baltensperger, 1991; Gallagher et al., 1990; Oppenheimer et al., 1998). The difficulties are due to various conditions that volcanic SO_2 encounters in the atmosphere. Sulphate, which is highly water soluble, is removed from the troposphere within less than a week by wet and dry deposition, but in dry conditions can remain longer, especially if it reaches higher tropospheric levels.

Table 2: Annual global volcanic sulphur emission fluxes into the atmosphere by different authors.

Authors	S (in 10^{12} g/yr)
Kellogg et al. (1972), excluded silent degassing (0.5% gas content)	0.75
Bartels (1972)	17.0
Friend (1973)	2.0
Stoiber and Jepsen (1973), based on only 5 central and south American volcanoes	5.0
Cadle (1975) excluded silent degassing, but used higher magma gas content (2.5%)	3.8
Naughton et al. (1976)	23.5
Granat et al. (1976)	3.0
Le Guern (1982)	5.0
Berresheim and Jaeschke (1983) 95% from silent, 5% from explosive, incl. $H_2S + SO_4$	7.6
Stoiber et al. (1987), 35 volcanoes, short time (1 year) 35% silent, 65% explosive	9.3
Lambert et al. (1988), uses ^{210}Po (50% of all from volcanic sources), but only for very few volcanoes $^{210}Po/SO_2$ ratio	25.0
Andres and Krasgnoc (1998), mainly explosive, only 1% silent	10.4
Graf et al. (1997), includes silent (fumaroles (5±2) post and extra-eruptive) (fuming (5±2) pre- and intra-eruption) and explosive (4±2 Mt S), incl. $H_2S + SO_4$	14.0 ± 6
Halmer et al. (2002)	9.0 ± 1.5

Sulphate aerosols in the atmosphere have effects altering the Earth's radiation balance and cloud microphysics, discussed in section 7. In addition, tropospheric sulphate aerosols have an impact on atmospheric chemistry. Acid rain, caused by scavenging of sulphate aerosols and damaging the vegetation, has been a problem for many years.

6. VOLCANIC EMISSIONS INTO THE STRATOSPHERE

Volcanic injections of sulphur into the stratosphere are sporadic and unpredictable. Explosive volcanic eruptions reach the stratosphere in general at least once every two years (Simkin et al., 1993). Toba 71-73 kyr ago is the largest eruption believed to have occurred since the beginning of the Quaternary period, ca. 2 million years ago. Rose and Chesner (1990) analysed the deposits and with simple assumptions estimated that Toba released about 6000 Tg of SO_2 into the atmosphere, but Zielinski et al. (1995, 1996) used ice core data to estimate a release of 1000-2000 Tg SO_2 and Scaillet et al. (1998) using experimental petrology methods estimated only about 70 Tg SO_2 (Oppenheimer, 2002).

Volcanic sulphur is not only directly injected into the stratosphere from explosive eruptions, but also from emissions of continuously degassing non-eruptive volcanoes and from small eruptive events (Graf et al., 1998). Volcanoes substantially contribute to the stratospheric sulphur burden (e.g., Bluth et al., 1997). The average input of volcanic sulphur over the last 200 years has been estimated as 1 Tg per year ranging from 0.3-3 Tg per year (Pyle et al., 1996). A minimum flux of 0.5-1.0 Tg S for the past 9000 years has been derived from ice core sulphate data. This mean flux, however, is highly variable due to singular explosive events. During periods of little volcanic activity, the background stratospheric load of sulphate is in the order of 0.15 Tg S (Kent and McCormick, 1984). In case of cataclysmic volcanic eruptions, this stratospheric sulphate mass can be increased for short periods of time by one to two orders of magnitude.

Table 3: The largest volcanic eruptions of the last 250 years. The Volcanic Explosivity Index (VEI) is a measure of explosivity (Newhall and Self, 1982, see section 8). Opacity is normalized to the Krakatau eruption and derived from geological data. The SO_2 data marked with a * are derived from the petrologic method, which gives a lower limit of the possible emission (see section 8). Table modified from Robock (2000).

Volcano	Year	VEI	Opacity	SO_2 [Tg]
Laki fissure, Iceland	1783	4	2.3	100*
Tambora, Indonesia	1815	7	3.0	130*
Cosiguina, Nicaragua	1835	5	4.0	
Askja, Iceland	1875	5	1.0	
Krakatau, Indonesia	1883	6	1.0	32*
Tarawera, New Zealand	1886	5	0.8	
Santa Maria, Guatemala	1902	6	0.6	13*
Ksudach, Kamchatka	1907	5	0.5	
Katmai, Alaska, USA	1912	6	0.5	12*
Agung, Indonesia	1963	4	0.8	5-13*
St. Helens, USA	1980	5	0.5	1
El Chichón, Mexico	1982	5	0.8	7
Pinatubo, Philippines	1991	6	1.0	17

Volcanic sulphur is emitted in the form of SO_2 and H_2S. H_2S is oxidized to SO_2 in the stratosphere, and H_2S has a chemical lifetime of only about three days (McKeen et al., 1984). From stoichiometric considerations it can be assumed that during oxidation one molecule of H_2O per H_2S is produced. This could be an important source for water in the stratosphere after volcanic eruptions. The effect of H_2S oxidation on the HO_x balance has been discussed by McKeen et al. (1984) and is still not known. SO_2 is transferred to sulphate by chemical reactions with an e-folding time of approximately 35 days in the dry stratosphere (Bluth et al., 1992). It is assumed that no net gain or loss of HO_x results from SO_2 oxidation (Read et al., 1993; McKeen et al., 1984). Two-dimensional model simulations of the Toba eruption (Bekki et al., 1996) showed that the injection of 200 Mt SO_2 could dehydrate the stratosphere, because approximately three molecules of H_2O are needed to convert SO_2 to sulphate, strongly reducing the HO_x-concentration.

Sulphate has residence times of a few years in the stratosphere. New particles are formed due to binary homogenous nucleation of sulphuric acid and water vapour. After the eruption of Mt. Pinatubo, 98% of the observed stratospheric aerosol was volatile (Deshler et al., 1992; Sheridan et al., 1992), which indicates that homogeneous nucleation is the most important process for stratospheric aerosol formation in the disturbed atmosphere. Deshler et al. (1993) found an increase of 1-2 orders of magnitude in the concentration of condensation nuclei in the volcanic layers. The aerosol size distribution is shifted to greater radii due to condensation of the vapour on existing aerosol particles. Model calculations that successfully reproduced Pinatubo sulphate aerosol (Weisenstein et al., 1997; Timmreck and Graf, 2000) were based on homogeneous nucleation.

Observations (Russell et al., 1996) and model studies (Zhao et al., 1995; Stenchikov et al., 1998; Timmreck and Graf, 2000) show that it takes about three months to build up the sulphate peak after big volcanic injections of SO_2 in the stratosphere. A volcanically enhanced stratospheric aerosol layer can be observed for about four years. Measurements of peak backscatter and mass, and columnar quantities (mass, backscatter and optical depth) indicate that the volcanic aerosol is removed with an e-folding time of approximately 1 ± 0.2 years (Jäger et al.1995; Ansmann et al., 1997; Deshler et al., 1997). The decay rates of vertical integrated quantities are strongly influenced by variations in the tropopause height and stratosphere-troposphere exchange processes, while the decay of peak parameters is smoother and primarily reflects gravitational removal by sedimentation. The relatively faster sedimentation of larger particles from the stratosphere leads to a rapid decrease in mass, but not in surface area (Ansmann et al., 1998). The e-folding time of surface area concentration is about 20 to 30% larger (16-17 months) than the decay of the aerosol backscatter and mass, since the ratio of surface area to mass increases with decreasing particle size. Similar decay rates have been observed after the eruption of Mt. Pinatubo and El Chichón

(Rosen et al., 1994; Barnes and Hoffmann, 1997). Chazette et al. (1995), however, found for the eruption of El Chichón a 25% slower decay of integrated backscatter in the lower stratosphere than for Pinatubo. This was possibly caused by the lofting of the particles to higher altitudes.

7. ATMOSPHERIC EFFECTS OF VOLCANIC EMISSIONS

Sulphate aerosols in the atmosphere have radiative effects that alter the Earth's radiation balance (e.g., Franklin, 1784; Charlson et al., 1991, 1992). While their optical depth is higher in visible wavelengths and the net effect in general is cooling, sulphate aerosols also absorb and emit in the longwave, heating the layer where they are and increasing the downward flux of radiation at the surface (for a review see Robock, 2000). The scattering of incoming solar radiation (direct effect) leads to cooling at the surface.

Sulphate aerosols in the troposphere act as cloud condensation nuclei and modify the radiative properties and the lifetime of clouds (indirect, or Twomey effect) (Twomey, 1974). The increase of the number of cloud droplets due to an increased number of condensation nuclei increases the albedo of clouds, enhancing surface cooling. It may also lead to changes in precipitation rate or rain suppression in deep convective clouds, thus changing the spatial and temporal distribution of latent heat release. This could have a significant effect on the global circulation, as was shown for aerosols in general by Nober et al. (2002).

Volcanic sulphur emissions in the troposphere have a disproportionate effect on the atmosphere, as shown by numerical experiments with an atmospheric general circulation model including a simplified sulphur cycle (Graf et al., 1997). Active volcanoes generally reach considerable elevations and most of their emissions, even during non-explosive events, are injected into the free troposphere, well above the planetary boundary layer. At this height, removal processes are slower and volcanic sulphur has longer atmospheric residence times than anthropogenic sulphur emitted from low elevations. In the mean, volcanic sulphur dominates the sulphate concentration in the middle and upper troposphere while anthropogenic emissions control the boundary layer (Graf et al., 1998). Table 4 shows the global sulphur budget and the radiative effect of different sulphur sources taken from Graf et al. (1997), who used a particular climate model for their calculations. Volcanoes contribute about 4.5 times less to the global sulphur emissions than anthropogenic emissions. But the atmospheric SO_2 burden is only 1.3 times less, and the global tropospheric sulphate burden is of similar size as the anthropogenic one. Hence, due to the longer life time of their sulphur emissions, volcanoes have a much higher relative radiative effect on global climate. Since the treatment of sulphur species, especially their vertical transport and cloud interactions, is not very realistic in current climate models, these results may be highly model-dependent. This is especially true for the transfer of troposphere sulphur to the stratosphere, which is discussed next.

The injection of large quantities of gases (SO_2, H_2O, H_2S, CO_2, HCl) and volcanic ash (mainly silicate particles) into the stratosphere can have an important impact on the global climate. In recent years, the understanding of the global volcanism-climate system has improved significantly, as the very detailed summaries by Robock (2000) and Zielinski (2000) show. But there is still more to be learned about the factors controlling the chemical, physical and optical properties of the volcanic aerosol, its radiative forcing and its role in the global climate system. A schematic representation of the atmospheric effects of volcanic eruptions is shown in figure 2.

Table 4: Global annual mean sulphur budget in Tg sulphur per year (from Graf et al., 1997) and top-of-the-atmosphere (TOA) radiative forcing in percentage of the total (101.8 Tg S/yr emission, 0.52 Tg S SO_2 burden, 0.78 Tg S SO_4 burden, -0.65 W/m^2 total direct global forcing). The efficiency is the relative sulphate burden divided by the relative source strength (i.e., column 4/column 2). DMS:

Source	Sulphur emission [%]	SO_2 burden [%]	SO_4^{2-} burden [%]	Efficiency	Direct TOA forcing [%]
Anthropogenic	65.6	46.1	37.1	0.56	40
Biomass Burning	2.5	1.2	1.6	0.64	2
Dimethyl sulphide (DMS), mainly from oceans	18.2	17.8	25.3	1.39	26
Volcanoes	13.7	34.9	36.0	2.63	33

Fine volcanic ash injected into the stratosphere is characterized by grain sizes in the micrometer range. They are efficiently removed by sedimentation from the stratosphere within about a month after the eruption (e.g., Pinto et al., 1989) and therefore have only a limited local impact on the climate system.

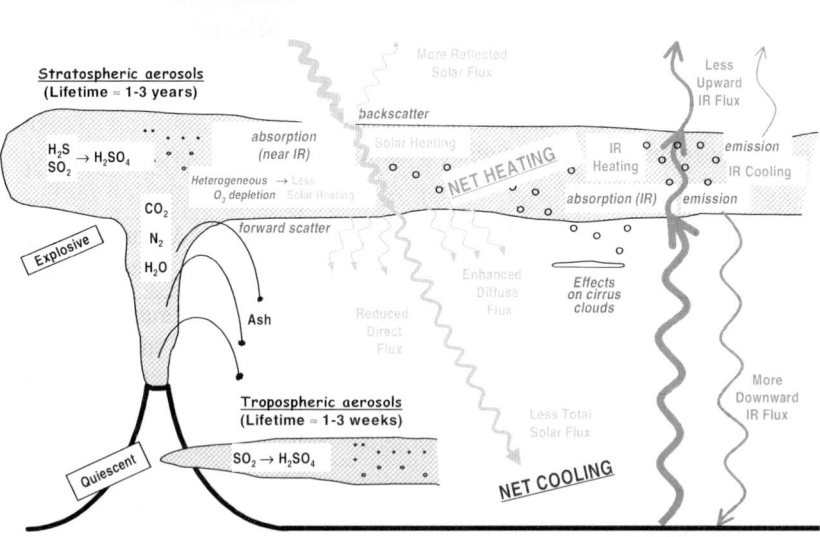

Figure 2 (Plate 15): Schematic diagram of volcanic inputs to the atmosphere and their effects. Reproduced with permission from Alan Robock (2000). (© American Geophysical Union).

Sulphur-containing volcanic gases, rather than ash, are responsible for the climatic effects of explosive volcanic eruptions. Sulphate aerosols in the stratosphere have radiative effects altering the Earth's radiation balance (Stenchikov et al., 1998). The scattering of incoming solar radiation (direct effect) leads to cooling at the surface and counteracts the greenhouse effect. The additional aerosol loading increases the temperature in the lower stratosphere through the absorption of near infrared and long wave radiation. According to the 1995 IPCC report (Schimel et al., 1996), the decadal mean radiative forcing due to volcanic aerosols has varied by as much as 1.5 W/m^2 since 1850, which can be large compared to the decadal-scale variation in any other known forcing. Volcanoes are therefore an important part of the global climate system that substantially contribute to the observed natural variability.

Stratospheric warming due to the presence of the volcanic aerosol in the stratosphere increases the pole-to-equator temperature gradient (e.g., Quiroz, 1984; Parker and Brownscombe, 1983; Angell, 1997a). This has a dynamical feedback on the tropospheric circulation, leading to abnormally warm winters over the northern hemisphere continents in years following the eruption (Groisman, 1985; Robock and Mao, 1992). The mechanism, including the production of stronger westerly winds in the lower stratosphere and their effect on tropospheric planetary waves was explained first by Graf et al. (1993, 1994) and Kodera (1996) to work via exaggeration of the North Atlantic Oscillation.

The effect of the geographical location of an explosive volcanic eruption on its climatic impact has not yet been quantified. Tropical volcanoes have the capability to affect the global climate system, because the cloud can be spread into both hemispheres. Mid to high latitude eruptions will primarily affect their own hemisphere (Graf and Timmreck, 2001). The dispersal of the volcanic cloud is also strongly dependent on the time of the year, and the eruption height. The location of the intertropical convergence zone and the phase of the Quasi Biennial Oscillation (QBO) play an important role for the transport of volcanic clouds.

The eruption of halogen species at the crater is rather significant (e.g., Varekamp et al., 1984; Westrich et al., 1992; Bureau et al., 2000). Their direct injection by explosive volcanic eruptions into the stratosphere could lead to catastrophic ozone loss (Prather, 1992). No severe increase of halogens was observed in the stratosphere after the eruption of Mt. Pinatubo (e.g., Mankin et al., 1992; Wallace and Livingston, 1992), however, after the eruption of El Chichón in 1982 a clear increased chlorine concentration was detected (Mankin and Coffey, 1984; Woods et al., 1985). The fraction of direct injection of halogen compounds depends on the magma composition and on the processes in the eruption column, as discussed in section 4.

Volcanic hydrated sulphate aerosols in the stratosphere can serve as sites for heterogeneous reactions which convert passive chlorine compounds (HCl, HOCl, ClNO$_3$) into active ones (ClO, Cl) (chlorine activation). After the Pinatubo eruption, the surface layer of the stratospheric sulphate aerosol was about 300 times higher than the usual value with peak concentrations higher than 3×10^{-7} cm^2/cm^3 (Jäger et al., 1995; Thomason et al., 1997). The critical value for ozone destruction of 10^{-7} cm^2/cm^3 was reached (Jäger et al.,

1995; Ansmann et al., 1993) for more than one year in northern hemisphere midlatitudes. Heterogeneous reactions deplete ozone in the presence of halogens like chlorine or bromine (e.g., Michelangeli et al., 1989; Hofmann and Solomon, 1989; Granier and Brasseur, 1992; Solomon et al., 1996). Since the human-induced increase of chlorine concentration in the stratosphere has peaked, the effect of ozone destruction at the surface of volcanic aerosol will probably decrease in the next few decades.

Column ozone reduction after the Mt. Pinatubo eruption ranged from about 2% in the tropics to about 7% in mid latitudes (Angell, 1997b; Solomon et al., 1998). Very large ozone losses were observed at high northern latitudes in February and in March. Randel et al. (1995) found losses of 10% in ozone total column in 1992 northward of 60°N and of 10-12% in 1993. Ozonesonde profiles show that the concentration did not decrease uniformly at all altitudes. Hofmann et al. (1993) found at 20°N an ozone decrease below 20-24 km and an ozone increase at about 26 km. Vertical profiles from the tropics (Grant et al., 1994) showed 10% losses of ozone below the peak and roughly 6% increases compared with SAGE data.

The observed ozone changes are a combined effect of heterogeneous chemistry and of perturbations in the heating rates and in the photolysis rates. The chemical composition of the atmosphere during the post-Pinatubo period has been analysed in several publications (e.g., Solomon et al., 1998; Toohey et al., 1995). Model studies, in particular for the Pinatubo episode, were carried out to investigate the impact of the volcanic aerosol contribution to stratospheric trace gas concentrations (e.g., Pitari and Rizzi, 1993; Bekki and Pyle, 1994; Kinnison et al., 1994; Solomon et al., 1996; Rosenfeld et al., 1997). A systematic analysis of the contribution of several components to stratospheric ozone destruction after the Pinatubo eruption was carried out by Tie et al. (1994) with a 2-dimensional radiative chemical model, which suggests that in the first year after the eruption the radiative-dynamical effects and in the second year the chemical effects were dominant. Rosenfeld et al. (1997) found with a 2-d interactive radiative-dynamical-chemical model, that 1-2% of the ozone depletion in the low latitudes after the Pinatubo eruption was due to the altered circulation (perturbation in the heating rate), and about 0.5% due to heterogeneous chemistry and perturbation in the photolysis rate each. However, global model studies considering the various interactions between aerosol microphysics, radiation, chemistry and dynamics, which are necessary to fully analyse the climate impact of a volcanic eruption, are still missing.

In the tropics and during tropopause folds at midlatitudes, volcanic sulphate aerosol particles can be transported vertically across the tropopause. Important lidar data by Ansmann et al. (1993) showed that in more than 50% of the observations the stratospheric aerosol layer penetrated the tropopause and influenced the formation and maintenance of cirrus clouds in the upper troposphere. Graf et al. (1997) suggested the importance of volcanic sulphate aerosol in the upper troposphere for cirrus formation. Unusual high cloud particle number concentration (600 L^{-1}), and extremely supercooled drops at 223-223 K have been observed in the year after the Mt. Pinatubo eruption (Sassen, 1992; Sassen et al., 1995). Song et al. (1996) suggested that the interannual variability of global high level clouds is related to explosive volcanism. The amount and persistence of such clouds increased by up to 10% after the eruptions of El Chichón and Pinatubo mainly at midlatitudes. These anomalies lasted for several years. Thus, violent volcanic eruptions lead to a change in radiative properties of cirrus clouds. Their impact on the climate is still not known, as it depends on changes in cloud microphysics. Scattering of solar radiation would lead to enhanced cooling, while absorption of terrestrial radiation would lead to warming.

8. VOLCANIC ACTIVITY INDICES

Multiple efforts have been made to compile a quantitative record of climatic impact of historical volcanic eruptions. When studying the effects of volcanic eruptions on climate, it is important to separate the volcanic forcing from the climatic response. Proxy records of past climate, such as tree rings and corals, measure climate change, not volcanism. Only if one already understands the climatic response, can these records be used to indicate past volcanism. Therefore, these records will be discussed in the next section as data to validate theories of the impact of eruptions on climate. Robock and Free (1995, 1996) described indices of past volcanism in detail and compare them, and Robock (2000) summarized their work. Here we briefly describe these indices. A perfect index would convey the radiative forcing associated with each explosive eruption. The radiative forcing is most directly related to the sulphur content of emissions that reach into the stratosphere, and not to the explosivity of the eruption. For all the indices, the problem of missing volcanoes and their associated dust veils becomes increasingly important the farther back in time they go. Even today, some indices may miss some southern hemisphere eruptions, as they may not be reported. Even in the 1980s, the December 1981 aerosols from the eruption of Nyamuragira were observed with lidar but were reported as the "mystery cloud" for several years until the source was identified by reexamining the TOMS satellite record (Krueger et al., 1996). The problem of missing an eruption does not exist for ice core records, except that associating acidity or sulphate peaks with particular eruptions may not be possible if the particular eruption is not known.

The first compilation of past volcanic eruptions, the dust veil index (DVI), was made by Lamb (1970, 1977, 1983). The methods used to create the DVI include historical reports of eruptions, optical phenomena, radiation measurements (for the period 1883 onward), temperature information, and estimates of the volume of ejecta. While the DVI has been criticized because of the circular reasoning used by including reports of temperature variations, for only a few eruptions between 1763 and 1882 was the northern hemisphere averaged DVI calculated based solely on temperature information. Robock (1981a) created a modified version of Lamb's DVI which excluded temperature information, and found that this had minor impacts on climate model simulations.

A second compilation is the volcanic explosivity index (VEI) (Newhall and Self, 1982). As the VEI is based only on volcanological data (Simkin et al., 1981; Simkin and Siebert, 1994), and included no atmospheric data, it cannot be used to estimate the height of the eruption columns. "Stratospheric injection" was the least reliable of 11 criteria for estimating VEI, and was never intended as a description of the eruption which had a VEI assigned from more reliable evidence. Therefore, "Since the abundance of sulphate aerosol is important in climate problems, VEI's must be combined with a compositional factor before use in such studies." (Newhall and Self, 1982). An example of a recent eruption for which VEI would be misleading, is the 1980 Mt. St. Helens eruption, with a large VEI of 5, but little stratosphere injection of sulphur and virtually no climatic impact (Robock, 1981b).

Despite its weakness, the VEI provides the most comprehensive geological record of past volcanism. Schnetzler et al. (1997) developed a volcanic SO_2 index (VSI) based on the VEI and 15-year satellite samplings of volcanic eruptions. These authors distinguished between subduction zone volcanoes and others, but they excluded silent degassing like fuming, fumarolic activity and dissipative emissions at the volcano flanks. Halmer et al. (2002) extended Schnetzler et al.'s work by including more published observations of sulphur emissions and reports of volcanic activity, also including silent degassing. Observed activity reports were compared with available sulphur measurements. This information was projected to include unmonitored active sub-aerial volcanoes. Halmer et al.'s modified VSI shows the contribution of different volcano types to the injection of sulphur into the atmosphere at different altitudes. Basaltic magmas can be important local sources, but only subduction zone volcanoes have the potential to significantly directly contribute to the stratospheric sulphur budget, because of their higher explosivities and source altitudes. In spite of the careful data review, the range of order of magnitudes of SO_2 emission "typical" for a specific geological and ambient setting shows big uncertainties. These do not only result from too sporadic and inaccurate measurements, but also to a large degree from the large variability of the processes involved.

A zonally averaged time series of the optical depth due to volcanic eruptions from 1850 on was compiled by Sato et al. (1993), and updated by Stothers et al. (1996). This record is based on optical data, since 1979 satellite measurements have been included. Based on the Sato et al. data and on radiative forcing calculations for the Pinatubo aerosol, Stenchikov et al. (1998) and Andronova et al. (1999) provided a data set of the instantaneous and adjusted radiative forcing by volcanic aerosol from 1850-1994.

Acidity in ice cores from the deposition of volcanic sulphate on glacial ice in the years following an eruption has been used to construct a record of past volcanism. However, each ice core record is extremely noisy and may have other problems (Robock and Free, 1995). For example, smaller local eruptions may have the same signal as distant larger eruptions, so estimating the global stratospheric aerosol loading from high latitude ice cores is problematic. Hammer et al. (1980) produced a 10,000 year data set using the Crete and Camp Century ice cores from Greenland. Crowley et al. (1993) reanalysed the Crete ice core data and constructed a 1420 year record of volcanic activity. The GISP2 ice core has been used to create a 2100 year estimate of stratospheric loading and optical depth, and a 110,000 record of explosive volcanism (Zielinski et al., 1995, 1996). Using multiple ices from both northern hemisphere ice sheets in Greenland and Ellesmere Island, and from Antarctica in the southern hemisphere, Robock and Free (1995) constructed a new Ice core Volcanic Index (IVI). Robock and Free (1996) extended this analysis back to 435 A.D., but the farther back one goes, the fewer ice core records are available. More cores are being drilled and analysed as time goes on, and they will provide valuable new information. The IVI, which correlates well with the VEI but to a lesser extent with the DVI, was used in energy balance model simulations of the past six centuries (Crowley and Kim, 1999; Free and Robock, 1999).

Recently a new ice core volcanic index, the Volcanic Aerosol Index VAI, was introduced by Robertson et al. (2001). It combines historical observations, ice core data from both hemispheres and satellite data to obtain an estimation of the stratospheric optical depth for the past 500 years. The VAI is latitude dependent, with a variable zonal band width of the stratospheric volcanic aerosol.

Proxy data suggest an increase in the frequency of volcanic eruptions on secular time scales. However, the number of highly explosive eruptions shows no such trend and the increase seems to be related to historical events and the number of observers.

9. CONCLUDING REMARKS

Volcanic emissions are highly variable in space and time. To better quantify the impacts of volcanic eruptions on climate, there is a need to better constrain their sulphur emissions. Important future investigations that are needed include exemplary long-term measurements at the sources, better satellite observations, and improved volcanic plume models that simulate the transformation rates of SO_2 into SO_4^{2-} in warm moist plumes and the contribution of H_2S, SO_4^{2-} and SO_2 associated with plume particles, since these species are not measured by the conventional methods. The fate of volcanic aerosols in the stratosphere and their influence on chemistry, microphysics and dynamics are still under discussion.

10. ACKNOWLEDGMENTS

The work of Christiane Textor was supported by the Bundesministerium für Bildung, Wissenschaft, Forschung und Technologie (BMBF) program AFO2000, grant 06-056-0180, and by the Volkswagen Foundation under the project EVA. The work of Claudia Timmreck was supported by the BMBF project KODYACS (Grant 07ATF43) and by the European Commission grant EVK2-CT-2001-000112 (project PARTS). The work of Alan Robock was supported by NASA grant NAG 5-9792 and NSF grant ATM-9988419.

11. REFERENCES

Andres, R. J., and W. I. Rose, Remote sensing spectroscopy of volcanic plumes and clouds, in Monitoring Active Volcanoes: Strategies, Procedures and Techniques, edited by B. McGuire, C. Kilburn, and J. Murray, pp. 301-314, UCL Press Limited, London, 1995.

Andres, R. J., and A. D. Kasgnoc, A time-averaged inventory of subaerial volcanic sulfur emissions, J. Geophys. Res., 103, 25.251-25.261, 1998.

Andronova, N. G., E. Rozanov, F. Yang, M. E. Schlesinger, and G. L. Stenchikov, Radiative forcing by volcanic aerosols from 1850 through 1994, J. Geophys. Res., 104, 16807-16826, 1999.

Angell, J. K., Stratospheric warming due to Agung, El Chichón and Pinatubo taking into account the quasi-biennial oscillation, J. Geophys. Res., 102, 9479948, 1997a.

Angell, J. K., Estimated impact of Agung, El Chichón and Pinatubo volcanic eruptions on global and regional total ozone after adjustment for the QBO, Geophys. Res. Lett., 24, 647-650,1997b.

Ansmann, A., U. Wandinger, and C. Weitkamp, One-year observations of Mount Pinatubo aerosol with an advanced Raman lidar over Germany at 53.5°N, Geophys. Res. Lett., 20, 711-714, 1993.

Ansmann, A., I. Mattis, U. Wandinger, F. Wagner, J. Reichardt, and T. Deshler, Evolution of the Pinatubo Aerosol: Raman lidar observations of particle optical depth, effective radius, mass, and surface area over central Europe at 53.4°N, J. Atmos. Sci., 54, 2630-2641, 1997.

Ansmann, A., I. Mattis , H. Jäger, and U. Wandinger, Stratospheric Aerosol Monitoring With Lidar: Conventional Backscatter versus Raman Lidar Observations of Mount- Pinatubo, Contributions to Atmospheric Physics. 213-222, 1998.

Antuña, J. C., A. Robock, G. L. Stenchikov, L. W. Thomason, and J. E. Barnes, Lidar validation of SAGE II aerosol measurements after the 1991 Mount Pinatubo eruption. J. Geophys. Res., 107, 2002.

Barnes, J.E., and D. J.Hofmann, Lidar measurements of stratospheric aerosol over Mauna Loa Observatory, Geophys. Res. Lett., 24, 1923-1926, 1997.

Bartels, O. G., An estimate of volcanic contributions to the atmosphere and volcanic gases and sublimates as the source of the radio isotopes ^{10}Be, ^{35}S, and ^{22}Na, Health Physics, 22, 387-392, 1972.

Bekki, S., and J. A. Pyle, A two-dimensional modeling study of the volcanic eruption of Mount Pinatubo, J. Geophys. Res., 99, 18,861-18,869, 1994.

Bekki, S., J. A. Pyle, W. Zhong, R. Toumi, J. D. Haigh und D. M. Pyle, The role of microphysical and chemical processes in prolonging the climate forcing of the Toba eruption, Geophys. Res. Lett., 23, 2669-2672, 1996.

Berresheim, H., and W. Jaeschke, The contribution of volcanoes to the global atmospheric sulfur budget, J. Geophys. Res., 88, 3732-3740, 1983.

Bluth, G. J. S. , Doiron, S. D., Schnetzler, C. C., Krueger, A. J., and L. S. Walter, Global tracking of the SO_2 clouds from the June, 1991 Mount Pinatubo eruptions, Geophys. Res. Lett., 19, 151-154, 1992.

Bluth, G. J. S., C. C. Schnetzler, A. J. Krueger, and L. S. Walter, The contribution of explosive volcanism to global atmospheric sulphur dioxide concentrations, Nature, 366, 327-329, 1993.

Bluth, G. J. S., C. J. Scott, I. E. Sprod, C. C. Schnetzler, A. J. Krueger, and L. S. Walter, Explosive SO_2 emissions from the 1992 eruptions of Mount Spurr, Alaska. U.S. Geological Survey Bulletin, 2139, 37-45, 1995.

Bluth, G. J. S., W. I. Rose, I. E. Sprod, and A. J. Krueger, Stratospheric loading of sulfur from explosive volcanic eruptions, Journal of Geology, 105, 671-684, 1997.

Bösenberg, J., M. Alpers, C. Böckmann, H. Jäger, V. Mathias, T. Trickl, U. Wandinger, and M. Wiegner, A lidar network for the establishment of an aerosol climatology, Proceedings of the 19th International Laser Radar Conference, (Annapolis, July 6-10), NASA/CP-1998-207671/PT1, 23-24,1998.

Bureau H., H. Keppler, and N. Metrich, Volcanic degassing of bromine and iodine: experimental fluid/melt partitioning data and applications to stratospheric chemistry, Earth and Planetary Science Letters, 183, 51-60, 2000.

Cadle, R. D., Volcanic emissions of halides and sulfur compounds to the troposphere and stratosphere, J. Geophys. Res., 80, 1650-1652, 1975.

Cadle, R. D., A comparison of volcanic with other fluxes of atmospheric trace gas constituents, Rev. Geophys. Space Phys., 18, 746-752, 1980.

Charlson, R. J., J. Langner, H. Rodhe, C. B. Leovy, and S. G. Warren, Perturbation of the Northern Hemisphere radiative balance by backscattering from anthropogenic sulfate aerosols, Tellus, 43, 152-163, 1991.

Charlson, R. J., S. E. Schwartz, J. M. Hales, R. D. Cess, J. A., Coakley, Jr., J. E. Hansen, and D. J. Hoffman, Climate forcing by anthropogenic aerosols, Science, 255, 423-430, 1992.

Chazette, P. C. David, J. Lefrere, S. Godin, J. Pelon, and G. Megie, Comparative lidar study of the optical, geometrical and dynamical properties of stratospheric postvolcanic aerosols, following the eruptions of El Chichón and Mount Pinatubo, J. Geophys. Res., 101, 23195-23207,1995.

Chin M., and D. D. Davis, Global sources and sinks of OCS and CS_2 and their distributions, Global Biogeochemical Cycles, 7, 321-337, 1993.

Crowley, T. J., T. A. Criste, and N. R. Smith, Reassessment of Crete (Greenland) ice core acidity/volcanism link to climate change, Geophys. Res. Lett., 20, 209-212, 1993.

Crowley, T. J., and K.-Y. Kim, Modeling the temperature response to forced climate change over the last six centuries, Geophys. Res. Lett., 26, 1901-1904, 1999.

Deshler, T., D. J. Hofmann, B. J. Johnson, and W. R. Rozier, Balloonborne measurements of the Pinatubo aerosol size distribution and volatility at Laramie, Wyoming during the summer of 1991, Geophys. Res. Lett., 19, 199-202, 1992

Deshler, T., B. J. Johnson, and W. R. Rozier, Balloonborne measurements of Pinatubo aerosol during 1991 and 1992 at 41°N: vertical profiles, size distribution, and volatility, Geophys. Res. Lett., 20, 1435-1438,1993.

Deshler, T., G. B. Liley, G. Bodeker, W. A. Matthews, and D. J. Hofmann, Stratospheric aerosol following Pinatubo, comparison of the north and south mid latitudes using in situ measurements, Adv. Space Res., 20, 2057-2061, 1997.

Facchini, M. C., S. Fuzzi, M. Kessel, W. Wobrock, W. Jaeschke, B. G. Arends, J. J. Möls, A. Berner, I. Solly, C. Kruisz, G. Reischl, S. Pahl, A. Hallberg, J. A. Ogren, H. Fierlinger-Oberlinninger, A. Marzorati, and D. Schell, The chemistry of sulfur and nitrogen species in a fog system: A multiphase approach, Tellus, 44B, 505-521, 1992.

Fiocco, G., M. Cacciani, A. G. di Sarra, D. Fuà, P. Colagrande, G. De Benedetti, P. Di Girolano, and R. Viola, The evolution of the Pinatubo stratospheric aerosol layer observed by lidar at South Pole, Rome, Thule: A summary of results, in: The Mount Pinatubo Eruption: Effects on The Atmosphere and Climate, NATO ASI Series, 42, 17-32, 1996.

Franklin, B., Meteorological imaginations and conjectures, Manchester Literary and Philosophical Society Memoirs and Proceedings 2, 122, 1784, reprinted in Weatherwise 35, 262, 1982.

Free, M., and Robock, A.: Global warming in the context of the Little Ice Age, J. Geophys. Res., 104, 19,057-19,070, 1999.

Friend, J. P., The global sulfur cycle, in Chemistry of the Lower Atmosphere, edited by S. I. Rasool, 177-201, Plenum, New York, 1973.

Fung, C. S., P. K. Misra, R. Bloxam, and S. Wong, A numerical experiment on the relative importance of H_2O_2 and O_3 in aqueous conversion of SO_2 to SO_4^{2-}, Atmos. Env., 25A, 411-423, 1991.

Gallagher, M. W., R. M. Downer, , T. W. Choularton, M. J. Gay, I. Stromberg, C. S. Mill, M/ Radojevic, B. J. Tyler, B. J. Bandy, S.A. Penkett, T. J. Davies, G. J. Dollard, and B. M. R. Jones, Case studies of the oxidation of sulphur dioxide in a hill cap cloud using ground and aircraft based measurements. J. Geophys. Res., 95, 18517-18537, 1990.

Gerlach, T.M., and Casadevall, T.J., Fumarole emissions at Mount St. Helens Volcano, June 1980 to October 1981: Degassing of a magma-hydrothermal system: Journal of Volcanology and Geothermal Research, 28, 141-160, 1986.

Gerlach, T.M., Present-day CO_2 emissions from volcanoes, EOS, 72, 249, 254-255, 1991.

Gerlach, T. M., H. R. Westrich, and R. B. Symonds, Preeruption vapor in magma of the climactic Mount Pinatubo eruption: Source of the giant stratospheric sulfur dioxide cloud, in Newhall, C. G., and R. S. Punongbayan, eds., Fire and mud: Eruptions and lahars of Mount Pinatubo, Philippines: Philippine Institute of Volcanology and Seismology, Quezon City, and University of Washington Press, Seattle, 415-433, 1996.

Graf, H-F., I. Kirchner, A. Robock, and I. Schult, Pinatubo eruption winter climate effects: Model versus observations, Climate Dynamics, 9, 81-93, 1993.

Graf, H.-F., J. Perlwitz, and I. Kirchner, Northern Hemisphere tropospheric mid-latitude circulation after violent volcanic eruptions, Contributions to Atmospheric Physics [Beitrage zur Physik der Atmosphäre.], 67, 3-13, 1994.

Graf, H.-F., J. Feichter, and B. Langmann, Volcanic sulfur emissions: Estimates of source strength and its contribution to the global sulfate distribution, J. Geophys. Res., 102, 10727-10738, 1997.

Graf, H.-F., B. Langmann, and J. Feichter, The contribution of Earth degassing to the atmospheric sulfur budget, Chemical Geology, 147, 131-145, 1998.

Graf, H.-F., M. Herzog, J. M. Oberhuber, and C. Textor, The effect of environmental conditions on volcanic plume rise, J. Geophys. Res., 104, 24,309, 1999.

Graf, H.-F., and C. Timmreck, A general climate model simulation of the aerosol radiative effects of the Laacher See eruption (10,900 BC), J. Geophys. Res., 106, 14,747- 14,756, 2001.

Granat, L., H. Rodhe, and R. O. Hallberg, The global sulphur cycle, in Nitrogen, Phosphorous and Sulphur-Global Cycles, edited by B. H. Svensson, and R. Söderlund, SCOPE Report 7, Ecological Bulletin, 22, 89-134, 1976.

Grant, W. B., E. V. Browell, J. Fishman. V. G. Brackett. R. E. Veiga, D. Nganga. A. Minga, B. Cros. C. F. Butler. M. A. Fenn, C. S. Long, and L. L Stowe, Aerosol-associated changes in tropical stratospheric ozone following the eruption of Mount Pinatubo, J. Geophys. Res., 99, 8197-8211, 1994.

Granier, C., and G. Brasseur, Impact of heterogeneous chemistry on model predictions of ozone changes, J. Geophys. Res., 97, 18015-18033, 1992.

Grgi, I., Hudnik, V., Bizjak, M., and J. Levec, Aqueous S(IV) oxidation-I. Catalytic effects of some metal ions, Atmos. Env., 25A, 1591-1597, 1991.

Groisman, P. Y., Regional climate consequences of volcanic eruptions (in Russian), Meteorol. Hydrol., 4, 39-45, 1985.

Halmer, M. M., H.-U. Schmincke, and H.-F. Graf, The annual volcanic gas input into the atmosphere, in particular into the stratosphere: a global data set for the past 100 years, Journal of Volcanology and Geothermal Research, 115, 511-528, 2002.

Hammer, C.U., H.B. Clausen, and W. Dansgaard, Greenland ice sheet evidence of post-glacial volcanism and its climate impact, Nature, 288, 230-235,1980.

Hansen, A. D. A., W. H. Benner, and T. Novakov, Sulfur dioxide oxidation in laboratory clouds, Atmos. Env., 25A, 2521-2530, 1991.

Herzog, M., H.-F. Graf, C. Textor, and J. M. Oberhuber, The effect of phase changes of water on the development of volcanic plumes, Journal of Volcanology and Geothermal Research, 87, 55-74, 1998.

Hitchman, M. H., M. McKay, and C. R. Trepte, A climatology of stratospheric aerosol, J. Geophys. Res., 99, 20,689-20,700, 1994.

Hofmann, D. J., J. M. Rosen, R. Reiter, and H. Jager, Lidar- and balloon-borne particle counter comparisons following recent volcanic eruptions, J. Geophys. Res., 88, 3777-3782, 1983.

Hofmann, D. J., and S. Solomon, Ozone destruction through heterogeneous chemistry following the eruption of El Chichón, J. Geophys. Res., 94, 5029-5041, 1989.

Hofmann, D. J., S.J. Oltmans, J.M. Harris, W.D. Komhyr, J.A. Lathrop, T. DeFoor, and D. Kuniyuki, Ozonesonde measurements at Hilo, Hawaii following the eruption of Pinatubo, Geophys. Res. Lett., 20, 1555-1558, 1993.

Jäger, H. Uchino, O. Nagai, T. Fujimoto, T. Freudenthaler, and V. Homburg, F., Ground-based remote sensing of the decay of the Pinatubo eruption cloud at three Northern Hemisphere sites, Geophys. Res. Lett., 22, 607-610, 1995.

Joos, F., and U. Baltensperger, A field study on chemistry, S(IV) oxidation rates and vertical transport during fog conditions. Atmos. Env., 25A, 217-230, 1991.

Kellogg, W. W., Cadle, R. D., Allen, E. R., Lazrus, A. L., and E. A. Martell, The sulfur cycle. Science, 175, 587-596, 1972.

Kent, G. S., and McCormick, M. P., SAGE and SAM-II measurements of global stratospheric aerosol optical depth and mass loading, J. Geophys. Res., 89, 5303-5314, 1984.

Kinnison, D. E., K.E. Grant, P.S. Connell, D.A. Rotman, and D. J. Wuebbles, The chemical and radiative effects of the Mount Pinatubo eruption, J. Geophys. Res., 99, 25,705-25,731, 1994.

Kjellström, E., A three-dimensional global model study of Carbonyl Sulfide in the troposphere and in the lower stratosphere, J. Atmos. Chem., 151-172,29, 1998.

Kodera K., M. Chiba, H. Koide, A. Kitoh, and Y. Nikaidou, Interannual variability of the winter stratosphere and troposphere in the Northern Hemisphere, Journal of the Meteorological Society of Japan, 74, 365-382,1996.

Krueger, A. J., L. S. Walter, P. K. Bhartia, C. C. Schnetzler, N. A. Krotkov, I. Sprod, and G. J. S. Bluth, Volcanic sulfur dioxide measurements from the Total Ozone Mapping Spectrometer (TOMS) instruments, J. Geophys. Res., 100, 14057-14076, 1995.

Krueger, A. J., C. C. Schnetzler, and L. S. Walter, The December 1981 eruption of Nyamuragira Volcano (Zaire), and the origin of the "mystery cloud" of early 1982, J. Geophys. Res., 101, 15,191-15,196, 1996.

Lamb, H. H., Volcanic dust in the atmosphere; with a chronology and assessment of its meteorological significance, Philosophical Transactions of the Royal Society of London Series A-Mathematical and Physical Sciences, 266, 425-533, 1970.

Lamb, H. H., Supplementary volcanic dust veil index assessments, Climate Monitoring, 6, 57-67, 1977.

Lamb, H. H., Update of the chronology of assessments of the volcanic dust veil index, Climate Monitoring, 12, 79-90, 1983.

Lambert, G., M.-F. Le Cloarec, and M. Pennisi, Volcanic output of SO_2 and trace metals: A new approach. Geochimica et Cosmochimica Acta, 52, 39-42, 1988.

Langmann, B., and H.-F. Graf, The chemistry of the polluted atmosphere over Europe: Simulations and sensitivity studies with a regional chemistry-transport-model, Atmos. Env. 31, 3239-3257, 1997.

Le Guern, F., Les débits de CO_2 et de SO_2 volcaniques dans l'atmosphère. Bulletin of Volcanology, 45, 197-202, 1982.

Mankin, W. G., and M. T. Coffey, Increased stratospheric hydrogen chloride in the El Chichón cloud, Science, 226, 170-172, 1984.

Mankin, W. G., M. T. Coffey, and A. Goldman, Airborne observations of SO_2, HCl, and O_3 in the stratospheric plume of the Pinatubo volcano in July 1991, Geophys. Res. Lett., 19, 179-182, 1992.

Mauldin, L. E., III, N. H. Zaun, M. P. McCormick, J. H. Guy, and W. R. Vaughn, Stratospheric Aerosol and Gas Experiment II Instrument: A Function Description, Optical Engeneering, 24, 307-312, 1985.

McClelland L., T. Simkin, M. Summers, E. A. Nielsen, and T. C. Stein (eds), Global Volcanism 1975-1985: The First Decade of Reports from the Smithsonian Institution's Scientific Event Alert Network (SEAN), Prentice Hall and American Geophysical Union, Englewood Cliffs, 657 pp., 1989.

McCormick, M. P., P. Hamill, T. J. Pepin, W. P. Chu, T. J. Swissler, and L. R. Master, Satellite studies of the stratospheric aerosol, Bulletin of the American Meteorological Society, 60, 1038-1046, 1979.

McCormick, M. P., SAGE II: An overview, Advances in Space Research, 7, 219-226, 1987.

McCormick, M. P., L. W. Thomason, and C. R. Trepte, Atmospheric effects of the Mt Pinatubo eruption, Nature, 373, 399-404, 1995.

McElroy, M. B., R. J. Salawitch, and K. Minschwaner, The changing stratosphere, Planetary and Space Science, 40, 373–401,1992.

McKeen, S. A., S. C. Liu, and C. S. Kiang, On the chemistry of stratospheric SO_2 from volcanic eruptions, J. Geophys. Res., 89, 4873-4881, 1984.

Menzies, R. T., and D. M. Tratt, Evidence of Seasonally Dependent Stratosphere-troposphere Exchange and Purging of Lower Stratospheric Aerosol from a Multiyear Lidar Data Set, J. Geophys. Res., 100, 3139-3148, 1995.

Michelangeli, D. V., M. Y. Allen, and L. Yuk, El Chichón volcanic aerosols: impact of radiative, thermal, and chemical perturbations, J. Geophys. Res., 94, 18429-18443, 1989.

Naughton J. J., V. A. Greenberg, and R. Goguel, Incrustations and Fumarolic Condensates at Kilauea Volcano, Hawaii - Field, Drill-Hole And Laboratory Observations, Journal of Volcanology and Geothermal Research, 1, 149-165, 1976.

Newhall,C., and S. Self, The Volcanic Explosivity Index (VEI): An estimate of explosive magnitude for historical volcanism, J. Geophys. Res., 87, 1281-1238, 1982.

Nober F., H.-F. Graf, and D. Rosenfeld, Sensitivity of the global circulation to the suppression of precipitation by anthropogenic aerosols, in print Global Planet. Change, 2002.

Oberhuber, J. M., M. Herzog, H.-F. Graf, and K. Schwanke, Volcanic plume simulation on large scales, Journal of Volcanology and Geothermal Research, 87, 29-53, 1998.

Oppenheimer C., P. Francis, and J. Stix, Depletion rates of SO_2 in tropospheric volcanic plumes, Geophys. Res. Lett., 25, 2671-2674, 1998.

Oppenheimer, C., Limited global change due to the largest known Quaternary eruption, Toba 74 kyr BP?, Quaternary Science Reviews, 21, 1593-1609, 2002.

Palais, J., Sigurdsson, H., Petrologic evidence of volatile emissions from major historic an pre-historic volcanic eruptions, In: Berger, A., Dickinson, R.E., Kidson, R.E., Understanding Climate Change, American Geophysical Union, Washington, DC, 31-53, 1989.

Parker, D. E., and J. L. Brownscombe, Stratospheric warming following the El Chichón volcanic eruption Nature, 301, 406-408, 1983.

Pinto, J. R., R. P. Turco, and O. B. Toon, Self-Limiting Physical and Chemical Effects in Volcanic Eruption Clouds, J. Geophys. Res., 94, 11165-11174, 1989.

Pitari, G., and V. Rizzi, An Estimate of the Chemical and Radiative Perturbation of Stratospheric Ozone Following the Eruption of Mt. Pinatubo, J. Atmos. Sci., 50, 3260-3276, 1993.

Prather, M., Catastrophic loss of stratospheric ozone in dense volcanic clouds, J. Geophys. Res., 97(D9), 10187-10191, 1992.

Pyle, D. M., P. D. Beattie, and G. J. S. Bluth, Sulphur emissions to the stratosphere from explosive volcanic eruptions, Bulletin of Volcanology, 57, 663-671, 1996.

Quiroz, R. S., Compact review of observational knowledge of blocking, with emphasis on the long-wave composition of blocks, IN: Annual Climate Diagnostics Workshop, 7th, Boulder, CO., Oct. 18-22, 1982, Proceedings, Wash., D.C., U.S. National Oceanic and Atmospheric Administration, March, 1983, p. 104-107, Climate Analysis Ctr., NMC/NWS/NOAA, Wash., D.C. M.G.A, 35, 8-254,1984.

Randel, W. J, F. Wu, J. M. Russell III, J. W. Waters, and L. Froidevaux, Ozone and temperature changes in the stratosphere following the eruption of Mount Pinatubo, J. Geophys. Res., 100, 16753-16764, 1995.

Rani, A., D. S. N. Prasad, P. V. S. Madnawat, and K. S. Gupta, The role of free fall atmospheric dust in catalysing autoxidation of aqueous sulphur dioxide, Atmos. Env., 26A, 67-673, 1992.

Read, W. G., L. Froidevaux, and J. W. Waters, Microwave Limb Sounder measurement of stratospheric SO_2 from the Mt. Pinatubo volcano, Geophys. Res. Lett., 20, 1299-1302, 1993.

Robertson, A., J. Overpeck, D. Rind, E. Mosley-Thompson, G. Zielinski, J. Lean, D. Koch, J. Penner, I. Tegen, and R. Healy, Hypothesized climate forcing time series for the last 500 years, J. Geophys. Res., 106, 14783-14803, 2001.

Robock, A., A latitudinally dependent volcanic dust veil index, and its effect on climate simulations, Journal of Volcanology and Geothermal Research, 11, 67-80, 1981a.

Robock, A., The Mount St. Helens volcanic eruption of 18 May 1980: Minimal climatic effect, Science, 212, 1383-1384, 1981b.

Robock, A., and J. Mao, Winter warming from large volcanic eruptions, Geophys. Res. Lett., 19, 2405-2408, 1992.

Robock, A., and M. P. Free, Ice cores as an index of global volcanism from 1850 to the present, J. Geophys. Res., 100, 11,549-11,567, 1995.

Robock, A., and M. P. Free, The volcanic record in ice cores for the past 2000 years, Climatic Variations and Forcing Mechanisms of the Last 2000 Years, edited by Philip D. Jones, Raymond S. Bradley, and Jean Jouzel, Springer-Verlag, Berlin, 533-546, 1996.

Robock, A., Volcanic eruptions and climate, Reviews of Geophysics, 38, 191-219, 2000.

Rose, W. I., and C. A. Chesner, Worldwide dispersal of ash and gases from the earth largest known eruption: Toba, Sumatra, 75 ka. Palaeogeography Palaeoclimatology Palaeoecology, 89, 269-275, 1990.

Rose, W. I., D. J. Delene, D.J. Schneider, G. J. S. Bluth, A. J. Krueger, I. Sprod, C. McKee, H. Davies, and G.G.J. Ernst, Ice in the 1994 Rabaul Eruption Cloud: Implications for volcano hazard and atmospheric effects, Nature, 375, 477-479, 1995.

Rose, W. I., G. J. S. Bluth, and G. G. J. Ernst, Integrating retrievals of volcanic cloud characteristics from satellite remote sensors - a summary, Philosophical Transactions of Royal Society, Series A, 358, 1585-1606, 2000.

Rosen, J. M., N. Kjome, R. L. McKenzie, and J. B. Liley, Decay of Mount Pinatubo aerosol at mid-latitudes in northern and southern hemispheres, J. Geophys. Res., 99, 25733-25739, 1994.

Rosenfeld, J. E., Considine, D. B., Meade, P. E., Bacmeister, J. T., Jackmon, C. H., and M. R. Schoeberl, Stratospheric effects of Mount Pinatubo aerosol studied with a coupled two-dimensional model, J. Geophys. Res., 102, 3649-3670, 1997.

Russell, P. B., J. M. Livingston, R. F. Pueschel, J. B. Pollack, S. Brooks, P. Hamill, J. Hughes, L. Thomason, L. Stowe, T. Deshler, and E. Dutton, Global to microscale evolution of the Pinatubo volcanic aerosol, derived from diverse measurements and analyses, J. Geophys. Res., 101, 18745-18763, 1996.

Sassen, K., Evidence for liquid-phase cirrus cloud formation from volcanic aerosol: Climate implications, Science, 257, 516-519, 1992.

Sassen, K., D. O'C. Starr, G. G. Mace, M. R. Poellot, S. H. Melfi, W. L. Eberhard, J. D. Spinhirne, E. W. Eloranta, D. E. Hagen, and J. Hallett, The 5-6 December 1991 FIRE IFO II jet stream cirrus case study: possible influences on volcanic aerosols, J. Atmos. Sci., 52, 97-123 1995.

Sato, M., J.E. Hansen, M. P. McCormick, and J. B. Pollack, Stratospheric aerosol optical depths 1850-1990, J. Geophys. Res., 98, 22,987-22994,1993.

Scaillet, B., B. Clemente, B. W. Evans, and M. Pichavant, Redox controls of sulfur degassing in silicic magmas, J. Geophys. Res., 103, 23,937-23,949, 1998.

Schimel, D., D. Alves, I. Enting, M. Heimann, F. Joos, D. Raynaud, T. Wigley, M. Prather, R. Derwent, D. Ehhalt, P. Fraser, E. Sauhueza, X. Zhou, P., Jonas, R. Charlson, H. Rodhe, S. Sadasivan, K. P. Shine, Y. Fouquart, V. Ramaswamy, S. Solomon, J. Srinivasan, D. Albritton, R. Derwent, I. Isaksen, M. Lal, and D. Wuebbles, Radiative forcing of climate change, In: Climate Change 1995: The Science of Climate Change, J. T. Houghton, L. G. M. Filho, B. A. Callander, N. Harris, A. Kattenberg, and K. Maskell (eds), 65-131, Cambridge University Press, Cambridge, UK, 1996.

Schmincke, H.-U.,Transfer von festen, flüssigen und gasförmigen Stoffen aus Vulkanen in die Atmosphäre, Zeitschrift für Umweltchemie und Ökotoxikologie, 5, 27-44, 1993.

Schnetzler, C. C., G. J. S. Bluth, A. J. Krueger, and L. S. Walter, A proposed volcanic sulfur dioxide index (VSI), J. Geophys. Res., 102, 20087-20091, 1997.

Seinfeld, J. H., and S. N. Pandis, Atmospheric Chemistry and Physics, From Air Pollution to Climate Change, John Wiley & Sons, New York, 1998.

Sheridan, P. J., R. C. Schnell, D. J. Hofmann, and T. Deshler, Electron microscope studies of Mt. Pinatubo aerosol layers over Laramie, Wyoming, during summer 1991, Geophys. Res. Lett., 19, 203-206, 1992.

Simkin T. L. Siebert, L. McClelland, D. Bridge, and C. Newhall, J. Latter, Volcanoes of the World: a regional directory, gazetteer, and chronology of volcanism during the last 10 00 years, Hutchinson Ross Pub Co, Stroudsburg, 1981.

Simkin, T., Terrestrial volcanism in space and time, Annual Review of Earth and Planetary Sciences, 21, 427-452, 1993.

Simkin T., and L. Siebert , Volcanoes of the World: a Regional Directory, Gazetteer, and Chronology of Volcanism During the Last 10,000 Years, (second edition), Geoscience Press, Tucson, 1994.

Solomon, S., R. W. Portmann, R. R. Garcia, L. W. Thomason, L. R. Poole, and M. P. McCormick, The role of aerosol variations in anthropogenic ozone depletion at northern midlatitudes, J. Geophys. Res., 101, 6713-6727, 1996.

Solomon, S., R. W. Portmann, R. R. Garcia, W. Randel, F. Wu, R. Nagatani, J. Gleason, L. Thomason, L. R. Poole, and M. P. McCormick, Ozone depletion at mid-latitudes: coupling of volcanic aerosols and temperature variability to anthropogenic chlorine, Geophys. Res. Lett., 25, 1871-1874, 1998.

Song, N., D. O'C. Starr, D. J. Wuebbles, A. Williams, and S. M. Larson, Volcanic aerosols and interannual variation of high clouds, Geophys. Res. Lett., 23, 2657-2660, 1996.

Stenchikov, G. L., I. Kirchner, A. Robock, H.-F. Graf, J. C. Antuña, R. G. Grainger, A. Lambert, and L. Thomason, Radiative forcing from the 1991 Mount Pinatubo volcanic eruption, J. Geophys. Res., 103, 13,837-13,857, 1998.

Stevermer, A., I. Petropavlovskikh, J. Rosen, and J. DeLuisi, Development of global stratospheric aerosol climatology: Optical properties and applications for UV, J. Geophys. Res., 105, 22763-22776, 2000.

Stoiber, R. E., and A. Jepsen, Sulfur dioxide contributions to the atmosphere by volcanoes, Science, 182, 577-578, 1973.

Stoiber, R. E., S. N Williams, and B. Huebert, Annual contribution of sulfur dioxide to the atmosphere by volcanoes. Journal of Volcanology and Geothermal Research, 33, 1-8, 1987.

Stothers, R. B., Major optical depth perturbations to the stratosphere from volcanic eruptions: Pyrheliometric period 1881-1960, J. Geophys. Res., 101, 3901-3920, 1996.

Symonds, R. B., W. I. Rose, and M. H. Reed, Contribution of Cl- and F-bearing gases to the atmosphere by volcanoes, Nature, 334, 415-418, 1988.

Symonds R. B., W. I. Rose, G. J. S. Bluth, and T. M. Gerlach, Volcanic gas studies: methods, results and applications, in: Volatiles in Magma, published by Mineral Society of America, M. R. Carrol and J. R. Holloway (eds), Reviews in Mineralogy, 30, 1-66, 1994.

Tabazadeh, A., and R. P. Turco, Stratospheric chlorine injection by volcanic eruptions: HCl scavenging and implications for ozone, Science, 260, 1082-1086, 1993.

Textor, C., Sachs, P.M., Graf, H.-F., and T. Hansteen, The 12,900 yr BP Laacher See eruption: estimation of volatile yields and simulation of their fate in the plume, In: Volcanic Degassing, Geological Society Special Publication 213, Clive Oppenheimer, David Pyle and Jenni Barclay (eds), 2003.

Textor, C., Graf, H.-F., Herzog, M., and J. M. Oberhuber, Injection of Gases into the Stratosphere by Explosive Volcanic Eruptions, J. Geophys. Res., in press, 2003.

Thomason, L. W., A diagnostic aerosol size distribution inferred from SAGE II measurements, J. Geophys. Res., 96, 22501-22508, 1991.

Thomason, L. W., L. R. Poole, and T. Deshler, A global climatology of stratospheric aerosol surface area density deduced from stratospheric aerosol and gas experiment II measurements: 1984-1994, J. Geophys. Res.,102, 8967-8976, 1997.

Tie, X., G. P. Brasseur, B. Briegleb, and C. Granier, Two-dimensional simulation of Pinatubo aerosol and its effect on stratospheric ozone, J. Geophys. Res., 99, 20,545-20,562, 1994.

Timmreck, C., and H.-F. Graf, A microphysical model to simulate the development of stratospheric aerosol in a GCM. Meteorologische Zeitschrift, 9, 263-282, 2000.

Toohey, D. W., A critical review of stratospheric chemistry research in the U.S.: 1991-1994, U.S. National Report to the International Union of Geodesy and Geophysics 1991- 1994, Reviews of Geophysics, 33, 759-773, 1995.

Turco, R. P., Toon, O. B., Whitten, R. C., Hamill, P., and R. G. Keese, The 1980 eruption of Mount St. Helens: Physical and chemical processes in the stratospheric cloud, J. Geophys. Res., 88, 5299-5319, 1983.

Twomey, S., Pollution and planetary albedo. Atmos. Env., 8, 1251-1265, 1974.

Uchino, O., et al. Observation of the Pinatubo volcanic cloud by lidar network in Japan, Journal of the Meteorological Society of Japan, 71, 285-295, 1992.

Varekamp, J.C., J.F. Luhr, and K.L. Prestegaard, The 1982 eruptions of El Chichón Volcano (Chiapas, México): Character of the eruptions, ash-fall deposits, and gasphase, and Geothermal Research, 23, 39-68, 1984.

Veiga, R. E., SAGE II measurements of volcanic aerosols, Technical Digest Series, 5, Optical Society of America, Washington, D.C., 467-470, 1993.

Wallace, L., and W. Livingston, The effect of the Pinatubo cloud on hydrogen chloride and hydrogen fluoride, Geophys. Res. Lett., 19, 1209, 1992.

Weisenstein, D. K., G. K. Yue, M. K. W. Ko, N.-D. Sze, J. M. Rodriguez, and C. J. Scott, A two-dimensional model of sulfur species and aerosols, J. Geophys. Res., 102, 13019-13053, 1997.

Westrich, H.R., and T.M. Gerlach, Magmatic gas source for the stratospheric SO2 cloud from the June 15, 1991 eruption of Mount Pinatubo, Geology, 20, 867-870, 1992.

Wilson, L., Explosive volcanic eruptions – III. Plinian eruption columns, Geophysical Journal of the Royal Astronomical Society, 45, 543-556, 1976.

Woods, D. C., R. L. Chuan, and W. I. Rose, Halite particles injected into the stratosphere by the 1982 El Chichón eruption, Science, 230, 170-172,1985.

Woods, A. W., The fluid dynamics and thermodynamics of volcanic eruption columns, Bulletin of Volcanology, 50, 169-193,1988.

Woods, A. W., Moist convection and the injection of volcanic ash into the atmosphere, J. Geophys. Res., 98, 17627-17636,1993.

Zielinski, G. A., Stratospheric loading and optical depth estimates of explosive volcanism over the last 2100 years derived from the GISP2 Greenland ice core, J. Geophys. Res., 100, 20937-20955, 1995.

Zielinski, G. A., P.A. Mayewski, L.D. Meeker, S. Whitlow, and M. Twickler, A 110,000 year record of explosive volcanism from the GISP2 (Greenland) ice core, Quaternary Research, 45, 109-118, 1996.

Zielinski, G. A., Use of paleo-records in determining variability within the volcanism-climate system, Quaternary Science Reviews, 19, 417-438, 2000.

Zhao, J., R. P. Turco, and O. B. Toon, A model simulation of Pinatubo volcanic aerosols in the stratosphere, J. Geophys. Res., 100, 7315-7328, 1995.

A satellite-based method for estimating global oceanic DMS and its application in a 3-D atmospheric GCM

Sauveur Belviso, Cyril Moulin, Laurent Bopp, Emmanuel Cosme, Elaine Chapman and Kazushi Aranami

1. INTRODUCTION

In order to assess in three-dimensional atmospheric models the climate effects of anthropogenic sulphate aerosols, it is necessary not only to compute spatial and temporal distributions of anthropogenic sulphate, but also to simulate spatially and temporally the emission, transport and transformation of natural sulphur gases and aerosols emitted at the Earth's surface. Jones et al. (2001) recently obtained a value of -1.9 W m^{-2} for the effect of anthropogenic sulphate aerosol on cloud albedo and on precipitation efficiency (the 'indirect' sulphate aerosol forcing effect), and demonstrated in a sensitivity test that doubling oceanic dimethylsulphide (DMS) emission fluxes reduced the indirect effect by over 25%. Thus, changes in marine DMS emissions appear to significantly affect estimates of the magnitude of anthropogenic sulphate forcing.

Estimates of annual global DMS emissions vary widely, but are expected to be in the range of 10 to 50 TgS yr^{-1} (Intergovernmental Panel on Climate Change (IPCC), 1996). The wide range results from uncertainties attached both to the global distribution of sea-surface DMS concentrations and to computing DMS air-sea exchange rates. Besides reducing uncertainties on present-day emissions estimates, it is also important to investigate the climate sensitivity of the marine DMS source. Ice core records of methanesulphonate and sulphate, some important atmospheric oxidation products of DMS, exhibit clear climatic variations over a full glacial cycle (Legrand et al., 1991 ; Hansson and Saltzman, 1993). Moreover, parameters driving DMS emissions are strongly dependent on climate variables such as sea-surface temperature, wind velocity, and irradiance, which affect atmospheric and oceanic physics, and control marine biology. Although considerable progress has been made in understanding the marine and atmospheric biogeochemical cycle of DMS, the impact of global warming on marine DMS emissions remains to be established, as well as the magnitude and sign of the climate-DMS feedback on indirect sulphate aerosol forcing. Central to the understanding of the links between climate change and the biogeochemical cycle of DMS is the investigation of the interannual variability of the marine source of DMS.

In this chapter, a new global distribution of sea-surface DMS concentrations based on satellite-derived data is presented and evaluated using field measurements. We use this new distribution in the three-dimensional Atmospheric General Circulation Model of the Laboratoire de Météorologie Dynamique (LMD-ZT), along with model time-step wind fields and a recent parameterization of the DMS mass transfer coefficient, to generate oceanic DMS emission fluxes in simulating the atmospheric sulphur cycle in remote areas of the southern hemisphere.

2. GLOBAL FIELDS OF SURFACE DMS

The modelling work of Archer et al. (2002) illustrates the complexity of oceanic DMS biogeochemistry and why no linear relationship exists to predict DMS concentration from a single biological parameter (chlorophyll α, for example). The principal precursor of DMS in oceanic surface waters is dimethylsulphoniopropionate (DMSP), which is primarily synthesised by phytoplankton. The transformation of DMSP to DMS and the accumulation of DMS in surface waters are intricately linked to food-web dynamics and physico-chemical processes, including photochemical degradation, vertical mixing, and sea to air flux (Archer et al., 2002, and references therein). To our knowledge, the study of Archer et al. (2002) is the first attempt to test our understanding of the cycling of DMS in the sea with such a detailed model. No attempts have been made yet to couple comprehensive ecosystem models of this type to atmospheric general circulation models to predict global DMS flux. Given the magnitude of computing resources and the time that would be required to complete even a one-month simulation, trying to couple such detailed models is currently impractical. For the foreseeable future, then, global scale distributions of DMS oceanic concentrations are likely to rely on statistical analyses of DMS datasets or parameterized empirical equations, such as described in this section.

2.1. Kettle et al. (1999)

A non-exhaustive summary of three-dimensional global climate change and chemical transport models which include a sulphur-cycle scheme is shown in Table 1. Many current models rely on the global database of sea surface DMS measurements assembled by Kettle et al. (1999).

Table 1. Marine sources of DMS in three-dimensional general circulation models (GCMs).

Name of GCM	References	Marine DMS
ECHAM3	Feichter et al., 1996	Bates et al., 1987
ECHAM4	Roeckner et al., 1999	Bates et al., 1992
NCAR CCM3	Barth et al., 2000	Benkovitz et al., 1994*
CCM1-GRANTOUR	Chuang et al., 1997	Spiro et al., 1992
CCCMA	Lohmann et al., 1999	Kettle et al., 1999**
GISS GCMII	Chin et al., 1996	Bates et al., 1987
GISS GCMII-prime	Koch et al., 1999	Kettle et al., 1999**
GOCART	Chin et al., 2000	Kettle et al., 1999**
LMD-ZT	Boucher and Pham, 2002	Kettle et al., 1999**
Hadley Center Climate Model	Jones et al., 2001	Kettle et al., 1999**
MIRAGE	Ghan et al., 2000 ab	Kettle & Andreae, 2000
COSAM GCM Intercomparison	Barrie et al., 2001	Kettle & Andreae, 2000

* The DMS emissions in each oceanic latitude band is distributed to the 1°x1° grid proportional to the chlorophyll concentration, while preserving the total emissions values of Bates et al., 1992.
** A. J. Kettle made his updated oceanic DMS concentration database (i.e., K&A2000) available on the web long before the Kettle and Andreae (2000) paper was submitted. It is likely that some of the models referred to in this table actually used K&A2000 but cited it as Kettle et al. 1999, because that was the only reference available at that time.

This database, initially derived from 15,617 measurements, was processed to create a series of climatological annual and monthly maps on a 1° latitude x 1° longitude grid. To form a first-guess global field of DMS sea surface concentrations, Kettle et al. divided the Earth's oceans into a series of 57 biogeochemical provinces. The average DMS concentration in each province was calculated, and in those instances where no data were available for a given climatological province, the average DMS concentration from an adjacent province was used.

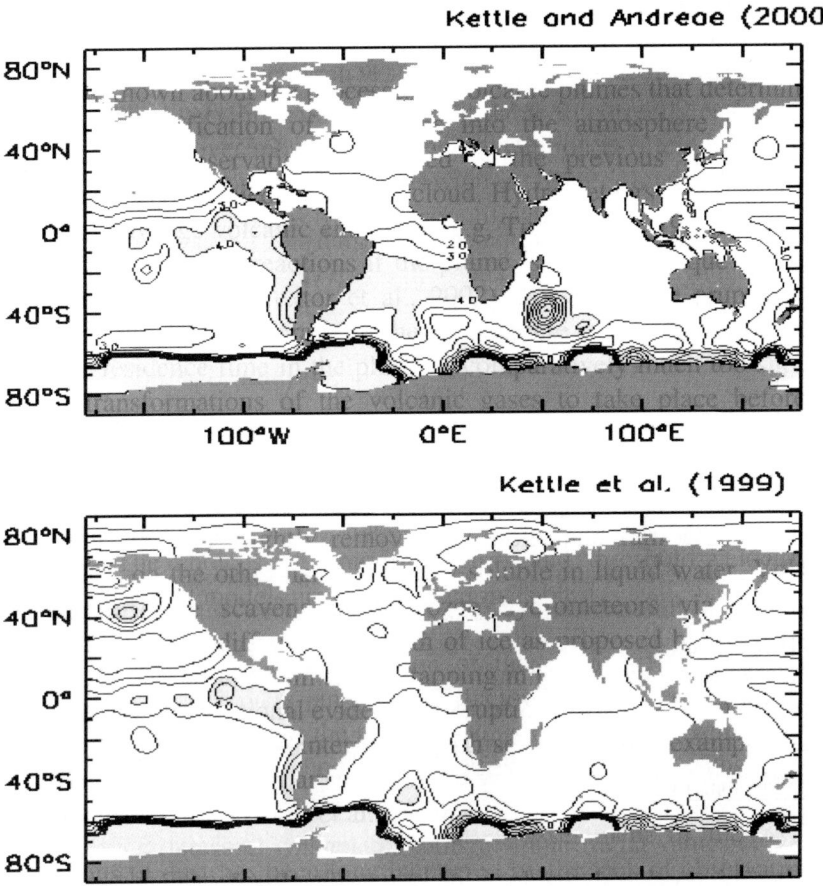

Figure 1 (Plate 16). Comparison of predicted oceanic DMS concentrations from the work of Kettle and Andreae, 2000 (upper panel) and Kettle et al., 1999 (lower panel) for the month of December. Shaded areas denote regions where DMS is higher than 4 nM. Isolines are every 1 from 1 to 10 nM. (© American Geophysical Union).

The southern ocean DMS field appears extremely sensitive to concentration changes in the south subtropical convergence (SSTC) and subantarctic (SANT) provinces, since the SSTC data are substituted into five adjacent provinces, and SANT data are substituted into all of the circumpolar Antarctic waters. Similarly, data from the North Atlantic drift province (NADR) are substituted in two oceanic provinces of the northern hemisphere.

This global database of sea surface DMS concentrations was recently updated (Kettle and Andreae, 2000, hereafter referred to as K&A2000), using observations collected in the SSTC, SANT and NADR provinces. New sea surface DMS data gathered by Sciare et al. (1999) and Belviso et al. (2000) were used by Kettle and Andreae in revising their 1999 data base to the K&A2000 version. As shown in Figure 1, the predicted DMS fields appear to be very dependent on the new measurements.

Substantial differences between the two global data sets were observed in the mid- and high- latitudes of both hemispheres for the month of December. This lack of stability suggests that it may not be appropriate to assimilate the original monthly fields of observed sea surface DMS (Kettle at al., 1999) to climatological fields, and demonstrates why it is extremely important to update this unique database as more observations become available.

2.2. Anderson et al. (2001)

In a recent study by Anderson et al. (2001), the K&A2000 database, which contains chlorophyll a as a recorded variable, was extended by merging nutrients and light from globally gridded fields to generate the CJQ index. Here, C is the chlorophyll concentration, J is the irradiance and Q is a nutrient term, a proxy of the algal growth rate. This index was shown to be significantly linearly correlated to DMS in the range 2.3-22 nmol l^{-1} (nM). However, DMS variability in low-concentration areas (high latitudes in the winter hemisphere, for example) is not resolved by this relationship.

2.3. Aumont et al. (2002)

Aumont et al. (2002) recently presented a model of the global distribution of sea surface DMS concentrations. The DMS parameterizations proposed were not based on mechanistic equations describing processes that control oceanic DMS production and removal, but instead were based on non-linear relationships relating DMS to the chlorophyll α content of surface waters and to the food-web structure of the ecosystem (i.e., the trophic state). These relationships were established from datasets obtained during several cruises carried out in contrasting areas of the world oceans (Figure 2, see also Aumont et al. (2002) for details). In the parameterizations, particulate dimethylsulphoniopropionate (pDMSP), a precursor compound of DMS, first is derived from surface chlorophyll a (Chl) concentrations. However, phytoplankton pDMSP production is highly specie-dependent; diatoms are poor DMSP producers whereas non-siliceous species are greater DMSP producers.

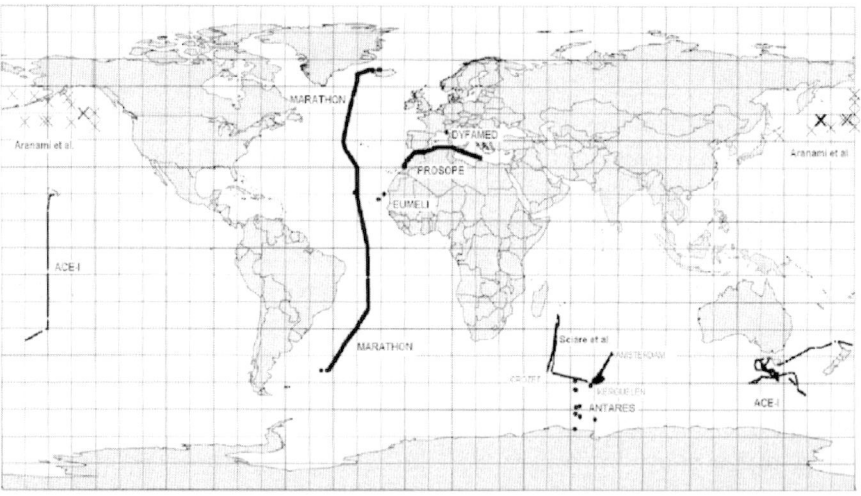

Figure 2 (Plate 17). Geographical coordinates of seawater samples used to establish the parameterizations of equations 1-3 (thick dots and lines: Mediterranean Sea (PROSOPE and DYFAMED projects), Atlantic Ocean (EUMELI and MARATHON projects) and Indian Sector of the Austral Ocean (ANTARES project)). Observations used to evaluate predicted sea surface DMSP and DMS levels (crosses and thin dots) are taken from the North Pacific Ocean (Aranami et al., 2001), the Central and South Pacific Ocean (cruise ACE-I, Bates et al., 1998), and the Indian Sector of the Austral Ocean (Sciare et al., 1999).

Thus, the pDMSP-Chl relationship uses the Fp-ratio, defined as the ratio of the diagnostic pigments fucoxanthin (of diatoms) and peridinin (of dinoflagellates) to total pigments (Claustre, 1994), to estimate the partition between non-siliceous and siliceous (diatoms) species. A linear relationship was used to estimate the contribution of diatoms to pDMSP. A non-linear function best accounted for the relationship between non-diatom Chl and pDMSP. Hence, the diagnosis of DMSP from Chl modulated by the Fp ratio is as follows:

$$pDMSP = (20 \times Chl \times Fp) + 13.64 + 0.10769/(1 + 24.97Chl(1-Fp))^{-2.5} \quad \text{(Eq. 1)}$$

The DMS-to-pDMSP ratio also is derived from the Claustre (1994) Fp-ratio. Aumont et al. (2002) found a significant correlation between these two ratios, the best fit relationship being:

$$DMS/pDMSP = 0.015316 + 0.005294/(0.0205 + Fp)$$
$$\text{for } Fp < 0.6 \qquad \text{(Eq. 2a)}$$

and

$$DMS/pDMSP = 0.569 \times Fp - 0.315$$
$$\text{for } Fp > 0.6 \qquad \text{(Eq. 2b)}$$

As evident from these equations, DMS can be estimated from a trophic status ratio (Fp) and the chlorophyll concentration of surface waters. This approach was first implemented in the global three-dimensional ocean carbon cycle model IPSL-OCCM2, using a proxy of the Fp-ratio directly predicted by the model (Aumont et al., 2002).

2.4. Use of SeaWiFs Satellite Data

The ocean colour Sea-viewing Wide Field-of-view Sensor (SeaWiFS) instrument was launched in August 1997 onboard the SeaStar spacecraft and is still operating after more than five years. SeaWiFS is a multispectral radiometer that measures the radiances scattered by the Earth-Atmosphere system at eight wavelenghts in the visible and near-infrared with a quasi-global daily coverage. Red and Near-infrared measurements (670, 765 and 865 nm) are used to estimate aerosol properties (optical thickness and Angström coefficient) for atmospheric correction of measurements in the rest of the visible spectrum (412, 443, 490, 510 and 555 nm). Once corrected from the atmospheric perturbation, these measurements are used to estimate the chlorophyll-α concentration in surface waters. SeaWiFS can pick out ocean colour features as small as 1 kilometre across.

2.4.1 Calculation of Global DMS Concentration Fields

The form of the relationships given in Equations 1 and 2 makes them suitable for use with ocean colour data from the satellite based Sea-viewing Wide Field-of-view Sensor. One key problem concerning the use of these equations in combination with ocean colour data is the prediction of the Fp ratio, since SeaWiFS does not yet provide any explicit speciation of phytoplankton.

Following the approach of Claustre (1994), Figure 3 shows that the Fp-ratio of sea surface waters is a highly significant ($r^2 = 0.89$), non-linear function of the chlorophyll concentration (Chl). The data shown in Figure 3 were obtained during the research cruises shown in Figure 2. The best fit relationships are:

$$Fp = 0.0168 + 0.481 Chl - 0.063(Chl)^2 \qquad \text{for Chl} < 4 \text{ mg m}^{-3}$$
(Eq. 3a)

and

$$Fp = 0.933 \qquad \text{for Chl} > 4 \text{ mg m}^{-3} \qquad \text{(Eq. 3b)}$$

Global fields of chlorophyll were obtained from one year (Oct. 1997 to Sept. 1998) of monthly composites of SeaWiFS chlorophyll available from NASA/GSFC/DAAC (http://eosdata.gsfc.nasa.gov/data/dataset/SEAWIFS/index.html). For our purposes, the data were regridded onto a one degree grid.

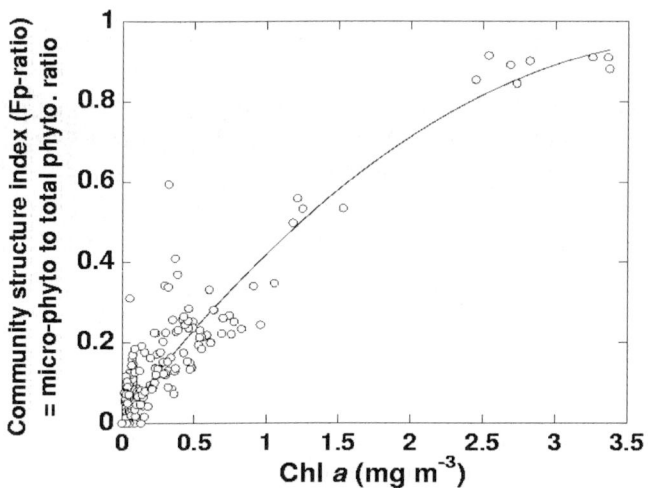

Figure 3. Fp-ratio versus sea surface concentrations of Chlorophyll α measured in the Mediterranean Sea, Atlantic Ocean and Indian Sector of the Austral Ocean. See Fig. 2 for geographic locations of sample collection.

Equations 3a and 3b were used to estimate the Fp-ratio from the SeaWiFS chlorophyll, and both parameters were then used in equations 1, 2a and 2b to retrieve monthly near-global gridded fields of DMS concentration. Figure 4 shows the monthly maps of sea-surface DMS for January and July.

Compared to the results of Kettle and Andreae (2000), our estimates show a much higher spatial variability, in particular in frontal and upwelling regions. This higher variability is expected since the distribution of sealife in the oceans is far from uniform. We thus expect this approach to strongly improve the capability to capture mesoscale DMS concentration variability. Note that no data are available at high latitude during winter in both hemisphere because SeaWiFS observations are limited to regions with sufficient solar irradiance.

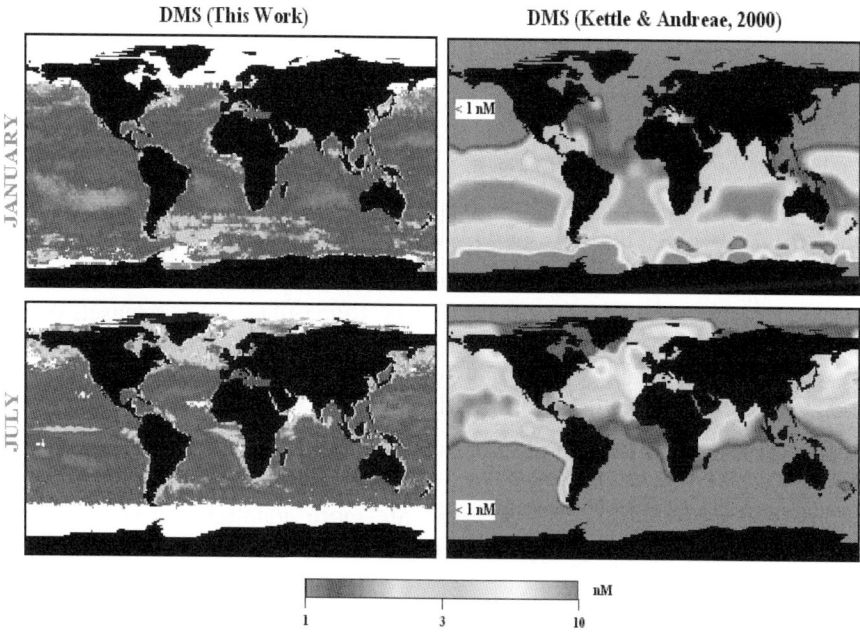

Figure 4 (Plate 18). Fields of sea surface DMS concentration (nM) for January (upper panels) and July (lower panels). This work (left column) and Kettle and Andreae (2000) (right column). (© American Geophysical Union).

2.4.2 Evaluation of SeaWiFS-derived DMS concentrations with temporally and spatially coincident observations

DMS concentrations and other oceanic constituents were measured in December 1997 and August 1998 during cruises of R/V Marion Dufresne

(Sciare et al., 1999; Sciare et al., unpublished). Ship transects are shown in Figure 2. Each trip was comprised of three legs: (1) from La Réunion Island (20°S, 56°E) to Crozet Island (46°S, 50°E); (2) from Crozet Island to Kerguelen Island (49°S, 69°E); and (3) from Kerguelen Island to Amsterdam Island (37°S, 77°E). Figure 5 compares the SeaWiFS-derived and K&A2000 DMS concentrations with observed sea-surface DMS concentrations in December 1997 (Sciare et al., 1999) and August 1998 (Sciare et al., unpublished). Use of weekly mean rather than monthly mean chlorophyll maps from SeaWiFS data may be preferable for evaluating predicted versus observed DMSP and DMS levels. Unfortunately weekly SeaWiFS data are not yet available.

Since DMSP was also measured (Sciare et al., unpublished), Figure 5 also compares predicted and observed DMSP concentrations. Note that Figure 5 shows observed total DMSP levels (tDMSP), i.e. the cumulated levels of particulate DMSP (pDMSP) and dissolved DMSP (dDMSP), whereas Equations 1-2 involve only pDMSP. Indeed, in subtropical and subantarctic waters of the Indian Ocean during December 1997 and August 1998, dDMSP accounted for 20 to 80% of tDMSP. Dissolved DMSP is a very labile compound usually exhibiting turnover times on the order of hours rather than days (e.g., Zubkov et al., 2001). Since dDMSP is released from phytoplankton by direct excretion, grazing or viral lysis, it is expected that measured dDMSP results from the turnover of pDMSP produced in the previous few days. Thus, it is more appropriate to compare predicted pDMSP from monthly mean chlorophyll maps and observed tDMSP than to compare directly predicted and observed pDMSP concentrations. In other words, observed tDMSP concentrations are more adapted to a comparison with predicted mean pDMSP because they provide a longer integration over time.

As seen in Figure 5, SeaWiFS data in combination with equations 1-3 slightly overestimate pDMSP in August 1998 inside and outside the chlorophyll patch. The SeaWiFS data predict a four- to five-fold enhancement of pDMSP in the patch, whereas observations show a two-fold increase. In August, observed DMS levels outside the chlorophyll patch exhibit fluctuations in the range 0.4-3 nM, but the median observed value of 0.95 nM is in fairly good agreement with the predicted baseline DMS level of 1.2 nM. Inside the patch, the median observed and predicted DMS concentrations are 1.1 and 1.6 nM, respectively. The estimates of K&A2000 are markedly lower relative to observations and the SeaWiFS-derived predictions, except at the beginning of the trip in the subtropical waters. It appears that use of SeaWiFS data in combination with equations 1-3 reduces the amplitude of the DMS fluctuations during the winter months in subantarctic waters of the Indian Ocean, and overestimates DMS by at most 50%.

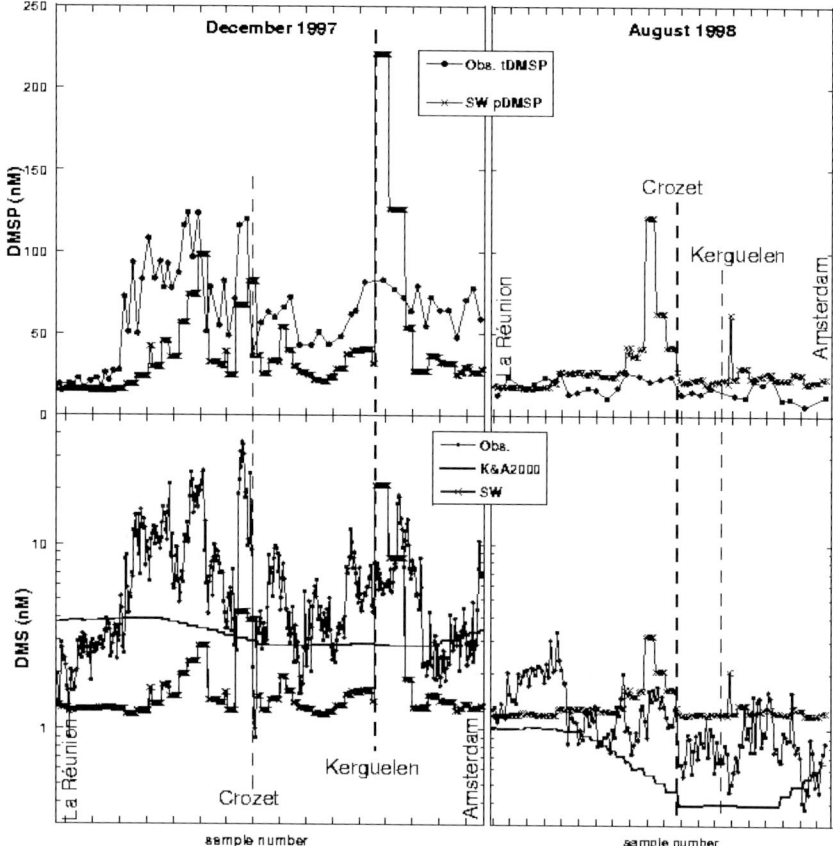

Figure 5. Comparison of spatially and temporally coincident DMSP (upper panel) and DMS (lower panel) in subantarctic water of the Indian Ocean for observed (Sciare et al., 1999 and Sciare unpublished) and predicted (SeaWiFS-derived and K&A2000) concentrations for December 1997 (left column) and August 1998 (right column).

Four major pDMSP peaks are predicted via the SeaWiFS data at the location of the chlorophyll patches crossed during the December 1997 trip. The predicted magnitudes are in close agreement with the field observations except near cruise end, where predicted pDMSP is three-fold higher than observed. Outside these chlorophyll patches, predicted pDMSP concentrations are similar (subtropical waters) or up to three-fold lower than observed. Four major DMS peaks are also predicted by the SeaWiFS data at the location of the chlorophyll patches. DMS field observations also were clearly enhanced at these locations. The SeaWiFS-derived magnitude of the first three DMS peaks, however, is considerably lower (5- to 10-fold) than observed, with better agreement on the fourth peak. Predicted baseline DMS, outside the chlorophyll patches, is at least twice and up to 10-fold lower than observed. However, as will be discussed in section 2.4.3, there is reason to believe that the DMS concentrations in this region may have been anomalously high during December 1997.

In summary, it appears that SeaWiFS-based predictions underestimate observed DMSP and DMS concentrations during the summer and may overestimate them during the winter. SeaWiFS-based predictions of both compounds thus have a reduced amplitude of seasonal variability relative both to observations and to the estimates of K&A2000. The comparisons indicate that DMS spatial variability is much better captured with SeaWiFS-based predictions than in the climatological data base of K&A2000 in the geographic areas covered by the ship cruises, at least for December 1997 and August 1998. The approach described here in employing SeaWiFS data and equations 1-3 results in an oceanic DMS baseline concentration of 1.1 nM, less than half the baseline of 2.3 nM reported by Anderson et al. (2001).

2.4.3 Evaluation of SeaWiFS-derived DMS concentrations using temporally non-coincident observations

We investigated the spatial distribution of predicted and observed DMS in the north, central and south Pacific Ocean at the same latitude, longitude and month of year but for different years, i.e., comparisons that are spatially coincident but temporally non-coincident. The location of the measurements is shown in Figure 2.

Since the field datasets contained chlorophyll as a recorded variable, we also can document the interannual variability in sea surface chlorophyll in the selected areas. The closer the agreement between measurements and SeaWiFS-derived predictions of chlorophyll, the closer should be the agreement between predicted and observed DMS. As shown in Figure 6, we found little difference in chlorophyll concentrations measured in July-

August-September 1997 (Aranami et al., 2001) and those derived from SeaWiFS data collected in July-August-September 1998.

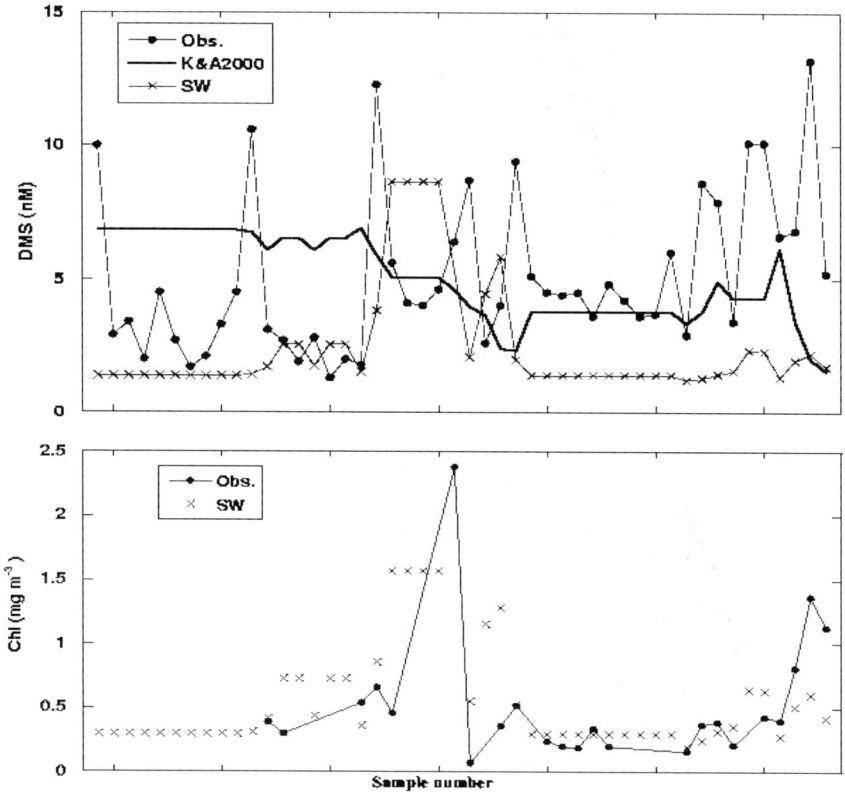

Figure 6. Comparison of spatially coincident but temporally non-coincident observed (Aranami et al., 2001) and predicted (SeaWiFS-derived and K&A2000) DMS (upper panel) and chlorophyll (lower panel) in surface waters of the North Pacific.

Both measured and predicted-baseline chlorophyll, and concentrations near the central chlorophyll patch, are similar. The amplitudes of measured and predicted DMS variations inside and near the central chlorophyll patch are in rather good agreement. The first fifteen measurements outside the patch fall much closer to SeaWiFS-based predictions than to the estimates of K&A2000. Elsewhere, the SeaWiFS-based predictions tend to underestimate the temporally non-coincident DMS observations. We found also little difference in chlorophyll measured in October-November-December 1995 (Bates et al., 1998) and predicted based on SeaWiFS data of October-November-December 1997 (Figure 7).

Exceptions occur at the beginning of the survey in the equatorial Pacific waters and at the end, where measured chlorophyll levels are two- to three-fold higher than the SeaWiFS-derived predictions. The discrepancy in equatorial Pacific waters is due to the relaxation of the equatorial upwelling associated with the 1997 El-Nino event. In these areas, the SeaWiFS-derived approach also predicts too little DMS. Outside these areas, such predictions differ markedly from the estimates of K&A2000. In general, observed baseline DMS shows closer agreement with SeaWiFS-derived baseline DMS than with the estimates of K&A2000. Table 2 shows the zonal distribution of observed and predicted average DMS concentrations in subtropical and subantarctic waters investigated during the ACE-I cruise (Bates et al., 1998).

Table 2. Zonal distribution of average oceanic DMS concentrations observed during cruise 128 and predicted from the data base of Kettle and Andreae (2000) and predicted from SeaWiFS satellite observations using the techniques described in this chapter.

Latitudinal bands	DMS observed (nM) mean (median)	K&A2000 DMS predicted (nM) mean (median)	SeaWiFS-derived DMS predicted (nM) mean (median)
20-30°S	1.1 (1.0)	0.6 (0.6)	1.4 (1.4)
30-40°S	2.4 (2.2)	2.1 (2.3)	1.4 (1.4)
40-50°S	1.8 (1.4)	2.6 (2.2)	1.6 (1.5)
50-60°S	0.8 (0.8)	2.0 (2.0)	1.3 (1.3)

Observations are in much better agreement with SeaWiFS-derived predictions than with the estimates of K&A2000, except in the latitude band 30-40°S where SeaWiFS-predicted DMS is underestimated by about 40% relative to observations. This contrasts with the temporally coincident results presented in Figure 5 for December 1997 in subantarctic Indian Ocean surface waters.

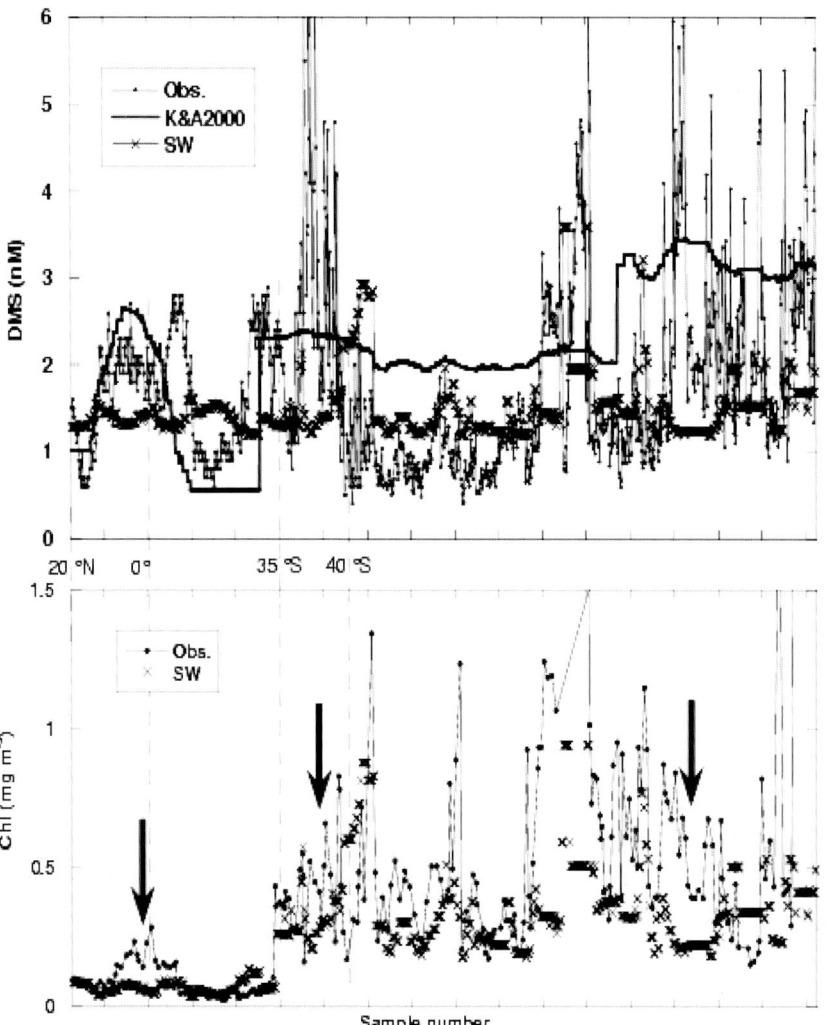

Figure 7. Comparison of spatially coincident but temporally non-coincident observed (Bates et al., 1998) and predicted (SeaWiFS-derived and K&A2000) DMS (upper panel) and chlorophyll (lower panel) in surface waters of the Central and South Pacific. Arrows indicate areas where observed and SeaWiFS-derived chlorophyll levels are markedly different.

In the corresponding latitude bands for December 1997, SeaWiFS-derived predictions are 45% (20-30°S), 82% (30-40°S) and 51% (40-50°S) lower than observed. Consequently, SeaWiFS-derived estimates appear to better reproduce the late austral spring DMS distribution in the Pacific than in the Indian Ocean. Since the zonal distribution of satellite chlorophyll is similar in both areas (data not shown, except in the latitude band 47-49°S corresponding to the Kerguelen Plateau where very high chlorophyll was observed with no equivalent in the Pacific along the trajectory of cruise ACE-I), and since the Indian Ocean DMSPs are rather well reproduced by SeaWiFS-derived data (Figure 5), the observed DMS levels in the area Crozet-Kerguelen-Amsterdam in December 1997 should be considered an anomaly. A situation where very high DMS levels are associated with very high concentrations of dDMSP and low pDMSP is typical of the senescence phase of a phytoplanctonic bloom. In the senescence phase, phytoplankton lysis releases to solution compounds present in the cytoplasm, where they undergo microbial degradation. Microbial degradation of DMSP and DMS is a matter of extensive investigation, but it is not within the scope of this chapter to summarize recent advances in this field. However, to maintain DMS and dDMSP at the very high levels observed in December 1997 we suggest, according to Kiene et al. (2000), that the bacterial sulphur demand in the area was low, a relatively high proportion of dDMSP was converted to DMS, and DMS consumption was low.

As indicated previously, the oceanic DMS data gathered by Sciare et al. (1999) in the area La Reunion-Crozet-Kerguelen were used by Kettle and Andreae in revising their 1999 data base to the K&A2000 version. Based on the evidence that DMS concentrations in this area were anomalously high during the measurement period, the substitution in K&A2000 of the Sciare et al. (1999) values into adjacent biogeochemical provinces, where no DMS data were available, may need to be revisited. Also, given the anomalous concentrations and the sensitivity of climatological global distributions (c.f., Figure 1) to such values, the need is clearly illustrated for more numerous, continuous, long-term surface measurements of oceanic DMS concentrations in the Southern Ocean and elsewhere.

3. USE OF GLOBAL OCEANIC DMS DISTRIBUTIONS IN ATMOSPHERIC MODELS

3.1. Air-sea exchange processes and parameterizations

The transfer of DMS from the world's oceans to the atmosphere is governed by parameters affecting air-sea exchange processes. In a simple conceptual model, the air-sea interface can be visualized with thin, stagnant films on both the water and air sides, with the interfacial flux governed by the DMS concentration gradient and by 'resistance' to mass transport in the air and water phases (Frost and Upstill-Goddard, 1999). Using typical DMS ocean and air concentrations, Henry's law coefficients, and basic equations of mass transfer, it can be demonstrated that for DMS the major resistance to mass transfer occurs on the aqueous side, and that DMS air concentrations are negligible compared to aqueous concentrations (C_w). The overall flux (F) of DMS across the air-sea interface thus can be mathematically represented as

$$F = -k_w C_w \quad \text{(Eq. 4)}$$

where k_w is a kinetic parameter which incorporates the aqueous phase resistance to mass transfer and represents the rate of approach to system equilibrium. k_w is typically referred to as the 'mass transfer velocity' or 'piston velocity' and has units of length per unit time. The minus sign in equation 4 indicates the direction of the DMS flux, from the ocean to the atmosphere.

Using mass transfer theory, k_w can be related to the ratio of the transfer coefficients for momentum and mass across the air-sea interface (Liss and Merlivat, 1986). This ratio is commonly referred to as the Schmidt number (Sc) and is mathematically expressed as

$$Sc = \nu / D \quad \text{(Eq. 5)}$$

where ν is the kinematic viscosity of the fluid (in this case, seawater) and D is the diffusivity of the gas of interest in this fluid (i.e., the diffusivity of DMS in seawater). Note that both ν and D are temperature and salinity dependent. If k_w can be determined for one gas, and is independent of gas solubility, then its value for other gases can be determined using the Schmidt number:

$$k_{w1} / k_{w2} = (Sc_1/Sc_2)^n \qquad \text{(Eq. 6)}$$

The value of the Schmidt number exponent n is considered to be a function of the thickness of the interfacial film across which the gas exchange takes place (Frost and Upstill-Goddard, 1999), and thus may have different values under different conditions.

Numerous empirical expressions for k_w as a function of wind speed have been developed over the years from field and laboratory experiments. Within current global three-dimensional atmospheric models, the most commonly used k_w parameterizations are those of Liss and Merlivat (1986) and Nightingale et al. (2000).

Liss and Merlivat (1986) proposed a three regime linear parameterization of k_w with the wind speed at a 10 m height above the surface (u_{10}). The linear coefficients varied according to regime. In contrast, Nightingale et al. (2000) postulated the existence of a continuous quadratic relationship between k_w and wind speed. Furthermore, they removed potential hidden stability effects from the parameterization by converting all field-measured-wind-speeds to an equivalent neutral wind speed at a 10 meter height above the surface (u_{10n}). This hidden stability effect can be illustrated by considering a situation where an air mass is advected over the ocean and where the water temperature (T_w) is greater than the air temperature (T_a). At the air-ocean interface, air parcels are warmed, begin to rise, and are replaced by cooler air parcels. Similarly water parcels are cooled, begin to sink, are replaced by warmer water parcels, and the process repeats itself. Conversely, consider a situation where $T_a > T_w$. At the air-ocean interface, air parcels are cooled and water parcels are warmed, reinforcing the tendency of the parcels to stay in their original positions. If in the two situations the air masses are moving at the same rate, intuitively one can see that the turbulence engendered in the first situation may enhance mass transfer, and that the possibility for this enhancement is missed if a k_w parameterization is based solely on u_{10}. The expression of Nightingale et al. (2000) thus is recommended for use in global atmospheric models where a wind-speed-dependent parameterization of k_w is needed.

The analysis of Nightingale et al. (2000) suggested that the following parameterization explains over 80% of the total variance in their combined observational data set:

$$k_w = 0.222 \, u_{10n}^2 + 0.333 \, u_{10n} \qquad \text{(Eq. 7)}$$

The above equation is for a gas with a Schmidt number of 600 in the marine environment, with a Schmidt number exponent of $n = -1/2$. Here u_{10n} is in m s^{-1} and k_w is in cm hr^{-1}.

Numerous global atmospheric models have utilized the approach described above in estimating DMS fluxes from the world's oceans, using everything from climatological monthly-averaged to instantaneous model time-step wind speeds in the parameterized k_w expressions. Recent work (Chapman et al., 2002) has demonstrated that substantial spatial and temporal variations in emissions fluxes occur when using different wind-speed-averaging periods. For example, a significant number of marine locations show DMS emissions fluxes that are 10-60% lower when calculated using monthly average wind speeds as opposed to 20-minute instantaneous model time-step winds. Time averaging eliminates the influence of extreme events on a solution. Use of time-averaged wind speeds with the continuous, quadratic Nightingale et al. (2000) k_w parameterization eliminates the contribution of sporadic high winds to DMS emission fluxes, with longer and longer averaging periods eliminating more and more events, leading to lower and lower flux estimates.

Once in the atmosphere, DMS emitted from the oceans will undergo transport, oxidation, and deposition processes. A model, most typically a three dimensional atmospheric general circulation model, incorporating such processes is needed to predict gas phase DMS concentrations if comparisons with field observations of gaseous DMS are to be made.

3.2. Three-dimensional atmospheric modeling

The Atmospheric General Circulation Model of the Laboratoire de Météorologie Dynamique (Paris, France) LMD-ZT (Boucher et al., 2002) was used to simulate the emission, transport and transformation of DMS and five other sulphur species. In the model version employed here, a 96 x 72 spatially variable grid zooms in on the mid- and high- southern latitudes, and specific parameterizations improve polar atmospheric physics (Krinner et al., 1997). The model simulates processes involving emissions, boundary layer mixing, advective and convective transport, dry and wet scavenging, and oxidation in the gaseous and aqueous phases. Gas-phase chemistry is based on the scheme first introduced by Pham et al. (1995). DMS is oxidized by OH and NO_3 radicals producing SO_2 and DMSO. No other reaction with an additional oxidant was considered here. We assume that MSA production proceeds via DMSO through the addition pathway of DMS oxidation. Aqueous-phase oxidation of SO_2 by O_3 and H_2O_2 is also considered. Dry deposition is parameterized through deposition velocities, which are prescribed for each chemical species and surface type (Cosme et al., 2002). The model is nudged to ECMWF analysis following the method described by Genthon et al. (2002). Meteorological data extending from October 1997 to September 1998, corresponding to the global distributions of SeaWiFS-derived DMS concentrations, are used for surface boundary conditions and nudging. Oceanic DMS concentrations are prescribed globally as monthly means with the constraints that (1) a minimum value of 0.2 nmol l^{-1} (Belviso et al., 2000) is assumed in regions where no SeaWiFS data are available (i.e., at high latitudes in wintertime due to low insolation), and (2) a maximum value of 50 nmol l^{-1} is also specified to eliminate the few unrealistic values obtained at very large Chl α content in coastal waters. DMS emissions from the oceans are simulated on-line using model-time-step wind speeds and the parameterization of Nightingale et al. (2000). Sea ice is assumed to produce a lid effect on air-sea DMS mass transfer. Two simulations were performed, one with the SeaWiFS-derived distribution of oceanic DMS concentrations and one with the K&A2000 distribution of oceanic DMS concentrations.

3.3. Results and discussion

3.3.1. Latitudinal distribution of DMS emissions

Table 3 summarizes the annual DMS emissions from both simulations, globally and in the southern Ocean south of 30°S. The global totals fall within the generally accepted DMS-emission-range of 10 to 50 Tg S yr^{-1} (IPCC, 1996). Southern Ocean emission totals represent approximately one-third of the global flux, regardless of whether the K&A2000 or SeaWiFS-derived DMS data base is used as input in the model. Latitudinal distributions based on the two oceanic DMS data bases are similar from 30° to 60°S, but differ markedly south of 60°S.

DMS emissions 325

Table 3. Latitudinal distribution of DMS emissions (TgS yr^{-1}) as produced by the LMD-ZT model. SEAWIFS and K&A2000 respectively refer to the simulations that used the oceanic DMS concentration maps derived from SeaWiFS data and from the data base of Kettle and Andreae (2000).

Reference	Global	30°S - 90°S	30°S - 40°S	40°S - 50°S	50°S - 60°S	60°S - 90°S
SEA WIFS	21.6	6.8	2.0	2.6	1.8	0.4
K&A2000	21.0	7.9	2.0	2.6	2.0	1.3

3.3.2. Seasonal variations of DMS emissions

Table 4 presents the primary statistics of monthly DMS emissions in the Southern Ocean. The simulation based on SeaWiFS-derived oceanic DMS concentrations displays a seasonality about 13 times less pronounced than the simulation based on K&A2000. Winter and summer DMS emission fluxes calculated with the SeaWiFS-derived DMS concentrations are approximately equal, whereas the summer emission flux is markedly higher than winter emission flux when K&A2000 DMS concentrations are used.

Table 4. Minimum, maximum, mean and standard deviation (SD) (in TgS yr^{-1}) of monthly DMS emissions in the southern ocean, between October 1997 and September 1998 as produced by the LMD-ZT model, using oceanic DMS concentrations derived from SeaWiFS data and from the data base of Kettle and Andreae (2000).

reference	Minimum	Maximum	Mean	SD
SEAWIFS	5.9	7.2	6.8	0.4
K&A2000	2.4	16.1	7.9	5.4

3.3.3. Atmospheric concentrations of sulphur compounds

Figure 8 compares time series of observed and simulated atmospheric concentrations of DMS and non-sea salt sulphate (nss-sulphate) at Amsterdam Island (77°30'E, 37°50'S), Halley (26°19'W, 75°35'S, Antarctica), and Dumont d'Urville (DDU, 140°1'E, 66°40'S, Antarctica). It should not be forgotten that model predictions are representative of the entire area within a given grid cell, whereas observations are at a very specific point. Also, the predicted values of both gas phase DMS and nss-sulphate depend strongly on the chemical mechanism and other sulphur (anthropogenic, volcanic, etc.) sources used in LMD-ZT model, and the accuracy with which the model simulates vertical mixing and other meteorological processes. Because of its relatively short atmospheric lifetime (about one day in summer), air concentrations of DMS are closely linked to localized oxidant fields, meteorological mixing processes, and emissions, whereas nss-sulphate, an end product of many atmospheric sulphur oxidation reactions, is likely to have been transported over long distances. Long-term surface measurements of sulphur compounds in the mid- and high- southern latitudes usually display a well marked seasonality, with a summer maximum and a winter minimum (e.g. Sciare et al., 2000; Minikin et al., 1998). This behaviour is captured by the model, as shown in Figure 8.

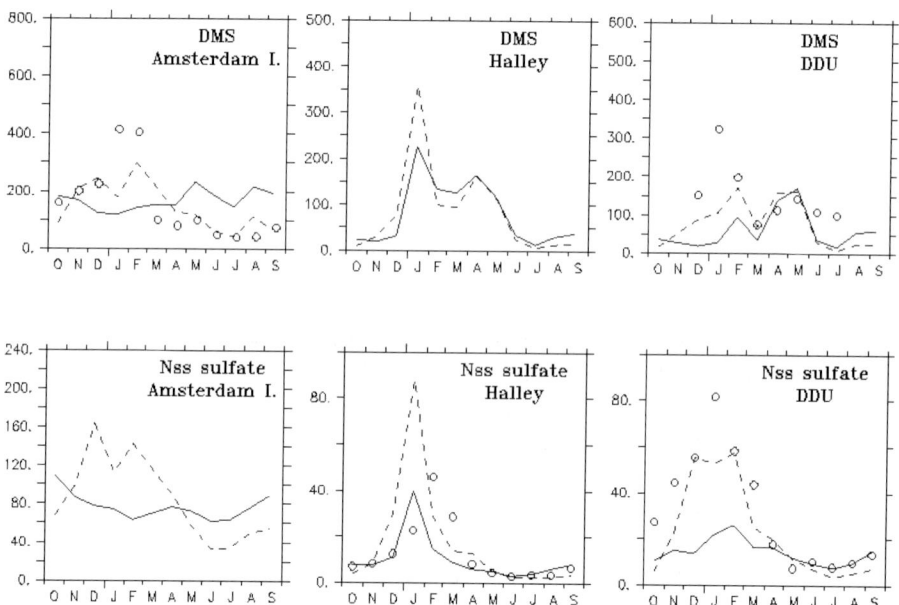

Figure 8. Seasonal variations in observed and model-predicted atmospheric surface DMS and nss-sulphate (pptv) at Amsterdam Island, Halley and Dumont d'Urville. Solid lines and dashed lines symbolize respectively model results using SeaWiFS-derived and K&A2000 oceanic DMS concentrations. Observations are drawn from Sciare et al. (2000) (Amsterdam Island), Minikin et al. (1998) (Halley) and Jourdain and Legrand (2001) (Dumont d'Urville).

Figure 8 also indicates that there are substantial differences between observed and simulated atmospheric DMS concentrations (up to factors of ~5), regardless of whether K&A2000 or SeaWiFS-derived oceanic DMS

DMS emissions 327

estimates are used. The SeaWiFS-derived DMS simulation produces atmospheric DMS levels at Amsterdam Island that are too high in austral winter and too low in austral summer relative to observations, while at Dumont d'Urville the simulations underestimate DMS and nss-sulphate during the austral summer. Conversely, at Halley simulations using the SeaWiFS-derived and K&A2000 DMS data bases both capture the seasonality of atmospheric nss-sulphate, although the observed nss-sulphate maximum of February occurs in January in both simulations.

Closer agreement between observed and simulated nss-sulphate occurs at Halley when the SeaWiFS-derived oceanic DMS data base is used. Substantial differences between observed and simulated atmospheric DMS concentrations have been noted in other works (e.g., Chin et al., 1996; Barth et al., 2000; Easter et al., 2002), and reflect the difficulties in accurately simulating localized wind fields, mixing and boundary layer phenomena, oxidant fields, and chemical processes in the as yet less-than-fully-understood atmospheric DMS reaction sequence, all of which influence the gas phase DMS concentrations predicted by any atmospheric general circulation model. Because of these modelling uncertainties, and the limited temporal period of comparison, it is not feasible for us to make any generalizations about which oceanic DMS concentrations data base (K&A2000 or SeaWiFS-derived) is 'better' to use. We note, however, that the use of K&A2000 apparently leads to more realistic atmospheric concentrations of DMS at Amsterdam Island. SeaWiFS-derived DMS concentrations, however, offer the opportunity to document the interannual variations of the marine source of DMS.

The SeaWiFS-derived oceanic DMS data also raise interesting questions for future research efforts. As noted previously, there is a lack of strong seasonality in SeaWiFS images of oceanic chlorophyll concentrations in the vicinity of Amsterdam Island (and thus the SeaWiFS-derived oceanic DMS levels), yet observed atmospheric DMS levels demonstrate a strong seasonality. Do oceanic DMS levels in this region show seasonality, or not? At Amsterdam Island, seawater DMS concentrations varied by a factor of 3.6 between winter and summer with mean concentrations of 0.4 and 1.4 nM, respectively (Nguyen et al., 1990). The maximum DMS concentration was 2 nM in December 1987. Mean DMS fluxes were in the range 1.3-3 $\mu mol\ m^{-2}\ d^{-1}$, respectively, so only roughly twice higher in summer than in winter. These results contrast with the more recent data of Sciare et al. (1999). Indeed, the mean concentration and the mean flux of DMS in

December 1997 were 9.1 nM and 16.4 µmol m^{-2} d^{-1}, respectively. The seawater DMS samples, however, were not collected in the vicinity of Amsterdam Island but in the latitude band 32-38°S about 2000 km to the west of Amsterdam Island. Although fluxes in the range 13-16 µmol m^{-2} d^{-1} appear to be consistent with long-term observations of summer atmospheric concentrations of DMS and rainwater concentrations of sulphate and methanesulphonate (Sciare et al., 1999), there is no evidence from seawater samples that DMS fluxes in summer closer to Amsterdam Island are indeed in such a range. If not, are atmospheric DMS variations due to localized seasonal differences in atmospheric oxidant fields and mixing processes, or to oceanic processes? For example, the flux of DMS from the ocean to atmosphere may be greater in winter than in summer because of stronger winter winds. If the removal of DMS by ventilation dominates other removal processes in winter, one can expect a feedback on winter sea surface DMS concentrations provided that DMS production from phytoplankton remains unchanged. Also, seasonal changes in phytoplankton speciation or changes in various DMSP and DMS turnover rates may be stronger in the subtropical Indian Ocean than elsewhere. There is currently very little information on the biogeochemistry of oceanic DMS in the area, and more work is clearly needed here and in improving the simulated behaviour of DMS in atmospheric models.

4. SUMMARY

Oceanic DMS concentration maps can be derived from SeaWiFS satellite data. DMS spatial variability is better captured with this approach relative to earlier works, although there appears to be a tendency to overestimate oceanic wintertime DMS in the mid-latitudes of the southern hemisphere. This tendency occurs because chlorophyll levels, used in deriving the DMS concentrations, remain high in winter at these latitudes. A baseline oceanic DMS level of 1.1 nM is attained with this technique. With these limitations in mind, nevertheless, we believe that the use of SeaWiFS-derived data should be continued to provide information on the interannual variability and potential climate effects on oceanic DMS.

The gridded SeaWiFS-derived oceanic DMS database can be used as input to three dimensional global atmospheric models. When coupled with parameterizations of the air-sea mass transfer coefficient, estimates of the flux of DMS from the ocean to the atmosphere can be obtained via model simulations. Using the three-dimensional atmospheric General Circulation Model of the Laboratoire de Météorologie Dynamique, model-time-step wind speeds, an atmospheric-stability-dependent parameterization of the mass transfer coefficient, and the SeaWiFS-derived oceanic DMS distributions, we estimate an annual southern ocean DMS emission of 6.8 Tg S yr^{-1}. This value represents approximately one-third of the annual global DMS marine emission, and underscores the importance of this region as a source of natural sulphur emissions.

5. REFERENCES

Anderson, T. R., S. A. Spall, A. Yool, P. Cipollini, P. G. Challenor, and M. J. R. Fasham, Global fields of sea surface dimethylsulfide predicted from chlorophyll, nutrients and light, J. Mar. Syst., 30, 1-20, 2001.

Aranami, K., S. Watanabe, S. Tsunogai, M. Hayashi, K. Furuya, and T. Nagata, Biogeochemical variation in dimethylsulfide, phytoplankton pigments and heterotrophic bacterial production in the Subarctic North Pacific during summer, J. Oceanogr., 57, 315-322, 2001.

Archer, S.D., F.J. Gilbert, P.D. Nightingale, M. V. Zubkov, A.H. Taylor, G.C. Smith, and P.H. Burkill, Transformation of dimethylsulfoniopropionate to dimethylsulfide during summer in the North Sea with an examination of key processes via a modelling approach, Deep-Sea Res. II, 49, 3067-3101, 2002.

Aumont, O., S. Belviso, and P. Monfray, Dimethylsulfoniopropionate (DMSP) and dimethylsulfide (DMS) sea surface distributions simulated from a global 3-D ocean carbon cycle model, J. Geophys. Res., 107, C4, 10.1029/1999 JC000111, 2002.

Barrie, L., Y. Yi, W.R. Leaitch, U. Lohmann, P. Kasibhatla, G.-J. Roelofs, J. Wilson, F. McGovern, C. Benkovitz, M.A. Meliere, K. Law, J. Prospero, M. Kritz, D. Bergmann, C. Bridgeman, M. Chin, J. Christensen, R. Easter, J. Feichter, C. Land, A. Jeuken, E. Kjellstrom, D. Koch, and P. Raschet al., A comparison of large scale atmospheric sulphate aerosol models (COSAM): Overview and highlights, Tellus, 53B, 625-645, 2001.

Barth, M. C., P. J. Rasch, J. T. Kiehl, C. M. Benkovitz and S. E. Schwartz, Sulfur chemistry in the National Center for Atmospheric Research Community Climate Model: Description, evaluation, features, and sensitivity to aqueous chemistry, J. Geophys. Res., 105, 1387-1415, 2000.

Bates, T. S., J. D. Cline, R. H. Gammon, and S. R. Kelly-Hansen, Regional and seasonal variations in the flux of oceanic dimethylsulfide to the atmosphere, J. Geophys. Res., 92, 2930-2938, 1987.

Bates, T. S., B. K. Lamb, A. Guenther, J. Dignon, and R. E. Stoiber, Sulfur emissions from natural sources, J. Atmos. Chem., 14, 315-337, 1992.

Bates, T. S., V.N. Kapustin, P.K. Quinn, D.S. Covert, D.J. Coffman, C. Mari, P.A. Durkee, W.J. DeBrutn, and E. Satzman, Processes controlling the distribution of aerosol particles in the lower marine boundary layer during the First Aerosol Characterization Experiment (ACE-1), J. Geophys. Res., 103, 16,369-16,383, 1998.

Belviso, S., R. Morrow, and N. Mihalopoulos, An Atlantic meridional transect of surface water dimethylsulfide concentrations with 10-15 km horizontal resolution and close examination of ocean circulation. J. Geophys. Res. 105, 14,423-14,431, 2000.

Benkovitz, C. M., C. M. Berkowitz, R. C. Easter, S. Nemesure, R. Wagener, and S. E. Schwartz, Sulfate over the North Atlantic and adjacent continental regions: Evaluation for October and November 1986 using a three-dimensional model driven by observation-derived meteorology, J. Geophys. Res., 99, 20,725-20,756, 1994.

Boucher, O., M. Pham, and C. Venkataraman, Simulation of the atmospheric sulfur cycle in the Laboratoire de Météorologie Dynamique General Circulation Model. Model description, model evaluation and global and European budgets, Note scientifique 23 de l'Institut Pierre Simon Laplace, Paris, France, 2002.

Chapman, E. G. , R. C. Easter, X. Bian, and S. J. Ghan, The influence of wind speed averaging on estimates dimethylsulfide emission fluxes, J. Geophys. Res., 107, 10.1029, doi: 2001JD001564, 2002.

Chin, M., D. J. Jacob, G. M. Gardner, M. S. Foreman-Fowler, and P. A. Spiro, A global three-dimensional model of tropospheric sulfate, J. Geophys. Res., 101, 18667-18690, 1996.

Chin, M., R. B. Rood, S.-J. Lin, J.-F. Muller, and A. M. Thompson, Atmospheric sulfur cycle simulated in the global model GOCART: Model description and global properties, J. Geophys. Res., 105, 24,671-24,687, 2000.

Chuang, C., J. E. Penner, K. E. Taylor, A. S. Grossman, and J. W. Walton, An assessment of the radiative effects of anthropogenic sulfate, J. Geophys. Res., 102, 3761-3778, 1997.

Claustre, H., The trophic status of various oceanic provinces as revealed by phytoplankton, pigment signatures, Limnol. Oceanogr., 39, 1206-1210, 1994.

Cosme, E., C. Genthon, P. Martinerie, O. Boucher and M. Pham, The sulfur cycle at high-southern latitudes in the LMD-ZT General Circulation Model, J. Geophys. Res., 107, 4690, 10.1029/2002JD002149, 2002.

Easter, R. C., S. J. Ghan, Y. Zhang, R. D. Saylor, E.G. Chapman, N. S. Laulainen, H. Abdul-Razzak, L. R. Leung, X. Bian and R. A. Zaveri, MIRAGE: Model description and evaluation of aerosols and trace gases, submitted for publication in J. Geophys. Res., 2002.

Feichter, J., E. Kjellstrom, H. Rodhe, F. Dentener, J. Lelieveld, and G-J. Roelofs, Atmos. Environ., 30, 1693-1707, 1996.

Frost, T. and R. C. Upstill-Goddard, Air-sea gas exchange into the millennium: Progress and uncertainties, in Oceanography and Marine Biology: an Annual Review, 37, 1-45, 1999.

Genthon, C., G. Krinner, and E. Cosme. Free and laterally-nudged antarctic climate of an atmospheric general circulation model, Mon. Wea. Rev., 130, 1601-1616, 2002.

Ghan, S., R. Easter, E. Chapman, H. Abdul-Razzak, Y. Zhang, R. Leung, N. Laulainen, R. Saylor and R. Zaveri, A physically based estimate of radiative forcing by anthropogenic sulfate aerosol, J. Geophys. Res., 106, 5279-5293, 2001a.

Ghan, S., N. Laulainen, R. Easter, R. Wagener, S. Nemesure, E. Chapman, Y. Zhang and R. Leung, Evaluation of aerosol direct radiative forcing in MIRAGE, J. Geophys. Res., 106, 5295-5316, 2001b.

Hansson, M. E. and E. S. Saltzman, The first Greenland ice core record of methanesulfonate and sulfate over a full glacial cycle, Geophys. Res. Lett., 20, 1163-1166, 1993.

IPCC, Climate Change, 1995: The science of climate change, edited by J. T. Houghton, L. G. Meira Filho, J. Bruce, H. Lee, B. A. Callander, E. Haites, N. Harris and K. Maskell, Canbridge University Press, 1996.

Jones, A., D. L. Roberts, M. J. Woodage, and C. E. Johnson, Indirect sulphate aerosol forcing in a climate model with an interactive sulphur cycle, J. Geophys. Res., 106, 20,293-20,310, 2001.

Jourdain B. and M. Legrand, Seasonal variations of atmospheric dimethylsulfide, dimethylsulfoxide, sulfur dioxyde, methanesulfonate and non-sea-salt sulfate aerosols at Dumont d'Urville (coastal Antarctica, December 1998-July 1999). J. Geophys. Res., 106, 14,391-14,407, 2001.

Kettle, A. J., M.O. Andreae, D. Amouroux, T.W. Andreae, T.S. Bates, H. Berresheim, H. Bingemer, R. Boniforti, M.A.J. Curran, G.R. DuTullio, G. Helas, C.B. Jones, M.D. Keller, R.P.Kiene, C. Leck, M. Levasseur, M. Maspero, P. Matrai, A.R. McTaggart, N. Mihalopoulos, B.C. Nguyen, A. Novo, J.P. Putaud, S. Rapsomanikis, G. Roberts, S. Schebeke, R. Simo, R. Staubes, S. Turner, ans G. Uher, A global database of sea surface dimethylsulfide (DMS) measurements and a procedure to predict sea surface DMS as a function of latitude, longitude, and month, Global Biogeochem. Cycles, 13, 399-444, 1999.

Kettle, A.J. and M. O. Andreae, Flux of dimethylsulfide from the oceans: A comparison of updated data sets and flux models, J. Geophys. Res., 105, 26,793-23,808, 2000.

Kiene, R. P., L. J. Linn, J. A. Burton, New and important roles for DMSP in marine microbial communities, J. Sea Res., 43, 209-224, 2000.

Koch, D., D. Jacob, I. Tegen, D. Rind, and M. Chin, Tropospheric sulfur simulation and sulfate direct radiative forcing in the Goddard Institute for Space Studies general circulation model, J. Geophys. Res., 104, 23,799-23,822, 1999.

Krinner, G., C. Genthon, Z.X. Li, and P. Le Van, Studies of the Antarctic climate with a stretched-grid general circulation model, J. Geophys. Res., 102, 13,731-13,745, 1997.

Legrand, M., C. Feniet-Saigne, E. S. Saltzman, C. Germain, N. I. Barkov, and V. N. Petrov, Ice-core record of oceanic emissions of dimethylsulfide during the last climatic cycle, Nature, 350, 144-146,1991.

Liss, P. S. and L. Merlivat, Air-sea gas exchange rates: Introduction and synthesis, in The Role of Air-Sea Exchange in Geochemical Cycling, edited by P. B. Menard, pp. 113-127. D. Reidel, Norwell, Mass., 1986.

Lohmann, U., K. von Salzen, N. McFarlane, H. G. Leighton, and J. Feichter, Tropospheric sulfur cycle in the Canadian general circulation model, J. Geophys. Res., 104, 26,833-26,858, 1999.

Minikin, A., M. Legrand, J. Hall, D. Wagenbach, C. Kleefeld, E. Wolf, E. C. Pasteur, and F. Ducroz. Sulfur-containing species (sulfate and methanesulfonate) in coastal Antarctic aerosol and precipitation, J. Geophys. Res., 103, 10,975-10,990, 1998.

Nightingale, P. D., G. Malin, C. S. Law, A. J. Watson, P. S. Liss, M. I. Liddicoat, J. Boutin and R. C. Upstill-Goddard, In situ evaluation of air-sea gas exchange parameterizations using novel conservative and volatile tracers, Glob. Biogeochem. Cycles, 14, 373-387, 2000.

Nguyen, B.C., N. Mihalopoulos, and S. Belviso, Seasonal variation of atmospheric dimethylsulfide at Amsterdam Island in the Southern Indian Ocean, J. Atmos. Chem., 11, 123-141, 1990.

Pham, M., J.-F. Müller, G. Brasseur, C. Granier, and G. Mégie, A three-dimensional study of the tropospheric sulfur cycle, J. Geophys. Res., 100, 26,061-26,092, 1995.

Roeckner, E., L. Bengtsson, J. Feichter, J. Lelieveld and H. Rodhe, Transient climate change simulations with a coupled Atmosphere-Ocean GCM including the tropospheric sulfur cycle, J. Clim., 12, 3004-3032, 1999.

Sciare, J., N. Mihalopoulos, and B. C. Nguyen, Summertime seawater concentrations of dimethylsulfide in the Western Indian Ocean: reconciliation of fluxes and spatial variability with long-term atmospheric observations, J. Atmos. Chem., 32, 357-373, 1999.

Sciare, J., M. Kanakidou, and N. Mihalopoulos, Diurnal and seasonal variation of atmospheric dimethylsulfoxide (DMSO) at Amsterdam Island in the southern Indian Ocean. J. Geophys. Res., 105, 17,257-17,265, 2000.

Spiro, P. A., D. J. Jacob, J. A. Logan, Global inventory of sulfur emissions with 1°x1° resolution, J. Geophys. Res., 97, 6023-6036, 1992.

Zubkov, M. V., B. M. Fuchs, S. D. Archer, R. P. Kiene, R. Amann, and P. H. Burkill, Linking the composition of bacterioplankton to rapid turnover of dissolved dimethylsulfoniopropionate in an algal bloom in the North Sea, Environ. Microbiol., 3, 304-311, 2001.

Sea-salt aerosol source functions and emissions

Michael Schulz, Gerrit de Leeuw and Yves Balkanski

1. INTRODUCTION

Sea spray aerosols are important for a wide variety of processes. Part of the current interest is their role in climate (Penner et al., 2001). Sea spray aerosol contributes to atmospheric cooling because they scatter incoming solar radiation. It is a natural component of the climate system and therefore can not be regarded as a forcing component. However, it is often neglected in global climate models and may be responsible for feedback effects. Latham and Smith (1990) suggested that a changing climate would alter surface winds and thus sea spray emissions. Although the sea spray aerosol number concentrations are not very high compared to those of anthropogenic aerosols such as ammonium sulphates, their role is significant because the oceans cover 70% of the Earth, whereas anthropogenic aerosols are rather locally produced. Sea-salt is the dominant submicrometer scatterer in most ocean regions and dominates the marine boundary layer particulate mass concentration in remote oceanic regions, with a significant fraction occurring in the submicrometer size range (IPCC, 2001). Sea-salt contributes 44% to the global aerosol optical depth. Estimates for top-of-atmosphere, global-annual radiative forcing due to sea-salt are -1.51 and -5.03 Wm^{-2} for low and high emission values, respectively (IPCC, 2001). Sea spray not only affects climate by scattering of solar radiation, it also acts as cloud condensation nuclei and thus contributes to the indirect aerosol effect (IAE). This effect accounts for the perturbation of the hydrological cycle due to the abundance of aerosol particles. The main contribution to the IAE, the most uncertain forcing mechanism in the prediction of climate change, comes from marine stratocumulus clouds. Boers et al. (1998) report an example of IAE related to changes in the distributions of natural cloud condensation nuclei (CCN) over the ocean, and O'Dowd et al. (1999ab) have demonstrated from aircraft observations, that sea spray particles can play a significant role in marine stratocumulus microphysics and chemistry. At very high wind speeds their role in the transfer of heat and water vapour from the ocean to the atmosphere becomes significant (e.g., Andreas et al., 1995).

Sea spray is also important in atmospheric chemistry. In this context it is interesting to note that sea spray particle composition may differ from one region to another. During their formation from bubbles (briefly discussed in section 2), they are enriched by organic material that is picked up by the bubbles rising in the water column and from the micro-layer at the sea surface (Monahan and Dam, 2001). Once formed, the aerosol droplets release halogens and they act as a surface for heterogeneous chemistry and can provide a significant sink for natural and anthropogenic trace gases (O'Dowd et al., 2000). For instance, sea spray provides a surface for the condensation of HNO_3, which influences the flux of nitrogen into the sea and thus eutrophication (Schulz et al., 1999; De Leeuw et al., 2001). At the coast line, sea spray is immediately available because of its production in the surf zone in very large amounts (De Leeuw et al., 2000) which are dispersed throughout the boundary layer as they are transported away from the source. The influence of the surf-produced sea spray aerosol on the HNO_3 fluxes was evaluated by Vignati et al. (2001), using a coastal aerosol transport model. The sea spray aerosol produced in coastal waters, both from breaking wind waves and in the surf zone, may also be important because of enrichment with pollutants, toxins, bacteria and viruses (occurring in coastal water due to, e.g., run-off, spills, etc., as well as natural causes such as red tide) which are transported to local beaches where they may affect the population. Sea-salt itself can be important for maintaining certain types of vegetation and natural habitats and coastal ecosystems over land. Sea spray aerosol also affects visibility and, because of the occurrence of super-micron droplets, infrared observation systems, which are often used for naval applications or the detection of drowning people after ship disasters.

To correctly assess the effects of sea spray, accurate estimates are required for the emission strength of sea spray aerosols for use in, e.g., global circulation models (GCM) and chemistry transport models (CTM). The emission strengths are commonly expressed as the flux of sea spray aerosols at or near the ocean surface, i.e. the number of droplets produced per unit of time and per square meter. The dependency on wind speed and eventually other factors such as fetch, whitecap cover or wave spectrum is combined to a sea-salt source function. The focus of studies on the source function depends on the application. For the role of sea spray in heat and mass transfer, the larger sea spray droplets are important and studies focus on very large particles, typically starting from a few μm in diameter. The critical diameter for this application is around 40 μm (Andreas, 1992). For infrared-propagation and heterogeneous chemistry, droplets with sizes from 1 to 20 μm are most important, and for visibility and climate applications the critical size is around 0.5 μm.

This contribution starts with a short overview on sea spray production mechanisms, as an introduction to a review of recent work on sea spray emission strengths described by the source function.

2. SEA SPRAY PRODUCTION MECHANISMS

Sea spray aerosol is mainly produced by two mechanisms, i.e. bubble-mediated production of film and jet droplets and, at elevated wind speeds, from direct tearing of droplets from the wave tops. Other production mechanisms are the spilling of droplets from breaking waves, splash droplets resulting from precipitation (e.g. Marks, 1990) or 'secondary droplets presumed to be created by the impact on the water's surface of their parent, primary film drops' (Spiel, 1998). The latter mechanism has not been quantified and no information is available with regard to its importance for the total sea spray aerosol concentrations. Below we will discuss in some detail bubble-mediated production and direct tearing, which have received most attention in the literature. Detailed studies on bubble-mediated production were made in particular by Blanchard and co-workers as reviewed in Blanchard (1963, 1983, 1989) and more recently by Spiel (1994ab, 1995, 1997, 1998), who studied very large bubbles (radius > 350 μm).

At wind speeds exceeding about 4 m s^{-1}, waves break and the air entrained in the water breaks up into bubbles. The bubbles are transported by various mechanisms such as turbulence produced by the breaking wave, Langmuir circulation, and currents. Small bubbles follow the motions of the fluid and may be entrained to depths several times the wave height, and Langmuir circulation may transport the bubbles to tens of meters below the surface. In contrast, the bubbles also tend to rise in the water due to their own buoyancy, which increases with the bubble volume. Hence the bubbles are separated in the water column and the bubble spectrum changes with time after the wave breaking event (Medwin and Breitz, 1989; Leifer and De Leeuw, 2001).

When the bubbles reach the water surface, they burst and produce two types of droplets. Film droplets are produced while the film cap opens. This mechanism has been studied in detail by Spiel (1998). During the opening of the film cap, the rim accelerates and obtains a corrugated character with areas of high and low density which accelerate centrifugally (forces are several g), and finally the high density areas reach such high momentum that they are torn off and ejected at small angles with respect to the horizontal. Most angles are negative and thus most film droplets will return back to the water surface where they may produce a "secondary" droplet. Film droplets are only produced by bubbles larger than circa 2.4 mm in diameter. The number of film droplets produced has been reported to be 100-1000 per bubble (Blanchard, 1963).

The opening bubble leaves a 'cavity' in the water surface. Due to hydrostatic forces, a vertical jet rises from the bottom of the cavity into the air where it breaks up into up to six droplets. This process has been described by e.g., Kientzler et al. (1954), MacIntyre (1972), Blanchard (1963, 1983) and Dekker and De Leeuw (1993). The number of these so-called jet droplets that are produced per bubble depends on the bubble size, bubbles larger than 3.4 mm in diameter produce no jet droplets; the smallest bubbles produce six jet droplets (Spiel, 1997).

The number of bubbles in oceanic spectra decrease very fast with size: presenting the bubble size distribution as a power law function ($dN/dD = CD^n$, where D is bubble diameter and C is a constant representing the number of bubbles for $D=1$ μm), the power n has been observed to vary between -2 and -6 (cf. the overview in De Leeuw and Cohen, 2001). Because most jet droplets are produced from small bubbles which occur in relatively high concentrations, and film droplets are produced only from bubbles larger than 2.41 mm (Spiel, 1998) for which the concentrations may be several orders of magnitude smaller, the jet droplet contribution to the size distribution is thought to be most important. This conclusion has been derived from oceanic spectra which are time-averaged 'background' spectra, whereas the laboratory experiments (Leifer and De Leeuw, 2001) show that the bubble spectra just after wave breaking have a very different shape, and thus the initial production is governed by film droplets. A detailed study of the evolution of the bubble spectra and the resulting sea spray aerosol spectra is required to decide which contribution is most important. This may seem an academic problem. However, the production mechanism may have consequences for the sea spray droplet composition, in particular the amount of organic material in the resulting aerosol particle and the distribution of the composition over the particle sizes. These chemical and physical properties in turn are important parameters for the aerosol life cycle and the effect of the aerosol particles on processes such as cloud formation and light scattering.

The second mechanism by which sea spray aerosol is formed is the direct disruption of spume droplets from the wave tops at wind speeds exceeding about 9 m s^{-1} (Monahan et al., 1983). The contribution of this mechanism to the aerosol size spectrum bears a large uncertainty, because in practice the sea spray droplets produced from either bubbles or by tearing from the wave tops cannot be discerned.

Mårtensson et al. (2002) undertook laboratory studies on the bubble-mediated production of sea spray droplets and observed droplets as small as 20 nm. The origin of such small droplets is not clear. Mårtensson et al. ascertained, that their measurement chamber was free of any detectable aerosol before the start of their experiment on bubble-mediated aerosol production. A possible explanation for the observation of nm-sized aerosol could be the secondary production by the splash of either film or jet droplets falling back into the water.

The size of the sea spray droplets is thought to vary according to the production mechanism. Commonly, particles with $r_{dry} \leq 0.5$ µm are referred to as film droplet range, particles with dry radii between 0.5 and 4 µm as jet droplet range, and particles with $r_{dry} \geq 4$ µm as spume droplet range (Guelle et al., 2001). r_{dry} corresponds to the radius of the particle at 0% RH. However, this size discrimination is not strict. For instance, Andreas (1998) classifies droplets with radii ranging roughly from 0.5 to 5 µm as film droplets, jet droplets have typical radii of 3 to 50 µm and spume droplets are larger than 20 µm. Although not mentioned in this context, Andreas (1998) likely refers to droplet radius r_o at formation, where high humidities prevail ($r_o \approx 4$ x $r_{dry,}$ see below). In practice, the various types of sea spray droplets cannot be separated, except, perhaps, from studies of single bubbles (Woolf et al., 1987). O'Dowd et al. (1997) classified the sea spray particles based on the observation of three different modes which have different physical behaviour, in particular as regards the variation of their concentrations with wind speed. The mode radii (r_m) were determined at $r_{m(dry)} = 0.1$ μm ($\sigma = 1.9$) for film droplets, at $r_{m(dry)} = 1$ μm ($\sigma = 2$) for jet droplets and for spume droplets at $r_{m(dry)} = 6$ μm ($\sigma = 3$). Keeping in mind that the size of a jet droplet is about one tenth of that of the parent bubble, and the largest bubbles that can produce a jet droplet have a diameter of 3.4 mm, the upper limit for the jet droplet diameter would be 340 μm. The lower limit is determined by the minimum size of the bubbles that would reach the surface before going into solution, which would result in minimum droplet sizes of the order of 1 μm.

3. SEA SPRAY SOURCE FUNCTIONS

The surface flux of sea spray aerosol, i.e. the number of droplets produced per unit surface area and per unit of time is described by the source function. The fluxes F are expressed in number of droplets in a certain size interval, i.e. μm^{-1} m^{-2} s^{-1}. It is noted that the particle radius may be given as dry radius (r_{dry}), i.e. the radius the sea-salt particle would have at relative humidity RH=0%, as the radius the particle would have when normalized to RH=80% (r_{80}), or as the radius at formation (r_o). The latter would be understood as a particle with the composition of sea water, in an environment with the equilibrium RH over sea water (RH=97%). The formulation depends on the way the source function has been derived. Formulations of Fitzgerald (1975) or Gerber (1985) have been used to relate the droplet sizes at various relative humidity. The equations of Fitzgerald and Gerber differ only for RH>97%. In this manuscript, the original formulations for the source functions will be used and the radius used will be indicated. As a rule of thumb, $r_o \approx 2r_{80} \approx 4r_{dry}$.

The sea spray source functions are usually formulated as function of the wind speed U_{10}, i.e. evaluated at 10 m above the water surface (unless otherwise mentioned), and expressed in m s^{-1}. When the wind speed is measured at another height than 10 m, U_{10} can be evaluated using micrometeorological expressions (cf. Stull, 1988):

$$U(z) = \frac{U_*}{\kappa} \ln\left(\frac{z}{z_0}\right) \qquad \text{(Eq.1)}$$

where U(z) is the wind speed at height z, U_* is the friction velocity, κ (= 0.4) is the Von Karman constant and z_0 is the roughness length for wind speed.

Eq.1 applies to neutral thermal stratification. For stable and unstable conditions, a correction factor needs to be applied, see e.g. Stull (1988) for further reference.

Sea spray source functions have been determined using different methods based on laboratory and field measurements or combinations thereof, as well as modelling. Laboratory measurements have been undertaken to determine the amounts of sea spray droplets produced from bubble plumes generated by breaking waves, weirs or aerators.

Source functions derived from field experiments are usually based on the assumption of a steady state, in which the surface production and the removal by dry deposition are in balance. Hoppel et al. (2002) showed that such condition can apply only for larger particles, the deposition velocity of which is determined by gravitation only. Recently, eddy correlation techniques have been applied to directly determine the particle fluxes. (Nilsson et al., 2001) and Reid et al. (2001) applied a technique in which the loading of the atmospheric column was observed.

An overview of different formulations of the sea-salt aerosol generation function published until 1993 has been presented by Andreas (1998). This overview clearly shows the wide variety of source functions resulting from different techniques, with predictions ranging over six orders of magnitude at any given radius. Andreas discounted most of these source functions based on various arguments and concluded, that "the bubble-only part of the Monahan et al. (1986) spray generation function is the best one available for predicting spray production by whitecap bubbles". Because the interest of Andreas was mainly the importance of sea spray droplets in transferring heat and moisture across the air-sea interface, for which spume droplets are probably more important, he undertook to formulate a source function including both types of droplets based on work by Smith et al. (1993).

Below an overview of current sea spray source functions is presented, starting with a summary of the currently most frequently used formulations of Monahan et al. (1986) and Smith et al. (1993), complemented by work published in the last decade. Other formulations published prior to 1998 will not be considered because they appear either too high or too low for reasons discussed by Andreas (1998). In global circulation models, the sea-salt contribution is often introduced as wind-speed dependent concentrations rather than surface fluxes, based on work by, e.g., Erickson et al. (1986), which is therefore also discussed.

3.1. Monahan et al., 1986 (M86)

The M86 sea-salt generation function derived by Monahan et al. (1986) is based on the combination of laboratory measurements and field observations of the whitecap cover as function of wind speed. The laboratory experiments consisted of measurements of the size distributions of sea spray droplets produced in the head space of a whitecap simulation tank during the decay of one artificially generated breaking wave. Accounting for the volume of the head space of the tank and the whitecap area, the total number of droplets produced per increment droplet radius per unit whitecap area was determined. A commonly used expression for the fraction of whitecap cover (W) has been derived by Monahan and co-workers (cf. Monahan and O'Muircheartaigh (1986) for a review):

$$W(U_{10}) = 3.84 \times 10^{-6} \, U_{10}^{3.41} \qquad (Eq.2)$$

Multiplication of this number with the oceanic whitecap cover (Eq. 2) and the whitecap decay time (3.53 s) yields the required source function:

$$\frac{dF_{M86}}{dr} = 1.373 \, U_{10}^{3.41} \, r^{-3} \left(1 + 0.057 \, r^{1.05}\right) \times 10^{1.19 e^{-B^2}} \qquad (Eq.3)$$

where U_{10} is the wind speed at a height of 10 m expressed in m s^{-1}, B=(0.380-log r)/0.650, and dF/dr is given in m^{-2} µm^{-1} s^{-1}. No mention is made in Monahan et al. as regards the nature of the radius. However, since the measurements were made in a closed head space in contact with the water surface, we assume that the air was saturated with water vapour and thus the droplets did not evaporate and retained their radius r_o. Because the measurements used to derive M86 were made with an optical particle counter, which is sensitive in the range of about 0.3 to 10 µm (Monahan et al., 1982), M86 applies only to that size range. However, often M86 is extrapolated to either side of the droplet spectrum. Andreas et al. (1995) pointed out that Monahan's formulation could well describe the production of particles below r_{dry} = 0.5 µm, and it was altogether evaluated to give satisfactory results in a global transport model (Guelle et al., 2001).

In an attempt to add also the spume component to their source function, Monahan et al. (1986) introduced a second mode. However, application of the latter in models describing effects of sea spray aerosols does not give realistic results (e.g., Burk, 1984; Stramska, 1987; Andreas, 1990, Smith et al., 1990). M86 explicitly mention that 'the short-comings of their equation as a description of the spume droplet production term are such that its use is not recommended'. Therefore, the spume droplet contribution as described by M86 is not further considered.

3.2. Smith et al., 1993 (S93)

S93 derived a sea spray source function from field measurements, using a budget method that considers the atmospheric concentrations as the result of the balance between surface production and removal. This aerosol source function contains contributions from both bubble-mediated droplet production and spume droplets:

$$\frac{dF_{S93}}{dr_{80}} = \sum_{i=1,2} A_i \exp\left[-f_i\left(\ln\frac{r_{80}}{r_{0i}}\right)^2\right] \quad \text{(Eq.4)}$$

where f_i and r_{oi} are constants (f_1=3.1, f_2=3.3, r_{01}= 2.1 µm and r_{02}= 9.2 µm) and the amplitudes A_i are wind speed-dependent coefficients given by:

$$\ln(A_1) = 0.0676\ U + 2.43$$
$$\ln(A_2) = 0.959\ U^{1/2} - 1.476 \quad \text{(Eq.5)}$$

where U is the wind speed measured at a height of 10 m above the beach, approximately 14 m above mean sea level. The aerosol particle size distributions and meteorological parameters used to derive S93 were obtained at the same height. Eq.4 is valid for 1µm ≤ r_{80} ≤ 25 µm and wind speeds at 14 m between 0 and 34 m s^{-1}.

Andreas (1998) uses Eq.1, assuming neutral stratification, to give a correction factor for the wind speed in Eq.5, i.e.:

$$U_{14} = U_{10}\left(1 + \frac{C_{DN10}^{0.5}}{0.4} \ln\left[\frac{14}{10}\right]\right) \quad \text{(Eq.6)}$$

where the neutral-stability drag coefficient C_{DN10} has the value of $1.2 \cdot 10^{-3}$ for $4 \text{ m s}^{-1} \leq U_{10} \leq 11 \text{ m s}^{-1}$, and $C_{DN10} = 0.49 + 0.065\, U_{10}$ for $U_{10} > 11 \text{ m s}^{-1}$. Taking this correction into account, S93 applies for U_{10} up to 32.5 m s^{-1} (Andreas, 1998).

3.3. Andreas, 1998 (A98)

Andreas (1998) was particularly interested in larger sea spray particles, because of their effect on the air-sea fluxes of water vapour and heat. Therefore, he needed a source function that describes both bubble-mediated production and direct production. The latter is included in field data, such as those used by S93 to formulate their source function. S93 also has about the expected dependence of the production rate on the third power of the friction velocity. Therefore, Andreas decided to formulate a new source function based on M86, which he concluded to be the best available bubble-mediated source function, and S93 which is needed to include spume droplets with radii up to 50 µm. Because Andreas had concerns about possible sampling bias in the data used by S93, he focused on droplets with initial radii r_o between 4 µm and 15 µm (2 µm $\leq r_{80} \leq$ 7.3 µm) for which the M86 and S93 source functions have approximately the same shape. Comparison of the integrated volume over this size range for wind speeds up to 20 m s^{-1} shows that S93 needs to be multiplied by a factor of 3.5 to also quantitatively coincide with M86. Furthermore, Andreas had a concern about the underestimation of the spume droplets by S93, and therefore used S93 only for radii up to r_{80}=10 µm. For larger radii, following earlier work presented in Andreas (1992), Andreas extrapolated with a function based on data obtained by Wu et al. (1984) at 0.2 m above the surface. It is noted that the measurements by Wu et al. (1984) were made in wind speeds of 6-8 m s^{-1} and the application of this data has been questioned by Katsaros and De Leeuw (1994). See also Andreas (1994) as regards this issue. The A98 source function is thus formulated as:

$$\frac{dF_{A98}}{dr_{80}} = \begin{cases} 3.5 * \dfrac{dF_{S93}}{dr_{80}} & \text{for } r_{80} \leq 10\,\mu m \\ C_1(U_{10})\, r_{80}^{-1} & \text{for } 10 \leq r_{80} \leq 37.5\,\mu m \\ C_2(U_{10})\, r_{80}^{-2.8} & \text{for } 37.3 \leq r_{80} \leq 100\,\mu m \\ C_3(U_{10})\, r_{80}^{-8} & \text{for } 100 \leq r_{80} \leq 250\,\mu m \end{cases} \quad \text{(Eq. 7)}$$

where S93 is used including the correction given by Eq.6. C_1, C_2, C_3 are wind speed dependent coefficients, which are computed in an iterative procedure, given in Andreas (1998).

3.4. Smith and Harrison, 1998 (SH98)

Smith and Harrison (1998) presented a modification of the S93 source function based on new experimental data, which is formulated as:

$$\frac{dF_{SH98}}{dr} = \sum_{i=1}^{2} A_i \exp\left(-f_i \ln\left[\frac{r}{r_{0i}}\right]^2\right) \quad \text{(Eq.8)}$$

where $r_{01} = 3\,\mu m$ and $r_{02} = 30\,\mu m$, $f_1 = 1.5$ and $f_2 = 1$, and coefficients A_1 and A_2 are approximated by $A_1 = 0.2\, U_{10}^{3.5}$ and $A_2 = 6.8\,10^{-3}\, U_{10}^{3}$.

3.5. Pattison and Belcher, 1999 (PB99)

Pattison and Belcher (1999) developed a model to simulate droplet trajectories over the sea for detailed studies on the behaviour of the droplets and their contribution to heat and mass transfer. The turbulent air flow over the waves was simulated, and droplet evaporation was accounted for. The rate at which the droplets are generated was determined by releasing large numbers of droplets (at an arbitrary height of 1 m above the surface), and tracking their subsequent motion and size. The results were compared with field data from De Leeuw (1986, 1990) and Smith et al. (1993). The droplet production rate derived from this exercise is:

$$P = 69.2 \, U_{10}^{3.25} \, D^{-2.74} \qquad (Eq.9)$$

where P is expressed in $m^{-2} \, \mu m^{-1} \, s^{-1}$ and the droplet diameter D in µm. The merit of this source function is discussed in Andreas et al. (2001).

3.6. De Leeuw et al., 2000 (DL00)

De Leeuw et al. (2000) determined a quantitative source function for sea spray aerosol produced by waves breaking in the surf zone from data collected with optical particle counters at two locations on the Californian coast. Three optical particle counters were used to measure profiles at the base of a pier; a fourth instrument was used at the end of the pier. Thus the surf-produced aerosol could be separated from the advected 'background' aerosol. Careful calibration and inter-comparisons of the instruments were made to avoid systematic errors. Concentrations of droplets with *in situ* diameters from 0.5 to 10 µm measured downwind from the surf, in wind speeds of up to 9 m s^{-1}, were 1-2 orders of magnitude higher than those upwind. Surf aerosol concentration gradients and plume heights vary with particle size and with wind speed. The concentrations were integrated over the plume height to determine surface fluxes. De Leeuw et al. (2000) proposed a single equation which approximately describes the surf source functions for wind speeds up to 9 m s^{-1}:

$$\frac{dF_N}{dD} = 1.1 \times e^{0.23 \times U_{10}} \times D^{-1.65} \qquad (Eq.10)$$

This simple model applies to particles with diameters D at formation between 1.6 and 20 µm. For comparison with other source functions, the difference in whitecap cover should be accounted for using e.g., W = 3.84x10^{-6} U$_{10}^{3.41}$ (Monahan and O'Muircheartaigh, 1986). The surf is considered a 100% whitecap.

3.7. Vignati et al., 2001 (V01)

Vignati et al. (2001) developed a coastal aerosol transport model (CAT) to evaluate the effect of sea spray production in the surf zone, the evolution of the aerosol composition in off-shore flow, and the effect of sea spray on the atmospheric concentrations of nitric acid over the coastal seas. Because all these processes are strongly connected with the production of sea spray, a source function was required. Although coastal effects were the focus, fetch dependent sea spray production was not considered and open ocean source functions were applied. Both M86 and S93 were included. As mentioned above, these source functions give similar effects for an intermediate size range. For the application in CAT, in particular for the evaluation of the surf zone effects for which only a limited range of wind speeds were available, the source functions were evaluated for a wind speed of 9 ms^{-1}. They are similar for particles with radii between 4 and 8 µm. For larger particles, S93 provides higher fluxes, for smaller particles the curves diverge rapidly. Comparison with surf zone source functions indicates that M86 applies better for smaller particles, whereas S93 appears to better describe the larger particles (De Leeuw et al., 2000). The comparison with experimental data from O'Dowd et al. (1997) shows that simulations using M86 overestimate the concentrations of the smallest particles. Therefore, V01 sought a practical solution in the form of an effective source function based on data presented in O'Dowd et al. (1997). These measurements were made on a ship on the North Atlantic, where coastal effects have no influence. CAT was used to iterate the source function until the data presented in O'Dowd et al. (1997) were reasonably well reproduced. The resulting source function is a sum of three log normal distributions:

$$\frac{dF_{V01}(\log r_{80})}{d\log r} = \sum_{i=1}^{3} \frac{N_i}{\sqrt{2\pi} \log \sigma_i} \exp\left(-\frac{(\log r_{80} - \log R_i)^2}{2\log^2 \sigma_i}\right) \quad \text{(Eq.11)}$$

with parameters shown in Table 1, and dlog r = 0.1. Note that with respect to the original publication the unit of N_i and the exponential factor of N_2 have been corrected. V01 is in good agreement with M86 for particles with radii larger than 0.1-0.2 µm, and with S93 for particles larger than about 10 µm. V01 applies to a relative humidity of 80%, and for wind speeds of 6-17 m s^{-1}.

Table 1. Parameters determined for the V01 source function (Eq. 11).

i	Number, N_i (cm^{-2}s^{-1})	Radius, R_i (μm)	Standard deviation, σ
1	$10^{(0.095U+0.283)}$	0.2	1.9
2	$10^{(0.0422U-0.288)}$	2	2
3	$10^{(0.069U-3.5)}$	12	3

3.8. Guelle et al., 2001 (G01)

The goal of Guelle et al. (2001) was to provide a source function for use in a GCM based upon previous work (M86, S93 and SH98), to represent particles with dry radii from 0.03 to about 64 μm. G01 presents different yearly model experiments to document the influence of different source functions on the global sea-salt loads of mass, aerosol surface and number. The source flux is divided into 7 bins for M86 (0.031- 4μm) and 4 bins for either SH98 or A98 (4-64 μm). Extensive comparison to data let the authors suggest that a composite source function of M86 and SH98 is probably the best solution. The work proposes to shift from M86 to SH98 at 4μm dry radius. The simulation shows that the fine sea-salt aerosol, produced with M86, is very nicely reproduced. Sea-to-land fluxes are validated herein by using wet deposition measurements in inner continental areas of the USA.

3.9. Reid et al., 2001 (R01)

Reid et al. (2001) undertook airborne measurements of the column burden of sea spray aerosol in off-shore wind, as the air mass was advected out to sea. They observed the development of the internal marine boundary layer over a fetch of 60 km. From the change in the column burden of sea-salt particles (dc_{col}), R01 derived the upward flux F_u:

$$\frac{dc_{col}}{dt} = F_u - F_d = F_u - V_d c_{sfc} \qquad (Eq.12)$$

where F_u is the upward flux and F_d is the downward dry deposition flux taken as the product of the dry deposition velocity V_d and the surface concentration c_{sfc}. Deposition velocities used were from Slinn and Slinn (1980). Effects of entrainment were negligible.

Because R01 had data in off-shore flow for only three wind speeds, 4, 8 and 12 m s^{-1}, where the lowest wind speed produced negligible sea spray aerosol, R01 did not undertake to parameterize their data and presented only the number for the two highest wind speeds.

3.10. Nilsson et al., 2001 (N01)

Nilsson et al. (2001) applied an eddy-covariance (EC) system to directly measure aerosol fluxes over the open ocean, from an ice breaker in the Arctic Ocean. The EC system consisted of a TSI 3010 condensation particle counter (CPC) and a Gill Solent 1012R research model ultrasonic anemometer. The CPC measured the number concentration (N) of aerosol particles with diameter D_p>10 nm, with a first order response time constant of 0.8 s. The vertical wind component (w) was sampled at 20.8 Hz.

The measured flux $\overline{N'w'}$, where the overbar denotes a time average and N' and w' are the turbulent fluctuations of N and w (e.g., $N = \overline{N} + N'$, $w = \overline{w} + w'$), is the sum of upward and downward fluxes. Hence the source flux was determined as:

$$F = \overline{N'w'} + \int v_d(D_p)\,\overline{N}(D_p)\,dD_p \quad \text{(Eq.13)}$$

where $v_d(D_p)$ is the dry deposition velocity (negative downward) and $\overline{N}(D_p)$ is the particle size distribution. N01 neglected entrainment effects because their contribution would only be a few percent. To obtain the surface flux from their measurements, the deposition term had to be subtracted from the observed flux. The deposition velocity $v_d(D_p)$ was calculated in terms of the aerodynamic resistance through the surface layer and the resistance through the quasi laminar sublayer (e.g., Seinfeld and Pandis, 1998). For the latter, the parameterization by Schack et al. (1985) was used. For the evaluation of the deposition term, measured values for the micro-meteorological parameters and the particle size distributions were used. This resulted in the following parameterization for the source flux:

$$\log F = k_F \overline{U_{10}} - m_F \qquad (Eq.14)$$

where $k_F = 0.20 \pm 0.06$ (r = 0.85) and $m_F = -1.71$. F is expressed in 10^6 m^{-2} s^{-1}. The U dependency can result both from increased bubble production with increased U (see discussion below), and from the increase in turbulent transport from the ocean surface to the measurement level with increased U. The wind-driven aerosol source flux consisted of a film drop mode centred at ~100 nm diameter and a jet drop mode centred at ~1 µm diameter.

3.11. Erickson et al., 1986 (E86)

E86 combined work from different authors on relationships between surface wind speed and sea-salt concentrations. It is as such not a source function but rather a parameterisation of the sea-salt aerosol load as a function of wind speed.

Sea-salt concentrations C_s in surface air (unit µg/m^3) are prescribed herein to depend only on surface wind speed (u in m/s) at a height of 15m. Concentrations can be directly measured with different techniques, whereas sea-salt emission fluxes are usually derived from such measurements (e.g. S93, A98, DL00, V01, R01). Concentrations often determine the effects of the aerosols and are therefore calculated in the models from either the fluxes and the governing transport processes or directly from the wind speed using relations such as derived by Erickson et al. (1986). This relationship is given for two different wind regimes:

$$\begin{aligned} C_s &= \exp^{(0.16u+1.45)} & u < 15 \text{ m/s} \\ C_s &= \exp^{(0.13u+1.89)} & u > 15 \text{ m/s} \end{aligned} \qquad (Eq.15)$$

To calculate important physical properties of the sea-salt aerosol such as deposition and optical effects, Erickson and Duce (1988) suggested to complement Eq.15 by calculating the mass median radius (unit [µm], defined for 80% relative humidity and 15 m above sea level) of the sea-salt size distribution as a function of wind speed:

$$\mathrm{mmr}_{80} = 0.422\ u + 2.12 \qquad \text{(Eq.16)}$$

Using equations (15) and (16), an emission flux can be derived in a model in an indirect way. Since emissions must balance deposition fluxes, the total emission flux can be calculated after a sufficiently long model run. Instantaneous emission fluxes are calculated after each time step, accounting for the loss of sea-salt due to dispersion and deposition from the surface layer air. A collection of different parameterizations of sea spray concentrations as function of wind speed published by Gong et al. (1997) shows that the concentrations vary by one order of magnitude, and significant differences are reported for the variations of the concentrations with wind speed. This implies that a single concentration-wind speed relationship such as presented by Eq.15 is not representative for the whole world, and an application in a GCM will yield spatially variable errors, which may or may not cancel. The derivation of more accurate source functions requires that other parameters than wind speed alone are accounted for.

4. COMPARISON OF SOURCE FUNCTIONS

Source functions have been compared by several authors (e.g., A98, V01 and G01). In Figure 1, the total sea-salt flux as a function of wind speed in the size range 3-5 µm, wet radius at 80% relative humidity, derived from four parameterizations is compared. Such a comparison is much dependent on the size range used. The comparison illustrates the similarity of the mass fluxes for particles in this size range calculated using M86 and SH98. It also illustrates the evolution of the Smith (1993) towards the Smith and Harrison (1998) function. Andreas (1998) predicts a considerable flux at low wind speeds, something which was debated e.g. in Guelle et al. (2001).

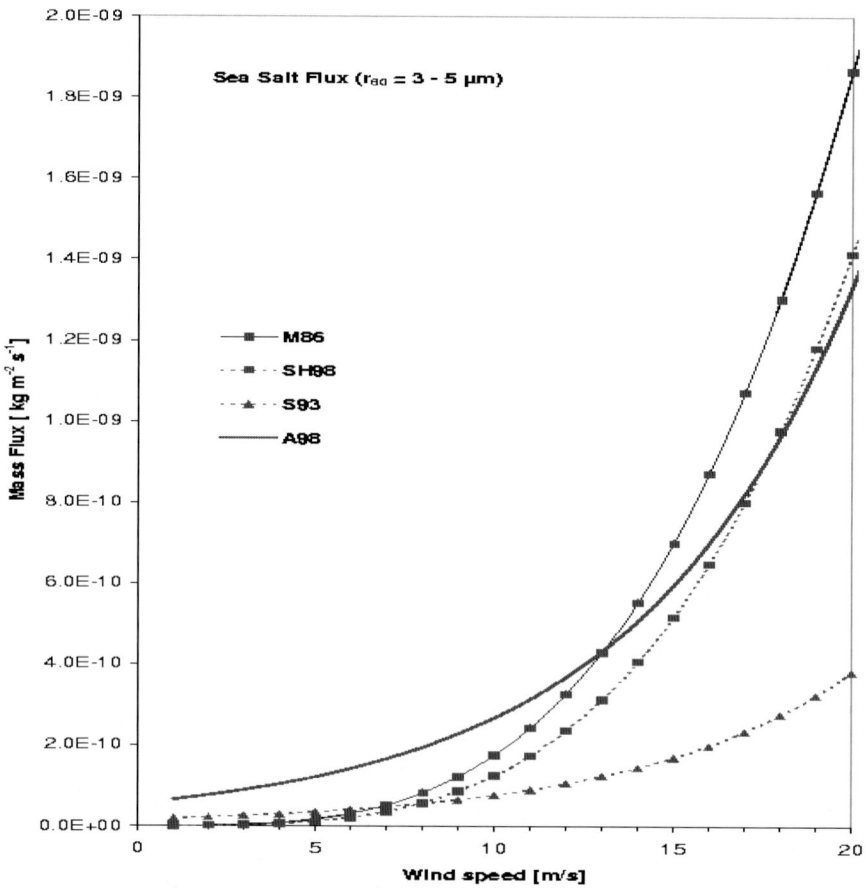

Figure 1. Sea-salt mass flux as a function of wind speed for wet particles at 80% relative humidity and four different source functions.

Figure 2 provides the size distributions for a given typical oceanic wind speed of 10 m/s. It shows that for larger particles, SH98 provides higher fluxes, for smaller particles the curves diverge rapidly. Comparison with surf zone source functions indicates that M86 applies better for smaller particles, whereas SH98 appears to better describe the larger particles (De Leeuw et al., 2000), which supports arguments presented by Andreas (1998) concerning the application of M86 and SH98. However, comparison with experimental data (Vignati et al., 2001) shows that simulations using M86 overestimate the concentrations of the smallest particles. It is noted that in this comparison M86 is extrapolated to sizes much smaller than those for which M86 was derived. SH98 and A98 do not apply to particles smaller than $r_{dry}=0.5$ µm. For very large particles (4 µm $\leq r_{dry} \leq$ 64 µm) the mass produced in A98 is greater than SH98. For very large particles the SH98 formulation produces the highest mass flux.

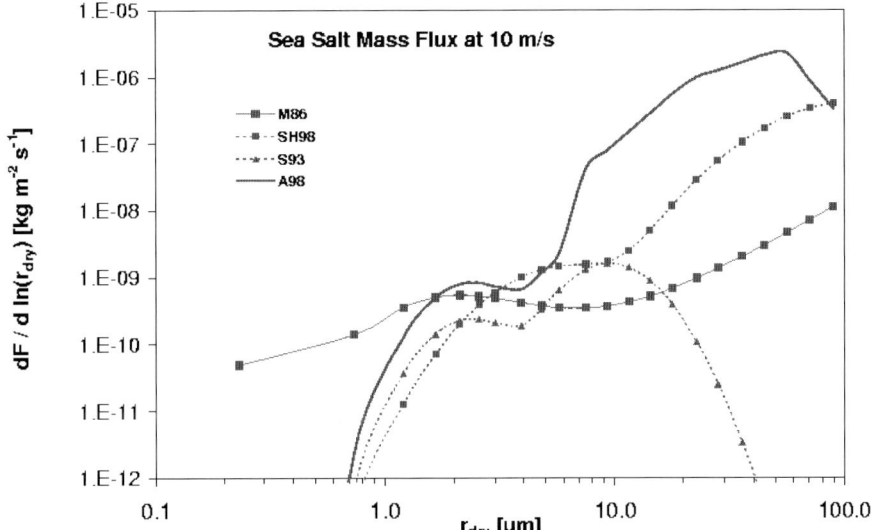

Figure 2. Mass size distribution for the same functions as in figure 1, at 10m/s surface wind speed.

5. GLOBAL EMISSIONS

Some recent estimates to illustrate the range of sea-salt fluxes to be expected with the source functions described above are presented in table 2. The compilation shows that only a few source functions have been used to derive a global sea-salt emission flux. But also those that seem to use the same source function are hardly comparable. The major reason is that the upper maximum of the size distribution is a critical parameter. The fact that a substantial amount of the emitted sea-salt mass is produced as coarse aerosol particles, renders the comparison difficult because of the different size ranges used in the models. We have split for illustration the total emissions of the work of (Guelle et al., 2001) into two contributions. It becomes clear that the coarse aerosol mode with a mass mean radius of approximately 5.5 μm dominates over the "fine" aerosol mode, described mainly by M86.

Table 2. Overview on global sea-salt emission flux estimates.

Reference	Source function used	Global sea-salt flux (Tg yr^{-1})	Size range for particle radius (μm)
Erickson and Duce, 1988	E86	10000-30000	Lognormal distribution, mmr vary 3-7,5 μm (E88)
Tegen et al., 1997	E86	5900	6 bins between 2-16μm using E88
Takemura et al., 2000	E86	3530	10 bins between 0.1-10 μm using E88
Spillane et al., 1986	M86	1300	
Gong et al., 1997	M86	11700	8 bins between 0.03 – 8 μm
Gong et al., 1998	M86	3300	8 bins between 0.03 – 8 μm
Erickson III et al., 1999	M86	2779	8 bins between 0.03 – 8 μm
Grini et al., 2002	M86+S93	6500	16 bins between 0.03-25 um
This study based on Guelle et al., 2001	M86+SH98	2680 17100	mode 1 (mmr ~ 1 μm) mode 2 (mmr ~ 5.5 μm)

E86: aerosol mass is split or used as given in eq. 5 (Erickson and Duce, 1988)

The studies which have used the concentration-wind speed relationship of E86 bear specific uncertainties. While surface concentrations might be well represented in these models, the total flux is dependent on the resolution and on the vertical dispersion and dry deposition fluxes. Any large flux removing sea-salt tracer from the surface layer within a time step implies large emission fluxes as compensation to maintain concentrations. The parameterisation itself had not taken into account that a fraction of the observed sea-salt concentration is due to advected sea-salt from long range transport. Thus, the small particle flux is rather overestimated by this parameterisation. It is interesting to note that the uncertainty of the removal flux in the model is dependent on the accuracy with which the wind data are available. Problems of surface wind speed data are translated into the sea-salt size distribution emitted.

Finally, the wind speed data used in the global estimates pose a problem, because different climate models and wind reanalysis products are expected to have rather different surface wind speed distributions over oceanic areas.

It is difficult to know which part of the difference in source emissions documented in table 2 is due to this effect.

6. DISCUSSION

The documented source functions are difficult to evaluate. Data on sea-salt size distributions are obtained with instruments that do not measure the complete range of particle sizes. Specific problems arise with all instruments used. Optical particle counters do not differentiate between sea-salt and other aerosol species. Impactor samples require sampling over a long period of time, during which the meteorological situation often changes. All devices miss a portion of the very coarse aerosol, since isocinetic sampling poses a problem in high winds. By lack of a common standard or a calibration facility in which results from different instruments can be compared, different sampling methods may yield different results. Coastal data may be contaminated with surf-produced aerosol. When surface concentrations are used to derive a source function, long-range transported aerosol contributes an unknown part to the observed concentration. It is unknown, how e.g. wave spectra influence the source function in a generalized way.

Validation of sea-salt aerosol concentrations has been attempted using surface station data and wet deposition measurements inland. The published estimates of total emissions indicate, that a large uncertainty is associated with the sea-salt aerosol component. Future work should include more detailed budgets of the sea-salt cycle for selected size ranges. Model simulations should be encouraged to simulate specific campaigns.

However, as discussed in various recent papers (Guelle et al., 2001; Grini et al., 2002 and Vignati et al., 2001), it seems to be most reasonable to combine different source functions if one is interested in a wide aerosol size range (e.g. M86 + SH98). The more difficult choice has to be made in the coarse particle range, where validation would be also more difficult. Recently, very fine particles have been suggested to be overlooked (Martensson et al., 2002), something that has not been tested in any large-scale model. It should also be noted that the application of any source function to small-scale problems, such as the surf-produced sea-salt aerosol, would require specific attention and adaptation.

7. REFERENCES

Andreas, E.L., Model estimates of the effects of sea spray on air-sea heat fluxes. In: Mestayer, P.G., E.C. Monahan and P.A. Beetham (Eds.), Modeling the fate and influence of marine spray. Marine Sciences Inst., Univ. of Connecticut, Avery Point Groton, CT 06340, pp. 17-28,1990.

Andreas, E.L., Sea spray and the turbulent air-sea heat fluxes, J. Geophys. Res., 97, 11 429-11 441, 1992.

Andreas, E.L., Reply, J. Geophys. Res., 99, 14345-14 350, 1994.

Andreas, E. L., J. B. Edson, E. C. Monahan, M. P. Rouault, and S. D. Smith, The spray contribution to net evaporation from the sea: A review of recent progress, Boundary Layer Meteorol., 72, 3-52, 1995.

Andreas, E. L., A new sea spray generation function for wind speeds up to 32 m s^{-1}, J. Phys. Oceanogr., 28, 2175-2184, 1998.

Andreas, E.L., M.J. Pattison and S.E. Belcher, "Production rates of seas-spray droplets" by M.J. pattisoin and S.E. Belcher: Clarification and elaboration, J. Geophys. Res., 106, 7157-7161, 2001.

Blanchard, D.C., Electrification of the atmosphere by particles from bubbles in the sea. In: M. Sears (Ed.), Progress in Oceanography, Vol I, Pergamon, NY, pp. 73-197, 1963.

Blanchard, D.C., The production, distribution, and bacterial enrichment of the sea-salt aerosol. In: P.S. Liss and W.G.N. Slinn (Eds.), Air-sea exchange of gases and particles, Reidel, pp. 407-454, 1983.

Blanchard, D.C., The size and height to which jet drops are ejected from bursting bubbles in sea water, J. Geophys. Res. 94, 10999-11002, 1989.

Boers, R., J.B. Jensen and P.B. Krummel, Microphysical and radiative structure of marine stratocumulus clouds over the Southern Ocean: Summer results and seasonal differences. Quart. J. Royal Met. Soc., 124, 151-168, 1998.

Burk, S.D.,The generation, turbulent transfer anddeposition of the sea-salt aerosol, J. Atmos. Sci., 41, 3040-3051, 1984.

Dekker, H.J., and G. de Leeuw, Bubble excitation of surface waves and aerosol droplet production: a simple dynamical model, J. Geophys. Res., 98, 10223-10232, 1993.

De Leeuw, G.,Vertical profiles of giant particles close above the sea surface, Tellus, 38B, 51-61, 1986.

De Leeuw, G., Profiling of aerosol concentrations, particle size distributions and relative humidity in the atmospheric surface layer over the North Sea, Tellus, 42B, 342-354, 1990.

De Leeuw, G., F.P. Neele, M. Hill, M.H. Smith and E. Vignati. Sea spray aerosol production by waves breaking in the surf zone, J. Geophys. Res., 105, 29397-29409, 2000.

De Leeuw, G., and L.H. Cohen. Bubble size distributions on the North Atlantic and the North Sea. in Gas Transfer and water Surfaces, edited by M.A. Donelan, W.M. Drennan, E.S. Salzman, and R. Wanninkhof, pp. 271-277, AGU, 2001.

De Leeuw, G., L.H. Cohen, L.M. Frohn, G. Geernaert, O. Hertel, B. Jensen, T. Jickells, L. Klein, G. J. Kunz, S. Lund, M.M. Moerman, F. Müller, B. Pedersen, K. von Salzen, K. H. Schlünzen, M. Schulz, C. A. Skjøth, L.L. Sorensen, L. Spokes, S. Tamm and E. Vignati, Atmospheric input of nitrogen into the North Sea: ANICE project overview. Continental Shelf Research 21, 2073-2094, 2001.

Erickson, D.J., J.T. Merrill, and R.A. Duce, Seasonal estimates of global atmospheric sea-salt distributions, J. Geophys. Res., 91, 1067-1072, 1986.

Erickson, D.J., and R.A. Duce, On the global flux of atmospheric sea salt, J. Geophys. Res., 93, 14079-14088, 1988.

Erickson III, D.J., C. Seuzaret, W.C. Keene, and S.L. Gong, A general circulation model based calculation of HCl and CLNO2 production from sea salt dechlorination: Reactive Chlorine Emissions Inventory, J. Geophys. Res., 104, 8347-8372, 1999.

Fitzgerald, J.W., Approximation formulas for the equilibrium size of an aerosol particle as a function of its dry size and composition and the ambient relative humidity, J. Appl. Meteor., 14, 1044-1049, 1975.

Gerber, H.E., Relative - humidity parameterization of the Navy Aerosol Model (NAM), Naval Research Laboratory, Washington D.C., NRL Report 8956, 1985.

Gong, S.L., L.A. Barrie, J.M. Prospero, D.L. Savoie, G.P. Ayers, J.-P. Blanchet, and L. Spacek, Modeling sea-salt aerosols in the atmosphere 2. Atmospheric concentrations and fluxes, J. Geophys. Res., 102, 3819-3830, 1997.

Gong, S.L., L.A. Barrie, J.-P. Blanchet, and L. Spacek, Modeling size-distributed sea-salt aerosols in the atmosphere. An application using Canadian climate models, in Air Pollution Modeling and its Applications XII, edited by S.-E. Gryning, and N. Chaumerliac, Plenum Press, New York, 1998.

Grini, A., G. Myhre, J.K. Sundet, and I.S.A. Isaksen, Modeling the annual cycle of sea salt in the global 3D model Oslo CTM2: Concentrations, fluxes and radiative impact, J. of Climate, 15, 1717-1730, 2002.

Guelle, W., M. Schulz, Y. Balkanski and F. Dentener, Influence of the source formulation on modeling the atmospheric global distribution of sea salt aerosol, J. Geophys. Res., 106, 27509-27524, 2001.

Hoppel, W.A., G.M. Frick and J.W. Fitzgerald, The surface source function for sea-salt aerosol and aerosol dry deposition to the ocean surface, Accepted for publication in J. Geophys. Res., 2002.

IPCC, Climate Change 2001: The Scientific Basis, Contribution of Working Group I to the Third Assessment Report of the Intergovernmental Panel on Climate Change, J.T. Houghton, Y. Ding, D.J. Griggs, M. Noguer, P.J. van der Linden, X. Dai, K. Maskell, and C.A. Johnson (eds), Cambridge University Press, U.K. and New York, NY., USA, 2001.

Katsaros, K.B., and G. de Leeuw,Sea spray and the turbulent air-sea heat fluxes – Comments, J. Geophys. Res., 99, 14339-14343, 1994.

Kientzler, C.F., A.B. Arens, D. Blanchard and A.H. Woodcock, Photographic investigation of the projection of droplets by bubbles bursting at the water surface, Tellus, 6, 1-7, 1954.

Latham, J. and M.H. Smith, Effect on global warming of wind-dependent aerosol generation at the ocean surface, Nature, 347, 372-373, 1990.

Leifer, I., and G. de Leeuw. Gas transfer at water interfaces. in Gas Transfer and water Surfaces, edited by M.A. Donelan, W.M. Drennan, E.S. Salzman, and R. Wanninkhof, pp. 303-309, AGU, 2001.

MacIntyre, F., Flow patterns in breaking bubbles, J. Geophys. Res., 77, 5211-5228, 1972.

Marks, R., Preliminary investigations on the influence of rain on the production, concentration, and vertical distribution of sea salt aerosol, J. Geophys. Res., 95, 22299-22304, 1990.

Mårtensson, M., E. D. Nilsson, G. de Leeuw, L.H. Cohen, and H-C Hansson, Laboratory simulations of the primary marine aerosol generated by bubble bursting, Accepted for publication in J. Geophys. Res., 2002.

Medwin, H. and N.D. Breitz, Ambient and transient bubble spectral densities inquiescent seas and under spilling breakers, J. Geophys. Res. 94, 12751-12759, 1989.

Monahan, E. C., K. L. Davidson and D. E. Spiel, Whitecap aerosol productivity deduced from simulation tank measurements, J. Geophys. Res. 87, 8898-8904, 1982.

Monahan, E.C., C.W. Fairall, K.L. Davidson and P.J. Boyle, Observed inter-relations between 10 m winds, ocean whitecaps and marine aerosols, Quart. J. R. Met. Soc. 109, 379-392, 1983.

Monahan, E. C., D. E. Spiel, and K. L. Davidson, A model of marine aerosol generation via whitecaps and wave disruption, in Oceanic Whitecaps and Their Role in Air-Sea Exchange, E. C. Monahan and G. Mac Niocaill. Eds., D. Reidel, 167-174, 1986.

Monahan, E.C. and I.G. O'Muircheartaigh, Whitecaps and the passive remote sensing of the ocean surface, Int. J. Remote Sensing, 7, 627-642, 1986.

Monahan E.C. and H.G. Dam, Bubbles: An estimate of their role in the global oceanic flux of carbon, J. Geophys. Res., 106, 9377-9383, 2001.

Nilsson, E.D., Ü. Rannik, E. Swietlicki, C.Leck, P.P. Aalto, J. Zhou and M. Norman,Turbulent aerosol fluxes over the Arctic Ocean 2. Wind-driven sources from the sea, *J. Geophys. Res., 106, 32139-32154, 2001.*

O'Dowd, C.D., M.H. Smith, I.E. Consterdine, and J.A. Lowe, Marine aerosol, sea salt, and the marine sulphur cycle: a short review, Atmospheric Environment 31, 73-80, 1997.

O'Dowd, C.D., J.A. Lowe and M.H. Smith, Coupling of sea-salt and sulphate interactions and its impact on cloud droplet concentration predictions, Geophy. Res. Lett., 26, 1311-1314, 1999a.

O'Dowd, C.D., J.A. Lowe and M.H. Smith, Observations and modelling of aerosol growth in marine stratocumulus, Atmos. Environ., 33, 3053-3062, 1999b.

O'Dowd, C.D., J.A. Lowe, N. Clegg N, M.H. Smith and S.L. Clegg, Modeling heterogeneous sulphate production in maritime stratiform clouds, J. Geophys. Res., 105 , 7143 – 7160, 2000.

Pattison, M.J. and S.E. Belcher, Production rates of sea-spray droplets, J. Geophys. Res.104, 18 397-18 407, 1999.

Penner, J.E., coordinator, Aerosols, their direct and indirect effects, in Contribution of Working Group I to the Third Assessment Report of the Intergovernmental Panel on Climate Change, J.T. Houghton, Y. Ding, D.J. Griggs, M. Noguer, P.J. van der Linden, X. Dai, K. Maskell, and C.A. Johnson (eds), Cambridge University Press, U.K. and New York, NY., USA, 2001.

Reid, J.S., H.H. Jonsson. M.H. Smith and A. Smirnov, Evolution of the vertical profile and flux of large sea-salt particles in the coastal zone, J. Geophys. Res., 106, 12,039-12,053, 2001.

Schack, C. J., S. E. Pratsinis, and S. K. Friedlander, A general correlation for deposition of suspended particles from turbulent gases to completely rough surfaces, *Atmos. Environ., 19*, 953-960, 1985.

Schulz M., J. van Beusekom, K. Bigalke, U. Brockmann, W. Dannecker, H. Gerwig, H. Grassl, C.-J. Lenz, K. Michaelsen, U. Niemeier, T. Nitz, E. Plate, T. Pohlmann, T. Raabe, A. Rebers, V. Reinhardt, M. Schatzmann, K.H. Schlünzen, R. Schmidt-Nia, T. Stahlschmidt, G. Steinhoff, K. von Salzen, The atmospheric impact on fluxes of matter and energy in the German Bight, Dt. Hydrogr. Z., 51, 133-154, 1999.

Seinfeld, J.H., and S. Pandis, Atmospheric Chemistry and Physics from Air Pollution to Climate Change, John Wiley and Sons, New York, 1326 pp., 1998.

Slinn, S.A. and W.G.N. Slinn, Predictions for particle deposition on natural waters, Atmospheric Environment, 14, 1013-1016, 1980.

Smith, M.H., M.K. Hill, P.M. Park and I.E. Consterdine, Aerosol concentrations and estimated fluxes ver the sea. In: Mestayer, P.G., E.C. Monahan and P.A. Beetham (Eds.), Modeling the fate and influence of marine spray. Marine Sciences Inst., Univ. of Connecticut, Avery Point Groton, CT 06340, pp. 17-28, 1990.

Smith, M. H., P. M. Park, and I. E. Consterdine, Marine aerosol concentrations and estimated fluxes over the sea, Q. J. R. Meteorol. Soc., 119, 809-824, 1993.

Smith, M. H., and N. M. Harrison, The sea spray generation function, J. Aerosol Sci., 29, Suppl. 1, S189-S190, 1998.

Spiel, D.E., The number and size of jet drops produced by air bubbles bursting on a fresh water surface, J. Geophys. Res., 99, 10,289-10,296, 1994a.

Spiel, D.E., The sizes of jet drops produced by air bubbles bursting on sea- and fresh water surfaces, Tellus, Ser. B, 46, 325-338, 1994b.

Spiel, D.E., On the births of jet drops from bubbles bursting on water surfaces, J. Geophys. Res., 100, 4,995-5,006, 1995.

Spiel, D.E., More on the births of jet drops from bubbles bursting on seawater surfaces, J. Geophys. Res., 102, 5,815-5,821, 1997.

Spiel, D.E., On the births of film drops from bubbles bursting on seawater surfaces, J. Geophys. Res., 103, 24,907-24,918, 1998.

Spillane, M.C., E.C. Monahan, P.A. Bowyer, D.M. Doyle, and P.J. Stabeno, Whitecaps and global fluxes, in Oceanic Whitecaps, pp. Monahan, E.C., Mac Niocaill G. (eds.), Reidel Publ.Comp., Dordrecht, 209-218, 1986.

Stramska, M., Vertical profiles of sea salt aerosol: A numerical model, Acta Geophys. Pol., 35, 87-100, 1987.

Stull, R., An Introduction to Boundary Layer Meteorology. Kluwer Academic Publishers, 666 pp., 1988.

Takemura, T., H. Okamoto, Y. Marujama, A. Numaguti, A. Higurashi, and T. Nakajima, Global three-dimensional simulation of aerosol optical thickness distribution of various origins, J. Geophys. Res., 105, 17853-17873, 2000.

Tegen, I., P. Hollrig, M. Chin, I. Fung, D. Jacob, and J. Penner, Contribution of different aerosol species to the global aerosol extinction optical thickness: Estimates from model results, J. Geophys. Res., 102, 23895-23915, 1997.

Vignati, E., G. de Leeuw and R. Berkowicz. Modeling coastal aerosol transport and effects of surf-produced aerosols on processes in the marine atmospheric boundary layer, J. Geophys. Res., 106, 20225-20238, 2001.

Wu, J., J. J. Murray, and R. J. Lai, Production and distributions of sea spray, J. Geophys. Res., 89, 8163-8169, 1984.

Woolf, D.K., P.A. Bowyer, and E.C. Monahan., Discriminating between the film drops and jet drops produced by a simulated whitecap, J. Geophys. Res., 92, 5142-5150, 1987.

Use of Isotopes

Valérie Gros, Carl A.M. Brenninkmeijer, Patrick Jöckel, Jan Kaiser, Dave Lowry, Euan G. Nisbet, Phillip O'Brien, Thomas Röckmann, and Nicola Warwick

1. INTRODUCTION

Isotope analysis is becoming more and more important in atmospheric chemistry, because it gives additional information about trace gas budgets and reaction pathways, complementary to concentration and flux measurements. The present chapter describes in detail how isotopes are used to obtain information about the sources of three important reactive atmospheric gases, namely methane (CH_4), carbon monoxide (CO) and nitrous oxide (N_2O). With respect to isotopes in CO_2, we refer to the following papers and the references therein: Ciais et al., 1995; Francey et al., 1995; Keeling, 1995; Bousquet et al., 1999; Rayner et al., 1999; Yakir and Sternberg, 2000.

The isotopic composition of a sample (SA) is usually expressed in terms of isotope ratios R (e.g., D/H, $^{13}C/^{12}C$, $^{17}O/^{16}O$, $^{18}O/^{16}O$ or $^{15}N/^{14}N$, where the isotope symbol is understood to represent its abundance). Since natural variations in isotope ratios are normally small, they are reported in the δ notation relative to a standard (ST), defining $\delta = (R_{SA}/R_{ST} - 1) \cdot 1000‰$. For the initial measurement, an arbitrary, laboratory-specific working standard can be chosen. However, in order to compare results from different laboratories, calibration of the working standard against a reference material of known isotopic composition is required. These reference materials define internationally agreed scales relative to which the isotopic composition of the sample is finally reported.

Atmospheric N_2 is the primary international standard for ^{15}N abundance (Mariotti, 1983; Mariotti, 1984). Hydrogen isotope ratios are reported relative to Vienna Standard Mean Ocean Water (VSMOW). For oxygen isotopes, VSMOW, CO_2 derived from Pee Dee Belemnite (VPDB-CO_2) and PDB itself are in use (VPDB) (Gonfiantini, 1978; IAEA, 1995). PDB was a fossil calcium carbonate from the Pee Dee formation in South Carolina, but the original material became exhausted decades ago. Therefore, the original PDB scale was replaced by a hypothetical VPDB (Vienna PDB) scale which is supposed to be identical to PDB. Carbon isotope ratios are also reported relative to VPDB. The VPDB scale was fixed by an internationally agreed assignment of exact δ values to NBS-19 (another carbonate reference material): $\delta^{13}C(NBS-19) \equiv +1.95‰$ and $\delta^{18}O(NBS-19) \equiv -2.20‰$ versus VPDB. A similar agreement was reached for interconversion between the two scales (Hut, 1987): $\delta^{18}O(VPDB) = 30.9‰$ versus VSMOW. VPDB-CO_2 is liberated from VPDB by treatment with 100% H_3PO_4 at 25 °C. The pertinent fractionation factor between the carbonate and the CO_2 gas was fixed at 1.01025, so that $\delta^{18}O(VPDB-CO_2) = 41.47‰$ versus VSMOW. Sometimes tropospheric O_2 is used as a reference for oxygen isotope ratios, since O_2 from background air at sites remote from local perturbations is known to have a relatively uniform enrichment of 23.8‰ versus SMOW (Coplen et al., 2002 and references therein). However, O_2 is not recommended as an international standard. Therefore, atmospheric N_2, VSMOW and VPDB are adopted as primary reference materials in the following text.

2. ISOTOPIC ANALYSIS OF METHANE

2.1. Introduction

Measurements of atmospheric methane have been made for over 30 years but the budget has still not been closed. The isotopic composition of methane has been measured for almost as long, but while many of the main methane sources were characterised isotopically during the 1980s, only in the last 5 to 10 years have long term high precision time series for background sites become available. New small sample, rapid, moderately high precision measurements are increasing the spatial coverage of isotope monitoring sites, providing valuable information to input into global models and to help verify inventories. Currently, several types of isotopic measurement are being carried out: 1) multi-sample time series to monitor global change; 2) spot sample population studies, coupled with back trajectory analysis, to identify distant and regional sources; 3) diurnal studies to characterise local source budgets; and 4) process studies of specific sources.

Use of Isotopes

The global background level of methane is still rising (see chapter 11) but the annual rate of this increase is not constant and there have been perturbations in the growth rate over the last 2 decades, most notably when the rate of change dropped below 0 ppb/yr in late 1992 (Dlugokencky et al., 1998). There is evidence that in parts of Europe at least, the emissions of methane are reducing: statistical emissions inventories suggest this, and measurements around the London region show this to be happening. It is important to know from independent atmospheric evidence, which sources are reducing or increasing emissions, and one way is to measure the isotopic signatures of the methane. There are 3 measurements that can be made: the ratios of the stable isotopes $^{13}C/^{12}C$ and $^{2}H/^{1}H$ (D/H), and the amount of the radiogenic isotope ^{14}C (in pmol). The $^{14}CH_4$ record is useful in the southern hemisphere as a tracer for non-fossil methane (Wahlen et al., 1989) but unfortunately in many parts of the northern hemisphere the emissions of $^{14}CH_4$ from PWR nuclear reactors are so large that the fossil/non-fossil methane emissions signature is smothered (e.g. Eisma et al., 1995).

The measurement of D/H ratio of methane has traditionally been a difficult and time-consuming experiment often requiring large volumes of air (up to $1m^3$; e.g. Levin et al., 1993; Bergamaschi et al., 1998a). These observations are useful for monitoring the changes in background air on seasonal to decadal time-scales (Quay et al., 1999; Bergamaschi et al., 2000a) and for identification of methane sources they can be useful when comparing measurements in carbon:hydrogen isotope space (e.g. Wahlen, 1993; Levin et al., 1993; Snover et al., 2000). The interpretation of methane sources using D/H data is dependent on the latitudinal and altitudinal variations in the D/H ratio of rainfall that contributes to water involved in methane-forming reactions. This can lead to wide D/H isotopic ranges for some widely distributed methane sources, and as yet, these processes are not fully understood.

The following sections will concentrate on the measurement and interpretation of ^{13}C data for methane, the most widely studied isotope, partly due to the relative ease of analysis and partly due to the less ambiguous interpretation of the data. For a recent comprehensive review of the global methane cycle including isotopes, see Breas et al. (2001).

2.2. Stable isotopic analysis of methane

An important aim of stable isotopic analysis of methane is to produce data with the highest precision possible. While high precision $\delta^{13}C$ data (replicates with 1σ <0.05‰) have many uses, data of low precision (1σ ~0.5‰) have limited use for source studies (see section 2.3.4). Traditionally the high precision analysis of $\delta^{13}C$ and δD in methane has involved the conversion of methane to CO_2 and water, the CO_2 being analysed directly and the water reduced to hydrogen as a second step.

The analysis of methane in background air by extraction and conversion has been approached in two different ways following the original method for background ^{13}C analysis of methane (Stevens and Rust, 1982; Quay et al., 1991): a) retain the methane in a molecular sieve trap at liquid nitrogen temperature allowing all additional species in the air or gas sample to pass through, then desorb the gas for conversion (e.g. Levin et al., 1993; Quay et al., 1999) and b) convert the methane dynamically using traps to remove all other carbon species before converting the methane in the air stream (Lowe et al., 1991; Tyler et al.,1994a; Sugawara et al., 1996; Lowry et al., 2001). Precision for $\delta^{13}C$ analysis of better than ±0.05‰ has been achieved using the dynamic extraction method (Lowe et al. 1999) but requires 5-60 litres of air (Lowe et al., 1991; Francey et al., 1999; Lowry et al. 2001), whereas D/H analysis by pre-concentration requires between 100 and 800 litres of air (Levin et al. 1993; Quay et al., 1999). CuO has been widely used as a reagent to oxidise the methane for source studies where methane concentrations are much higher than background ones with a reported reproducibility of ±0.13 for $\delta^{13}C$ and ±1.5‰ for δD (e.g. Revesz et al., 1995). Additional methods for methane analysis include the tuneable diode laser (TDL), which allows small samples to be analysed for ^{13}C albeit at low precision (±0.5‰, Bergamaschi et al., 1994; Bergamaschi and Harris, 1995).

In recent years continuous flow GC-IRMS (ConFlo) methods have been used for gas analysis (Merritt et al., 1995; Hilkert et al., 1999), and precisions of 0.1‰ and 1‰ are now being achieved for $\delta^{13}C$ and δD, respectively (Miller et al., 2002; Rice et al., 2001). The advantages of these ConFlo techniques are the much smaller air sample size (20mL is possible) and a more rapid analysis time. While most investigators looking at mechanistic source studies of methane emissions have already switched to this faster, easier technique, it still has to be proven for long term isotopic monitoring of methane at background air sampling stations.

2.3. Isotopic characterisation of anthropogenic and wetland methane sources

Methane sources have a wide range of $\delta^{13}C$ and δD signatures depending on the processes of formation (Table 1). The isotopic signatures of specific sources also vary according to the local conditions (e.g. Levin et al., 1993, 1999a). However, some generalisations can be made. Isotopically the biomass burning (pyrogenic) and fossil fuel (thermogenic) sources ($\delta^{13}C$ mostly of −52 to −16‰, δD mostly of −250 to −80‰) are distinct from the natural and agricultural (biogenic) sources ($\delta^{13}C$ −75 to −50‰, δD −360 to −270‰). At the upper end of the biogenic range are the landfill and waste sources ($\delta^{13}C$ −60 to −50‰, δD −320 to −270‰).

Although there is a similar trend in D/H as for $\delta^{13}C$, with the biogenic sources showing the greatest depletion and the pyrogenic sources the least depletion, there are large ranges reported for δD of the main sources. Better constraints on source ranges and relative ease of measurement have resulted in a much larger $\delta^{13}C$ data set for methane sources, but D/H are particularly useful in some case studies where $\delta^{13}C$ data would prove to be ambiguous (Wahlen, 1993). Such examples include CH_4 derived by different methanogenic pathways, distinguishing bacterial from early mature thermogenic gas (Whiticar, 1993), and identifying latitudinal variation in individual source types.

2.3.1. Pyrogenic sources

The pyrogenic sources show the smallest depletions in ^{13}C ($\delta^{13}C$ −30 to −10‰). Emissions are mostly from biomass burning, but there are minor emissions from vehicle fuel combustion processes and wood stoves (Chanton et al., 2000).

The $\delta^{13}C$ of biomass burning emissions is related to the signature of the source vegetation. C3 plants (approx. 95% of total), including nearly all trees and bushes, use the enzyme rubisco to make a three-carbon compound as the first stable product of photosynthesis. They preferentially select ^{12}C and hence strongly fractionate the stable carbon isotopes from the atmospheric CO_2 source (−8‰).

Table 1. Global methane emissions estimates, $\delta^{13}C$ and δD characteristics of anthropogenic and wetland source emissions to atmosphere

Methane sources	Emission, Tg/yr	$\delta^{13}C$, ‰	δD, ‰
Biomass burning	40±20		−210±16
C4 vegetation-dominated		−17±3	
C3 vegetation-dominated		−26±3	
Fossil fuel related	100±30		−175±10
Vehicle emissions	<5	−16±6 (1)	
		−28±3 (2)	
Coal mines	35±10	−35±3	
Natural gas	60±20		
North Sea		−34±3	
Siberia		−50±3	
Waste management systems	90±30		−293±20
Landfills	40±20	−53±2	
Domestic sewage treatment	25±5	−57±3	
Animal waste	25±5	−58±3	
Rice cultivation	80±60	−62±3	−323±18
Enteric fermentation	80±20		−305±9
C4 vegetation diet		−49±4	
C3 vegetation diet		−70±4	−358±15
Wetlands	160±50		−322±30
Florida Everglades		−55±3	
Eastern Canada		−63±3	
Western Siberia		−67±2	
Total methane emissions	540±100	−52.9	−283±13

Emission and $\delta^{13}C$ data are derived from Stevens (1988), Wahlen et al. (1989), Wahlen (1993), Levin et al. (1993 1999a), Sugawara et al. (1996), Hein et al. (1997), Bergamaschi et al. (1998ab), Quay et al. (1999), Rata (1999), Chanton et al. (2000), Allen et al. (2001), Bilek et al. (2001) and Nisbet (2001). δD data are from Snover et al. (2000) and references therein, except C3 enteric fermentation, which is from Bilek et al. (2001). See Breas et al. (2001) for comprehensive details of source studies prior to 2000. The divergent vehicle emission isotopic estimates are based on (1) new estimates for American automobiles (Chanton et al. 2000), and (2) estimates for German vehicles reported in the work of Levin et al. (1999a). Natural gas figures are based on gas in the distribution system, not well samples. Total methane emissions estimates are subject to large errors, more than 50% of which is due to poor estimates of natural emissions.

C3 plant matter typically has $\delta^{13}C$ between –33 and –22‰. C4 plants (including savanna grasses) evolved during periods of lower atmospheric CO_2 and developed a pre-concentration mechanism for carbon dioxide. These plants fractionate carbon much less than C3 plants, and hence have isotopic ratios only slightly depleted in ^{13}C relative to carbon dioxide in ambient air ($\delta^{13}C$ around –19 to –10‰). Burning emissions have a wide range of $\delta^{13}C$ values between the C4 plants (–20 to –15‰, e.g. Rata, 1999; Chanton et al., 2000) and the C3 plants (–30 to –22‰), but with a range within ±3‰ for any specific burn area. Biomass burning produces relatively D-enriched methane but ranging from –230‰ to –10‰, with recent work giving values of –225 to –184‰ for C3 fires (Snover et al., 2000) and -136 to –99‰ for C4 fires (Rata, 1999). These signatures are largely controlled by a terrestrial plant δD range of –140 to –20‰ with fractionations of up to 150‰ during burning (Snover et al., 2000).

Minor emissions of methane from vehicles are around –28±3‰ for $\delta^{13}C$ based on experiments in Heidelberg (Levin et al., 1993, 1999a). These emissions are related to the $\delta^{13}C$ of the hydrocarbon mix of the fuel used locally and the distribution and age of vehicle types.

The isotopic values of methane produced by combustion processes are considered to represent a 2-stage process (Chanton et al., 2000). CH_4 released from the vegetation during burning has a more ^{13}C-depleted signature than the source, by 2-3‰, which is followed by subsequent partial combustion of light hydrocarbons during the flaming phase of burning. This second phase can result in $\delta^{13}C$ values less depleted than the source vegetation if combustion is very efficient, due to preferential consumption of $^{12}CH_4$. Engines and wood stoves are high efficiency closed systems, which can produce very little methane. For example, the vehicles which have less efficient combustion give higher methane emissions which are more depleted in ^{13}C. The vehicles which are most efficient give values close to –10‰, but their emissions are so low that these vehicles are a negligible source of methane (Chanton et al., 2000). Forest or grass fires are less efficient, the smoldering fires producing more methane with ^{13}C depleted by 2-3‰ compared to the fuel and the flaming phase (Chanton et al., 2000). These processes are partly dependent on burn temperature, which is in excess of 800°C for the flaming phase.

δD of biomass burning methane is thought to be controlled by high temperature pyrolysis involving methyl radical extraction of hydrogen from organic material, a process that might have a large kinetic isotope effect (KIE) (Snover et al., 2000).

2.3.2. Thermogenic sources

The thermogenic sources comprise most of the fossil fuels and have isotopic emissions to the atmosphere with $\delta^{13}C$ mostly between –52‰ and –30‰. Emissions from coal mining have reported averages in the range –38 to –35‰ (e.g. Hein et al., 1997; Allen et al., 2001), close to the average for thermogenic natural gas leaks. North Sea gas (thermogenic origin) has values in the supply lines of –34±3‰ (e.g. Lowry et al., 2001), although well values are mostly in the range –40‰ to –25‰. These values are significantly less depleted in ^{13}C than the mix of Siberian gases supplied to most European countries east of the Rhine, and which averages close to –50‰ when used in the Heidelberg system, SW Germany (Levin et al., 1999a). At other sites in the network supplied with west Siberian gas the values are –50.5±2‰ (Levin et al., 1999a; Nisbet et al., 2001). This represents mixing of the shallow West Siberian gas, which is dominantly biogenic (–64 to –58‰), and deeper reserves, which are thermogenic (–50 to –38‰, Grace and Hart, 1986).

2.3.3. Biogenic sources

The methane isotopic composition of biogenic sources is largely controlled by methanogenic and methanotrophic bacteria and like the burning is normally a 2-stage process involving methane production and consumption. Methanogens normally produce methane by either CO_2-reduction or acetate fermentation. This results in methane that is depleted in ^{13}C or D relative to the initial source material. Isotopic analysis has been used to distinguish between the two processes because CO_2-reduction produces methane that is relatively enriched in D (δD -170 to –250‰) but depleted in ^{13}C ($\delta^{13}C$ –60 to –110‰) whereas acetate fermentation produces methane that is relatively depleted in D (δD < –250‰) but enriched in ^{13}C ($\delta^{13}C$ –50 to –65‰) (e.g. Happell et al. 1994). The second stage involves methane oxidation by methanotrophs or in the water column prior to release to the atmosphere; the residual methane emitted by for example landfills, rice paddies or wetlands being enriched in the heavier isotope.

Emissions to atmosphere from landfills have a tightly constrained range of $\delta^{13}C$ from –57 to –50‰ with averages varying from –55 to –51‰

(Wahlen, 1993; Levin et al., 1993; Bergamaschi et al., 1998ab; Lowry et al., 2001) dependent on the region of the world and the landfill practices in operation. Below ground and in gas recovery systems the values tend to be more depleted (–62 to –53‰; e.g. Bergamaschi and Harris, 1995) and extreme methane oxidation in topsoil cover layers can produce areas of low emission relatively enriched in ^{13}C (up to –35‰, Liptay et al., 1998; Chanton and Liptay, 2000; Lowry et al., 2001). Bacterial oxidation is generally greater during higher summer temperatures resulting in residual CH_4 emitted to atmosphere with $\delta^{13}C$ close to –40‰, compared to a higher proportion of methane released during winter with $\delta^{13}C$ around –52‰ (Chanton and Liptay, 2000).

The natural and agricultural biogenic sources have average $\delta^{13}C$ for emissions mostly in the range –65 to –62‰, but with large regional and seasonal variations. Emissions from ruminants are mostly in the range –65±5‰ (e.g. Bilek et al., 2001), but this is strongly controlled by diet. Emissions from those animals eating a C4-grass dominated diet (–12‰) range from –56 to –49‰ (Stevens, 1988; Levin et al. 1993) while those with a C3-grass dominated diet (–29‰) give emissions of –75‰. The dominant process is CO_2-reduction but elevated H_2 results in δD values similar to those produced by acetate fermentation (–358‰, Bilek et al., 2001).

Rice paddy methane averages around –62‰ (e.g. Tyler et al., 1994a; Bergamaschi, 1997; Chanton et al., 1997; Bilek et al., 1999), although direct emissions can be lower than this (–65‰). During flooding methane oxidation is enhanced. Any methane component not oxidised in the water when the paddies are flooded is enriched relative to the average, around –55‰ (Bergamaschi, 1997; Chanton et al., 1997).

The northern wetland emissions average –64‰ (e.g. Lowe et al., 1999), and flooded tropical wetlands average –55‰ (Happell et al., 1994, and references therein) although values can vary within the broad biogenic range depending on latitude and the degree of oxidation in soil horizons or surface water before emission (e.g. Happell et al., 1994; Hornibrook et al., 1997; Waldron et al., 1999). Wetlands also produce very D-depleted methane ranging from –380‰ to –250‰ (e.g. Levin et al., 1993; Wahlen, 1993; Waldron et al., 1999; Snover et al., 2000). Up to 60-70% of methane produced by wetlands can be oxidized before emission, the residual methane emitted being enriched in ^{13}C relative to the methane produced at –80 to –70‰. The degree of methane oxidation varies between geographical wetland zones. Northern wetlands south of 60-62°N generally have a longer emission season than those further north, and often greater oxidation, which may in part explain a ^{13}C-enrichment of approximately 4‰ recorded for the Hudson Bay wetlands of Canada compared to the western Siberian wetlands. Methane ebullation, occurring as rapid, short-lived events, possibly following thawing, may result in pulses of less oxidized, more ^{13}C-depleted methane than these averages, which are not detected during flux experiments (Waldron et al., 1999).

2.3.4. Precision of source isotope measurements

Emissions to atmosphere from most sources can be constrained to within ±3‰ on a national or regional scale. Detailed source studies often find wider ranges of values in localised areas or small variations at the diurnal to seasonal time-scale. Sub-ground surface analysis also reveals wide ranges for many sources, such as landfill, wetlands, sea-floor sediments and gas fields. The key here is that landfill and wetland methane emissions are oxidised to varying degrees in the soil horizons before being emitted to the atmosphere and are isotopically homogenized. With natural gas emissions most of the leaks are from the pipeline network or pumping stations. The gas in the pipelines (similarly with landfill gas extraction systems) is homogenised and so atmospheric emissions regionally have only a narrow range of values, such as the North Sea gas supply at –34±3‰ and the west Siberian emissions at –50±3‰.

For studies of sources larger than point emissions to be easily utilized in regional and global isotopic models it is important to estimate the integrated emission to atmosphere from the whole source site. As such, sampling downwind in the emissions plume from major sources or source regions, as well as in the middle of them, is essential. Examples of such experiments are being carried out in west Siberia to distinguish the wetland and natural gas emissions, with some success in distinguishing the characteristics of the two end members in this mix at –67‰ and –50‰, respectively (Nisbet, 2001). Similarly, on a more local scale it has been possible to distinguish the natural gas supply from landfill emissions in the London region of the UK (–34‰ and –52‰, respectively, Lowry et al., 2001). In addition to sampling such plumes the gas supply in London was measured each month from 1997 to 1999 for confirmation, giving an average of –34.2±1.4‰ (n=20).

2.4. Isotopic characteristics of methane in background air

Before the interpretation of excess methane in urban environments can begin, it is necessary to understand the seasonal changes in isotopic ratios observed at background stations. The background $\delta^{13}C$ ratio averaged –47.1‰ for 1998 and 1999 (Miller et al., 2002) but this value varies with latitude, generally in the range –47.9 to –46.8‰. During 1998 and 1999 the annual average difference between Barrow, Alaska and the South Pole was 0.6‰; in the same period the average for three northern hemisphere sites was –47.2‰ and the average for three southern hemisphere sites was –46.9‰ (Miller et al., 2002). These values are very different from the weighted average of all sources of approximately –53‰ (e.g. –52.3‰, Tyler et al., 1999a; –52.9‰; Lowe et al., 1999; –53.6‰, Snover et al., 2000). Most of this difference is due to the fractionation effect from methane destruction in the atmosphere by reaction with OH or Cl (e.g. Cantrell et al., 1990; Saueressig et al., 1995; Gupta et al., 1997). This fractionation with OH has been estimated at –5.4‰, although the most recent study suggests that it may be much lower (–3.9‰, Saueressig et al., 2001). Some of the discrepancy can also be accounted for by the soil absorption sink, which has a fractionation of ~22‰ (Tyler et al., 1994b). Lowe et al. (1999) and Allen et al. (2001) suggested that this second methane sink, at –69‰ shifts the methane mix to –52.0‰ prior to fractionation through removal by OH. An additional 0.6‰ of the offset may be caused by the lag time between the emission of methane and the isotopic equilibration at background levels (Lassey et al., 2000).

Long-term isotopic records have been maintained for four Pacific sites: Point Barrow, Alaska; Olympic peninsula, Washington; Mauna Loa, Hawaii (all since 1988); and Samoa (since 1990) (Quay et al., 1999). Other coastal background sites with long-term records are Baring Head, New Zealand (since 1989) and Scott Base, Antarctica (since 1991; Lowe et al., 1994, 1997); Alert, Canadian Arctic (since 1991; Levin et al., 1999b); Izaña, Tenerife (since 1991; Levin et al., 1999a; Bergamaschi et al., 2000a); Montana de Oro, California (since 1995; Tyler et al., 1999a); and Mace Head, Ireland (since 1995; Lowry et al., 1998, 2001). Data are also available for Cape Grim, Tasmania from 1992-1996, with measurements on archived air back to 1978 (Francey et al., 1999). Time series for continental background sites are rare, the most notable being for Schauinsland, Germany (Levin et al., 1999a) and Niwot Ridge, Colorado (Tyler et al., 1999b). With the advent of the new continuous flow analysis techniques more background sites are coming on line, so that Miller et al. (2002) were able to report 1998-99 data for Point Barrow, Niwot Ridge, Mauna Loa, Samoa, Cape Grim and the South Pole.

The methane mixing ratio in the northern hemisphere was increasing by 6-8ppb/yr on average (Dlugokencky et al., 1998; Quay et al., 1999) over the 1995-2000 period (see Chapter 11). An enrichment in ^{13}C at the rate of ~0.02‰/yr was also observed at the coastal background sites during the same period (Quay et al., 1999; Lowry et al., 2001). This global trend of enrichment was originally inferred by Stevens (1988), who combined data for several sites over the 1978-1987 period. Conversely, the continental time series showed a depletion of up to 0.1‰/yr between 1989 and 1995 (Gupta et al., 1996; Tyler et al., 1999b; Levin et al., 1999a).

In addition to these general long-term trends (Figure 1) there is a seasonal cycle. The amplitude of this cycle varies from 0.7-0.8‰ at high northern latitudes (Alert, Levin et al., 1999b) to 0.3‰ at Tenerife (Bergamaschi et al., 2000), and down to 0.1‰ in tropical regions and 0.2‰ in Antarctica (Lowe et al., 1999). In New Zealand the seasonality is ~0.3‰ (Lowe et al., 1994,1997; Lassey et al., 2000). The mid-latitudinal sites are most prone to variations in the seasonal cycle, since they receive winds from many directions during the year. For example, the northerly limit of movement of the Azores high-pressure system is affected by the position of the Polar Front. In some summers these high-pressure systems rarely reach Mace Head, and air arriving from Canada dominates. This has a significant effect on the seasonal isotopic pattern, with the result that isotopic values recorded at Mace Head can vary between those recorded at Alert and Tenerife, depending on wind direction. In air from the south-west, where high pressure cells originating at low latitudes are sampled, the influence of methane removal by OH is increased, which has the effect of enriching the ^{13}C of the background methane signal. Conversely, in air from the west/north-west, source emissions collected by the depressions traversing the Atlantic from Canada (e.g. wetlands) are observed at Mace Head, which has the effect of depleting the ^{13}C of the background signal.

Use of Isotopes

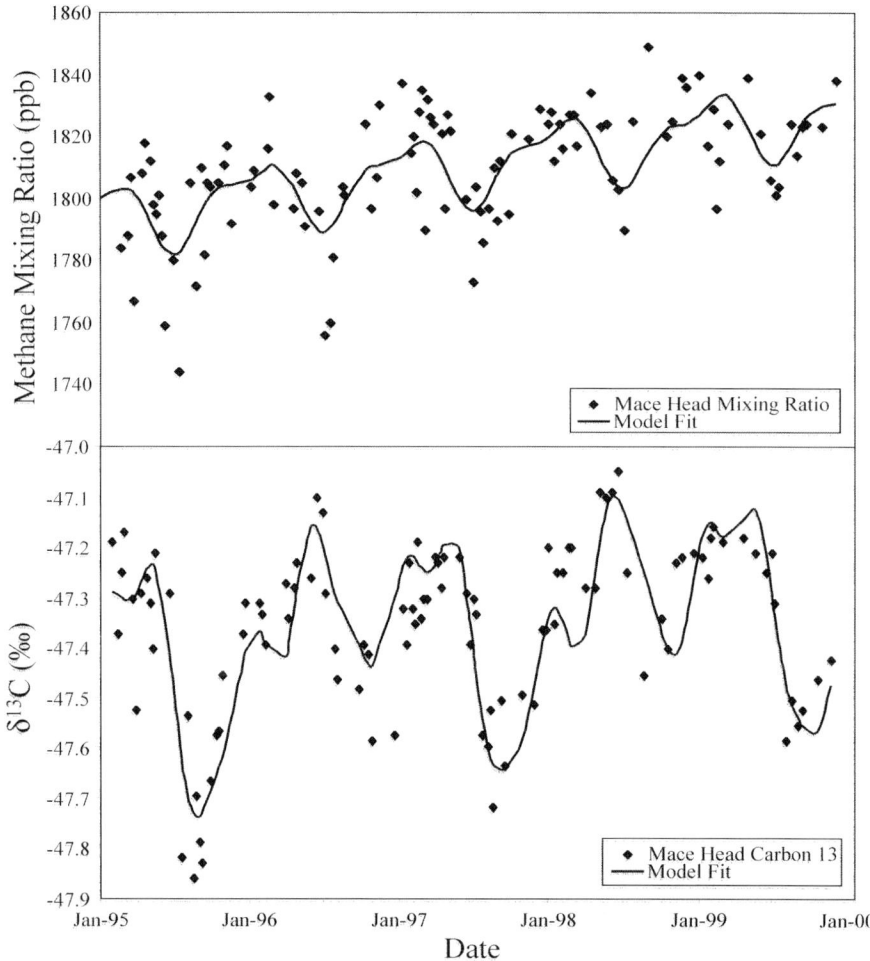

Figure 1. Mace Head mixing ratio (relative to NOAA standards) and isotopic records from 1995-2000 derived from tank samples collected for isotopic analysis. The curve fits (O'Brien, unpublished data) highlight the northern hemisphere growth rate of methane over this period of 6-8ppb/yr and the changing amplitude of the seasonal isotopic cycle.

The isotopic data from the network of background stations will greatly help to improve global models of methane emission and removal. Many sources are so widespread as to be difficult to sample in close proximity, and high precision isotopic background records, which are occasionally influenced by emissions plumes, can provide important information. For example, isotopic sampling to detect the seasonal and inter-annual variation in ^{13}C on Ascension Island (mid South Atlantic) caused by African biomass-burning emissions is currently underway.

δD isotopic time-series for background stations are rare (e.g., Quay et al., 1999; Bergamaschi et al., 2000a), although there are some spot samples from earlier decades (e.g. Wahlen et al., 1989). Values are in the range –100 to –70‰, with seasonal cycles of 3.5-10‰, depending on latitude. There is also an inter-hemispheric difference of ~10‰, being heavier in the north (Wahlen 1993; -91 and –81‰, Quay et al., 1999), possibly caused by greater biomass burning in the southern hemisphere (relatively enriched in δD), with more biogenic methane in the Northern hemisphere. The average δD for total sources has been estimated at –283 ±13‰ with the D/H fractionation during reaction with OH calculated as –163 to –255‰ (see Snover et al., 2000, and references therein).

Many recent studies have involved latitudinal or vertical sampling transects. For aircraft sampling the vertical transects have so far provided only limited isotopic data (e.g. Wahlen et al., 1989ab; Brenninkmeijer et al., 1995; Sugawara et al., 1996, 1997; Tyler et al., 1999ab; Mak et al., 2000). These suggest that there is a gradual shift from tropospheric methane levels of ~1800 ppb to ~900ppb moving upward from the tropopause, with an associated shift in δ^{13}C from –47.5‰ to –39‰ (e.g. Sugawara et al., 1997). Similarly the Trans-Siberian railroad transect (Bergamaschi et al., 1998a) provided only limited methane isotopic results, but both this study and the aircraft sampling of Sugawara et al. (1996) implied a light (δ^{13}C of –70 to –65‰) wetland source in Siberia, since confirmed by ground plume studies (Nisbet, 2001). Data from ship transects have provided information about the seasonal latitudinal gradients across the equatorial regions (e.g. Quay et al., 1999), most notably the New Zealand-California Pacific transect (Lowe et al., 1998; Bergamaschi et al., 2001). While in June there is very little inter-hemisphere difference, at –47.2‰, by late September / early October the difference is at its maximum, with δ^{13}C values of –47.0‰ for the southern hemisphere and –47.4‰ for the northern hemisphere. In general there is a 0.23‰ depletion in ^{13}C in the northern hemisphere relative to the southern hemisphere (Quay et al., 1999).

2.5. Using isotopes in air samples to identify sources

The background mixing ratio and isotopic trends for many sites in the Pacific and Atlantic regions are now known, and background patterns for many intermediate sites could be interpolated with a reasonable degree of accuracy. Any deviation from these background isotopic trends is usually interpreted as resulting from changes in emissions rather than perturbations in OH. These background trends can then be used as a starting point to understand emissions within broad latitudinal bands. For example the Mace Head record is probably typical of the North Atlantic background from 45-70°N and the dominant air movement is from west to east, so this baseline will apply to much of Northern Europe.

One of the keys to understanding the source puzzle is the use of back trajectories, normally recording air movements for 4 or 5 days before arrival at the sampling site. This provides some idea of the source regions causing any anomalies in methane mixing ratio. The isotopes then often provide the evidence of which source type is responsible for the excess methane. The isotopic signature of the source or source mix ($\delta^{13}C_{source}$) causing the methane mixing ratio to increase above a background with isotopic signature $\delta^{13}C_{background}$ is approximated by the lever rule (e.g. Thom et al., 1993):

$$\delta^{13}C_{source} \simeq (\delta^{13}C_{air} \cdot mr_{air} - \delta^{13}C_{background} \cdot mr_{background})/\Delta mr \qquad (Eq.1)$$

where Δc is the difference in mixing ratio (mr) between the selected background ($mr_{background}$) and the sampled air mass (mr_{air}). $\delta^{13}C_{air}$ shifts from $\delta^{13}C_{background}$ depending on the isotopic signature of additional methane from emissions sources that are mixed with background air (see section 3.3. for more details on this approach).

2.5.1 Regional sources

The reasons for the perturbations in methane growth rate from 1991 to 1992 (see chapter 11) are still unclear, although many suggestions have been made (see Lowe et al., 1997, and references therein), including reductions in natural gas leaks, wetland emissions, or biomass burning, or changes in OH due to the eruption of Mt. Pinatubo. The most compelling evidence comes from the southern hemisphere isotopic record (Lowe et al., 1997) which suggests that the ^{13}C depletion at the time of growth rate slowing can only be explained by a source with $\delta^{13}C$ of –22‰, and this coincides in parts of the

Amazon Basin and Africa with increased biomass burning in 1991 followed by drought (much reduced burning) in 1992. On a similar note it has been possible using the Mace Head record to estimate the source causing the excess methane in Canadian air masses compared to the baseline signal for air masses arriving from the SW within one week of the Canadian air. The shifts are small, resulting in errors often greater than ±10‰ on the calculation but consistently give $\delta^{13}C$ values of −64 to −62‰ (Lowry et al., 1998), within the range for Hudson Bay wetlands and Canadian Tundra (e.g. Wahlen et al., 1989).

When the air masses cross more densely populated regions the understanding of contributing sources of methane becomes more difficult. The ability to sample background and 'polluted' air masses at regular intervals and over long time series then becomes important. Information can still be gained from monitoring the isotopic change in a regional emissions plume over time by using excess over background. For example changes in the regional European source mix can be monitored in this way using samples collected at Mace Head. Air masses crossing central Europe and collected at Mace Head between 1996 and 2000 show a shift in isotopic value of the source mix for the excess methane from $\delta^{13}C$ of −54.1 to −55.6‰. This implies a reduction over the period in the proportion of fossil sources (e.g. coal) in the European region. The rate of change of this shift is likely to slow down as the change from coal to gas use slows, and best practice gas pipelines and landfill gas extraction systems are implemented. The shift observed using the Mace Head data confirms and broadens the implications of the isotopic shift from −47.4 to −52.9‰, observed in the source mix for the Heidelberg region between 1992 and 1995 (Levin et al., 1999a) using the Izaña station as background. This was attributed to reduction of coal mining emissions and a change in the source of the gas supply from North Sea to Siberia. This change in source mix in Europe by over 7‰ in 12 years confirms to some extent what inventory statistics such as CORINAIR 94 and EMEP have been suggesting: that major reductions have been occurring in landfill and coal emission. The isotopic shift over the 1986-1998 period suggested by the inventories would be approximately 3‰. Exchanging a coal source for a natural gas source has much the same isotopic characteristics. Many European countries, however have been replacing an enriched coal source with a much more depleted Russian gas source and therefore a greater rate of ^{13}C depletion in the emissions mix would be expected.

2.5.2. Local sources

When studying emissions on a more local scale in the urban environment the trajectories do not provide accurate information on the arrival direction of air masses at the sampling site, and local meteorological measurements are required. Urban studies have tended to concentrate on pollutants and CO_2 rather than methane. Fossil fuel and waste make up 30-35% of global CH_4 emissions and 45% of methane emissions in the European Community. These source categories are those which are easiest to reduce without causing significant social upheaval, but the emissions, particularly from landfills are poorly constrained. The isotopes provide one way of distinguishing emissions plumes from these sources and monitoring changes in emission over the long term.

Many studies have looked at the isotopes of methane emissions from major anthropogenic sources in industrial and urban areas such as Levin et al. (1993) on major German sources, Bergamaschi et al. (1998b), Liptay et al. (1998) and Chanton and Liptay (2000) on landfill sites, and Chanton et al. (2000) on vehicle emissions. There have been few studies analysing the isotopes in plumes some distance from these sources though. Moriizumi et al. (1998) used the isotopes of methane in urban air (Nagoya, Japan) at different times of day to calculate the contributions of fossil to non-fossil methane. Levin et al. (1999a) used a more comprehensive study of methane isotopes in the Heidelberg region to verify the Germany emissions inventories. Lowry et al. (2001) used a site at the western edge of London to verify London emissions and question recent revisions in UK landfill emissions estimates. Both the German and UK studies relied heavily on the availability of background records of methane isotopes and concentration in their interpretation of local data, utilising the Mace Head and Tenerife Atlantic background records and the Schauinsland continental interior background record, as well as using the baseline records from their own urban sites in the calculations.

In one diurnal experiment from summer 1995, Levin et al. (1999a) showed that the source mix for Heidelberg was −53.0‰ confirming the long term trend based on excess over background. Heidelberg is only a relatively small urban conurbation. London on the other hand is one of the largest cities in Europe with ~8 million people. In such a region the landfill emissions are almost 80% of the total, while gas leaks contribute most of the remainder (Lowry et al., 2001).

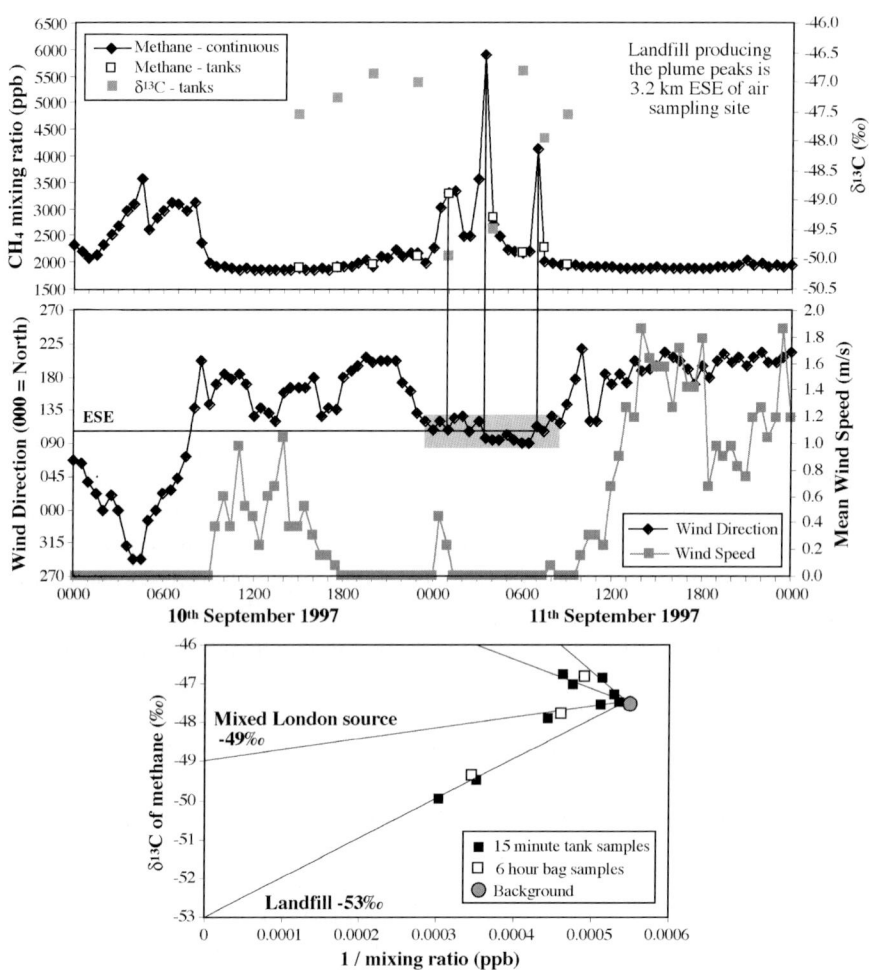

Figure 2. Diurnal experiment in September 1997 at the RHUL site in west London. Tanks were filled with air every 2 hours for isotopic analysis. Bags were filled slowly with air over an 18-hour period to give 6-hour integrated samples. This experiment recorded three significant methane peaks on an otherwise low build-up. The peaks were encountered each time the air movement was from the ESE. These landfill emissions were sampled throughout the 6-hour filling period of one bag, as shown by its position on the –53‰ source line.

Both main sources are well characterised isotopically from the distribution and extraction systems, from source plume studies and from plume sampling up to 4km away (gas –34‰, landfill –52‰). In addition to these sources there is a wider regional mix of emissions that is identified as a gradual build-up under inversion conditions (Figure 2). The calculated $\delta^{13}C$ for this mix was –48.7±0.3‰ during 1996 and 1997. This is significantly different to the –53‰ recorded at Heidelberg during 1995, but can be readily explained by the difference in source mixes. The main differences are the bigger proportion of agricultural emissions in the Heidelberg regional inventory and the use of Russian gas. In general though, the larger the urban region within a country the bigger the shift toward ^{13}C-enrichment in the source mix. More data from studies such as these is needed to verify global models of methane emissions as well as inventories. To be accurate, using isotopes for these calculations requires that there are significant isotopic differences between the main sources. The degree of certainty of the solution is also greatly increased if both sources that are enriched and sources that are depleted in ^{13}C relative to the background signal are present in the mix. In London this is the case with background at –47.2‰, gas at –34‰ and landfill at –52‰. Similarly promising results are also being obtained for the 2 main sources in west Siberia with a 17‰ difference between gas at –50‰ and wetland at –67‰ (Nisbet, 2001). This means that in countries supplied solely by a thermogenic gas supply (such as the UK) it is easiest to distinguish the proportions of fossil fuel to landfill emissions.

2.6. Can isotopes be used to verify methane emission inventories?

Official inventory estimates are calculated for many of the world regions as part of International Agreements, as detailed in chapter 1 (e.g., European Union - CORINAIR 94, Europe - EMEP). Such estimates depend largely on statistical calculations that scale up from the 'per-unit' results of studies of specific source types to obtain a regional bulk estimate. There is, as yet, no independent external validation of emissions estimates. Isotopes provide one method of placing some constraint on the accuracy of inventories.

The use of isotopes of methane in correlation with concentration data and regional and national emissions statistics is still in its infancy (e.g. Lowry et al., 2002). Statistical inventories have estimated emissions for each source category for each year that records are available. In Germany and the UK for instance, there are good constraints on the isotopic signatures of the main emission sources and the isotopic signature of the urban plume has been calculated. Thus the combination of emissions totals for the region being studied and the isotopic ratio of each source allow the calculation of what the mix should be. In both the Heidelberg and London studies the statistic-based estimates did not agree with atmospheric calculations and landfill emission estimates (which are considered to be the least well constrained of the statistical estimates at ±40% errors) were suspected to be in error (see Levin et al., 1999a; Lowry et al., 2001). These studies revealed though that calculation of source proportions by isotopic analysis could, when combined with data from background methane measurement stations, be used in the long term to monitor the progress of reduction strategies.

Applying the calculations to the EMEP 1996 inventory for European countries suggests that the European source area should have a mixed source plume with $\delta^{13}C$ of –53.1‰, close to the Heidelberg plume average, but higher than the calculation of –54.1‰ based on European air masses sampled at Mace Head. Small offsets such as this are accounted for by the lack of reporting of natural emissions (–65±4‰) by many European countries. The isotopic comparison suggests that the European inventories provide a good estimate of the source mix, but the isotopes can only suggest which source emissions may be in error. Alone they cannot provide quantitative information on amounts of source reduction but with longer time-series comparison they should be able to provide relative increases or reductions in emissions categories at local, regional and continental scale.

2.7. Use of methane isotopes in global modelling

Global modelling of methane isotopes varies from simple two-box hemispheric models (e.g. Lassey et al., 1993; Quay et al., 1999) to three-dimensional (Hein et al., 1997; Warwick et al., 2002) transport models. As detailed in chapter 12, inverse modelling allows the background site records to be individually modelled (e.g. Hein et al., 1997). The monthly or latitudinal emission and isotopic contribution from each source type can be modelled (Bergamaschi et al., 2000a, 2001), and estimates of the proportions of each source contributing to the observed seasonal or latitudinal profiles can be calculated. Much of the two dimensional modelling (such as the Oslo global tropospheric photochemical model) has concentrated on the inter-hemispheric isotopic differences (e.g. Gupta et al., 1996; Tyler et al., 1999a). The isotopic implications from Global Circulation Model simulations show that there is net transport of CH_4 from the Northern Hemisphere to the

Southern Hemisphere (e.g. Hein et al 1997; Quay et al., 1999). A complication added to the models is that the methane budget is at present unbalanced. Global isotopic equilibration following significant perturbations in source emissions or OH levels may be very slow (decades) and much slower than the response of the methane growth rate (Tans, 1997). Only a major change in emissions from a source which is isotopically very different from background (e.g., biomass burning or wetlands) will cause an immediately observed shift in the background isotopic records (Lassey et al., 2000). How can we tell if the observations represent a large change in relative source strengths or a relaxation following a previous change in the sources, and more importantly can models distinguish between the two? It may be possible to answer these questions using isotopic models with a long spin-up time. So far only the 1 box model of Lassey et al. (2000) has such a lead-in with sources varying over time. Currently all 3-D models using $\delta^{13}C$ are run to steady state. A 100-year spin up in a 3-D model may take many months to compute and only if accurate source parameters were available for the start date. The 3D models can be used though to simulate small spatial and temporal variations in $\delta^{13}C$ (e.g. Allen et al., 2001). They suggested that the discrepancies between measured and modelled $^{13}CH_4$ could be due to the existence of a Cl sink in the marine boundary layer.

While there are many global distribution models of methane concentrations (e.g. Houweling, 2000 and references therein), the addition of $\delta^{13}C$ to such models has only recently produced realistic spatial distributions of $\delta^{13}C$. The TOMCAT Eulerian chemical transport model (Warwick et al., 2002) predicts the isotopic seasonality at all stations studied and models the general global isotopic distribution. The isotopic characterization of each source type allows modelling of the proportions of each source contributing the excess methane for each site. Good fits of the seasonality of isotopic data have been obtained for Point Barrow, Alaska and Scott Base, South Pole (Figure 3). So far the data fitting has been more problematic for mid-high latitude northern hemisphere stations where the model predicts that the isotopes should be more depleted in ^{13}C than actually recorded. The main discrepancies are most notable between the modelled prediction of an isotopically-depleted plume over Northern Europe and the actual site measurements within this region.

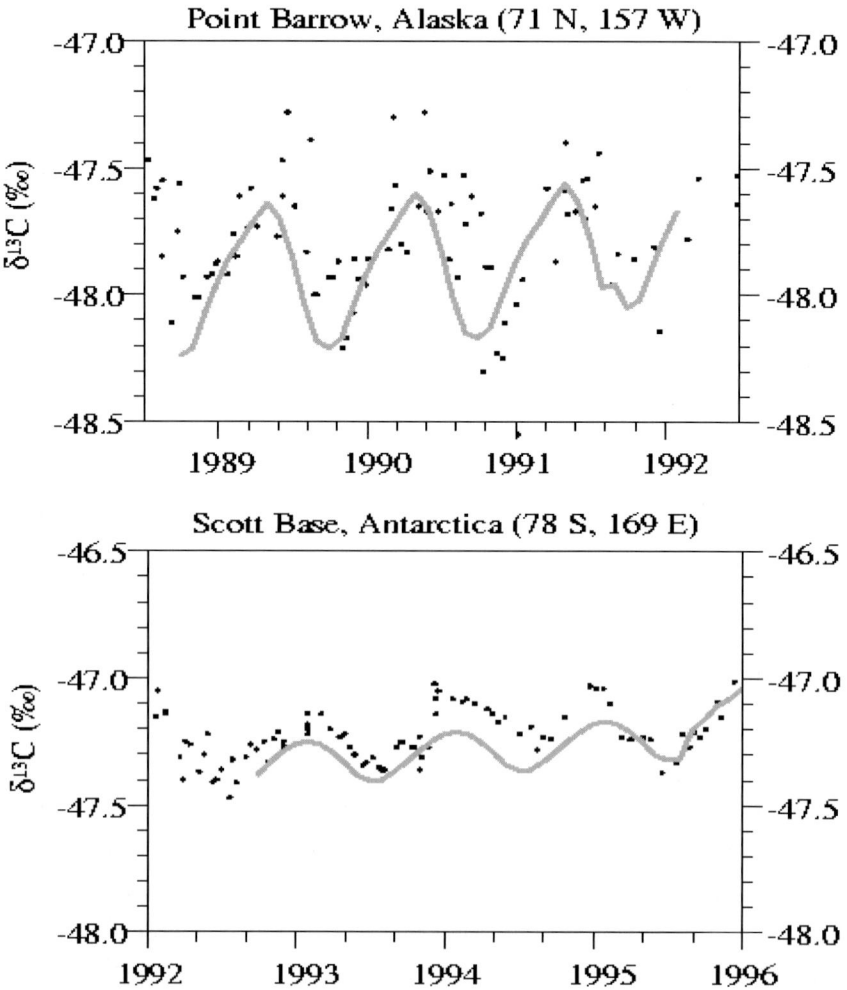

Figure 3. Results of TOMCAT 3D modelling of individual site isotopic records for Point Barrow and Scott Base. The dots are actual measurements, the curves represent the modelled fits (see Warwick et al., 2002, for further details). © American Geophysical Union.

The most likely cause is the difficulty of estimation of the northern wetland emission. The global emissions models also have difficulties in reproducing the seasonal cycles at these northern latitude stations (Houweling et al., 2000), which are very sensitive to the parameterization of wetland emissions (e.g., Walter and Heimann, 2000). Reduction in emissions from this wetland source was considered as a possible cause for the rapid decrease in methane growth rate in 1992 (Hogan and Harriss, 1994) and as a major contributor to the significant increase in growth rate in 1998 (Dlugokencky et al., 2001). Only continuing emissions experiments in Siberia and Canada, and monitoring at distant sites downwind (e.g. at Mace Head) will help to solve this problem. Isotopes will play an important role in understanding these emissions.

2.8. Future requirements of methane isotope analysis

To understand methane sources and sinks, and quantify them in terms of the Kyoto emissions accounting, high quality measurements are needed, both of mixing ratios and isotopes. These measurements must be in sustained multi-year time series. Ground based measurements are logistically and economically the most viable foundation for a long term observing system. A network of high precision and high accuracy ground based stations is a prerequisite for an observing system that can properly constrain modelling studies, especially to assess seasonality of different sources by using the isotopic record.

Can we learn much more from methane isotopes? A key to this will be a greatly expanded spatial coverage of high precision isotopic measurements, coupled to wind field analysis, trajectory studies and global modelling. This approach may help to clearly identify regions where the global isotopic models are in error, and point to areas requiring detailed isotopic source study. The growing knowledge linking natural source emission anomalies to annual climate variability, such as with the Northern wetlands, could lead to maps linking percentage methane oxidation with isotopic distribution of methane emitted, thus providing further constraints on model input parameters. It is now possible to work from background to source, using both marine and continental receptor sites to identify the causes of perturbations in the methane growth rate and use isotopes to discriminate between changes in emissions from specific sources, identifying them by character, seasonality, and location. The next decade will likely provide many answers.

3. ISOTOPIC ANALYSIS OF CARBON MONOXIDE

3.1. Introduction

Carbon monoxide (CO) is one of the atmospheric gases for which isotopic analysis is most useful. Its comparatively short lifetime, the isotopic fractionation occurring during its removal from the atmosphere by OH, as well as the distinct isotopic signatures of different source categories, all contribute to clear isotopic signals. Consequently $^{13}C(CO)$ and $^{18}O(CO)$ do show useful variations, yet this is not the whole story: it has been discovered recently that an unusual isotope effect occurs in the CO + OH reaction. This effect has yet to be explained, but it results in a small excess of ^{17}O in the unreacted CO, relative to nearly all other oxygen-containing substances in the Earth system. This novel isotope effect gives an additional, independent isotope signal, which may prove to be very useful in future research. The radioactive isotope ^{14}CO is largely (75%) of cosmogenic origin, making this species a tracer *par excellence* both for OH and for transport. CO and ^{14}CO can effectively be treated as two independent tracers. In the following section we present the main features of the use of isotopic information for CO, and describe how current analytical developments will make isotope analysis much more accessible and therefore useful in the near future.

3.1.1. Analytical methods

The ^{13}C and ^{18}O isotopic analyses are based on the oxidation of CO to CO_2 (Stevens 1972; Brenninkmeijer 1993; Brenninkmeijer et al., 2001), followed by mass spectrometry of the ion beams at m/z 44, 45 and 46. The measurement uncertainty is less than 0.1‰ for $\delta^{13}C$ and less than 0.2‰ for $\delta^{18}O$. Since mass 45 is used to establish the $^{13}C/^{12}C$ ratio, it cannot be used to measure the $^{17}O/^{16}O$ ratio. The CO_2 is therefore further converted to O_2, where mass 33 represents an unambiguous measure of ^{17}O. A new method for this conversion has been developed (Brenninkmeijer and Röckmann, 1998). However, another new method (Assonov and Brenninkmeijer, 2001) circumvents this problem, but gives a lower precision (0.45‰ rather than 0.2‰). Using the conventional off-line method, at least tens of litres of air (at standard temperature and pressure) are required for isotopic assay.

In the future, use will be made of the GC-IRMS technique for isotope measurements, in which the separation is by gas chromatography, coupled to an isotope ratio mass spectrometer (Mak and Yang, 1998). Less than one litre of air is sufficient for an analysis by this technique, and the (automated) sample preparation/analysis time is reduced to typically ¼ hour. However, this does not solve the problem of ^{14}CO analysis. Due to the very low abundance of this molecule (about 10 molecules cm^{-3}), large volumes of air (at least 100 litres) are required for measuring ^{14}CO in the atmosphere. The analysis technique using accelerator mass spectrometry (AMS) allows determination of ^{14}CO with an overall uncertainty of 2-3% (Rom et al., 2000).

3.1.2. ^{14}CO

The main source of atmospheric ^{14}CO is production by cosmic radiation. High-energy cosmic rays (mainly protons) produce neutrons in the atmosphere. Most of these neutrons diffuse, are slowed down, and thermalise, before they are eventually captured by nitrogen nuclei forming ^{14}C (^{14}N (n,p) ^{14}C). The recoil ^{14}C atom is rapidly oxidised to ^{14}CO, with a yield of approximately 95% (MacKay et al., 1963). In this way, a natural tracer is produced throughout the atmosphere, almost equally partitioned between the stratosphere and the troposphere, with a natural maximum in the polar regions caused by the influence of the geomagnetic field on the primary cosmic ray particles. The average source strength is 1.6-2 molecules per second per square-centimetre of the Earth's surface, corresponding to a total production of approximately 13-16 kg ^{14}CO per year. Since the cosmic ray flux reaching the atmosphere is modulated by the solar wind intensity, the cosmogenic ^{14}CO production rate oscillates with a phase of 11 years (solar cycle) with higher production rates during times of low solar activity. The secondary ("biogenic") contribution to ^{14}CO, comprising 20-25% of the total source, consists of ^{14}C recycled from the biosphere, entering or evolving in the atmosphere by oxidation of natural methane and higher hydrocarbons, and by biomass burning. The use of fossil fuel does not contribute to atmospheric ^{14}CO, as geological production times vastly exceed the ^{14}C half-life of about 5730 years.

The significance of ^{14}CO is that it constitutes a natural tracer that can be used to assess the hydroxyl radical (OH) abundance, because ^{14}CO + OH is its main sink reaction. The average tropospheric lifetime of ^{14}CO is about 2-3 months. However, tropospheric ^{14}CO also depends on its rate of transport from the stratosphere. Even before the discovery of the important role of OH in the troposphere, Weinstock (1969) estimated the residence time of CO using ^{14}CO measurements, or more precisely the specific activity of CO measured by MacKay et al. (1963). Volz et al. (1981) applied the ^{14}CO concept in a systematic manner and concluded that the abundance and seasonality of ^{14}CO is in accordance with that of the OH used in a two-dimensional (2D) atmospheric chemistry model.

Three independent estimates of the primary cosmogenic ^{14}CO source distribution exist (Lingenfelter, 1963; O'Brien et al., 1991; Masarik and Beer 1999), which differ mostly with respect to the vertical gradient of the production rate. However, Jöckel et al. (1999) showed that this uncertainty is not a problem for the ^{14}CO methodology. The effect of solar variation is also well understood, and can be taken into account when ^{14}CO observations of different epochs are to be compared (Jöckel et al., 2000). This prepares the way for compiling a ^{14}CO climatology, i.e., a zonally averaged seasonal cycle at the surface comprising 1088 ^{14}CO observations from 4 institutes (Jöckel and Brenninkmeijer, 2002). Jöckel et al. (2002) used this climatology for the evaluation of two three-dimensional (3D) atmospheric models and explained the observed inter-hemispheric asymmetry of atmospheric ^{14}CO with reference to inter- hemispheric differences in the exchange rate between the stratosphere and the troposphere. For a summary of the literature of ^{14}CO observations, we refer to Jöckel et al. (2002) and Jöckel and Brenninkmeijer (2002).

3.1.3 The stable isotopes

Different sources of CO have different δ values, and our current knowledge is depicted in Figure 4. The δ value of ^{13}C essentially follows that of the precursor molecule or compound, and as a whole the ^{13}C content of atmospheric CO reflects that found in organic plant matter. The combustion of coal, oil, wood, and biomass burning in general gives a typical δ value of -26‰. In this case a difference between C3 and C4 plants shows an effect which can be of importance when considering biomass burning. The CO produced in methane oxidation has a very negative δ^{13}C value, because methane is mainly a product of bacterial activity and is strongly depleted in ^{13}C (about -45‰). In summary, the main information obtained from measurements of ^{13}C in CO is the partitioning of CO between source types, which is mainly observable in the seasonal cycle.

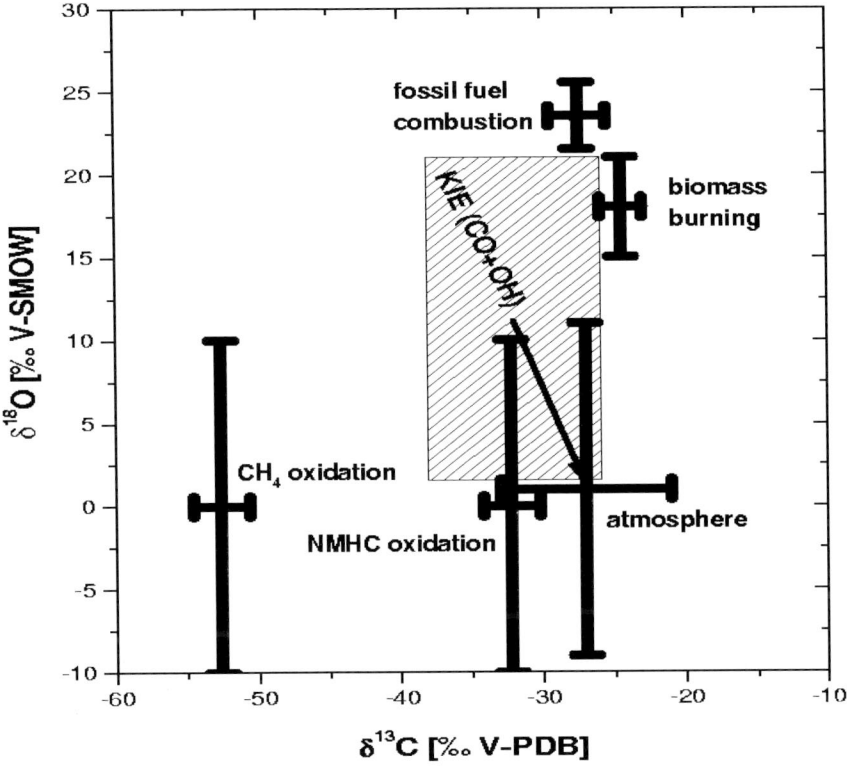

Figure 4. Estimated isotopic composition for the main sources of CO. Data are derived from Stevens et al. (1972), Brenninkmeijer, (1993), Conny et al. (1997), Conny, (1998), Stevens and Wagner, (1989), Quay et al. (1991), Cantrell et al. (1990) and Brenninkmeijer and Röckmann, (1997).

For oxygen, the interesting and useful difference is between CO from combustion processes and CO from other sources. Here, high temperature combustion gives an ^{18}O content simply reflecting that of atmospheric oxygen (~ 23‰). Biomass burning, at a lower temperature, gives rise to fractionation and isotope-exchange processes, and to a range of values of 10-18‰. At the lower end of the range is CO from hydrocarbon oxidation. The actual value is not yet known, and estimates range from -10 to 10‰. Figure 4 also shows the impact of the kinetic isotope effect (KIE) that occurs when CO diminishes in the atmosphere, mainly through its reaction with OH. This effect leads to a gradual enrichment in ^{13}C (^{12}CO reacts faster by 5‰ at 1 bar) and a depletion in ^{18}O ($C^{18}O$ reacts faster by 10‰ at 1 bar). The values reported here were measured by Röckmann et al. (1998). Other measurements of the relative reaction rates of CO isotopes with OH can be found in Feilberg et al. (2002).

^{17}O in CO deviates from the mass-dependent relation $\delta^{17}O = 0.52\ \delta^{18}O$; this is because when CO reacts with OH, the remaining fraction becomes progressively enriched in $C^{17}O$ (Röckmann et al., 1998). This excess in $C^{17}O$ can be expressed as $\Delta^{17}O = \delta^{17}O - 0.52\ \delta^{18}O$. It is close to zero for each of the main sources of CO. The small source of CO from the ozonolysis of unsaturated hydrocarbons produces CO that has an excess of ^{17}O, inherited from that of ozone.

3.2. Understanding and applying CO inter-annual variations in terms of source changes

One important goal of monitoring CO and its isotopes at selected stations is to observe and understand their seasonal variations in terms of the impact of sources and sinks. Measurements of CO and its isotopic composition performed at different latitudes have shown, for remote stations, a clear seasonal cycle (Brenninkmeijer, 1993; Kato et al., 2000; Mak and Kra, 1999, Tyler et al., 1999b; Gros et al., 2001; Röckmann et al., 2002). A maximum in late winter and a minimum in late summer is observed for CO, ^{14}CO and $\delta^{18}O$, and this seasonal cycle is mainly controlled by the same sink of CO, i.e. OH reaction. The observed maxima and minima in $\delta^{13}C$ are slightly shifted (delayed by one to two months) due to the competition between sources (the methane oxidation source, with its maximum contribution in summer, is very depleted in ^{13}C) and sinks (the reaction of CO with OH tends to enrich the remaining CO in ^{13}C). Up to now $\Delta^{17}O$ has only been measured regularly at the Spitsbergen station (79°N). As its seasonal cycle is uniquely controlled by OH, it is clearly anti-correlated with CO (the remaining CO becoming enriched in ^{17}O) (Röckmann et al., 2002).

Use of Isotopes

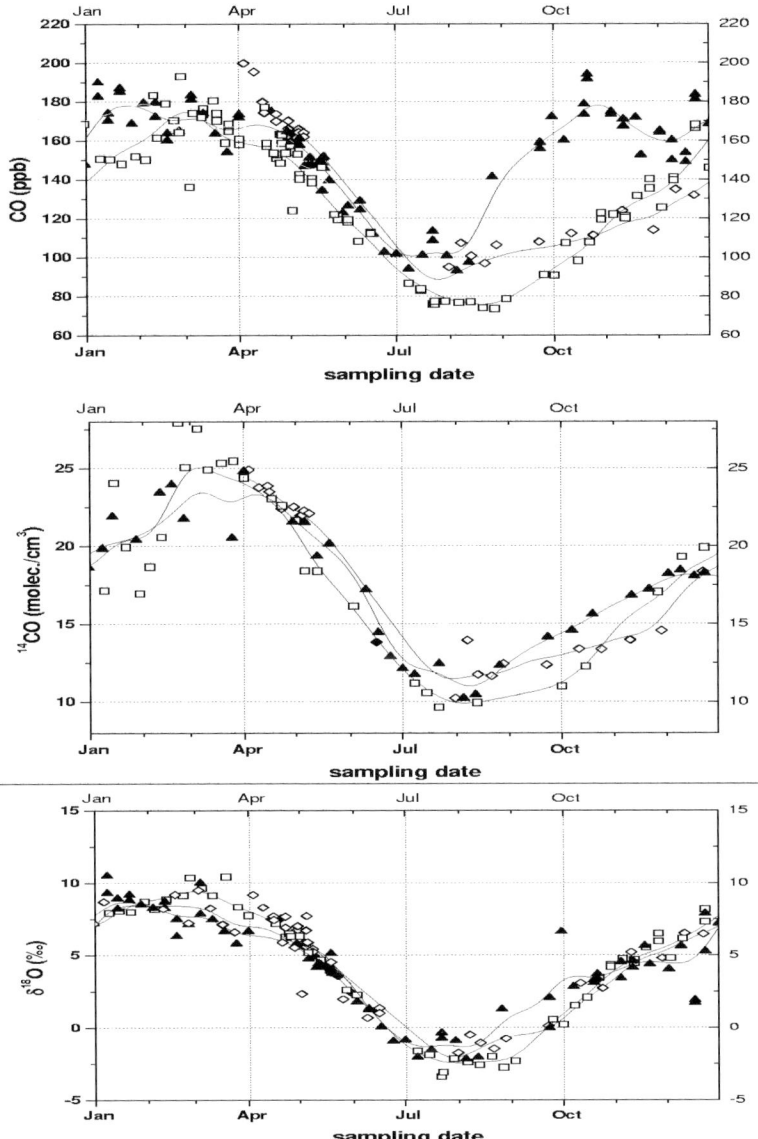

Figure 5. Seasonal cycle of CO (a), ^{14}CO (b) and δ^{18}O (c) observed at Spitzbergen. Open diamonds, open squares and solid triangles stand for 1996 1997 and 1998 respectively. Curve fittings are represented by solid lines. Adapted from Röckmann et al. (2002).

Once the first step of observing and interpreting the seasonal cycles of CO and its isotopes is completed, the next goal is to interpret the inter-annual variations observed in CO. Significant inter-annual variations and trends of CO have indeed been observed in the last decade (Novelli et al., 1998 and references therein). The origin of these variations is currently unclear, as the impacts of the different sources and sinks cannot be separated by standard CO measurements alone.

As systematic measurements of CO and its isotopic composition have begun only recently at most participating stations, more measurements are clearly required in order to interpret the data collected so far. However, in order to illustrate the principle, we present briefly here the measurements collected at the Spitsbergen station. Figure 5 shows the CO, ^{14}CO and δ^{18}O measurements for 1996, 1997 and 1998. Particularly striking in Figure 4 is the large and highly unusual increase in CO observed during the second part of 1998. The fact that a small but significant increase is observed concomitantly in the ^{14}CO and δ^{18}O signals suggests that an enhancement in biomass burning is the most likely cause for this phenomenon, as this is the only CO source enriched in both ^{14}C and ^{18}O (Röckmann et al., 2002).

3.3. Identification and characterization of CO sources at local and regional scale

The measurement of CO and its isotopes provides information about the origin of CO. This requires taking into account all the processes affecting CO and its isotopic composition. Typically, one has to consider the impact of sources (which have specific isotopic signatures), transport (i.e., dilution with air having a different isotopic signature) and photochemistry (KIE from the reaction of CO with OH). A quantitative interpretation of such a complex signal therefore often requires the use of two- or three-dimensional models (see 3.4). In some cases, justified assumptions allow the formulation of simpler approaches as already mentioned in section 2.5 and as explained here in more detail. For instance, studies performed in the vicinity of sources are based on the assumption that most of the freshly emitted CO has not yet reacted with OH, and therefore that the KIE can be neglected. In that case, the observed CO results from the mixing of the emitted CO with the surrounding background air, i.e.:

$$[CO]_{obs} \simeq [CO]_{bg} + [CO]_s \tag{Eq.2}$$

where the subscripts 'obs', 'bg' and 's' represent the observed CO, the background CO and CO freshly emitted by the local source 's', respectively. It is important to note that 's' can either represent one single source (CO observed near an isolated industrial installation) or a mixed source (for example, CO measured in a city, where the sources are mixed - cars, trucks, industry, heaters etc.).

Neglecting the KIE, the isotopic ratio (either $\delta^{13}C$ or $\delta^{18}O$) is given by:

$$\delta_{obs} \simeq (\delta_s [CO]_s + \delta_{bg} [CO]_{bg}) / [CO_{obs}] \qquad (Eq.3)$$

and by combining with

$$[CO]_s \simeq [CO]_{obs} - [CO]_{bg} \qquad (Eq.4)$$

the following expression can be derived:

$$\delta_{obs} \simeq ([CO]_{bg} (\delta_{bg} - \delta_s)) / [CO]_{obs} + \delta_s \qquad (Eq.5)$$

Consequently, the intercept of a plot of δ_{obs} against the inverse of $[CO]_{obs}$ gives the isotopic signature of the polluting source. Although this is only an approximation, this method is still valuable and especially for $C^{18}O$, since signatures associated with combustion sources are very distinct from background values.

Kato et al. (1999) used this approach to determine the average signature of the pollution in the city of Mainz, Germany. Using the results of 40 measurements, the authors observed a very clear linear relationship between the measured $\delta^{18}O$ and 1/CO. They determined that the average isotopic signature of the pollution source was 20.7 ± 0.3‰, indicating that most of the CO in Mainz came from car exhausts.

However, this concept is probably not suitable for more complex environments, where many processes affect CO. For example, Huff and Thiemens (1998) failed to observe a strong correlation between $\delta^{18}O$ and CO concentration in La Jolla, California. They attributed this result to the geographical location of the city, since in this case local sources are mixed with air of different origins (clean air from the Pacific Ocean and more polluted air from the land).

This approach has nevertheless been used successfully in a recent study by Gros et al. (2002) to distinguish between local and non-local pollution events in a semi-rural environment. This study used the information provided by isotope measurements to derive qualitative and quantitative information about pollution affecting Schauinsland, a site located in the Black Forest of Germany. A short intensive sampling campaign was performed in August 2000 and the samples obtained were analysed in the laboratory for CO and its isotopic composition ($\delta^{13}C$, ^{14}CO, $\Delta^{17}O$ and $\delta^{18}O$). The results of the campaign are presented in Figure 6. A quite high degree of variability was observed in the CO data, which varied between 100 and 180 ppb over these few days. By interpreting the CO mixing ratios alone, it would not be possible to derive information about the cause of this variability (i.e. the impact of sources or meteorology).

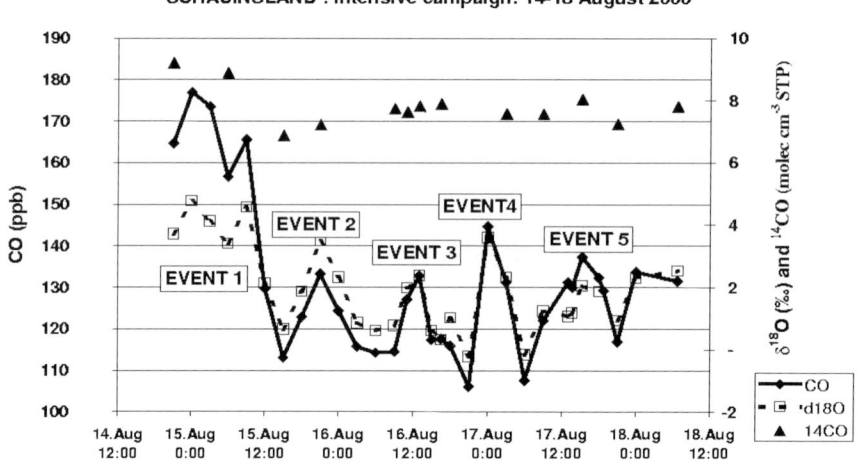

Figure 6. CO, $\delta^{18}O$ and ^{14}CO observed at Schauinsland in August 2000 Adapted from Gros et al., 2002, with permission from Elsevier.

Although the five pollution events associated with significantly elevated CO mixing ratios look superficially similar, the study of the isotopes demonstrated that some of the events had not only very different geographical origins, but also originated from different sources. The simultaneous enhancement of CO, $\delta^{18}O$ and ^{14}CO during Event 1 unambiguously indicated that the contaminating source was biomass burning, whereas the absence of significant elevations of ^{14}CO during the rest of the campaign indicated that all the other events were due to fossil fuel combustion. Using the concept discussed above, it was then possible to determine that Event 2 was of local origin. Finally, a detailed study performed with a box model, taking into account all the isotopic information (source signatures, KIE, observations), allowed additional conclusions to be reached. This study suggested: Event 1 was due to long-range transport of biomass burning emissions originating in North America; Event 3 was an aged pollution plume; and Event 4 was a fresh pollution plume, probably originating in the nearby Rhine valley (Gros et al., 2002).

As the $\delta^{13}C$ signatures associated with combustion sources are close to background values, they can not be used to characterize those specific sources. Conversely, since CO produced from methane oxidation has a signal very distinct from background values, $\delta^{13}C$ variations can be used to track this source, as shown by Röckmann et al. (1999). These authors collected several air samples at Spitsbergen during the Arctic springs of 1995, 1996 and 1997, and showed that drops of $\delta^{13}C$ values of 0.6-0.8‰ were observed at the same time as ozone depletion events. This observation could be attributed to the presence of chlorine during such ozone episodes. Although only a small fraction of CO observed at Spitsbergen is formed in the reaction $CH_4 + Cl$, this source of CO could be identified, since the very high kinetic isotope effect of 70‰ associated with the reaction $CH_4 + Cl$ leads to a significant depletion in ^{13}C values.

3.4 Modelling

3.4.1 Box models

We have seen in the previous section that box models for calculation of the isotopic signatures can be used to determine the nature and origin of observed CO for specific events at a measurement station. They can also be used to study seasonal variations of CO and isotopes and therefore help in determining the relative contributions of CO sources. For example, Mak and Kra (1999) used a chemistry box model to determine the relative strengths of CO sources contributing to CO observed at Montauk Point, Long Island. The authors calculated directly the methane-oxidation contribution (based on methane data and OH fields) and solved the set of three equations (for CO, $\delta^{13}C$ and $\delta^{18}O$) involving three unknown parameters (the contributions of fossil fuel combustion, biomass burning and non-methane-hydrocarbon oxidation). From these calculations, they were able to show that fossil fuel combustion was the dominant source, even during the summer and fall periods. In a different approach, Kato et al. (2000) studied the seasonal variation of CO and its isotopes at Happo (Japan) by using a box model for the entire northern hemisphere and the estimated source contributions from global models. They compared the observed and modelled seasonal cycles of CO, $\delta^{18}O$, $\delta^{13}C$ and ^{14}CO, and analysed the discrepancies in terms of source contributions. They concluded that an enhanced CO production from biomass burning or other CO sources was required to reproduce the observed spring maximum of CO.

3.4.2 Two-dimensional and three-dimensional modelling

The intermediate step between the simple approach offered by box models and the complexity of 3D global models is the use of 2D or 3D regional models. Manning et al. (1997) used a 2D global model in order to derive a budget for CO in the southern hemisphere, focusing on $\delta^{13}C$. They prescribed latitudinal and seasonal distributions of CO sources, and a priori estimates of the source strength and $\delta^{13}C$ of the CO produced. Optimising the source strength by minimizing the differences between observed and modelled values, they were able to reproduce the observations of CO and $\delta^{13}C$ at a site in New Zealand with their 2D model. The strongest constraint resulting from the $\delta^{13}C$ observations concerned the role of the methane oxidation source, as CO issued from this source is highly depleted in ^{13}C and consequently isotopically very different from CO produced by other sources. Since the modelled $\delta^{13}C$ was significantly higher than the measurements, Manning et al. (1997) concluded that either a smaller CO yield per molecule of CH_4 was required (about 0.7), or that a ^{13}C enrichment must occur in the oxidation of CH_4 leading to CO production. Very recently a slightly smaller constant has been measured (Saueressig et al., 2001) for the fractionation occurring during the CH_4+OH reaction (3.9‰ instead of 5.4‰), leading to a fractionation of -51.1‰ instead of -52.6‰ for CO produced from methane, which is more consistent with the conclusions of Manning et al. (1997).

Using a transport and dispersion 3D model focused on Brazil, Conny et al. (1997) showed that variations in CO isotopic composition monitored at four different stations in Brazil during the biomass-burning season were due to isotopically distinct sources and changing meteorology. They concluded that it is incorrect to assume that the CO source contributions calculated from single-day air samples represent a season and a region. However, their conclusions are limited to areas in the vicinity of important sources, especially the biomass-burning source, which is very variable in space and time.

3.4.3. Global modelling

To our knowledge, there has been only one study of the stable isotopic composition of CO on a global scale (Bergamaschi et al., 2000b) and only one series of 3D global modelling studies of atmospheric ^{14}CO (Jöckel et al., 1999, 2000, 2002, see section 3.1.2). Bergamaschi et al. (2000b) performed an inverse modelling study in order to derive information on the global budget of CO and they used the isotopic composition of CO (δ^{13}C and δ^{18}O) in order to put additional constraints on its sources. The isotopic signatures of the main CO sources and the KIE values were implemented in the TM2 model. With an inversion technique, the observations performed at five stations (Spitsbergen, Alert, Izaña, Baring Head and Scott Base) were used to optimise source contributions. Two important conclusions were derived from this study. First, the model reproduced simultaneously CO, δ^{13}C and δ^{18}O within two standard deviations of the observed data, indicating that our understanding of the processes governing the isotopic variations (source signatures, KIE) was consistent with the observations. Second, the inputs of the isotopes brought additional constraints to the model. In particular, the δ^{13}C observations suggested a CO yield from CH_4 oxidation of about 86%, decreasing the total amount of CO produced by this route. However, there are some problems inherent in the inversion technique. The most important of these is that the inversion technique will always implicitly correct systematic errors by changing the a-priori estimates of the free parameters, and this effect is especially significant for short-lived species like CO (see chapter 12 for further details on this technique). Consequently, some of the inferred source signatures do require experimental verification.

3.5 Carbon monoxide isotope analysis in the future

As we have endeavoured to show in this section by means of some examples, isotopic analysis of CO can indeed give useful additional insight into the cycle of CO (see summary in Table 2). All the information presented here has been gleaned from a modest number of isotope analyses that nevertheless represent a tremendous amount of work in the field and in the laboratory. Soon CO isotope analyses will be conducted on small air samples with almost the same ease as concentration measurements. Even the newly developed method for ^{17}O can be modified to allow "on-line" analyses. This all is due to the possibility of coupling gas chromatography with accurate isotope ratio mass spectrometry. We will therefore see an enormous increase in the number of isotope analyses in the future, which will make it very worthwhile to incorporate isotopic information into model simulations.

Table 2. Main use of CO isotope information

^{13}C	1.	Clear seasonal cycle. Partitioning of CO contribution from all sources relative to the methane oxidation source. Low values from OH + CH_4.
	2.	Estimation of CO from Cl + CH_4 in the ozone hole, or during low ozone events. Very low values from Cl + CH_4.
	3.	Determination of biomass-burning type: C3 or C4 plants.
	4.	Possible detection of CO from burning of natural gas, although this source is very small. Low ^{13}C value.
^{18}O	1.	Contribution of CO from fossil fuel combustion. High ^{18}O values for CO from fossil fuel burning.
	2.	Photochemical age of CO: large inverse isotope fractionation in the reaction with OH, $C^{18}O$ reacts 1% faster than $C^{16}O$.
^{17}O	1.	Fraction of CO from the ozonolysis of unsaturated hydrocarbons in certain environments.
	2.	Photochemical age of CO. All sources except the above have a value of zero. CO gradually gains excess ^{17}O during its residence in the atmosphere; this effect is diluted by all in situ sources.
^{14}C	1.	Determination of the OH-based oxidative capacity of the troposphere. In theory, a detailed picture of OH distribution and trends may be obtained.
	2.	Assessment of stratosphere to troposphere transport.
	3.	Estimation of biomass-burning contributions.

4. ISOTOPIC ANALYSIS OF NITROUS OXIDE

4.1. Introduction

The use of stable isotopes to characterise nitrous oxide (N_2O) sources is still in its early stages. Constraints on the global budget result primarily from the extent of isotopic fractionation in the stratospheric N_2O sink reactions, namely photolysis and reaction with $O(^1D)$. New techniques for isotope analysis on N_2O as well as a number of new N_2O source measurements have only recently become available and may help to pin down the isotopic signature of N_2O sources. Here we summarise the existing information on the isotopic composition of the main N_2O sources (nitrification and denitrification in soils and oceans) and of smaller contributions from industry or transportation and from atmospheric in-situ sources.

The first measurements of the ^{15}N content of atmospheric (N_2O) were already carried out in the early 1970s (Moore, 1974). Further important foundations for the field were laid in the 1980s by the work of Yoshida and co-workers (1984, 1988. 1989) for ^{15}N and Wahlen and Yoshinari for ^{18}O (1985), and in the early 1990s with the first dual isotope measurements (^{15}N and ^{18}O) by Kim and Craig (1990, 1993). N_2O from different sources can have strongly varying isotopic composition, forming the basis of today's isotope studies. One aim of isotope research on atmospheric trace gases is to construct atmospheric budgets for individual isotopic N_2O species based on trace gas emissions from different sources with characteristic and sufficiently different isotopic composition. To this end, isotope measurements provide more observables for a certain species than measurements of mixing ratios alone, but also introduce more unknowns, namely the isotopic source signatures of the individual sources. Of course, isotope fractionations in the sinks of trace gases also have to be taken into account.

In the case of N_2O, it became clear early that tropospheric N_2O is isotopically enriched relative to most known terrestrial sources. This enrichment is due to isotope fractionation in the stratospheric sink reactions, which has been thoroughly verified in the recent past and appears to be well understood now. On the source side, however, investigations of the global N_2O isotope budgets suffer from the same limitations that are hindering the construction of a reliable atmospheric N_2O cycle: Because of the long atmospheric lifetime of about 120 years, the tropospheric N_2O reservoir is globally very well mixed. Temporal and spatial gradients remote from local sources are barely detectable. Annual emissions represent only a small fraction of the atmospheric reservoir, so that the atmospheric concentration and isotopic composition change only slowly compared to the average background conditions. Furthermore, in contrast to the uniform and relatively simple sink distribution, N_2O sources at the Earth's surface are widespread and have a complex and variable distribution (see Kroeze et al. (1999) and Olivier et al. (1998) for estimates of the global N_2O budget).

Emissions from the main sources, oceans and soils, depend on many parameters like temperature, substrate availability, redox state, moisture, soil texture, etc. Therefore, many measurements have to be carried out to make representative estimates on the global scale. In the past, this has been a major limitation in isotope ratio research on N_2O, because of the large air samples that were required and the laborious and time-consuming analytical techniques. Nevertheless, due to the unique information that can be derived from knowledge of the isotopic composition of N_2O, the number of isotope studies on the sources and sinks of N_2O is steadily increasing in recent years. This trend will certainly continue in the future, because new continuous-flow techniques considerably reduce the time as well as the sample size required for an isotope ratio determination (Brand, 1995; Röckmann et al., 2003a).

4.2. Measurement techniques

Various experimental techniques have been developed to carry out isotope ratio measurements on atmospheric N_2O. Two optical absorption methods, Fourier Transform Infrared (FTIR) absorption spectroscopy and tuneable diode laser (TDL) absorption spectroscopy are described in detail in Wahlen and Yoshinari (1985), Esler et al. (2000), Zhang et al. (2000) and Uehara et al. (2001). Still the most common and most precise methods are based on isotope ratio mass spectrometry (IRMS).

An overview of various pathways to determine individual isotope signatures of N_2O by IRMS is shown in Figure 7. N_2O can be converted to N_2 and CO_2 to determine the ^{15}N and ^{18}O isotope content using established methods for isotope ratio measurement on these reaction products (Kim and Craig, 1990; Yoshida et al., 1984; pathways 1 and 2 in Figure 7). This method is still widely used today for calibration, i.e. to relate the heavy isotope content of N_2O to international standards in the form of N_2 and CO_2. Kim and Craig (1993) introduced pure N_2O into the mass spectrometer to determine ^{15}N and ^{18}O on N_2O at m/z 44, 45 and 46 directly (pathway 3a), thus circumventing the need for chemical conversion. Fragmentation of N_2O following electron impact yields a significant fraction of NO^+ fragment ions, which is the basis for the recently developed technique to measure the intramolecular distribution of ^{15}N in N_2O (pathway 3b in Figure 7). Due to its asymmetric structure (NNO), the two nitrogen atoms are not equivalent, and different fractionations are to be expected at the two positions. The NO^+ fragment ion essentially retains the central N atom (position 2), interfered only by minor scrambling of about 8% (Brenninkmeijer and Röckmann, 1999; Kaiser, 2002; Toyoda and Yoshida, 1999). Isotope ratio measurement of this fragment therefore yields $\delta^{15}N$ at the central position. The isotopic composition of the terminal nitrogen atom (position 1) can then be calculated as the difference between position 2 and the average fractionation if the absolute ^{15}N distribution in the mass spectrometer reference gas is known.

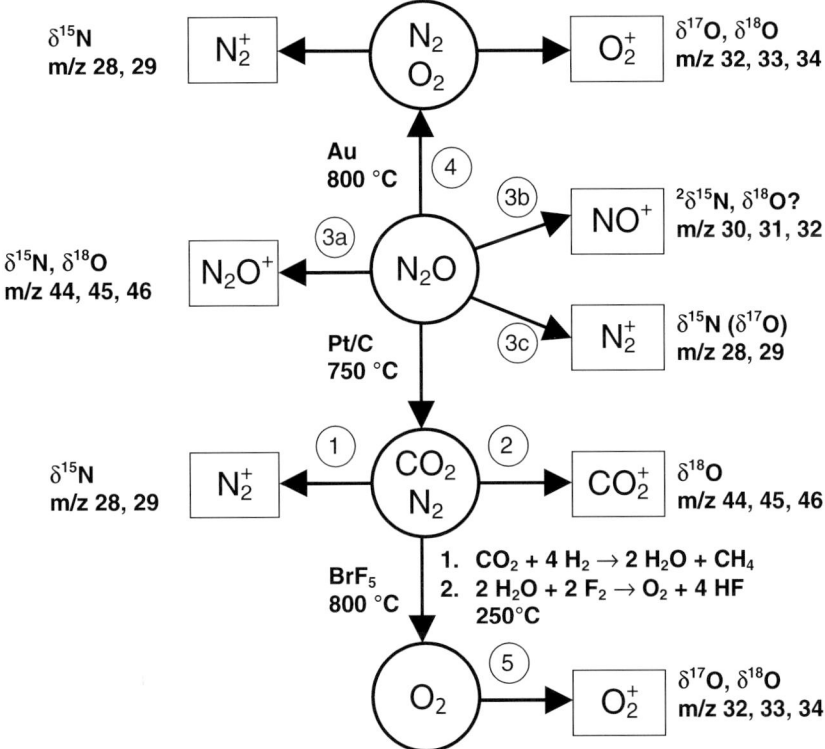

Figure 7. Overview of methods to determine isotope fractionation signals in N_2O using isotope ratio mass spectrometry; modified from Brenninkmeijer and Röckmann (1999). The numbering of reaction pathways refers to the main text

The N_2^+ fragment (m/z 28 and 29) can be used similarly to determine the average ^{15}N fractionation (Brenninkmeijer and Röckmann, 1999) which allows the determination of $\delta^{17}O$ (Kaiser et al., 2003a). Unfortunately, the achievable precision is only about 0.025‰. Since ^{15}N is about twenty times more abundant than ^{17}O, this limits the precision of $\delta^{17}O$ to about 0.5‰. Therefore, conversion of N_2O to O_2 is usually preferred for ^{17}O measurements. Two methods have been developed recently, conversion over gold to N_2 and O_2 (pathway 4, Cliff and Thiemens, 1994) or conversion to CO_2 and subsequent conversion to O_2 (pathway 5, Röckmann et al., 2001b). Using a combination of these techniques, all four isotopically mono-substituted N_2O species are accessible. Apart from the ^{17}O techniques, all methods are already available as continuous-flow-methods.

Isotope ratios of a sample (SA) are reported relative to the standards (ST) atmospheric N_2 and VSMOW in the δ notation, where δ is defined as $\delta^{15}N = (^{15}N/^{14}N)_{SA}/(^{15}N/^{14}N)_{ST} - 1$, or $\delta^{18}O = (^{18}O/^{16}O)_{SA}/(^{18}O/^{16}O)_{ST} - 1$. $^1\delta^{15}N$ and $^2\delta^{15}N$ correspond to the terminal and central N-atom in the linear NNO molecule, respectively. Alternative nomenclatures use $\delta^{15}N^\beta/\delta^{15}N^\alpha$ or $\delta(546)/\delta(456)$ (Toyoda and Yoshida, 1999; Yung and Miller, 1997). The average $\delta^{15}N$ and $\delta^{18}O$ values of tropospheric N_2O versus atmospheric N_2 and VSMOW are (6.7±0.1)‰ and (44.6±0.2)‰, respectively (Kaiser et al., 2002a). Yoshida and Toyoda (2000) found similar numbers, but with larger uncertainties: $\delta^{15}N$=(7.0±0.6)‰ and $\delta^{18}O$=(43.7±0.9)‰. They also measured individual ^{15}N signatures as $^1\delta^{15}N = -2.35‰$ and $^2\delta^{15}N = 16.35‰$. Preliminary results from Kaiser (2002) indicate even larger differences between both nitrogen positions.

4.3. Isotopic source signatures

4.3.1. Biological sources

The main sources of atmospheric N_2O are microbial nitrification and denitrification processes in soils and oceans. The production processes themselves are not fully understood, however, and isotope studies have been carried out in the past both to study individual nitrification and denitrification pathways (often using isotopically labelled substrates) and to determine the isotopic source signature of N_2O emitted from soils and oceans. Although the details of production pathways are beyond the scope of this review, it has become clear in the recent past that knowledge about the relevant nitrification and denitrification reactions, as well as the isotopic composition of the substrates available for N_2O production are very helpful for the interpretation and understanding of isotope measurements on N_2O sources (Stein and Yung, 2003).

Generally, there are two pathways for N_2O formation, denitrification according to

$$NO_3^- \rightarrow NO_2^- \rightarrow NO \rightarrow N_2O \rightarrow N_2 \qquad (Eq.6)$$

and nitrification according to

$$NH_4^+ = NH_3 \rightarrow NH_2OH \rightarrow NO \rightarrow NO_2^- \rightarrow NO_3^- \qquad (Eq.7)$$
$$\searrow \quad \downarrow \quad \nearrow$$
$$N_2O$$

Whereas production of N_2O via denitrification seems straightforward, several different steps in the nitrification sequence have been proposed to produce N_2O.

The production pathways are associated with normal kinetic isotope effects in which the light isotope generally reacts faster, leading to enrichments in the remaining substrate and to isotopically depleted reaction products. Whether an individual compound in the reaction sequence is depleted or enriched depends on the isotopic fractionations in its production and its consumption. Thus, although N_2O produced is generally depleted relative to the ammonium or nitrate precursors (Barford et al., 1999; Mandernack et al., 2000; Mariotti et al., 1981; Wada et al., 1996; Webster and Hopkins, 1996; Yoshida 1988; Yoshida et al., 1984; Yoshinari and Wahlen, 1985), it can be isotopically enriched again in the final step of the denitrification sequence (reduction to N_2). The enrichment depends on the environmental conditions, and a large range of fractionation constants have been found in pure culture and soil incubation studies (Barford et al., 1999; Mandernack et al., 2000; Wahlen and Yoshinari, 1985; Yamazaki et al., 1987 and references therein). Similarly, isotopically enriched N_2O may be produced via enriched intermediate compounds (NH_2OH or NO) in the nitrification sequence (Kim and Craig, 1990).

4.3.1.1. Soil emissions

Production of N_2O due to nitrification and denitrification reactions in soils represents the largest source of atmospheric N_2O (Kroeze et al., 1999). Only few investigations on the isotopic composition of soil N_2O have been carried out, although a number of studies have reported measurements on pure cultures. The available soil data are shown in Figure 8. Early isotope measurements on soil emissions revealed pronounced depletions in ^{18}O and ^{15}N (Kim and Craig 1993; Mariotti et al., 1981; Wahlen and Yoshinari, 1985). Emissions from soils are regarded as an important end-member in the construction of a global N_2O isotope budget.

Only in the past three years has this sparse data base been expanded, providing a bit more insight into the great temporal and spatial variability of the isotopic composition of soil N_2O. Pérez et al. (2000) found very variable values depending on soil type, soil texture, substrate availability and water content ($\delta^{15}N = -34$ to $2‰$, $\delta^{18}O = 20$ to $42‰$). Hot spots with high emissions were associated with low $\delta^{15}N$. Generally, N_2O emissions from tropical rain forests were isotopically not as light as those reported by Kim and Craig (1993), but fertilised soil was shown to produce very light N_2O. In a follow-up study, Pérez et al. (2001) concluded that N_2O from fertilised systems with a flux-weighted average $\delta^{15}N$ of $(-37.9\pm8.1)‰$ is more depleted in ^{15}N than N_2O from more N-limited forest soils. Also Yamulki et al. (2000) found very ^{15}N-depleted N_2O emitted from soils with urine amendment (-40 to -21‰), but in a later study N_2O emitted from such soils was considerably less depleted in ^{15}N (-20 to 0‰) and ^{18}O (30 to 50‰).

In landfill cover soils, Mandernack et al. (2000) even reported enrichments relative to tropospheric background N_2O: $\delta^{15}N$ is in a range from -5 to 19‰, and $\delta^{18}O$ from 43 to 58‰, but their inorganic substrate nitrogen pool was also enriched in ^{15}N.

In the above investigations, relatively enriched N_2O is usually attributed to partial reduction of the formed N_2O to N_2 before it can escape to the atmosphere, to substrate limitation, or to different pathways for N_2O production in nitrification (e.g. via NO_2^- and NH_2OH).

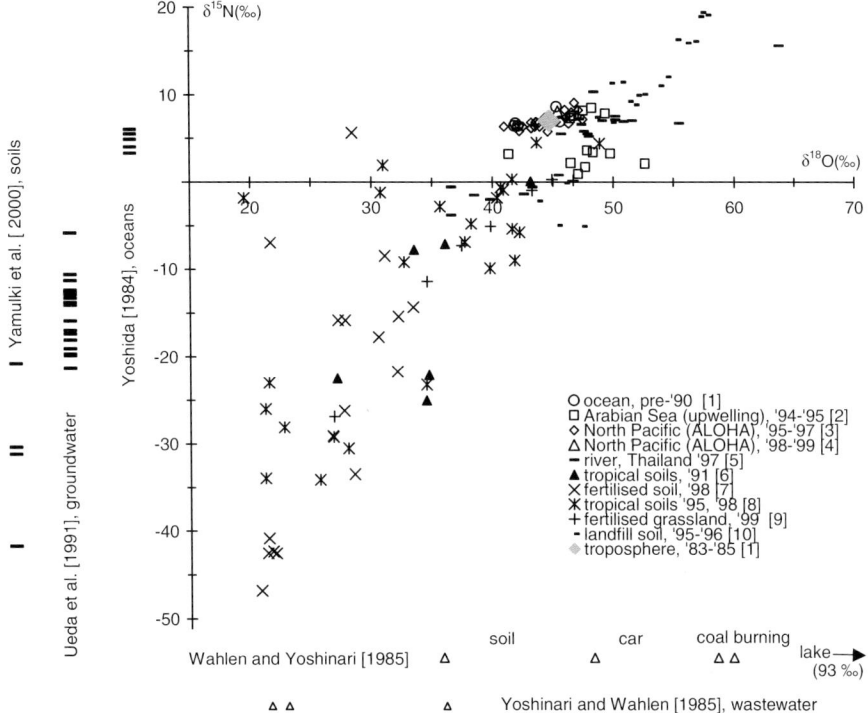

Figure 8. The $\delta^{15}N$ and $\delta^{18}O$ source signatures of N_2O emitted from various sources. Unless indicated, data have been taken from (1) Kim and Craig (1990), (2) Naqvi et al. (1998), (3) Dore et al. (1998), (4) Ostrom et al. (2000), (5) Boontanon et al. (2000), (6) Kim and Craig (1993), (7) Pérez et al. (2001), (8) Pérez et al. (2000), (9) Yamulki et al. (2001), (10) Mandernack et al. (2000).

4.3.1.2. Oceanic emissions

The emission of N_2O from oceans are not homogeneous on a global scale. Especially two areas have been identified which contribute a substantial fraction to the oceanic N_2O source: the eastern tropical North Pacific and the Arabian Sea (Naqvi and Jayakumar, 2000). These have been the focus of most of the studies on the isotopic composition of oceanic N_2O and the data are shown in Figure 8. We note that recent studies indicate that coastal areas and estuaries may be at least as important as open oceans for N_2O emissions (Kroeze and Seitzinger, 1998), but no isotope investigations have been reported from there yet.

^{15}N depleted N$_2$O in the shallow water column of the eastern tropical Pacific was first measured by Yoshida et al. (1984) and attributed to nitrification. Kim and Craig (1990) confirmed the observed depletions, and found similar depletions in ^{18}O. Strongly oxygen deficient layers were characterised by ^{15}N and ^{18}O enrichments in N$_2$O in. This enrichment in the oxygen minimum zone has been observed in numerous subsequent studies (Naqvi et al., 1998; Yoshida et al., 1989), with highest values of $\delta^{15}N = 37\%o$ and $\delta^{18}O = 83\%o$ in the oxygen minimum zone of the Arabian Sea (Yoshinari et al., 1997). The exact cause for this enrichment is not yet understood and may be different in different environments. The very strong ^{15}N enrichments in anoxic waters are most likely produced via denitrification studies. Other evidence, generally in more oxic waters, suggests that the enriched N$_2$O is likely produced via nitrification (Kim and Craig, 1990; Ostrom et al., 2000). Here, the isotopic enrichment of N$_2$O may be produced via the enrichment of intermediate products in the nitrification sequence, i.e. hydroxylamine (NH$_2$OH) or NO. Couplings of nitrification and denitrification in the production of N$_2$O have been suggested (Naqvi, 1991; Ostrom et al., 2000). These results show that no process can be singled out that accounts for the isotope enrichment of N$_2$O found in oxygen deficient layers of various parts of the oceans: Two recent publications show evidence for two different processes contributing to N$_2$O production at different depths at the same geographical location (Naqvi et al., 1998; Ostrom et al., 2000).

The layers of strong N$_2$O isotope enrichments just described do not extend to the ocean surface, but are usually capped by a layer exhibiting much lower abundance of the heavy isotopes. This surface layer represents the direct boundary to the atmosphere and therefore the isotopic composition of N$_2$O in this layer is indicative of the isotopic signature of the oceanic N$_2$O emissions. In addition to the early data by Kim and Craig (1990) and Yoshida et al. (1984; 1989), recent data have confirmed that a slight depletion of ^{15}N relative to the atmospheric value and a slight depletion or slight enrichment of ^{18}O is the typical isotope signature for N$_2$O in the surface layer of the Arabian Sea (Naqvi et al., 1998; Yoshinari et al., 1997) and the extratropical North Pacific (Dore et al., 1998). However, the production mechanism of this relatively depleted N$_2$O in the surface layer is not yet understood. Whereas Dore et al. (1998) attribute the N$_2$O production in the surface layer to nitrification, Naqvi and Jayakumar (2000) suggest a nitrification-denitrification coupling according to NH$_4^+$ → NH$_2$OH → NO → N$_2$O.

Although investigations of the underlying mechanisms reveal an increasingly complex picture, the new data from the surface oceans obtained in the past few years generally support the concept of an oceanic N_2O source with an isotope signature close to that of atmospheric N_2O provided already in 1993 by Kim and Craig. The recognition that processes close to the surface determine the isotopic composition of the oceanic N_2O source, implies that this source signature may be subject to considerable variability on short and long time scales with possible implications for global change (Dore et al., 1998; Naqvi et al., 1998).

4.3.1.3. Other aquatic sources

Early measurements of N_2O produced by nitrification in a wastewater facility (Yoshinari and Wahlen, 1985) showed depletions of about 20‰ in ^{18}O relative to ambient N_2O, the low values indicating production via nitrite (NO_2^-). Low $\delta^{15}N$ values between -21.3 and -5.8‰ were found in groundwater (Ueda et al., 1991), which was also attributed to production via nitrification.

The very high $\delta^{18}O$ value (+93‰ versus VSMOW) found in a single N_2O sample extracted from lake water (Wahlen and Yoshinari, 1985) was attributed to denitrification. In another denitrification environment (mesotrophic lake), surprisingly low values of ^{15}N were found (Wada et al., 1996). Boontanon et al. (2000) observed significantly different isotope values in N_2O from the Bang Nara River in Thailand depending on oxygen availability, with high isotope enrichments under anaerobic conditions ($\delta^{15}N$ = 15.6‰ and $\delta^{18}O$ = 63.8‰) and low values under aerobic conditions ($\delta^{15}N$ = –3.8 to –0.6‰ and $\delta^{18}O$ = 36.6 to 39.8‰). Intermediate values ($\delta^{15}N$ = 0.8 to 10.3‰ and $\delta^{18}O$ = 45.5 to 48.4‰) were attributed to the occurrence of simultaneous nitrification and denitrification processes. These isotope data can help in the identification of N_2O formation processes, but there are not sufficient data to assess the importance for the global N_2O isotope budget reliably.

4.3.2. Emissions from industry and transportation

Information about the source signatures of N_2O from industrial sources is extremely sparse. One sample of N_2O from automobile exhaust had a $\delta^{18}O$ value similar to tropospheric N_2O and two samples from coal combustion were about 14‰ enriched (Wahlen and Yoshinari, 1985). The ^{15}N content of coal and oil shales is relatively close to atmospheric N_2, as summarised by Rigby and Batts (1986). N_2O from nylon production was partly characterised isotopically ($\delta^{18}O$) when this source was first identified (Thiemens and Trogler, 1991), yielding $\delta^{18}O$ values between 17 and 27‰ relative to VSMOW, i.e., much lower than tropospheric N_2O. Based on these data, Rahn and Wahlen (2000) assume the source signature from industrial emissions to be close to atmospheric N_2 and O_2.

4.3.3. Atmospheric sources

The existence of possible atmospheric sources of N_2O due to non-standard chemistry in the stratosphere has been repeatedly postulated over the past decade (McElroy and Jones, 1996; Prasad et al., 1997; Wingen and Finlayson-Pitts, 1998), and the need for such sources was often argued to be substantiated by isotope data. Although isotope enrichments due to photolysis had already been reported in 1990 (Yoshida et al., 1990), subsequent UV photolysis experiments at 185 nm that yielded negligible enrichments (Johnston et al., 1995) were seen as evidence for the existence of additional sources that would explain the isotope enrichment of tropospheric N_2O relative to its surface emissions. A number of recent studies have now confirmed strong isotope enrichments in the stratosphere that cause the observed enrichment of the tropospheric reservoir (Griffith et al., 2000; Rahn and Wahlen, 1997; Röckmann et al., 2001a; Toyoda et al., 2001) and clearly identified UV photolysis as the dominant origin (Kaiser et al., 2002a, 2002b, 2003b; Rahn et al., 1998; Röckmann et al., 2000, 2001a).

After its discovery in 1997 (Cliff and Thiemens, 1997), the anomalous ^{17}O excess of atmospheric N_2O has also been discussed as evidence for additional N_2O sources, in particular such involving ozone, which is known to possess an exceptionally large ^{17}O excess (Krankowsky et al., 1995). Such a source has recently been identified (Röckmann et al., 2001b). The isotope anomaly can be transferred from O_3 to NO_2 via the well-established O_3-NO_x cycle, and in a second step NO_2 reacts with NH_2 to produce N_2O that carries a large ^{17}O excess. This mechanism is a well established reaction sequence in the atmosphere and does not require non-standard chemistry. However, recent measurements (Lindholm and Hershberger, 1997; Park and Lin, 1997) of the product branching ratio in the reaction of $NH_2 + NO_2$ resulted in smaller yields than assumed in previous studies (Dentener and Crutzen, 1994; Kohlmann and Poppe, 1999), so that this pathway is likely too small to quantitatively explain the ^{17}O anomaly observed in tropospheric N_2O (Kaiser, 2002). We mention that a possible increase of the anomaly in the stratosphere has been suggested (Cliff et al., 1999). As the above isotope transfer mechanism is located close to the surface, this stratospheric trend – if it exists – would indicate additional N_2O sources in the stratosphere. Such a source could be the reaction of $N_2 + O(^1D) + M \rightarrow N_2O + M$ as proposed by Estupiñán et al. (2002).

4.4. Position-dependent ^{15}N fractionation

Since the development of new methods to resolve the ^{15}N fractionation at both positions in the N_2O molecule, the position dependent ^{15}N fractionations in the stratospheric sink have been studied in considerable detail and are relatively well understood. In contrast, published measurements on position resolved ^{15}N source signatures are very sparse, and the presently available data from soil emissions, together with some stratospheric data, are plotted in Figure 9. The stratospheric data show a tight correlation which is satisfactorily explained by the irreversible removal of N_2O through photolysis and reaction with $O(^1D)$. Interestingly, the general correlation – although with a smaller slope – appears to continue to the region of isotopically depleted N_2O emitted from soils. This shows that the residual N_2O sources (i.e., mainly oceans) should lie to the lower right of tropospheric N_2O in a $^2\delta^{15}N - ^1\delta^{15}N$ diagram (Figure 9) in order to balance the enrichment induced in the stratosphere. However, Toyoda et al. (2002) found that oceanic N_2O rather lies to the upper left of tropospheric N_2O in a $^2\delta^{15}N$ versus $^1\delta^{15}N$ diagram. Clearly, more data from soil measurements are required to resolve this apparent paradox.

Yoshida and Toyoda (2000) showed that in urban tropospheric air the central nitrogen atom is on average 18.7‰ more enriched than the terminal nitrogen atom, but they also reported a large variability in this difference between $^2\delta^{15}N$ and $^1\delta^{15}N$ (14.5 to 25.5‰). Lower values were attributed to anthropogenic sources, higher values to soil emissions.

The slightly negative slope defined by these data (Figure 9) is in disagreement with the clearly positive slope derived from the few available soil data. This indicates that the spread is likely not caused by soil emissions as originally suggested. Toyoda et al. (2002) explain it by the influence of oceanic N_2O emissions. Alternatively, a possible shift in N_2O production processes (nitrification, denitrification) may be mirrored in a changing $^2\delta^{15}N$ versus $^1\delta^{15}N$ relationship (Yamulki et al., 2001, Stein and Yung, 2003).

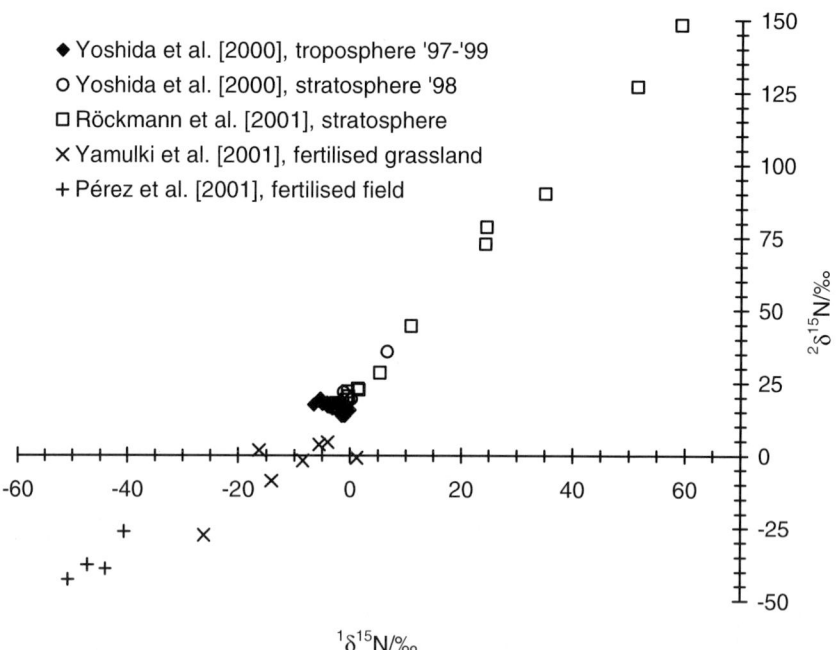

Figure 9. Individual ^{15}N signatures from stratospheric and soil N_2O emitted (data have been digitised from Pérez et al. (2001), Röckmann et al. (2001a), Yamulki et al. (2001) and Yoshida and Toyoda (2000)).

Regarding the intramolecular distribution of ^{15}N, one should keep in mind that $^{15}N^{14}NO$ and $^{14}N^{15}NO$ are independent molecules. Although linear combinations of two δ values like the "site preference" $^2\delta^{15}N$-$^1\delta^{15}N$ defined by Yoshida and Toyoda (2000) may in some cases be illustrative (Yamulki et al., 2001; Yoshida and Toyoda, 2000), they can also conceal the actual physical fractionation processes. This is particularly true for the stratospheric sink processes like photolysis. Due to the fact that the fractionation constant at position 2 is larger than at position 1, $^2\delta^{15}N - {^1\delta^{15}N}$ trivially increases with increasing degree of photolysis. Thus, changes in this difference do not necessarily imply changing physical processes. For N_2O emitted from surface sources, it is important to keep in mind that fundamental fractionations occur for the individual molecules $^{15}N^{14}NO$ and $^{14}N^{15}NO$, and ^{15}N cannot redistribute between the positions within the molecule, as may be misleadingly implied by the term "site preference". By the same token, average $\delta^{15}N$ values can be misleading (Kaiser et al., 2003a). However, average $\delta^{15}N$ values are still in common use since they were the first isotopic signatures of N_2O to be analysed and are still the easiest to obtain by isotope ratio mass spectrometry.

4.5. The global isotope budget of N_2O

Global isotope budgets of atmospheric trace gases can be calculated when the isotopic source signatures and the fractionation in the sinks (and possible isotope exchange reactions) are known. In the case of N_2O, the sink part is relatively simple, since its only known sinks are photolysis and reaction with $O(^1D)$ in the stratosphere. The isotopic composition of stratospheric N_2O has been characterised relatively well over the past years, so that the impact of the sinks on the N_2O isotope budget is now quite well understood. As shown above, much larger uncertainties exist on the source side. Among the two major source terms, the isotopic signature of the oceanic N_2O source seems to be better defined (from the surface water data summarised above), whereas N_2O emitted from soils exhibits large variability in its isotopic composition, and it is presently difficult to derive a reliable globally averaged soil source signature.

Increasing isotopically depleted N_2O surface emissions are expected in the future from growing anthropogenic application of fertilisers in agriculture (Kroeze and Seitzinger, 1998), which should decrease the heavy isotope abundance in N_2O similar to the "Suess effect" known for CO_2 (Rahn and Wahlen, 2000). Due to the long atmospheric lifetime of N_2O, changes in its tropospheric isotope signature by variations in the emission balance take a long time to become measurable (Kaiser et al., 2003a), but measurements of the isotopic composition of N_2O trapped in firn-air and ice provide hints for changes since pre-industrial times (Röckmann et al., 2003b; Sowers, 2001). Especially the firn air data have confirmed the expected "Suess effect" for N_2O and allowed to determine the current trend in the isotopic composition of tropospheric N_2O.

As the isotopic composition of tropospheric N_2O is the parameter in the N_2O isotope budget that is constrained best, this value together with reliable estimates of the stratospheric enrichment can be used to estimate the average isotope signature of the surface sources when the respective fluxes are known. This approach was first used by Kim and Craig (1993), who concluded that the source strength of isotopically depleted N_2O from the surface is not sufficient to balance the enrichment produced in the stratospheric sinks. More recent calculations (Yoshida and Toyoda, 2000) employing new stratospheric data show better agreement between revised flux estimates (Kroeze et al., 1999) and the new isotope data from soil emissions (Pérez et al., 2000; 2001). Most recently, Röckmann et al. (2003b) concluded from their firn air data that the imbalance between (isotopically depleted) N_2O sources and stratospheric sinks is not sufficient to account for the presently decreasing isotope ratios. In addition, even the global average isotope ratio of the N_2O source flux must have decreased to explain the N_2O "Suess effect" quantitatively. The global N_2O isotope budget derived from stratospheric and firn air measurements as well as the available data for the dominant sources (soils and oceans) is not in contradiction with conventional estimates of the N_2O budget, but it is difficult to impose more stringent constraints on source fluxes from the isotope data, because of the large variability of the isotopic composition of N_2O emissions (Kaiser, 2002).

4.6. Nitrous oxide isotope analysis in the future

One reason to carry out isotope measurements on atmospheric N_2O at natural isotope abundance is to understand the various N_2O production pathways in water and in soils. It appears that in order to gain reliable information about individual pathways, it is not sufficient to determine the abundance of ^{15}N and ^{18}O in N_2O alone. Additional information about the isotopic composition of the substrate and about the environmental conditions helps in the interpretation of the N_2O data, as has been shown in recent studies (Ostrom et al., 2000; Pérez et al., 2000). Complex mechanisms for N_2O production like coupled nitrification-denitrification have been invoked to explain isotope ratio measurements. Often isotope information is most useful when regarded in the context of other parameters that are conventionally measured to study N_2O production. The combination of isotope investigations and other evidence can lead to a better understanding of N_2O production at the Earth's surface (Pérez et al., 2001).

Regarding the global N_2O budget, isotope ratio measurements have not yet reached a level where they can be used to put additional constraints on the budget. In fact, the opposite is true, and at present independent knowledge on the N_2O cycle is used in a top-down approach to derive flux-weighted globally representative isotopic source signatures for the surface sources (Yoshida and Toyoda, 2000). In particular the large variability in the isotopic composition of N_2O emitted from soils represents a challenge. With the modern continuous-flow techniques, there is hope that the data-base on N_2O emitted from surface sources can be expanded considerably in the future.

5. ACKNOWLEDGMENTS

NERC (UK) are acknowledged for funding Atlantic monitoring and global modelling of methane, the Norsk Polar Institute for funding a sampling campaign in Ny-Alesund, and INTAS for funding projects in St.Petersburg and Siberia. Thanks go to collaborators on these projects, Ingeborg Levin, Kim Holmen and all at RCRSA, St. Petersburg. The European Community is acknowledged for support via the Marie Curie Fellowship of V. Gros (EVK2-CT-1999-50003) and the project CO-OH-Europe (ENV4-CT96-0318). Thanks to all the partners involved in this project and to G. Salisbury for his help in proof-reading the chapter.

6. REFERENCES

Allen W., M.R. Manning, K.R. Lassey, D.C. Lowe and A.J. Gomez, Modelling the variation of $\delta^{13}C$ in atmospheric methane: Phase ellipses and the kinetic isotope effect, Global Biogeochem. Cycles, 15, 467-481, 2001.

Assonov, S., and C.A.M. Brenninkmeijer, A new method to determine the ^{17}O isotopic abundance in CO2 using oxygen isotope exchange with a solid oxide., Rapid Communications in Mass Spectrometry, 15, 2426-2437, 2001.

Barford, C.C., J.P. Montoya, M.A. Altabet, and R. Mitchell, Steady-state nitrogen isotope effects of N_2 and N_2O production in Paracoccus denitrificans, Appl. Environ. Microbiol., 65, 989-994, 1999.

Bergamaschi P., M. Schupp and G.W. Harris, High precision direct measurements of $^{13}CH_4/^{12}CH_4$ and $^{12}CH_3D/^{12}CH_4$ ratios in atmospheric methane sources by means of a long path tunable diode laser absorption spectrometer, Applied Optics, 33, 7704-7716,1994.

Bergamaschi P and G.W. Harris, Measurements of stable isotope ratios ($^{13}CH_4/^{12}CH_4$; $^{12}CH_3D/^{12}CH_4$) in landfill methane using a tunable diode laser absorption spectrometer, Global Biogeochem. Cycles, 9, 439-447, 1995.

Bergamaschi P., Seasonal variations of stable hydrogen and carbon isotope ratios in methane from a Chinese rice paddy, J. Geophys. Res., 102, 25,383-25,393, 1997.

Bergamaschi P., C.A.M. Brenninkmeijer, M. Hahn, T. Rockmann, D.H. Scharffe, P.J. Crutzen, N.F. Elansky, I.B. Belikov, N.B.A. Trivett and D.E.J. Worthy, Isotope analysis based source identification for atmospheric CH_4 and CO sampled across Russia using the Trans-Siberian railroad, J. Geophys. Res., 103, 8227-8235, 1998a.

Bergamaschi P., C. Lubina, R. Konigstedt, H. Fischer, A.C. Veltkamp and O. Zwaagstra Stable isotopic signatures ($\delta^{13}C$, δD) of methane from European landfill sites, J. Geophys. Res., 103, 8251-8265.

Bergamaschi P., M.Braunlich, T. Marik and C.A.M. Brenninkmeijer, Measurements of the carbon and hydrogen isotopes of atmospheric methane at Izaña, Tenerife: Seasonal cycles and synoptic scale variations, J. Geophys. Res., 105, 14,531-14,546, 2000a.

Bergamaschi, P., R. Hein, C.A.M. Brenninkmeijer, and P.J. Crutzen,. Inverse modelling of the global CO cycle: 2. Inversion of $^{13}C/^{12}C$ and $^{18}O/^{16}O$ ratios, J. Geophys. Res., 105, 1929-1945, 2000b.

Bergamaschi P., D.C. Lowe, M.R. Manning, R. Moss, T. Bromley and T.S. Clarkson, Transects of atmospheric CO, CH_4, and their isotopic composition across the Pacific: Shipboard measurements and validation of inverse models, J. Geophys. Res., 106, 7993-8011, 2001.

Bilek R.S., S.C. Tyler, R.L. Saa and F.M. Fisher, Differences in CH_4 oxidation and pathways of production between rice cultivars deduced from measurements of CH_4 flux, and $\delta^{13}C$ of CH_4 and CO_2, Global Biogeochem. Cycles, 13, 1029-1044, 1999.

Bilek R.S., S.C. Tyler, M. Kurihara, and K. Yagi, Investigation of cattle methane production and emission over a 24 hour period using measurements of $\delta^{13}C$ and δD of emitted CH_4 and rumen water, J. Geophys. Res., 106, 15,405-15,413, 2001.

Boontanon, N., S. Ueda, P. Kanatharana, and E. Wada, Intramolecular stable isotope ratios of N_2O in the tropical swamp forest in Thailand, Naturwissenschaften, 87, 188-192 2000.

Bousquet, P., P. Peylin, P. Ciais, M. Ramonet, and P. Monfray, Inverse modelling of annual atmospheric CO_2 sources and sinks. Part 2 : sensitivity study, J. Geophys. Res., 104, 26,179-26,193, 1999.

Brand, W.A., PreCon: A fully automated interface for the pre-GC concentration of trace gases in air for isotopic analysis, Isotopes Environ. Health Stud., 31, 277-284 1995.

Breas O., C. Guillou, F. Reniero and E. Wada, The global methane cycle: Isotopes and mixing ratios, sources and sinks, Isotopes Environ. Health Stud., 37, 257-379, 2001.

Brenninkmeijer, C.A.M., Measurement of the abundance of ^{14}CO in the atmosphere and the $^{13}C/^{12}C$ and $^{18}O/^{16}O$ ratio of atmospheric CO, with application in New-Zealand and Antarctica, J. Geophys. Res., 98, 10,595-10,614, 1993.

Brenninkmeijer C.A.M., D.C. Lowe, M.R. Manning, R.J. Sparks and P.F.J. van Velthoven, The ^{13}C, ^{14}C, and ^{18}O isotopic composition of CO, CH_4, and CO_2 in the higher southern latitudes lower stratosphere, J. Geophys. Res., 100, 26,163-26,172, 1995.

Brenninkmeijer, C.A.M., and T. Röckmann, Principal factors determining the $^{18}O/^{16}O$ ratio of atmospheric CO as derived from observations in the southern hemispheric troposphere and lowermost stratosphere, J. Geophys. Res., 102, 25477-25485, 1997.

Brenninkmeijer, C.A.M., and T. Röckmann, A rapid method for the preparation of O_2 from CO_2 for mass spectrometric analysis of $^{17}O/^{16}O$ ratios, Rapid Communications in Mass Spectrometry, 12, 479-483, 1998.

Brenninkmeijer, C.A.M., and T. Röckmann, Mass spectrometry of the intramolecular nitrogen isotope distribution of environmental nitrous oxide using fragment-ion analysis, Rapid Commun. Mass Spectrom., 13, 2028-2033, 1999.

Brenninkmeijer, C.A.M., C. Koeppel, T. Röckmann, D.S. Scharffe, M. Bräunlich, and V. Gros, V., Absolute measurements of the abundance of atmospheric carbon monoxide, J. Geophys. Res., 106, 10,003-10,010, 2001.

Cantrell, C.A., R.E. Shetter, A.H. McDaniel, J.G. Calvert, J.A. Davidson, D.C. Lowe, S.C. Tyler, R.J. Cicerone and J.P. Greenberg, Carbon kinetic isotope effect in the oxidation of methane by the hydroxyl radical, J. Geophys. Res., 95, 22,455-22,462, 1990.

Chanton J.P., G.J. Whiting, N.E. Blair, C.W. Lindau and P.K. Bollich, Methane emission from rice: Stable isotopes, diurnal variations and CO_2 exchange, Global Biogeochem. Cycles, 11, 15-27, 1997.

Chanton J. and K. Liptay K., Seasonal variation in methane oxidation in a landfill cover soil as determined by an in situ stable isotope technique, Global Biogeochem. Cycles, 14, 51-60, 2000.

Chanton J.P., C.M. Rutkowski, C.C. Schwartz, D.E. Ward and L. Boring, Factors influencing the stable carbon isotopic signature of methane from combustion and biomass burning, J. Geophys. Res., 105, 1867-1877, 2000.

Ciais, P., P.P. Tans, M. Trolier, J.W.C. White, and R.J. Francey, A large Northern hemisphere Terrestrial CO_2 sink indicated by the $^{13}C/^{12}C$ ratio of atmospheric CO_2, Science, 269, 1098-1102, 1995.

Cliff, S.S., C.A.M. Brenninkmeijer, and M.H. Thiemens, First measurement of the $^{18}O/^{16}O$ and $^{17}O/^{16}O$ ratios in stratospheric nitrous oxide: A mass-independent anomaly, J. Geophys. Res., 104, 16171-16175, 1999.

Cliff, S.S., and M.H. Thiemens, High-precision isotopic determination of the $^{18}O/^{16}O$ and $^{17}O/^{16}O$ ratios in nitrous oxide, Annal. Chem., 66, 2791-2793, 1994.

Cliff, S.S., and M.H. Thiemens, The $^{18}O/^{16}O$ and $^{17}O/^{16}O$ ratios in atmospheric nitrous oxide: A mass-independent anomaly, Science, 278, 1774-1776, 1997.

Conny, J.M., M. Verkouteren, and L.A. Curie, Carbon 13 composition of tropospheric CO in Brazil: A model scenario during the biomass burn season, J. Geophys. Res., 102, 10,683-10,693, 1997.

Conny, J.M., The isotopic characterization of carbon monoxide in the troposphere, Atmos. Environ., 32, 2669-2683, 1998.

Coplen, T.B., J.A. Hopple, J.K. Böhlke, H.S. Peiser, S.E. Rieder, H.R. Krouse, K.J.R. Rosman, T. Ding, J. Vocke, R. D., K.M. Révész, A. Lamberty, P. Taylor, and P. De Bièvre, Compilation of minimum and maximum isotope ratios of selected elements in naturally K.A. occurring materials and reagents, 131 pp., US Geological Survey, Reston, Virginia, 2002.

Dentener, F., and P.J. Crutzen, A three-dimensional model of the global ammonia cycle, J. Atmos. Chem., 331-369, 1994.

Dore, J.E., B.N. Popp, D.M. Karl, and F.J. Sansone, A large source of atmospheric nitrous oxide from subtropical North Pacific surface waters, Nature, 396, 63-66, 1998.

Dlugokencky E.J., K.A. Masarie, P.M. Lang and P.P. Tans, Continuing decline in the growth rate of the atmospheric methane burden, Nature, 393, 447-450, 1998.

Dlugokencky E.J., B.P. Walter, K.A. Masarie, P.M. Lang and E.S. Kasischke, Measurements of an anomalous global methane increase during 1998, Geophys. Res. Lett., 28, 499-502, 2001.

Eisma R, A.T. Vermeulen and. K. van der Borg, $^{14}CH_4$ emissions from nuclear power plants in Northwestern Europe, Radiocarbon, 37, 475-483, 1995.

Esler, M.B., D.W.T. Griffith, F. Turatti, S.R. Wilson, T. Rahn, and H. Zhang, N_2O concentration and flux measurements and complete isotopic analysis by FTIR spectroscopy, Chemosphere: Global Change Sci., 2, 445-454, 2000.

Estupiñán, E.G., R.E. Stickel, J.M. Nicovich, and P.H. Wine, Investigation of N_2O production from 266 and 532 nm laser flash photolysis of $O_3/N_2/O_2$, J. Phys. Chem. A, 106, 5880-5890, 2002.

Feilberg, K., S.R. Sellevag, C.J. Nielsen, D.W.T. Griffith, and M.S. Johnson., CO + OH→CO2 + H: the relative reaction rate of five CO isotopologues, Phys. Chem. Chem. Phys., 4, 4687- 4693, 2002.

Francey, R. J., P.P. Tans, C.E. Allison, I.G. Enting, J.W.C. White, and M. Trolier, Changes in oceanic and carbon uptake since 1982, Nature, 373, 326-330, 1995.

Francey R.J., M.R. Manning, C.E. Allison, S.A. Coram, D.M. Etheridge, R.I. Langenfelds, D.C. Lowe and L.P. Steele, A history of $\delta^{13}C$ in atmospheric CH_4 from the Cape grim Air Archive and Antarctic firn air, J. Geophys. Res., 104, 23,631-23,643, 1999.

Gonfiantini, R., Standards for stable isotope measurements in natural compounds, Nature, 271, 534-536, 1978.

Grace J.D. and G.F. Hart, Giant gas fields of Northern West Siberia, AAPG Bulletin 70, 830-852, 1986.

Griffith, D.W.T., G.C. Toon, B. Sen, J.-F. Blavier, and R.A. Toth, Vertical profiles of nitrous oxide isotopomer fractionation measured in the stratosphere, Geophys. Res. Lett., 27, 2485-2488, 2000.

Gros, V., M. Bräunlich, T. Röckmann, P. ., Jöckel, P. Bergamaschi, C.A.M. Brenninkmeijer, W. Rom, W. Kutschera, A. Kaiser, H.E. Scheel, M. Mandl, J.V.D. Plicht, and G. Possner, Detailed analysis of the isotopic composition of CO and characterization of the air masses arriving at Mt. Sonnblick (Austrian Alps), J. Geophys. Res., 106, 3179-3193, 2001.

Gros, V., P. Jöckel, C.A.M. Brenninkmeijer, T. Röckmann, F. Meinhardt, R. and Graul, Characterization of pollution events observed at Schauinsland using CO and its stable isotopes, Atmos. Environ., 36, 2831-2840, 2002.

Gupta M, S.C. Tyler and R.J. Cicerone, Modelling atmospheric $\delta^{13}CH_4$ and the causes of recent changes in atmospheric CH_4 amounts, J. Geophys. Res., 101 , 22,923-22,932, 1996.

Gupta M.L., M.P. McGrath, R.J. Cicerone, F.S. Rowland and M. Wolfsberg, $^{12}C/^{13}C$ kinetic isotope effects in reactions of CH_4 with OH and Cl, Geophys. Res. Lett., 24, 2761-2764, 1997.

Happell J.D., J.P. Chanton and W.S. Showers, The influence of methane oxidation on the stable isotopic composition of methane emitted from Florida swamp forests, Geochim. Cosmochim. Acta, 58, 4377-4388, 1994.

Hein R., P.J. Crutzen and M. Heimann, An inverse modelling approach to investigate the global atmospheric methane cycle, Global Biogeochem. Cycles, 11, 43-76, 1997.

Hilkert A.W., C.B. Douthitt, H.J. Schluter and W.A. Brand, Isotope Ratio Monitoring Gas Chromatography / Mass Spectrometry of D/H by High Temperature Conversion Isotope Ratio Mass Spectrometry, Rapid Communications in Mass Spectrometry, 13, 1226-1230, 1999.

Hogan K.B. and R.C. Harriss, Comment on "A dramatic decrease in the growth rate of atmospheric methane in the Northern hemisphere during 1992" by E.J. Dlugokencky et al., Geophys. Res. Lett., 21, 2445-2446, 1994.

Hornibrook E.R.C., F.J. Longstaffe and W.S. Fyfe, Spatial distribution of microbial methane production pathways in temperate zone wetland soil: Stable carbon and hydrogen isotope evidence, Geochim. Cosmochim. Acta, 61, 745-753, 1997.

Houweling S., F. Dentener, J. Lelieveld, B. Walter and E. Dlugokencky, The modelling of tropospheric methane: How well can point measurements be reproduced by a global model?, J. Geophys. Res., 10, 26,137-26,160, 2000.

Huff, A.K., M.H. and Thiemens, $^{17}O/^{16}O$ and $^{18}O/^{16}O$ isotope measurements of atmospheric carbon monoxide and its sources, Geophys. Res. Lett., 25, 3509-3512. 1998.

Hut, G., Consultants' Group Meeting on stable isotope reference samples for geochemical and hydrological investigations, 42 pp., IAEA, Vienna, 1987.

IAEA, Reference and intercomparison materials for stable isotopes of light elements, 165 pp., International Atomic Energy Agency, Vienna, 1995.

Jöckel, P., M.G. Lawrence, and C.A.M. Brenninkmeijer, Simulations of cosmogenic ^{14}CO using the three-dimensional atmospheric model MATCH: Effects of ^{14}C production distribution and the solar cycle, J. Geophys. Res., 104, 11,733-11,743, 1999.

Jöckel, P., C.A.M. Brenninkmeijer, and M.G. Lawrence, Atmospheric response time of cosmogenic ^{14}CO to changes in solar activity, J. Geophys. Res., 105, 6737-6744, 2000.

Jöckel, P., C. A. M. Brenninkmeijer, M. G. Lawrence, A. B. M. Jeuken, and P. F. van Velthoven. Evaluation of stratosphere - troposphere exchange and the hydroxyl radical distribution in three-dimensional global atmospheric models using observations of cosmogenic ^{14}CO, J. Geophys. Res., 107, 4446, 101029/2001JD001324, 2002.

Jöckel, P., and C.A.M. Brenninkmeijer, The seasonal cycle of cosmogenic ^{14}CO at the surface level: A solar cycle adjusted, zonal average climatology based on observations, J. Geophys. Res., 4656, 10.1029/2001JD001104, 2002.

Johnston, J.C., S.S. Cliff, and M.H. Thiemens, Measurement of multioxygen isotopic ($\delta^{18}O$ and $\delta^{17}O$) fractionation factors in the stratospheric sink reactions of nitrous oxide, J. Geophys. Res., 100, 16801-16804, 1995.

Kaiser, J., Stable isotope investigations of atmospheric nitrous oxide, Ph. D. thesis, Johannes Gutenberg-Universität, Mainz (also available as a book from http://www.dr.hut-verlag.de, ISBN 3-934767-90-7, or at http://archimed.uni-mainz.de/pub/2003/0004), 2002.

Kaiser, J., C.A.M. Brenninkmeijer, and T. Röckmann, Intramolecular ^{15}N and ^{18}O fractionation in the reaction of N_2O with $O(^1D)$ and its implications for the stratospheric N_2O isotope signature, J. Geophys. Res., 107, 4214, 10.1029/2001JD001506, 2002a.

Kaiser, J., T. Röckmann, and C.A.M. Brenninkmeijer, Temperature dependence of isotope fractionation in N_2O photolysis, Phys. Chem. Chem. Phys., 4, 4220-4230, 10.1039/B204837J, 2002b.

Kaiser, J., T. Röckmann, and C.A.M. Brenninkmeijer, Complete and accurate mass-spectrometric isotope analysis of tropospheric nitrous oxide, Journal of Geophysical Research, in press, 2003a.

Kaiser, J., T. Röckmann, C.A.M. Brenninkmeijer, and P.J. Crutzen, Wavelength dependence of isotope fractionation in N_2O photolysis, Atmos. Chem. Phys., 3, 303-313, 2003b.

Kato, S., Akimoto, H., Bräunlich, M., Röckmann, T., and Brenninkmeijer, C.A.M., Measurements of stable carbon and oxygen isotopic composition of CO in automobile exhausts and ambient air from semi-urban Mainz, Germany, Geochemical Journal, 33, 73-77, 1999.

Kato, S., J. Kajii, H. Akimoto, M. Bräunlich, T. Röckmann, and C.A.M. Brenninkmeijer, Observed and modelled seasonal variation of ^{13}C, ^{18}O and ^{14}C of atmospheric CO at Happo, a remote site in Japan, and a comparison with other records, J. Geophys. Res., 105, 8891-8900, 2000.

Keeling, R. F., The atmospheric oxygen cycle: the oxygen isotopes of atmospheric CO_2 and O_2 and the O_2/N_2 ratio, Reviews of Geophysics, 1253-1262, 1995.

Kim, K.-R., and H. Craig, Two-isotope characterization of nitrous oxide in the Pacific Ocean and constraints on its origin in deep water, Nature, 347, 58-61, 1990.

Kim, K.-R., and H. Craig, Nitrogen-15 and oxygen-18 characteristics of nitrous oxide: A global perspective, Science, 262, 1855-1857, 1993.

Kohlmann, J.-P., and D. Poppe, The tropospheric gas-phase degradation of NH_3 and its impact on the formation of N_2O and NO_x, J. Atmos. Chem., 32, 397-415, 1999.

Krankowsky, D., F. Bartecki, G.G. Klees, K. Mauersberger, K. Schellenbach, and J. Stehr, Measurement of heavy isotope enrichment in tropospheric ozone, Geophys. Res. Lett., 22, 1713-1716, 1995.

Kroeze, C., and S. Seitzinger, Nitrogen input to rivers, estuaries and continental shelves and related nitrous oxide emissions in 1990 and 2050: a global model, Nutr. Cycl. Agroecosyst., 52 195-212, 1998.

Kroeze, C., A. Mosier, and L. Bouwman, Closing the N_2O budget: A retrospective analysis 1500-1994, Global Biogeochem. Cycles, 13, 1-8, 1999.

Lassey K.R., D.C. Lowe, C.A.M. Brenninkmeijer and A.J. Gomez, Atmospheric methane and its carbon isotopes in the southern hemisphere: Their time series and an instructive model, Chemosphere, 26, 95-109, 1993.

Lassey K.R., D.C. Lowe and M.R. Manning, The trend in atmospheric methane $\delta^{13}C$ and implications for isotopic constraints on the global methane budget, Global Biogeochem. Cycles, 14, 41-49, 2000.

Levin, I., P. Bergamaschi, H. Dorr and D. Trapp, Stable isotopic signature of methane from major sources in Germany, Chemosphere, 26, 161-177, 1993.

Levin I., H. Glatzel-Mattheier, T. Marik, M. Cuntz, M. Schmidt and D.E. Worthy, Verification of German methane emission inventories and their recent changes based on atmospheric observations, J. Geophys. Res., 104, 3447-3456, 1999a.

Levin I., B. Kromer, C. Poss, T. Marik, H. Sartoris, W. Weiss and M. Maiss, Quasi-continuous observations of atmospheric trace substances and their isotopic composition at Alert. Canadian Baseline Program: Summary of Progress to 1998, Environment Canada, Toronto, 5.1-5.4, 1999b.

Lindholm, N., and J.F. Hershberger, Product branching ratios of the $NH_2(X^2B_1) + NO_2$ reaction, J. Phys. Chem. A, 101, 4991-4995, 1997.

Lingenfelter, R. E., Production of carbon 14 by cosmic-ray neutrons, Rev. Geophys.,1 ,35-55, 1963.

Liptay K., J. Chanton, P. Czepiel and B. Mosher, Use of stable isotopes to determine methane oxidation in landfill cover soils, J. Geophys. Res., 103, 8243-8250, 1998.

Lowe, D.C., C.A.M. Brenninkmeijer , S.C. Tyler and E.J. Dlugokencky, Determination of the isotopic composition of atmospheric methane and its application in the Antarctic, J.Geophys. Res., 96, 15455-15467, 1991.

Lowe D.C., C.A.M. Brenninkmeijer, G.W. Brailsford, K.R. Lassey, A.J. Gomez and E.G. Nisbet, Concentration and ^{13}C records of atmospheric methane in New Zealand and Antarctica: Evidence for changes in methane sources. J. Geophys. Res., 99, 16,913-16,925, 1994.

Lowe D.C., M.R. Manning, G.W. Brailsford and A.M. Bromley, The 1991-1992 Atmospheric Methane anomaly: Southern hemisphere ^{13}C decrease and growth rate fluctuations, Geophys. Res. Lett., 24, 857-860, 1997.

Lowe D.C., W. Allan, M.R. Manning, T. Bromley, G. Brailsford, D. Ferretti, A. Gomez, R. Knobben, R. Martin, Z. Mei, R. Moss, K. Koshy and M. Maata, Shipboard determinations of the distribution of 13C in atmospheric methane in the Pacific, J. Geophys. Res., 104 , 26,125-26,135, 1999.

Lowry D., P. O'Brien, E.G. Nisbet. And N.D. Rata, $\delta^{13}C$ of atmospheric methane: an integrated technique for constraining emission sources in urban and background air. In: Isotope Techniques in the Study of Environmental Change. Proceedings of IAEA Symposium SM-349, 57-67, 1998.

Lowry D., C.W. Holmes, N.D. Rata, P. O'Brien and E.G. Nisbet, London methane emissions: Use of diurnal changes in concentration and $\delta^{13}C$ to identify urban sources and verify inventories, J. Geophys. Res., 106, 7427-7448, 2001.

Lowry D., C.W. Holmes, E.G. Nisbet and N.D. Rata, Can EC and UK national methane emission inventories be verified using high precision stable isotope data? Proceedings: Study of Environmental Change using Isotope Techniques, IAEA-CSP-13/P, IAEA, Vienna, 399-409, 2002.

McElroy, M.B., and D.B.A. Jones, Evidence for an additional source of atmospheric N_2O, Global Biogeochem. Cycles, 10, 651-659, 1996.

MacKay, C., M. Pandow, and R. Wolfgang, On the chemistry of natural radiocarbon, J. Geophys. Res., 68, 3929-3931, 1963.

Mak, J.E., and W.B. Yang, Technique for analysis of air samples for C-13 and O-18 in carbon monoxide via continuous-flow isotope ratio mass spectrometry, Analytical Chemistry, 70, 5159-5161, 1998.

Mak, J.E., and G. Kra, The isotopic composition of carbon monoxide at Montauk Point, Long Island, Chemosphere Glob. Change Sci., 1, 205-218, 1999.

Mak J.E., M.R. Manning and D.C. Lowe, Aircraft observations of $\delta^{13}C$ of atmospheric methane over the pacific in August 1991 and 1993: Evidence of an enrichment in $^{13}CH_4$ in the Southern hemisphere, J. Geophys. Res., 105, 1329-1335, 2000.

Mandernack, K.W., T. Rahn, C. Kinney, and M. Wahlen, The biogeochemical controls of the $\delta^{15}N$ and $\delta^{18}O$ of N_2O produced in landfill cover soils, J. Geophys. Res., 105, 17709-17720 2000.

Manning, M.R., C.A.M. Brenninkmeijer and W. Allan, Atmospheric carbon monoxide budget of the southern hemisphere: Implications of $^{13}C/^{12}C$ measurements, J. Geophys. Res., 102, 10,673-10,682, 1997.

Mariotti, A., J.C. Germon, P. Hubert, P. Kaiser, R. Letolle, A. Tardieux, and P. Tardieux, Experimental determination of nitrogen kinetic isotope fractionation some principles; illustration for the denitrification and nitrification processes, Plant Soil, 62, 413-430. 1981.

Mariotti, A., Atmospheric nitrogen is a reliable standard for natural ^{15}N abundance measurements, Nature, 303, 685-687, 1983.

Mariotti, A., Natural ^{15}N abundance measurements and atmospheric nitrogen standard calibration, Nature, 311, 251-252, 1984.

Masarik, J. and J. Beer, Simulation of particle uxes and cosmogenic nuclide pro-duction in the Earth's atmosphere, J. Geophys. Res., 104, 12 099-12 111, 1999.

Merritt D.A., J.M. Hayes and D.J. Des Marais, Carbon isotopic analysis of atmospheric methane by isotope-ratio-monitoring gas chromatography-mass spectrometry, J. Geophys. Res., 100 , 1317-1326, 1995.

Miller J.B., K.A. Mack, R. Dissly, J.W.C. White, E.J. Dlugokencky and P.P Tans, Development of analytical methods and measurements of $^{13}C/^{12}C$ in atmospheric CH_4 from the NOAA Climate Monitoring and Diagnostics Laboratory Global Air Sampling Network, J. Geophys. Res., 107, 10.1029/2001JD000630, 2002.

Moore, H., Isotopic measurement of atmospheric nitrogen compounds, Tellus, 26, 169-174 1974.

Moriizumi J., K. Nagamine, T. Iida and Y. Ikebe, Carbon isotopic analysis of atmospheric methane in urban and suburban sources: Fossil and non-fossil methane from local sources. Atmos. Env., 32, 2947-2955, 1998.

Naqvi, S.W.A., N_2O production in the ocean, Nature, 349, 373-374, 1991.

Naqvi, S.W.A., T. Yoshinari, D.A. Jayakumar, M.A. Altabet, P.V. Narvekar, A.H. Devol, J.A. Brandes, and L.A. Codispoti, Budgetary and biogeochemical implications of N_2O isotope signatures in the Arabian Sea, Nature, 394, 462-464, 1998.

Naqvi, S.W.A., and D.A. Jayakumar, Ocean biogeochemistry and atmospheric composition: Significance of the Arabian Sea, Curr. Sci., 78, 289-299, 2000.

Nisbet E.G., I. Levin, G.P. Wyers, and A.F. Roddy (eds.), Quantification of the west European methane emissions budget by atmospheric measurements. Report EUR 17511 EN, European Commission, Luxembourg, 24pp, 1998.

Nisbet E.G. (ed.), Russian Emissions of Atmospheric Methane: Study of Sources. Final Report of INTAS 97-2055 1998-2001, 2001.

Novelli, P.C., K.A. Masarie, and P.M. Lang, Distributions and recent changes of carbon monoxide in the lower troposphere, J. Geophys. Res., 103 19,015-19,033, 1998.

O'Brien, K., A. de la Zerda Lerner, M. A. Shea, and D. F. Smart, The production of cosmogenic isotopes in the earth's atmosphere and their inventories. In C. P. Sonett, M. S. Giampapa, and M. S. Matthews, editors, The sun in time, pages 317-342. The University of Arizona Press, Tucson, Arizona, 1991.

Olivier, J.G.J., A.F. Bouwman, K.W. van der Hoek, and J.J.M. Berdowski, Global air emission inventories for anthropogenic sources of NO_x, NH_3 and N_2O in 1990, Environ. Pollut., 102, 135-148, 1998.

Ostrom, N.E., M.E. Russ, B. Popp, T.M. Rust, and D.M. Karl, Mechanisms of nitrous oxide production in the subtropical North Pacific based on determinations of the isotopic abundances of nitrous oxide and di-nitrogen, Chemosphere: Global Change Sci., 2, 281-290, 2000.

Park, J., and M.C. Lin, A mass spectrometric study of the $NH_2 + NO_2$ reaction, J. Phys. Chem. A, 101, 2643-2647, 1997.

Pérez, T., S.E. Trumbore, S.C. Tyler, E.A. Davidson, M. Keller, and P.B. De Camargo, Isotopic variability of N_2O emissions from tropical forest soils, Global Biogeochem. Cycles, 14, 525-535, 2000.

Pérez, T., S.E. Trumbore, S.C. Tyler, P.A. Matson, I. Ortiz-Monasterio, T. Rahn, and D.W.T. Griffith, Identifying the agricultural imprint on the global N_2O budget using stable isotopes, J. Geophys. Res., 106, 9869-9878, 2001.

Popp, B.N., M.B. Westley, S. Toyoda, T. Miwa, J.E. Dore, N. Yoshida, T.M. Rust, F.J. Sansone, M.E. Russ, N.E. Ostrom, and P.H. Ostrom, Nitrogen and oxygen isotopomeric constraints on the origins and sea-to-air flux of N_2O in the oligotrophic subtropical North Pacific gyre, Global Biogeochem. Cycles, 16, 1064, 10.1029/2001GB001806, 2002.

Prasad, S.S., E.C. Zipf, and X. Zhao, Potential atmospheric sources and sinks of nitrous oxide: 3. Consistency with the observed distributions of the mixing ratios, J. Geophys. Res., 102, 21537-21541, 1997.

Quay, P.D., S.L. King, J. Stutsman, D.O. Wilbur, L.P. Steele, I. Fung, R.H. Gammon, T.A. Brown, G.W. Farwell, P.M. Grootes, and F.H. Schmidt, Carbon isotopic composition of atmospheric CH_4: fossil and biomass burning source strength, Global Biogeochem. Cycles, 5, 25-47, 1991.

Quay P., J. Stutsman, D. Wilbur, A. Snover, E. Dlugokencky and T. Brown, The isotopic composition of atmospheric methane, Global Biogeochem. Cycles, 13, 445-461, 1999.

Rahn, T., and M. Wahlen, Stable isotope enrichment in stratospheric nitrous oxide, Science, 278, 1776-1778, 1997.

Rahn, T., H. Zhang, M. Wahlen, and G.A. Blake, Stable isotope fractionation during ultraviolet photolysis of N_2O, Geophys. Res. Lett., 25, 4489-4492, 1998.

Rahn, T., and M. Wahlen, A reassessment of the global isotopic budget of atmospheric nitrous oxide, Global Biogeochem. Cycles, 14, 537-543, 2000.

Rata N.D., Development of new cryogenic extraction techniques for studying stable isotope ratios in atmospheric methane, PhD thesis, University of London, 265pp, 1999.

Rayner, P., I. Enting, R.J. Francey and R.L. Langefelds, Reconstructing the recent carbon cycle from atmospheric CO_2, $d^{13}C$ and O_2/N_2 observations, Tellus 51B, 213-232, 1999.

Révész K., T.B. Coplen, M.J. Baedecker, P.D. Glynn and M. Hult, Methane production and consumption monitored by stable H and C isotope ratios at a crude oil spill site, Bemidji, Minnesota, Applied Geochemistry, 10, 505-516, 1995.

Rice, A.L., A.A. Gotoh, H.O. Ajie, and S.C. Tyler, High-precision continuous-flow measurements of $\delta^{13}C$ and δD of atmospheric CH_4, Analytical Chemistry, 73 (18), 4104-4110, 2001.

Rigby, D., and B.D. Batts, The isotopic composition of nitrogen in Australian coals and oil shales, Chemical Geology (Isotope Geoscience Section), 58, 273-282, 1986.

Röckmann, T., C.A.M. Brenninkmeijer, G. Saueressig, P. Bergamaschi, J. Crowley, H. Fischer, and P.J. Crutzen, Mass independent fractionation of oxygen isotopes in atmospheric CO due to the reaction CO + OH, Science, 281, 544-546, 1998.

Röckmann, T., C.A.M. Brenninkmeijer, P.J. Crutzen, and U. Platt, Short term variations in the $^{13}C/^{12}C$ ratio of CO as a measure of Cl activation during tropospheric ozone depletion events in the Arctic, J. Geophys. Res., 104, 1691-1697, 1999.

Röckmann, T., C.A.M. Brenninkmeijer, M. Wollenhaupt, J.N. Crowley, and P.J. Crutzen, Measurement of the isotopic fractionation of $^{15}N^{14}N^{16}O$, $^{14}N^{15}N^{16}O$ and $^{14}N^{14}N^{18}O$ in the UV photolysis of nitrous oxide, Geophys. Res. Lett., 27, 1399-1402, 2000.

Röckmann, T., J. Kaiser, C.A.M. Brenninkmeijer, J.N. Crowley, R. Borchers, W.A. Brand, and P.J. Crutzen, The isotopic enrichment of nitrous oxide ($^{15}N^{14}NO$, $^{14}N^{15}NO$, $^{14}N^{14}N^{18}O$) in the stratosphere and in the laboratory, J. Geophys. Res., 106, 10403-10410, 2001a.

Röckmann, T., J. Kaiser, J.N. Crowley, C.A.M. Brenninkmeijer, and P.J. Crutzen, The origin of the anomalous or "mass-independent" oxygen isotope fractionation in atmospheric N_2O, Geophys. Res. Lett., 28, 503-506, 2001b.

Röckmann, T., P. Jöckel, M. Bräunlich, V. Gros, G. Posnert and C.A.M. Brenninkmeijer, Using 14C, 13C, 18O and 17O isotopic variations to provide insights into the high northern latitude surface CO inventory, Atmos. Chem. Phys., 2, 147-159, 2002.

Röckmann, T., J. Kaiser, C.A.M. Brenninkmeijer, and W.A. Brand, Gas-chromatography, isotope-ratio mass spectrometry method for high-precision position-dependent ^{15}N and ^{18}O measurements of atmospheric nitrous oxide, Rapid Communications in Mass Spectrometry, submitted, 2003a.

Röckmann, T., J. Kaiser, and C.A.M. Brenninkmeijer, The isotopic fingerprint of the pre-industrial and the anthropogenic N_2O source, Atmos. Chem. Phys., 3, 315-323, 2003b.

Rom, W., C.A.M. Brenninkmeijer, C.B. Ramsey, W. Kutschera, A. Priller, S. Puchegger, T. Röckmann, and P. Steier, Methodological aspects of atmospheric (CO)-C-14 measurements with AMS, Nuclear Instruments and Methods in Physics Research Section B- Beam Interactions With Materials and Atoms, 172, 530-536, 2000.

Saueressig G., P. Bergamaschi, J.N. Crowley, H. Fischer and G.W. Harris, Carbon kinetic isotope effect in the reaction of CH_4 with Cl atoms, Geophys. Res. Lett., 22, 1225-1228, 1995.

Saueressig G., J.N. Crowley, P. Bergamaschi, C. Brühl, C.A.M. Brenninkmeijer and H. Fischer, Carbon 13 and D kinetic isotope effects in the reactions of CH_4 with $O(^1D)$ and OH: New laboratory measurements and their implications for the isotopic composition of stratospheric methane, J. Geophys. Res., 106 , 23,127-23,138, 2001.

Snover A.K., P.D. Quay and W.M. Hao, The D/H content of methane emitted from biomass burning, Global Biogeochem. Cycles, 14, 11-24, 2000.

Sowers, T., N_2O record spanning the penultimate deglaciation from the Vostok ice core, J. Geophys. Res., 106, 31903-31914, 2001.

Stein, L.Y., and Y.L. Yung, Production, isotopic composition, and atmospheric fate of biologically produced nitrous oxide, Annu. Rev. Earth Planet. Sci., 110502.080901, 2003.

Stevens, C.M., L. Krout, D. Walling, A. Venters, A. Engelkemeir, and L.E. Ross, The isotopic composition of atmospheric carbon monoxide, Earth and Planetary Science Letters, 16, 147-165, 1972.

Stevens C.M. and F.E. Rust, The carbon isotopic composition of atmospheric methane, J. Geophys. Res., 87 , 4879-4882, 1982.

Stevens, C.M., Atmospheric methane, Chem. Geol., 71, 11-21, 1988.

Stevens, C.M., and A.F. Wagner, The role of isotope fractionation effects in atmospheric chemistry, Z. Naturforschung, 44a, 376 – 384, 1989.

Suguwara S., K. Nakazawa, G. Inoue, T. Machida, H. Mukai, N.K. Vinnichencko and V.U. Khattatov, Aircraft measurements of the stable carbon isotopic ratio of atmospheric methane over Siberia, Global Biogeochem. Cycles, 10, 223-231, 1996.

Suguwara S., T. Nakazawa , Y. Shirakawa, K. Kawamura, S. Aoki, T. Machida and H. Honda, Vertical profile of the carbon isotopic ratio of stratospheric methane over Japan. Geophys. Res. Lett., 24, 2989-2992, 1997.

Tanaka, N., D.M. Rye, R. Rye, H. Avak, and T. Yoshinari, High precision mass spectrometric analysis of isotopic abundance ratios in nitrous oxide by direct injection of N_2O, Int, J. Mass Spectrom. Ion Processes, 142, 163-175, 1995.

Tans, P.P., A note on isotopic ratios and the global atmospheric methane budget, Global Biogeochem. Cycles, 11, 77-81, 1997.

Thiemens, M.H., and W.C. Trogler, Nylon production: An unknown source of atmospheric nitrous oxide, Science, 251, 932-934, 1991.

Thom M., R. Bosinger, M. Schmidt and I. Levin I., The regional budget of atmospheric methane of a highly populated area, Chemosphere 26, 143-60, 1993.

Toyoda, S., and N. Yoshida, Determination of nitrogen isotopomers of nitrous oxide on a modified isotope ratio mass spectrometer, Anal. Chem., 71 , 4711-4718, 1999.

Toyoda, S., N. Yoshida, T. Urabe, S. Aoki, T. Nakazawa, S. Sugawara, and H. Honda, Fractionation of N_2O isotopomers in the stratosphere, J. Geophys. Res., 106, 7515-7522 2001.

Toyoda, S., N. Yoshida, T. Miwa, Y. Matsui, H. Yamagishi, U. Tsunogai, Y. Nojiri, and N. Tsurushima, Production meachnism and global budget of N_2O inferred from its isotopomers in the western North Pacific, Geophys. Res. Lett., 29, 10.1029/2001GL014311, 7, 1-4, 2002.

Tyler S.C., G.W. Brailsford, K. Yagi K. Minami and R. Cicerone, Seasonal variations in methane flux and $\delta^{13}CH_4$ values for rice paddies in Japan and their implications, Global Biogeochem. Cycles, 8, 1-12, 1994a.

Tyler S.C., P.M. Crill and K. Yagi K, $^{13}C/^{12}C$ Fractionation of methane during oxidation in a temperate forested soil, Geochim. Cosmochim. Acta, 58, 1625-1633, 1994b.

Tyler S.C., H.O. Ajie, M.L. Gupta, R.J. Cicerone, D.R. Blake. and E.J. Dlugokencky, Stable carbon isotopic composition of atmospheric methane: A comparison of surface level and free tropospheric air, J. Geophys. Res., 104, 13,895-13,910, 1999a.

Tyler, S.C., G.A. Klouda, G.W. Brailsford, A.C. Manning, J.M. Conny, and A.J.T. Jull, Seasonal snapshots of the isotopic (^{14}C, ^{13}C) composition of tropospheric carbon monoxide at Niwot Ridge, Colorado, Global Change Science, 1, 185-203, 1999b.

Ueda, S., N. Ogura, and E. Wada, Nitrogen stable isotope ratio of groundwater N_2O, Geophys. Res. Lett., 18 , 1449-1452, 1991.

Uehara, K., K. Yamamoto, T. Kikugawa, and N. Yoshida, Isotope analysis of environmental substances by a new laser- spectroscopic method utilizing different pathlengths, Sens. Actuator B-Chem., 74, 173-178, 2001.

Volz, A., D.H. Ehhalt, and R.G. Derwent, Seasonal and latitudinal variation of 14CO, and the tropospheric concentration of OH radicals, J. Geophys. Res., 86, 5163-5171, 1981.

Wada, E., N. Yoshida, T. Yoshioka, M. Yoh, and Y. Kabaya, The abundance of ^{15}N in N_2O in aquatic ecosystems with emphasis on denitrification, Mitt. Internat. Verein. Limnol., 25, 115-123, 1996.

Wahlen, M., and T. Yoshinari, Oxygen isotope ratios in N_2O from different environments, Nature, 313, 780-782, 1985.

Wahlen M., N. Tanaka, R. Henry, T. Yoshinari , R. Fairbanks, A. Shemesh and W. Broeker, ^{13}C, D and ^{14}C in methane, EOS Trans AGU, 68, 1220, 1987.

Wahlen M., N. Tanaka, R. Henry, B. Deck, J. Zeglen, J.S. Vogel, J. Southton, A. Shemesh, R. Fairbanks and W. Broeker, Carbon-14 in methane sources and in atmospheric methane: The contribution from fossil carbon, Science, 245, 286-290, 1989.

Wahlen M., The global methane cycle, Annual review of Earth and Planetary Science, 21, 407-426, 1993.

Waldron S., A.J. Hall and A.E. Fallick, Enigmatic stable isotope dynamics of deep peat methane, Global Biogeochem. Cycles, 13, 93-100, 1999.

Walter, B.P. and M. Heimann, A process-based, climate sensitive model to derive methane emissions from natural wetlands: Applications to five wetland sites, sensitivity to model parameters and climate, Global Biogeochem. Cycles, 14, 745-765, 2000.

Warwick N.J., S. Bekki, K.S. Law, E.G. Nisbet and J.A.Pyle, The impact of meteorology on the inter-annual growth rate of atmospheric methane, Geophys. Res. Lett., 29, 1947, 10.1029/2002GL015282, 2002.

Webster, E.A., and D.W. Hopkins, Nitrogen and oxygen isotope ratios of nitrous oxide emitted from soil and produced by nitrifying and denitrifying bacteria, Biol. Fertil. Soils, 22, 326-330 1996.

Weinstock, B., Carbon monoxide: Residence time in the atmosphere, Science, 166 ,224-225, 1969.

Whiticar, M.J., Stable isotopes and global budgets. In: Atmospheric Methane: Sources, Sinks and Role in Global Change, NATO ASI Series I, 13, edited by M.A.K. Khalil, 139-167, Springer Verlag, New York, 1993.

Wingen, L.M., and B.J. Finlayson-Pitts, An upper limit on the production of N_2O from the reaction of $O(^1D)$ with CO_2 in the presence of N_2, Geophys. Res. Lett., 25, 517-520, 1998.

Yakir D., Sternberg L. daS.L. 2000. The use of stable isotopes to study ecosystem gas exchange, Oecologia, 123, 297-311.

Yamazaki, T., N. Yoshida, E. Wada, and S. Matsuo, N_2O reduction by Azotobacter vinelandii with emphasis on kinetic nitrogen isotope effects, Plant and Cell Physiology, 28, 263-272 1987.

Yamulki, S., I. Wolf, R. Bol, B. Grant, R. Brumme, E. Veldkamp, and S.C. Jarvis, Effects of dung and urine amendments on the isotopic content of N_2O released from grasslands, Rapid Commun. Mass Spectrom., 14, 1356-1360, 2000.

Yamulki, S., S. Toyoda, N. Yoshida, E. Veldkamp, B. Grant, and R. Bol, Diurnal fluxes and the isotopomer ratios of N_2O in a temperate grassland following urine amendment, Rapid Commun. Mass Spectrom., 15, 1263-1269, 2001.

Yoshida, N., A. Hattori, T. Saino, S. Matsuo, and E. Wada, $^{15}N/^{14}N$ ratio of dissolved N_2O in the eastern tropical Pacific Ocean, Nature, 307, 442-444, 1984.

Yoshida, N., ^{15}N-depleted N_2O as a product of nitrification, Nature, 335, 528-9, 1988.

Yoshida, N., H. Morimoto, M. Hirano, I. Koike, S. Matsuo, E. Wada, T. Saino, and A. Hattori, Nitrification rates and ^{15}N abundances of N_2O and NO_3^- in the western North Pacific, Nature, 342, 895-897, 1989.

Yoshida, N., H. Morimoto, and S. Matsuo, UV photolysis and microbial reduction as major sinks of nitrous oxide with emphasis on kinetic nitrogen isotope discrimination, Eos Trans. AGU, 71, 933-934, 1990.

Yoshida, N., and S. Toyoda, Constraining the atmospheric N_2O budget from intramolecular site preference in N_2O isotopomers, Nature, 405, 330-334, 2000.

Yoshinari, T., and M. Wahlen, Oxygen isotope ratios in N_2O nitrification at a wastewater treatment facility, Nature, 317, 349-350, 1985.

Yoshinari, T., M.A. Altabet, S.W.A. Naqvi, L. Codispoti, A. Jayakumar, M. Kuhland, and A. Devol, Nitrogen and oxygen isotopic composition of N_2O from suboxic waters of the eastern tropical North Pacific and the Arabian Sea — Measurement by continuous-flow isotope-ratio monitoring, Mar. Chem., 56, 253-264, 1997.

Yung, Y.L., and C.E. Miller, Isotopic fractionation of stratospheric nitrous oxide, Science, 278, 1778-1780, 1997.

Zhang, H., P.O. Wennberg, V.H. Wu, and G.A. Blake, Fractionation of $^{14}N^{15}N^{16}O$ and $^{15}N^{14}N^{16}O$ during photolysis at 213 nm, Geophys. Res. Lett., 27, 2481-2484, 2000.

Determination of emissions from observations of atmospheric compounds

Claire E. Reeves, Derek. M. Cunnold, Richard G. Derwent, Edward Dlugokencky, Sandrine Edouard, Claire Granier, Richard Ménard, Paul Novelli, David Parrish

1. INTRODUCTION

In previous chapters, many examples are presented in which emission estimates have been derived from flux measurements at specific sites and then extrapolated to larger areas by, for example, knowledge of the spatial extent of the conditions represented by the measurement site. This can be thought of as a "bottom up" approach. In the present and in the following chapter, "top-down" approaches will be discussed, in which measured atmospheric spatial and temporal distributions of a trace gas are used to determine the global budget of chemical species and their emissions.

Over the past decades, measurements at a number of globally-distributed surface locations have been used to provide information on emissions for a range of atmospheric trace gases. This chapter describes several examples of global datasets of atmospheric observations, as well as a few examples of the use of these observations. A number of surface measurement networks are described in section 2 in terms of how the networks were developed, which trace compounds are measured and by which techniques. Section 3 relates to remote sensing and specifically discusses measurement of trace gases and aerosols by instruments on board satellites. Section 4 describes the methodology of using observations to analyse atmospheric budgets and determine emissions, with examples presented in section 5. This chapter does not extend to inverse modelling using complex chemical transport models (CTMs), which is covered in chapter 12, but shows how observations, often combined with models, can provide constraints on emissions estimates.

2. GLOBAL MEASUREMENTS OF TROPOSPHERIC CONSTITUENTS

2.1. Observations of Methane and Carbon Monoxide

Growing concern that anthropogenic emissions would lead to 'greenhouse' warming of the lower atmosphere renewed scientific interest in human-induced climate change during the early 1970's (Wang et al., 1976; Hansen et al., 1981). In response, monitoring programs in the United States, Europe and Australia expanded to include measurements of tropospheric CO_2. Arguments that a potential increase in the abundance of atmospheric methane (CH_4) would also result in atmospheric warming (Wang et al., 1976; Ramanathan et al, 1985), called for global study of CH_4. In response, several existing programs began measuring methane along with CO_2 during the late 1970's and early 1980's. Carbon monoxide (CO), while not a greenhouse gas, nonetheless plays an important role in atmospheric chemistry and climate change. Monitoring of CO, first initiated at Cape Point, South Africa in 1978 (Brunke et al., 1990), was begun at a number of locations around the world during the 1980's and 1990's.

The World Meteorological Organization (WMO) has managed a network of atmospheric observing sites since the 1950's under the Global Ozone Observing System and the Background Air Pollution Monitoring Network. The two programs were combined in 1989 under the Global Atmospheric Watch Program (GAW) (WMO, 1997). GAW serves as an umbrella organization which establishes and co-ordinates the scientific operations needed to determine regional and global distributions of natural and anthropogenic constituents. Included in GAW was the development of six new global observatories (Table 1) (Figure 1). The approximately 20 'Global Stations' are located in remote locations free of local or regional pollution, and many monitor atmospheric CO and CH_4 (http://www.wmo.ch/web/arep/exlinks.html) using standard methods (e.g. gas chromatography (GC) plus flame ionization detection (FID), and GC + hot mercuric oxide reduction or NDIR, respectively (WMO, 2001).

Table 1: WMO GEF/GAW observatories – 2003[#]

Name	Agency[**]	Latitude	Longitude	Altitude
Arembepe, Brazil[*]	INMET	12.77	-38.17	0
Alert, Nunavut, Canada	AES	82.45	-62.50	210
Baring Head Station, New Zealand	NIWA	-41.41	174.87	80
Barrow, Alaska, United States	NOAA	71.32	-156.60	11
Bukit Koto Tabang, Indonesia[*]	MGA	0.20	100.3	865
Cape Grim, Tasmania, Australia	CSIRO	-40.68	144.68	94
Cape Point, South Africa	SAWS	-34.35	18.48	210
Lauder, New Zealand	NIWA	-45.04	169.68	390
Mace Head, County Galway, Ireland	AEROCE	53.33	-9.90	25
Mauna Loa, Hawaii, United States	NOAA	19.53	-155.58	3397
Minamitorishima, Japan	JMA	24.30	-153.97	18
Mt. Kenya, Kenya[*]	KMA	0.00	37.20	----
Mt. Waliguan, Peoples Rep. of China[*]	CAMS	36.29	100.90	3810
Neumayer, Antarctica	AWI	-70.65	-8.25	42
Ny-Alesund, Svalbard, Norway	MISU	78.90	11.88	475
Pallas-Sammaltunturi, Finland	FMI	67.97	24.12	560
Tamanrasset, Algeria[*]	AMS	23.78	5.52	1377
Tenerife, Canary Islands, Spain	INM	28.30	-16.48	2360
Tutuila, American Samoa	NOAA	-14.25	-170.57	42
South Pole, Antarctica, United States	NOAA	-89.98	-24.80	2810
Ushuaia, Argentina[*]	SMN	-54.87	-68.48	20
Zugspitze, Germany	IFU	47.42	10.98	2962

[#] http://www.wmo.ch/web/arep/gaw/stations.html
[*] Denotes the six GAW/GEF sites.
[**] Operating agency of the GAW site: AEROCE = Aerosol Oceanic Chemistry Experiment; AES = Environment Canada/Atmospheric Environment Service; AMS = Algeria Meteorological Service; AWI = Alfred-Wegener-Institute; CAMS = Chinese Agency of Meteorological Sciences; CSIRO = Commonwealth Science and Industry Research Organization; FMI = Finnish Meteorological Institute; IFU = Fraunhofer Institute for Atmospheric Environmental Research; INM = National Meteorological Institute of Spain; INMET = Brazilian National Institute of Meteorology; KMA = Meteorological Agency of Kenya; MGA = Meteorological and Geophysical Agency of Indonesia; MISU = University of Stockholm Meteorological Institute; NOAA = US National Oceanic and Atmospheric Administration/CMDL; SAWB = South African Weather Bureau; SWN = Argentine National Weather Service

The U.S. National Oceanic and Atmospheric Administration (NOAA) Geophysical Monitoring for Climate Change laboratory (NOAA/GMCC) was created in 1971 to develop a global network of sampling sites that would better define changes in tropospheric CO_2. Under GMCC a cooperative air sampling program was created which consisted of eight sites in 1978 and had expanded to 22 globally-distributed monitoring locations by 1984 (Conway et al., 1988). Sampling sites were primarily chosen to represent the background marine boundary layer, free from local and regional-scale pollution (Table 2) (Figure 1).

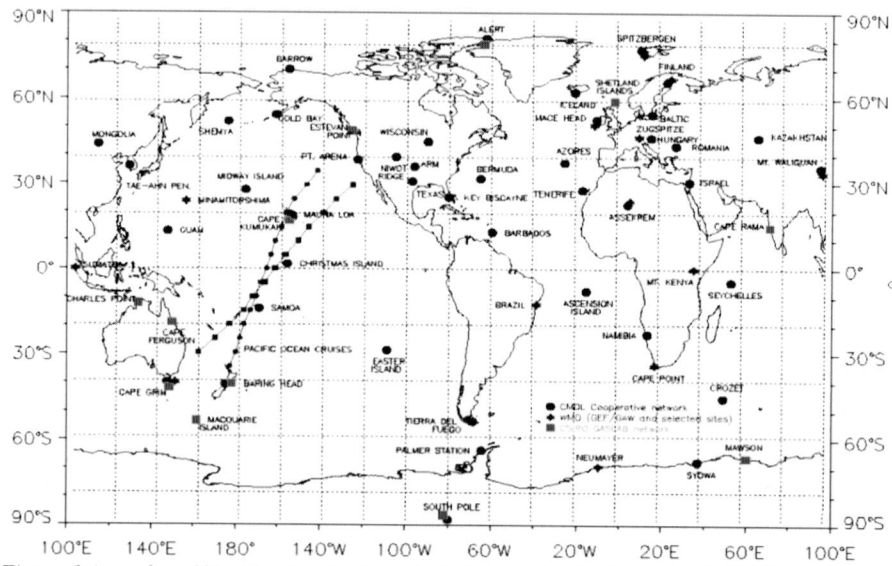

Figure 1 (see plate 19). The distribution of 'global monitoring' sites in 2003. Large circles indicate sites that comprise the NOAA/CMDL Carbon Cycle Group Cooperative Air Sampling Network (the small connected symbols show the shipboard sampling tracts). Squares show the CSIRO sampling sites, and stars indicate sites in the WMO GAW. Several locations are used by two or more networks.

Beginning 1983 CH_4 was analyzed in all network air samples using GC-FID, and the global distribution of methane was quickly established (Steele et al., 1987). In 1987, CO analysis (by GC plus hot HgO detection) was gradually added to those of CO_2 and CH_4. All sites in the network were monitored for the three carbon gases by 1993 (Novelli et al., 1994]). GMCC was combined with the NOAA Climate Research Division to form the Climate Monitoring and Diagnostics Laboratory (CMDL) in 1990, and additional expansion of the network included locations representing regionally-polluted air masses. By 2002 the NOAA/CMDL measurements of CO and CH_4 had grown to include 50 locations in the marine boundary layer, 8 (free troposphere (mountaintop) sites, 7 continental sites in areas of regional pollution, and regular shipboard sampling of latitudinal transects across the Pacific Ocean and the South China Sea (Table 2, http://www.cmdl.noaa.gov/ccgg/index.html). All data are archived at the WMO World Data Center for Greenhouse Gases (WDCGG), Tokyo, Japan, and at the Carbon Dioxide Information Analysis Center, Oak Ridge, TN, USA.

Table 2: NOAA GMCC/CMDL Sampling Sites 1968-2003[#]

Code	Name	Latitude	Longitude	Altitude	First	Last**
1. Cooperative Air Sampling Network						
ALT	Alert, Nunavut, Canada	82.45	-62.50	210	1985	2003+
AMS	Amsterdam Island, France	-37.95	77.53	150	1979	1990
ASC	Ascension Island, UK	-7.92	-14.42	54	1979	2003+
ASK	Assekrem, Algeria	23.18	5.42	2728	1995	2003+
AVI	St. Croix, Virgin Islands, USA	17.75	-64.75	3	1979	1990
AZR	Terceira Island, Azores, Portugal	38.77	-27.38	40	1979	2003+
BAL	Baltic Sea, Poland	55.50	16.67	7	1992	2003+
BHD	Baring Head Station, New Zealand	-41.41	174.87	80	1999	2003+
BME	St. Davids, Bermuda, UK	32.37	-64.65	30	1989	2003+
BMW	Tudor Hill, Bermuda, UK	32.27	-64.88	30	1989	2003+
BRW*	Barrow, Alaska, USA	71.32	-156.60	11	1971	2003+
BSC	Black Sea, Constanta, Romania	44.17	28.68	3	1994	2003+
CBA	Cold Bay, Alaska, USA	55.20	-162.72	25	1978	2003+
CGO	Cape Grim, Tasmania, Australia	-40.68	144.68	94	1984	2003+
CHR	Christmas Island, Kiribati	1.70	-157.17	3	1984	2003+
CMO	Cape Meares, Oregon, USA	45.48	-123.97	30	1982	1998
COS	Cosmos, Peru	-12.12	-75.33	4600	1979	1985
CRZ	Crozet Island, France	-46.45	51.85	120	1991	2001
EIC	Easter Island, Chile	-27.15	-109.45	50	1994	2003+
FLK	Falkland Islands, UK	-51.70	-57.87	51	1980	1982
GMI	Mariana Islands, Guam	13.43	144.78	2	1978	2003+
GOZ	Dwejra Point, Gozo, Malta	36.05	14.18	30	1993	1999
HBA	Halley Station, Antarctica	-75.58	-26.50	10	1983	2003+
HUN	Hegyhatsal, Hungary	46.95	16.65	344	1993	2003+
ICE	Heimaey, Vestmannaeyjar, Iceland	63.25	-20.15	100	1992	2003+
ITN	Grifton, North Carolina, USA	35.35	-77.38	505	1992	1999
IZO	Tenerife, Canary Islands, Spain	28.30	-16.48	2360	1991	2003+

Table 2 (Cont'd). NOAA GMCC/CMDL Sampling Sites 1968-2003[#]

KCO	Kaashidhoo, Republic of Maldives	4.97	73.47	1	1998	1999
KEY	Key Biscayne, Florida, USA	25.67	-80.20	3	1972	2003+
KPA	Kitt Peak, Arizona, USA	32.00	-112.00	2083	1982	1989
KUM	Cape Kumukahi, Hawaii, USA	19.52	-154.82	3	1971	2003+
KZD	Sary Taukum, Kazakhstan	44.45	77.57	412	1997	2003+
KZM	Plateau Assy, Kazakhstan	43.25	77.88	2519	1997	2003+
LEF	Park Falls, Wisconsin, USA	45.93	-90.27	868	1994	2003+
MBC	Mould Bay, Nunavut, Canada	76.25	-119.35	58	1980	1997
MCM	McMurdo Station, Antarctica, USA	-77.83	166.60	11	1985	1987
MHD	Mace Head, County Galway, Ireland	53.33	-9.90	25	1991	2003+
MID	Sand Island, Midway, USA	28.22	-177.37	4	1985	2003+
MKO	Mauna Kea, Hawaii, USA	19.83	-155.47	4220	1978	1984
MLO*	Mauna Loa, Hawaii, USA	19.53	-155.58	3397	1969	2003+
NMB	Gobabeb, Namibia	-23.57	15.03	408	1997	2001
NWR	Niwot Ridge, Colorado, USA	40.05	-105.58	3475	1967	2003+
NZL	Kaitorete Spit, New Zealand	-43.83	172.63	3	1982	1985
OPB	Pacific Ocean, N/A	-99.99	-999.99	10	1995	2000
OPC	Pacific Ocean, N/A	-99.99	-999.99	10	1993	1999
OPD	Pacific Ocean, N/A	-99.99	-999.99	10	2002	2002
OPE	Pacific Ocean, N/A	-99.99	-999.99	10	2002	2003+
OPW	Olympic Peninsula, Washington, USA	48.25	-124.42	488	1984	1990
OXK	Ochsenkopf, Germany	50.08	11.80	1356	2003	2003+
PAC	Pacific Ocean, N/A	-99.99	-999.99	10	1986	1993
PAL	Pallas-Sammaltunturi, GAW Station, Finland	67.97	24.12	560	2001	2003+
PAW	Pacific Ocean, N/A	-99.99	-999.99	10	1990	1993
PSA	Palmer Station, Antarctica, USA	-64.92	-64.00	10	1978	2003+
PSM	Point Six Mountain, Montana, USA	47.03	-113.98	2462	1978	1982
PTA	Point Arena, California, USA	38.95	-123.73	17	1999	2003+
RPB	Ragged Point, Barbados	13.17	-59.43	45	1987	2003+
SCS	South China Sea, N/A	-99.99	-999.99	15	1991	1998
SEY	Mahe Island, Seychelles	-4.67	55.17	3	1980	2003+
SGI	Bird Island, South Georgia, UK	-54.00	-38.05	30	1989	1992
SGP	Southern Great Plains, Oklahoma, USA	36.62	-97.48	----	2002	2003+
SHM	Shemya Island, Alaska, USA	52.72	174.10	40	1985	2003+
SIO	La Jolla, California, USA	32.83	-117.27	14	1968	1986

Table 2 (Cont'd). NOAA GMCC/CMDL Sampling Sites 1968-2003[#]

SMO*	Tutuila, American Samoa	-14.25	-170.57	42	1972	2003+
SPO*	South Pole, Antarctica, USA	-89.98	-24.80	2810	1975	2003+
STC	Ocean Station C, USA	54.00	-35.00	6	1968	1973
STM	Ocean Station M, Norway	66.00	2.00	7	1981	2003+
SUM	Summit, Greenland	72.58	-38.48	3238	1997	2003+
SYO	Syowa Station, Antarctica, Japan	-69.00	39.58	11	1986	2003+
TAP	Tae-ahn Peninsula, Republic of Korea	36.73	126.13	20	1990	2003+
TDF	Tierra Del Fuego, La Redonda Isla, Argentina	-54.87	-68.48	20	1994	2003+
UTA	Wendover, Utah, USA	39.90	-113.72	1320	1993	2003+
UUM	Ulaan Uul, Mongolia	44.45	111.10	914	1992	2003+
WIS	Sede Boker, Negev Desert, Israel	31.13	34.88	400	1995	2003+
WKT	Moody, Texas, USA	31.32	-97.62	708	2001	2003+
WLG	Mt. Waliguan, Peoples Republic of China	36.29	100.90	3810	1990	2003+
ZEP	Ny-Alesund, Svalbard (Spitsbergen), Norway	78.90	11.88	475	1994	2003+
	2. *Aircraft Program*					
CAR	Carr, Colorado, USA	40.10	-104.10	2200→7800	1993	2003+
FLT	Fortaleza, Brazil	-4.15	-38.28	200→4200	1999	2003+
HAA	Molokai, Hawaii, USA	21.40	-157.20	200→7800	1999	2003+
HFM	Harvard Forest, Massachusetts, USA	42.50	-71.2	800→7900	2003	2003+
EPC	Estavan Point, Canada	49.38	-126.53	400→5200	1993	2003+
LEF	Park Falls, Wisconsin, USA	45.90	-38.30	250→5500	1998	1999
	Moscow, Russia	35.90	35.50	300→6000	1995	1996
PFA	Poker Flats, Alaska, USA	65.10	-147.5	2100→7500	1999	2003+
RTA	Rarotonga, Cook Islands	-21.20	-159.8	400→7600	2000	2003+
SAN	Santarem, Brazil	-2.85	-54.95	200→4400	2000	2003+

[#] http://www.cmdl.noaa.gov/ccgg/iadv/ccggmdb_list_available.php
* Locations of observatories
** 2003+ indicates on-going programs

The Commonwealth Science and Industry Research Organization (CSIRO), Division of Atmospheric Research, Australia, maintains a cooperative network of sampling locations from which atmospheric constituents are monitored (Langenfelds et al., 2002). Beginning in the early 1970's, in situ measurements of CO_2 were made at Cape Grim, Tasmania. Grab samples of air from Macquarie Island and Mawson, Antarctica were also measured at Cape Grim (Beardsmore et al., 1978).

Table 3: CSIRO GASLAB Sampling Sites 1976-2003[#]

Code	Name	Latitude	Longitude	Altitude	First	Last*
ALC	Alert, Nunavut, Canada	82.45	-62.50	210	1989	2003+
BHZ	Baring Head Station, New Zealand	-41.41	174.87	80	1991	----
BWU	Barrow, United States	71.32	-156.60	11	1983	1993
CFA	Cape Fergusen, Australia	-19.28	147.06	2	1989	2003+
CGA	Cape Grim, Tasmania, Australia	-40.68	144.68	94	1988	2003+
CIA	Cape Grim, in situ CO_2, Tasmania, Australia	-40.68	144.68	94	1976	2003+
CRI	Cape Rama, India	15.08	73.83	60	1993	2003+
CSA	Cape Schanck, Australia	-38.50	144.88	108	1994	2003+
CPU	Cheeka Peak, United States	48.25	-124.75	488	1989	1993
DAA	Darwin (Charles Point)/Jabiru, Australia	-12.42	130.57	3	1987	2003+
EPC	Estavan Point, Canada	49.38	-126.53	39	1993	2003+
FRC	Fraserdale, Australia	49.9	-81.6	200	1990	1993
MAA	Mawson, Antarctica	-67.6	62.9	32	1977	1993
MLU	Mauna Loa, United States	19.53	-155.58	3397	1984	2003+
MQA	Macquarie Island, Australia	-54.48	158.97	12	1986	2003+
SIS	Shetland Islands, Scotland	60.17	-1.17	30	1992	2003+
SMU	Matatula Point, American Samoa	-14.23	-170.56	42	1984	1990
SPU	South Pole, Antarctica	-89.99	-102.00	2810	1992	2003+
AIA	Aircraft	-38 → -41	140 →141	150→8000	1992	2000

[#] Francey et al., 1996; http://cdiac.ornl.gov/trends/atm_meth/csiro/csiro_gaslabch4.html
* 2003+ indicates on-going programs

Beginning in 1980, CH_4 in flask samples was measured using GC + FID, and CO was analyzed by methanization followed by GC + FID (Fraser et al,

1986). The CSIRO Global Atmospheric Sampling Laboratory (GASLAB) was developed in the late 1980's from the earlier trace gas monitoring programs at Cape Grim (Francey et al., 1996). GASLAB includes both surface in situ and flask sampling programs, air sampling from aircraft and from ships. All air samples are measured at Cape Grim. GASLAB uses the more precise GC + HgO reduction method for CO (Francey et al., 1996). GASLAB measures trace gases at sites in the Southern Hemisphere at regional locations often not studied by other networks. (Table 3) (Figure 1). CSIRO also conducts air sampling from oceanic shipboard programs in the Antarctic and has maintained a 30 year program of vertical trace gas sampling above southern Australia (Fraser et al., 1986; Francey et al., 1996).

Long term intercomparisons of field data have proved to be a rigorous method for comparing data sets (Masarie et al., 2000). Differences in reference scales, drift in standard gases, and differences in calibration procedures have all been identified through these comparisons. The absence of a reliable reference scale for CO has complicated global measurements. During the 1990's CO in air standards were provided to the community by NOAA/CMDL. It was discovered in 2002 that the reference scale had been drifting upward over time and a correction was applied to all standards produced in that laboratory (Novelli et al., 2003). Intercomparisons of measurements, particularly through round robin experiments, continue to show differences between laboratories, and it remains difficult to combine CO data sets from different labs with confidence to better than ~ ±10%.

In contrast, measurements of CH_4 are more easily combined. Methane measurement records from many international laboratories have been integrated and extended to produce a globally consistent cooperative data product called GLOBALVIEW. GLOBALVIEW (http://www.cmdl.noaa.gov/globalview/index.html) was developed, coordinated and maintained by NOAA/CMDL. The data product includes synchronized smoothed time series derived from continuous and discrete land-surface, ship, aircraft, and tower observations. It also provides a marine boundary layer reference matrix used in the data extension process; uncertainty estimates; and extensive documentation. It is specifically intended as a tool for use in modeling studies. Unfortunately access is restricted to active data contributors.

2.2. Observations of halocarbons

Measurements of chlorofluorocarbons were made, because of their use as tracers of atmospheric motions, in the early 1970s (Lovelock et al., 1973).

Table 4a: Location of surface network stations

Station	Latitude	Longitude	Altitude (masl)	Network *
Adrigole, Ireland (AD)	52N	10W	50	A
Alert, NWT, Canada (ALT)	82N	63W	210	C
American Samoa (SMO)	14S	171W	77	C
Barrow, Alaska, USA (BRW)	71N	157W	8	C
Cape Grim, Tasmania, Australia (CGO)	41S	145E	94	A, C
Cape Matatula, American Samoa (AS)	14S	171W	42	A
Cape Meares, USA (CM)	45N	124W	30	A
Harvard Forest, Massachusetts, USA (HFM)	43N	71W	340	C
Kumukahi, Hawaii, USA (KUM)	20N	155W	3	C
Mace Head, Ireland (MH)	53N	10W	25	A
Mauna Loa, Hawaii, USA (MLO)	20N	156W	3397	C
Niwot Ridge, Colorado, USA (NWR)	40N	106W	3475	C
Palmer Station (PSA)	65S	64W	10	C
Ragged Point, Barbados (RP)	13N	59W	45	A
South Pole (SPO)	90S	25W	2837	C
Trinidad Head, California, USA (THD)	41N	124W	120	A, C
Wisconsin, USA (LEF)	46N	90W	868	C

* ALE/GAGE/AGAGE sites (A), CMDL HATS sites (C)

The outgrowths of these pioneering efforts have been formalized through programs of regular measurements such as the Atmospheric Lifetime Experiment (ALE), the Global Atmospheric Gases Experiment (GAGE) and the Advanced GAGE (AGAGE) global network programs originated in 1978, and collect continuously high frequency gas chromatographic measurements of six anthropogenic gases: $CFCl_3$ (CFC-11), CF_2Cl_2 (CFC-12), $CF_2ClCFCl_2$ (CFC-113), methyl-chloroform (CH_3CCl_3), chloroform ($CHCl_3$) and carbon tetrachloride (CCl_4). Methane and nitrous oxide are also measured at these stations. The measurements of a few other important species have been added at select sites in recent years, HFC-134a (CF_3CH_2F), HCFC-141b (CH_3CFCl_2), HCFC-142b (CH_3CF_2Cl), as well as H_2 and CO. Table 4a indicates the location of the stations belonging to this surface network along with those of the NOAA CMDL Halocarbons and other Atmospheric Trace Species (HATS), and Table 4.b indicates the period each species was observed by ALE/GAGE/AGAGE.

Table 4b: Periods of observation at ALE/GAGE/AGAGE sites

Species	CG	AS	RP	TH	CM	AD	MH
CFC-11	1978-2002	1978-2002	1978-2002	1995-2002	1979-89	1978-83	1987-2002
CFC-12	1978-2002	1978-2002	1978-2002	1995-2002	1980-89	1978-83	1987-2002
CFC-113	1982-2002	1985-2002	1985-2002	1995-2002	1984-89	N/A	1987-2002
CH_3CCl_3	1978-99	1978-99	1978-99	1995-99	1979-89	1978-83	1987-99
CCl_4	1978-2002	1978-2002	1978-2002	1995-2002	1979-89	1979-83	1987-2002
$CHCl_3$	1993-2002	1996-2002	1996-2002	1995-2002	N/A	N/A	1994-2002
HFC-134a	1998-2002	N/A	N/A	N/A	N/A	N/A	1994-2002
HCFC-141b	1998-2002	N/A	N/A	N/A	N/A	N/A	1994-2002
HCFC-142b	1998-2002	N/A	N/A	N/A	N/A	N/A	1994-2002
N_2O	1978-2002	1978-2002	1978-2002	1995-2002	1980-89	1978-83	1987-2002
CH_4	1986-2002	1987-2002	1985-2002	1995-2002	1985-89	N/A	1987-2002
CO	1996-2002	N/A	N/A	N/A	N/A	N/A	1996-2002
H_2	1993-2002	N/A	N/A	N/A	N/A	N/A	1996-2002

Systematic measurements of halogen-containing gases at a number of globally distributed sites began in 1978. Some of these have been operated continuously at a number of sites yielding measurements of mole fractions in ambient air samples at roughly hourly intervals. More spatially extensive measurements have been made by collecting paired air samples weekly (or bi-weekly) in evaluated flasks that have then been transported back to

Boulder, Colorado for analysis by CMDL. A gradual increase in detector capabilities used in association with the gas chromatographs, from electron capture and flame ionization detectors to mass spectrometers, have provided a substantial expansion in the number of measurable halogenated compounds in recent years. Details of the instrumentation and the measurement procedures are given for the AGAGE network in Prinn et al. (2000) and for the CMDL Halocarbon network in Elkins et al. (1996) and Hall et al. (2001).

Calibration differences between the networks are typically at the 1% level; for a few of the newly measured species the differences approach 5% but these differences are expected to decrease over time as they have for the other species. In addition to intercomparison of standards, there are a few sites where more than one group of investigators have made measurements. This has allowed calibration differences between the networks to be accurately assessed and calibration procedures to be improved. The measurements from each of the networks may be found at the Department of Energy-Carbon Dioxide Information Analysis Center (DOE-CDIAC) World Data Center (email to cpd@ornl.gov, Dataset No. DB-100) for AGAGE and the data from CMDL are available through the CMDL web site (http://www.cmdl.noaa.gov).

An additional source of information on the historical record of the increases in trace gas concentrations since the 1970s is obtainable from the archived air bank at CSIRO (Commonwealth Scientific and Industrial Research Organisation) in Australia (Langenfelds et al., 1996; Fraser et al., 1996). Each of the tanks in the bank contain air which was collected south of Melbourne at Cape Grim under baseline conditions (i.e. when the wind direction was such that the air was expected to be uncontaminated by regional pollution effects). Several such samples were collected annually starting in 1978 and since 1990 an increased number of samples have been collected and archived on a regular basis. With the enhancements in detector technology in recent years, analysis of this archive has provided historical information on previously undetectable (or previously uninteresting) gases (e.g. Oram et al., 1998). The archive has also provided an important test of the consistency of network calibration over the past two decades (e.g. Cunnold et al., 1994).

More recently air trapped in unconsolidated snow (firn) has been extracted and analysed to extend the historical records back as far as 1900 (Butler et al., 1999; Sturges et al., 2001). These data are extremely useful as they can indicate the natural background concentrations of various trace gases prior to major, known, industrial production of these compounds. Furthermore, the increase in concentration over the last century can be used to quantify the anthropogenic contribution to the budget. Atmospheric trends of halocarbons from air trapped in ice cores are expected to become available in the near future which will extend the historical records further back in time.

2.3. Observations of N_2O

The air samples collected for the CMDL Carbon Cycle Greenhouse Gases (CCGG) group at the sites listed in Table 2 are also analysed for nitrous oxide (N_2O) and sulfur hexafluoride (SF_6) (Tans et al., 2001). In addition the CMDL Halocarbon and Other Atmospheric Trace Species (HATS) group have made measurements of N_2O at a limited number of sites (Point Barrow, Niwot Ridge, Mauna Loa, American Samoa) using flasks back to the 1970s and in-situ instruments since 1987 (Hall et al., 2001). An intercomparison between the measurements of these 2 groups is being undertaken. The 2 data sets complement each other in that the HATS record extends further back in time, whilst the CCGG network has the advantage of sampling from many more sites thus allowing a better spatial picture of the distribution of N_2O. The global growth rate of N_2O from 1978 to 2000 was 0.74 ± 0.01 ppb yr^{-1} whilst for 1996-2000 it was 0.85 ± 0.06 ppb yr^{-1}, suggesting that atmospheric concentrations of N_2O were increasing more rapidly near the end of the 20th century (Hall et al., 2001).

The ALE/GAGE/AGAGE community has been measuring N_2O from its network of sites (Table 4) since 1978 (Prinn et al., 1990; 2000). The mean concentrations from the 5 sites were between 306.4 and 307.5 ppb for the period 1978 and 1998 and showed an increasing trend of 0.69 to 0.74 ppb yr^{-1}, consistent with that reported by CMDL. Further the AGAGE N_2O data are sufficiently precise to enable the detection of an annual cycle in qualitative agreement with that predicted by models (Prinn et al., 2000).

2.4. Summary

As a summary of the observations available within the different surface networks, Table 5 reports the species measured by the CMDL carbon cycle group (C), the AGAGE team (A), WMO/GAW is (W), GASLAB is (G) and the CMDL Halocarbons and Other Atmospheric Trace Species group (H).

Table 5. Summary of available observations

Species	Chemical Formula	Network
methane	CH_4	A, C, W, G
carbon monoxide	CO	A, C, W, G
nitrous oxide	N_2O	A, H, C, W, G
carbon dioxide	CO_2	C, G
hydrogen	H_2	A, H, C, G
methyl chloroform	CH_3CCl_3	A, H
carbon tetrachloride	CCl_4	A, H
CFC-11	$CFCl_3$	A, H
CFC-12	CF_2Cl_2	A, H
CFC-113	CCl_2FCClF_2	A, H
HFC-134a	CH_2FCF_3	A, H
HCFC-22	CHF_2Cl	A, H
HCFC-141b	CH_3CFCl_2	A, H
HCFC-142b	CH_3CF_2Cl	A, H
Halon-1211	$CBrClF_2$	A, H
Halon-1301	$CBrF_3$	A, H
methyl chloride	CH_3Cl	A, H
methyl bromide	CH_3Br	A, H
chloroform	$CHCl_3$	A, H
sulfur hexafluoride	SF_6	H, C

3. OBSERVATIONS OF ATMOSPHERIC COMPOUNDS BY REMOTE SENSING

Remote sensing is characterized by the fact that the measuring instrument and the quantity being measured are at different locations. The information is carried at a distance through electromagnetic radiation, and it is from the physics of the interaction of radiation with matter, in the form of emission, absorption, and scattering, that it can reveal the atmospheric gases and particles and their abundances. All methods in use on satellites today are passive; the source of radiation is natural and it is usually the sun or the thermal radiation of the atmosphere itself. The radiation received at the sensor is then the result of the radiative interaction with matter that occurred all along the ray path, and basically provides a weighted integral of atmospheric properties, although the weighting function (vertical sensitivity) may vary with wavelength. It is precisely when the weighting function changes significantly across a strong absorption band (or line) of a molecule, that it is possible to extract profile information along the ray path about the absorbent molecule.

Through time a wide variety of instruments has been developed using many different principles of measurements, which are reviewed for example in the books of Kidder and Vonder Haar (1995) and Stephens (1994) as well as Burrows (1999), Brasseur et al. (1999) and King et al. (1999).

3.1. Passive remote sensing of trace gases

Passive remote sensing of gases in trace amounts in the troposphere requires high sensitivity and resolving power. Historically, trace gases were first retrieved in the stratosphere by limb sounding. By looking at the earth limb, and measuring at a range of elevations above the earth surface, a vertical sounding is obtained. The atmospheric path is maximized and therefore small amounts of trace gases can be measured. Since the density of air decreases roughly exponentially with height, the contribution arises primarily around the tangent point, and thus limb sounding offers quite a good vertical resolution, typically of the order of two kilometres. The horizontal resolution is, in counterpart, not so good being a few hundred kilometres. In the troposphere, the probability of a cloud occluding the path becomes very high and for all practical purposes limb sounding is only used in the upper troposphere and stratosphere. The Upper Atmosphere Research Satellite (UARS) launched in September 1991 has been one of the first large missions for chemical measurements; it used four limb sounding instruments. The UARS was particularly successful in validating new techniques and, with a duration of nearly ten years, its lifetime far exceeded expectations.

Tropospheric sounding, however, requires nadir view; although some of the more advanced solar occultation limb sounders, can measure down to 5 km in altitude. Nadir view is useful to distinguish molecules that are "strong" absorbers, i.e. that have large optical depths such as O_3, H_2O, CH_4, CO_2, O_2, with "weak" absorbers such as N_2O, CO, BrO, SO_2, ClO, OClO, and HCHO. To detect minor trace gases or weak absorbers, the instrument needs to measure intensity changes in optical depth of 10^{-3} to 10^{-4}, and thus requires high sensitivity. A list of species observed by current and planned satellite experiments is given in Table 6.

Table 6 (see plate 20): Species observed by current and planned satellite experiments

Sensor	TOMS	GOME	IMG	MOPITT	SAGE III	SCIA-MACHY	MIPAS	AIRS	TES	MLS	HRDLS	OMI
Launch		04/95	10/96	12/99	12/01	02/02	02/02	05/02	01/04	01/04	01/04	01/04
O_3					>6 km	T	>8 km	P	P	>8 km	>7 km	T
OClO												
CO				P		T			P	>8 km		
CO_2			P			T						
CH_4						T	>8 km				>7 km	
NO									>5 km			
NO_2						T			>5 km			
N_2O						T	>8 km				>7 km	
H_2O		P				T	>8 km	P	P	>5 km	>7 km	
HNO_3							>8 km		P			
HCOH												
HCN										>8 km		
BrO												
SO_2												
CFC-11											>7 km	
CFC-12											>7 km	

Near-real time P : vertical profile information
Research mode T : tropospheric column by combining nadir and limb soundings
Under special conditions only. Profile information from limb sounders (e.g. > 8 km)

The sensitivity to detect small concentrations in remote sensing can be achieved in several ways. In the case of limb sounding, sensitivity was obtained by having a long atmospheric path; the weak absorption or emission of tenuous molecules is accumulated along the path and can then be detected at the receiver. In general high sensitivity is achieved by having a large signal-to-noise ratio. The signal depends on the amount of radiation reaching the instrument and thus on the intensity of the source of radiation. The sun is the most intense source of radiant energy, and the scattering of solar radiation by molecules or particles in nadir view yields a very effective method for the remote sensing of trace gases. Several instruments have been developed on this idea, e.g. TOMS, GOME, SCIAMACHY, and OMI. The remote sensing is limited to daytime, and uses absorption lines in the UV, visible and near IR. Another method to increase the signal-to-noise ratio is noise reduction as used in thermal infrared remote sensing. The noise is normally given by the root sum square of the shot noise and various other electronic noise sources, and it can be reduced by lowering the temperature. Cooling devices, such as cryogenic coolants are often used. However, the cryogenic coolant sublimates and eventually runs out so that the expected lifetime of such instruments is limited.

Remote sensing in the thermal infrared is not very sensitive in the lowest layers of the atmosphere, because the boundary layer has about the same temperature as the ground and so emits as much radiation as it absorbs. The remote sensing of minor trace gases, whether in the UV visible or thermal infrared, is characterized by weak absorption lines. Even if the radiance is resolved across the line as with Michelson interferometers, the variation of transmittance remains small and cannot provide very different weighting functions and thus profile information. Strong absorbers such as ozone and water vapour are not in this category. To obtain tropospheric column amount of minor trace gases, a combination of limb and nadir sounding have been proposed. Higher resolution instruments such as Michelson interferometers can provide vertical profile information of minor gases, but it is generally limited to a few layers in the troposphere. Dramatically improved vertical resolution can only be achieved with active remote sensing techniques. These techniques use a ranging system, such as with lidars and radars, where a pulse of energy is created, and upon timing the return signal provides a measure of the distance between the instrument and the target. Vertical resolution of a few tens to hundred of meters can be achieved with such systems.

A list of current and future satellite instruments (up to December 2004) along with their characteristics and the species retrieved is presented in Table 7. Only instruments providing tropospheric measurements are listed in the table. Many more instruments will provide measurements in the stratosphere, but this is outside of the scope of this volume.

Table 7. Current and future satellite instruments for trace gases

Instrument / Satellite Period of Operation/orbit	Product	Coverage	Vertical Resolution
TOMS; 1987-1993 Meteor–3; 1991-1994 ADEOS; 1996-1997 Earth Probe; 1996-2002 Triana; late 2004	O_3 SO_2	Global, Daylight	
GOME / ERS-2: 1995-current Sun synchronous 10:30 AM	O_3, NO_2, O_3 Profile, BrO, OClO, SO_2, HCOH	Global, Daylight	Column/ 5 km
IMG / ADEOS : 1996-1997 Sun synchronous ; 10:30 AM	H_2O, CO_2, N_2O, CO, O_3 CH_4, HNO_3	Global	Column/ Profile
MOPITT / EOS Terra; 1999-current Sun synchronous, 10:30 AM	CO CH_4	Global	5 layers/ Column
SAGE III /Meteor 3 2001-current ISS, launch 2004	O_3 H2O	50-80°N- 35-60°S ISS: Nearly global	Limb profile Above 6 km
SCIAMACHY / Envisat 2002-current Sun synchronous 10:00 AM	O_3, NO_2, NO_3, BrO, N_2O CO, CO_2, CH_4 NO, OClO, ClO, HCHO, SO_2	Global, Daylight	Profile/ Column
MIPAS / Envisat 2002-current Sun synchronous - 10:00 AM	O_3, H_2O, CH_4, N_2O, HNO_3	Global	Limb profile
AIRS /EOS Aqua 2002-current Sun synchronous 1:30 PM	O_3 CO, CH_4	Global	Profile/ column
TES /EOS Aura Launch 2004 Sun synchronous 1:45 PM	O_3, CO, CH_4 HNO_3 H_2O NO, NO_2	Global	Profile
MLS / EOS Aura Launch 2004 Sun synchronous 1:45 PM	H_2O CO, O_3, HCN	Global	Limb profile
HIRDLS /EOS Aura Launch 2004 Sun synchronous 1:45 PM	O_3 N_2O, H_2O, CH_4, CFC-11, CFC-12	Global ·· ··	Limb profile > 7 km ··
OMI EOS Aura Launch 2004 Sun synchronous 1:45 PM	O_3, SO_2, NO_2, BrO, OClO	Global, Daylight	Column

3.2. Passive remote sensing of aerosols

As seen in the previous chapters, aerosols originate from a wide variety of sources: volcanoes, sea-salt particles from the oceans, wind-blown mineral particles such as desert dust, sulphates and nitrates derived from gas to particle conversions, organic material, carbonaceous substances from biomass burning and industrial combustion. The radiative properties of aerosols depend on their shape, size distribution, chemical composition, and total amount.

Passive remote sensing of aerosols is generally obtained by measuring the reflected sunlight. Paradoxically, this was first accomplished using AVHRR and TOMS, two multi-channel radiometers that were not designed for this application in the first place. In the last few years, an approach to retrieve aerosols in the near UV has emerged as a by-product of the recent improvement (Version 7) in the TOMS ozone retrieval. As detailed in Chapter 6, a TOMS's Aerosol Index (AI) has been developed that is a measure of the change of spectral contrast in the near UV due to radiative effects of aerosols in a Rayleigh scattering atmosphere. Recently new remote sensing techniques designed specifically for the aerosol remote sensing problem were developed which greatly increase the information content. One new technique, called multi-angle observations, provides observations of the same location on the earth's surface at different observation angles; for example, the ATSR-2 launched in April 1995 aboard the ERS-2 consists of a conical scanning radiometer that provides curved swaths at two measurement angles, one nadir and one at 56° forward viewing angle from four spectral bands. The POLDER instrument aboard ADEOS acquired eight months of data between November 1996 and June 1997. A follow-on instrument is planned for the launch of ADEOS-2 in November 2002. POLDER uses a wide-angle imaging system and an array detector that can measure up to 14 view angles 60° forward to 60° backward on height spectral bands. The MISR on EOS Terra has a very high resolution footprint down to 275 m^2 and view nine angles 70.5° forward to 70.5° backward in four spectral bands. Table 8 gives a list of the major instruments and their characteristics.

Table 8. Current and future satellite instruments for aerosols; (Reproduced and adapted from King et al. 1999)

Instrument/ Satellite Launch date or Time coverage	Spatial Resolution	Comments
AVHRR / NOAA-7, -9, -11, -L, Metop-1 Since 1979	1.1 km (local mode) 4.4 km (global)	Long data set (since 1981) Aerosol optical thickness retrieved over oceans
TOMS /Meteor-3; 8/22/91 – 11/24/94 ADEOS; 7/25/96 – 6/28/97 Earth Probe; 7/25/96 – 4/5/02 Triana; Late 2004	50 km	Relatively long data set (since 1978) Low spatial resolution
ATSR-2 /ERS-2 April 21, 1995	1 km	Narrow swath, takes 6 days for global coverage
OCTS /ADEOS 11/26/96 – 6/30/97	0.7 km	Cloud screening using three thermal channels
POLDER / ADEOS; 11/26/96 – 6/30/97 ADEOS-II; Nov. 2002 or later	7 × 6 km	Polarization more sensitive to aerosol refractive index than radiance
SeaWiFS / OrbView-2 August 1, 1997	1.1 km(local) 4.5 (global)	Tilt capability to avoid sun glint
MISR /EOS Terra 18 Dec., 1999	1.1 km	In flight calibration
MODIS / EOS Terra; 18 Dec., 1999 EOS Aqua; May 4, 2002	0.25 – 1 km	Wide spectral range High calibration accuracy
AATSR / Envisat 28 Feb., 2002	1 km	
MERIS / Envisat 28 Feb., 2002	0.3 – 1.2 km	
SEVERI / MSG August 2002	2 km IR 3 km VIS	Provide in addition H_2O, O_3 (total column)
GLI /ADEOS-II Nov. 2002 or later	0.25 -1 km	Wide spectral coverage
OMI /EOS Aura 2004	13 km (local) 13 × 24 km (global)	High calibration accuracy
CALIPSO / Proteus April, 2004	330 m (horizontal) 30 m (vertical)	Lidar

Another development in aerosol remote sensing consists in polarization measurements. The microphysical properties such as the effective particle radius and refractive index can be detected from polarization measurements. With the instrument POLDER aboard ADEOS polarization measurements are made in addition to multi-angle measurements providing simultaneous and almost independent retrievals of aerosol optical thickness and effective radius. To help capture aerosol types, and even aerosol size parameters, narrow band radiometers over a wide spectral range have been developed. MODIS aboard EOS Terra, is one of those. Scattering by aerosols is weakly dependent on wavelengths; actually this property is often used to identify the presence of aerosols in remote sensing.

Various satellite sensors have now been developed for the remote sensing of aerosols. Measurement of the radiative and microphysical properties of aerosols can now be derived from many different methods, including single- and multiple-channel reflectance, multi-angle reflectance, contrast reduction and polarization, some of which are still in experimental stage (see King et al. (1999) for a succinct yet clear description of the methods). To derive the optical thickness from the radiance measurements requires necessary supplemental information, namely the single scattering albedo, the aerosol scattering phase function. Single scattering albedo and phase function are typically estimated from aerosol climatology for a given location. An assumption on the aerosol distribution is also required. Monomodal and bimodal distributions are mostly used.

4. OBSERVATION-BASED BUDGET ANALYSES

4.1. General methodology

Different methodologies have been developed to evaluate the budgets of chemical species from observations. The main approach discussed here uses atmospheric chemical transport models (CTMs) to simulate the concentration field of a gas and through comparison with measured atmospheric concentrations the distribution and temporal trends of the sources that best reproduce the observations are then determined. Another method is to ratio the measured concentrations of 2 trace gases, one of which has a well-known emission strength, to determine the emission strength of the other. Depending on the particular application the CTMs used in the determination of emissions and the way in which they are employed will vary. However, there are many principals that are generally applicable to most of the modelling approaches. These relate to the model grid size, and simulation of atmospheric transport and sink processes.

4.2. Model grids and atmospheric transport

The modelling of the measurements to derive emission information has ranged from using a single box to represent the entire atmosphere to using the most sophisticated three-dimensional (3-D) models. For compounds that are well mixed throughout the atmosphere a first approximation of global emissions can be derived simply from the measured atmospheric burden and growth rate, along with an estimate of the lifetime of the gas. For species with atmospheric lifetimes longer than approximately 1 year, useful estimates of global and hemispheric emissions can be obtained using two-dimensional (2-D) (latitude-altitude) atmospheric models with for example 12 boxes (Cunnold et al., 1994), 108 boxes (Oram et al., 1995) and 288 boxes (Fraser et al., 1999). It is however necessary to validate the transport rates in these models. This is mostly done using measured concentrations of CFC-11 and CFC-12, which are inert in the troposphere, have well known emissions and thus their modelled distributions are primarily a function of the transport terms.

The success of the 12-box model (Cunnold et al., 1994), for example, which uses interannually invariant transport rates, is indicated by its ability to simulate the evolution of the latitudinal gradient of CFCs as their production and release have declined (Figure 2). It is equally encouraging that using these transport rates for many other gases, which are measured at the ALE/GAGE/AGAGE (AGA) sites, the model generally simulates their multi-year average hemispheric gradients within 10%. This model should therefore provide useful estimates of the semi-hemispheric proportions for gases, such as anthropogenically-released gases, which have emission distributions somewhat similar to the CFCs but, not surprisingly, for methane, which has an emission distribution very different from the CFCs, the CFC-based transport rates can lead to errors in the semi-hemispheric emission proportions. The hemispheric emission proportions predicted by the model for chloroform (O'Doherty et al., 2001) and nitrous oxide (Prinn et al., 1990) respectively are nevertheless expected to be reasonably accurate.

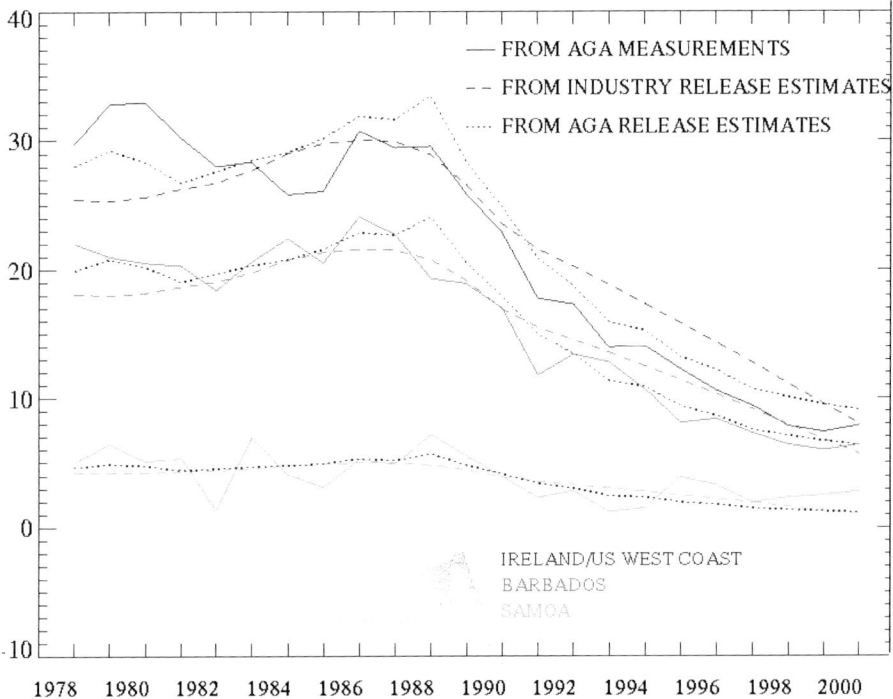

Figure 2 (see plate 21). ALE/GAGE/AGAGE measured differences between annual mean CFC-12 mole fractions in Ireland, Barbados and Samoa (full lines, top to bottom) and those in Tasmania between 1978 and 2000. These are compared against the differences calculated with the 12-box 2-D model (Cunnold et al., 1994) using the emissions estimated by McCulloch et al. (2003) (dashed lines), emissions estimated from the AGA measurements (dotted line), and a lifetime of 100 years.

4.3. Atmospheric lifetimes

Emission estimates for compounds with lifetimes in excess of approximately 1 year can be produced using these models provided that the atmospheric destruction rates (i.e. lifetimes) are accurately known. For compounds such as the CFCs, the only significant loss results from photodissociation in the stratosphere. This loss rate has been calculated using 2-D and 3-D (altitude-longitude-latitude) stratospheric models (e.g. Golombek and Prinn, 1986) and in some cases directly from stratospheric measurements based on correlations against other long-lived species (Volk et al., 1997). Because of the necessary tropospheric emphasis of the models used to estimate emissions, the calculated stratospheric loss rates are typically imposed at the upper levels of the models. Emission estimates then have some sensitivity to the accuracy of the model simulations of the rate of mass exchange between the troposphere and the stratosphere.

The hydroxyl (OH) distribution, which controls the lifetimes of many of the trace gases of interest, can be calculated from reasonably well understood photochemical reactions which determine its production, primarily via the photodissociation of ozone in the presence of water vapor, and loss rates (e.g. reaction with hydrocarbons). These OH concentrations can be constrained by observations of other trace gases that have well known emissions and reaction rates with OH. Industry estimated emissions of methyl chloroform (Midgley and McCulloch, 1995) are considered to be reasonably accurate because there were only a few companies that produced it and its release into the atmosphere occurs within a year of production. Therefore the global mean concentration of OH in the AGAGE model (Cunnold et al., 1994) has been adjusted to simulate the AGA methyl chloroform measurements. The accuracy of the estimate of global mean OH using this method is reported to be 14% (Prinn et al., 2001). The most recent independent calculations of OH by Spivakovsky et al. (2000) have simulated the methyl chloroform measurements within approximately 10%. Other models (e.g. Oram et al., 1995) have used OH fields that have been constrained by measurements of ^{14}CO (Volz et al., 1981), which is produced by cosmic rays at a reasonably well-known rate.

For many species the reaction with OH dominates the loss. For these gases the emission calculations are dependent on the accuracy of the rate coefficients (and their temperature dependencies) for their reactions with OH. These have typically been measured in laboratories with accuracies of approximately 10%.

For some trace gases, such as the halons, photodissociation in the troposphere is also an important loss mechanism, which therefore requires the use of radiative transfer schemes and knowledge of absorption cross-sections and quantum yields to simulate their destruction (e.g. Fraser et al., 1999). For other molecules, such as methyl chloroform and carbon tetrachloride, the ocean can also be a sink.

4.4. A priori emissions

In some approaches an a priori emission distribution is used as discussed below for the regional emission estimates. Comparison of concentrations predicted by a model using a priori emission distributions and strengths can also be used to identify regions where additional sources may be found (e.g. Krol et al., 2003). A priori emission estimates are obviously required for the reference trace gas when the measured concentration ratio method is applied.

4.5. Regional emissions

Regional emissions have been estimated by comparing measured concentrations with those inferred from back trajectories and an a priori emission distribution based on population density (Simmonds et al., 1996; Derwent et al., 1998a). More recently this approach has been replaced by a Lagrangian, mesoscale model in which particles are released continuously from all the major source regions and their trajectories calculated (Derwent et al., 1998b; Ryall et al., 1998). Advection occurs via the three resolved wind components as well as by three randomized wind components, which are intended to represent particle dispersion. The strength of pollution event is estimated from the number of particles reaching the measurement site. The strength of the source assigned to each particle is originally based on an a priori emission distribution, which is then uniformly multiplied by a factor determined by the correlation between pollution events as measured versus those modelled.

Regional emissions are also often determined using the ratio between the concentrations of trace gases during pollution events (e.g. Schmidt et al., 2001).

4.6. Atmospheric measurement requirements

This "top-down" approach requires suitable measurement programmes that provide sufficient information on the distribution and trends of trace gases. The major measurement networks and programmes that have been used in the determination of emission estimates are described above in sections 2 and 3.

Houweling et al. (2000) have discussed the difficulties associated with using point measurements in models, all of which have limited spatial resolution. This problem is ameliorated by the use of remote, relatively clean sites and by the removal of regional pollution effects in the time series analyzed. The measurements by the networks that make continuous in situ observations are, at times, influenced by local or regional pollution events (e.g. air from the European continent). These events are very useful for assessing regional emissions, but are filtered out using a statistical procedure (e.g. O'Doherty et al., 2001; Cunnold et al., 2002) when estimating global, or hemispheric, emissions. This filtering is largely unnecessary for the flask sample observations in which air samples are collected under specified wind conditions which characteristically are unpolluted (e.g. Langenfelds et al., 1996). For estimating regional emissions the pollution events can be identified by trajectory, or wind sector, analysis (Simmonds et al., 1998), which results in residual baseline concentrations not too different from the statistical procedure.

Estimating emissions from atmospheric measurements depends crucially on the accuracy of the measurements. Uncertainties in the total global emission estimates arise from measurement imprecisions and absolute calibration uncertainties. Measurement imprecisions tend to dominate the precision of the annual estimates of release, but these may be largely filtered out by smoothing the release estimates over 3 (or more) years, leaving the more systematic errors to typically dominate the uncertainties. For species with atmospheric lifetimes longer than approximately 10 years, 2-D and 3-D model comparisons (e.g. Cunnold et al., 2002) suggest that globally-averaged tropospheric concentrations derived from the surface networks (after the removal of regional pollution effects) have an accuracy of approximately 1%. This is supported by comparisons of CMDL and AGAGE global averages, which are based on different measurement sites especially in the Northern Hemisphere where much of the spatial variability occurs.

5. MEASUREMENT CONTRAINTS ON EMISSIONS OF ATMOSPHERIC COMPOUNDS

This section illustrates how measured concentrations of atmospheric compounds have been used to constrain emission estimates. It is not intended to be an exhaustive review, but rather, through a series of examples, a demonstration of a number of the different approaches and applications used.

5.1. Emissions of CH$_4$

As discussed in section 2.1, NOAA CMDL has made high-precision measurements of atmospheric methane from a globally-distributed network of air sampling sites since 1983. Figure 3 shows zonally-averaged atmospheric CH$_4$ mole fractions at the earth's surface for 1992-2001. The plot is based on measurements from about 50 sites. Though the plot is not very quantitative by itself, it qualitatively demonstrates the main features in the global distribution of atmospheric CH$_4$. These features include the long-term increase (because CH$_4$ sources and sinks are not in balance), seasonal variations (because of the photochemical sink, which is strongest during summer), and the strong latitude gradient in CH$_4$ (because ~2/3 of emissions are in the northern hemisphere). From the matrix of CH$_4$ values used to define this plot, the global mean surface concentration and budget of atmospheric methane has been quantified.

The globally, annually averaged CH$_4$ surface mixing ratio for 2001 is 1751 nmol mol^{-1}, compared to an average value of 1610 nmol mol^{-1} in 1983. This corresponds to a burden of 4840 Tg (where 1 Tg = 10^{12}g) CH$_4$, based on a model derived conversion factor of 2.767 from surface mole fraction in ppb to global burden of CH$_4$ in Tg [Fung et al, 1991]. The globally averaged CH$_4$ growth rate decreased from ~14 nmol mol^{-1} yr^{-1} in 1984 (source-sink ~ 40 Tg CH$_4$) to near zero in 2000 and 2001 (Figure 4). Relative uncertainties in the burden and source/sink imbalance are small, likely less than 5%.

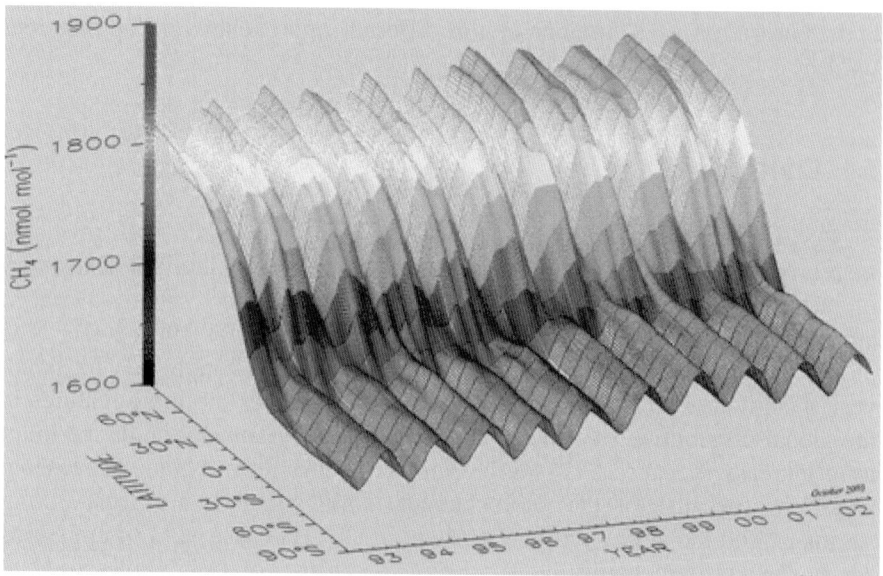

Figure 3 (see plate 22). Smoothed, zonally-averaged representation of the global distribution of CH_4 in the boundary layer for 1992-2001. Grid spacing is 10° latitude by 1 week. It is constructed from curves fitted to 43 sites in the CMDL cooperative air sampling network.

With an estimate of the global CH_4 lifetime, global emissions can be calculated using the following mass balance equation: $Q = d[CH_4]/dt + [CH_4]/\tau$, where Q is emissions, $[CH_4]$ is the global CH_4 burden in Tg CH_4, and τ is the CH_4 lifetime. The CH_4 burden and its rate of increase are determined from the observations, and τ is from independent studies. Figure 5 shows global emissions as a function of time, calculated with a lifetime of 8.9 years, which is assumed constant over the period of measurements. From 1984-2001, global emissions averaged ~540 Tg yr^{-1}, in good agreement with Cunnold et al. [2002], with no significant trend. Uncertainty in total global emissions (possibly ±15%) is larger than that in the burden and source/sink imbalance, and it is mostly because of the uncertainty in CH_4 atmospheric lifetime. A conclusion drawn from this analysis is that, if the CH_4 lifetime has been constant over the period of observations, then global CH_4 emissions are constant and the global CH_4 burden is approaching steady state [Dlugokencky et al., 1998, Cunnold *et al.*, 2002].

Determination of Emissions from Observations

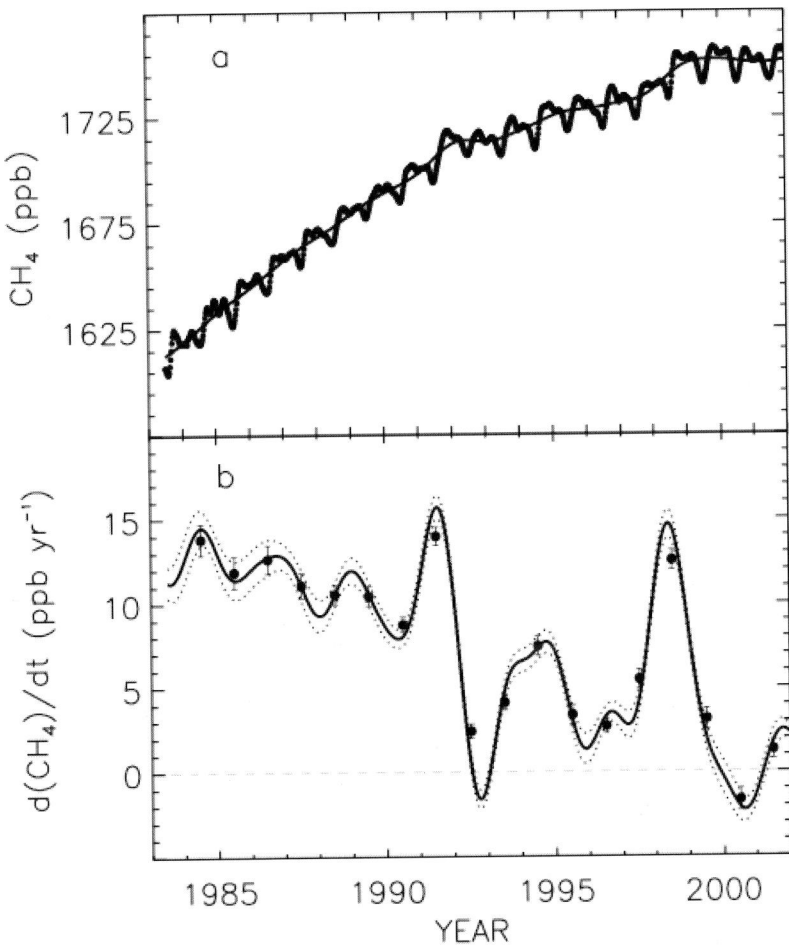

Figure 4 a. Symbols are globally averaged CH_4 mole fractions. Solid line is a deseasonalized trend curve fitted to the global averages. *b.* Instantaneous growth rate for globally averaged atmospheric CH_4 (solid line) with $\pm 1\sigma$ uncertainties (dashed lines). The growth rate is the derivative as a function of time of the solid line in a. Circles are annual increases, calculated from the trend line in Figure 1a as the increase from January 1 in one year to January 1 in the next. Uncertainties were determined with a nonparametric statistical technique (bootstrap).

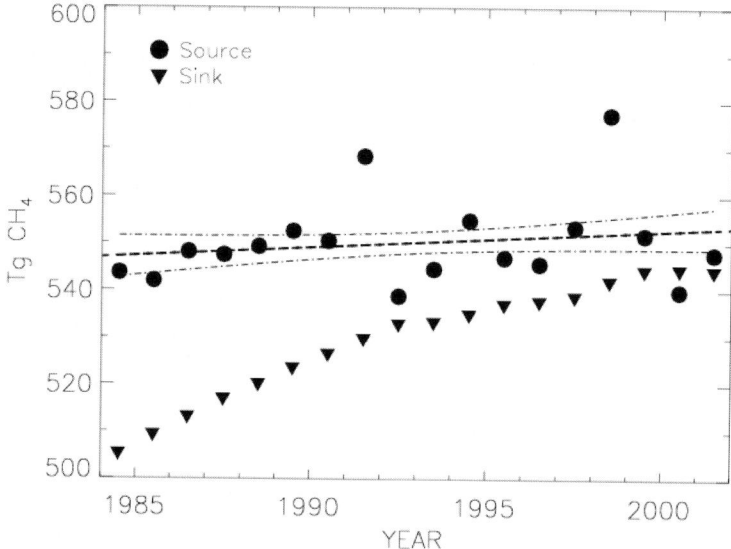

Figure 5. Global CH_4 emissions (circles) and sink (triangles) calculated with a 1-box mass balance assuming constant CH_4 lifetime and CMDL measurements of the global burden and annual increase. A straight line fitted to the annual emissions gives a slope of 0.4 ± 0.4 Tg yr^{-1}.

In cases where large, observed changes in CH_4 growth rate or spatial gradients can be linked to an "event", the measurements can be used to quantify variations in emissions from specific sources. The zonal distribution of CH_4 growth rate is shown in Figure 6 as a contour plot. Strong growth rate anomalies during 1992 at high northern latitudes (negative) and 1998 at high northern and southern tropical latitudes (positive) have been linked to interannual variations in temperature and soil moisture content in wetland regions, thereby affecting CH_4 emissions, and emphasizing the strong link between wetland CH_4 emissions and climate. For the 1998 anomaly, observations suggested that global emissions were greater than average by 24 Tg CH_4; a process based model, which included soil-temperature and precipitation anomalies, was used to calculate CH_4 emission anomalies from wetlands of +24.6 Tg CH_4, split nearly equally between high-northern latitudes and the southern tropics [Dlugokencky et al., 2001].

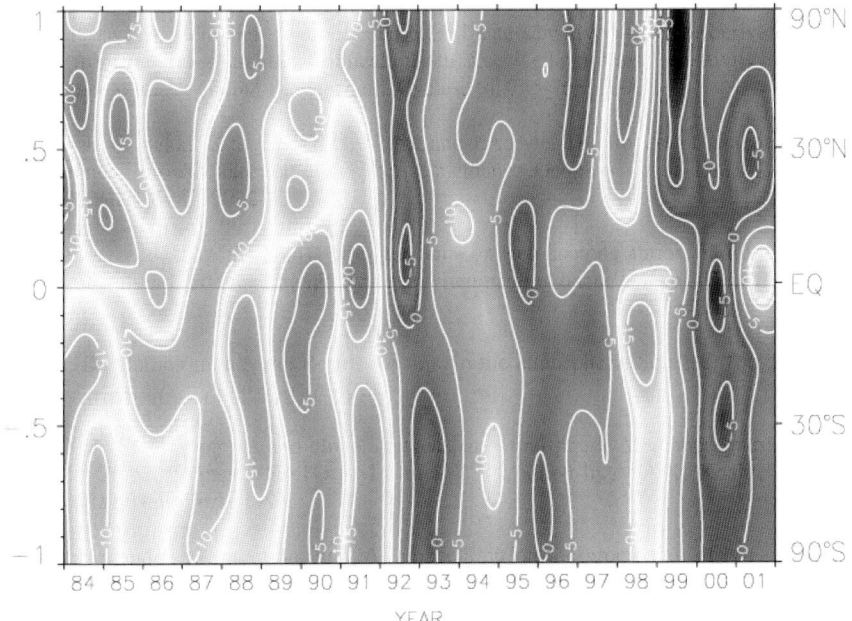

Figure 6 (see plate 23). Contour plot showing spatial and temporal variations in the growth rate of atmospheric CH_4 for 1984-2001. Cool colors (blue, violet, and black) represent regions and periods of low growth rate; warm colors (yellow, orange, and red) identify high growth rate.

The excellent agreement between process model and observations is likely somewhat fortuitous, since global process models of wetland emissions are in early stages of development. But the result emphasizes the usefulness of the measurements in constraining models of CH_4 emission processes. The large positive CH_4 growth rate anomaly during 1991 in the tropics was also observed in atmospheric CO and is likely related to changes in the oxidative capacity of the troposphere after the eruption of Mt. Pinatubo [Dlugokencky et al., 1996]. Sulphur dioxide and subsequent sulphate aerosol reduced the flux of UV radiation to the troposphere, reducing the production of $O(^1D)$ by 12%, which reduced hydroxyl radical (OH) concentration. Reducing [OH] results in an increase in CH_4 growth rate because its main sink, reaction with OH, has been reduced.

In addition to these easily applied observational constraints, the measurements have also provided constraints on studies of the global CH_4 budget using 3-D models of atmospheric transport and chemistry. Fung et al. (1991) effectively used the latitudinal distribution of CH_4 and its seasonal cycle at individual sampling sites in combination with measurements of the isotopic composition of atmospheric CH_4 to produce 7 potential scenarios of CH_4 emissions by source sector, 6 of which met the observational constraints. They suggested that the existing network of observations could not produce a unique emission scenario and that further observations were needed, particularly closer to strong emission regions such as South and Southeast Asia.

In summary, the measurements provide important constraints on the global CH_4 budget; uncertainties associated with these constraints cover a wide range, from the global burden and trend, with relatively small uncertainties, to speculative scenarios of changes in specific sources that can only be tested with models. Improvements in using measurements to constrain the global CH_4 budget will come by expansion of the current network of low-frequency sampling, addition of high-frequency measurements downwind of strong CH_4 source regions, and improvements to models.

5.2. Emissions of N_2O

Prinn et al. (1990) used inverse theory and a 9-box model to examine 10 years of ALE/GAGE N_2O measurements and from this concluded that the cause of the trend of increasing concentrations appeared to be a combination of a growing tropical source and a growing northern mid-latitude source. Concentrations measured in CMDL samples collected from northern hemispheric sites are 1-3 ppb higher than samples collected at the South Pole, which is consistent with about two-thirds of total N_2O emissions occurring in the northern hemisphere (Tans et al., 2001). Khalil and Rasmussen (1992) used a simple one-dimensional model and data from the CMDL and ALE/GAGE networks along with ice core data to determine that anthropogenic sources make a one-third contribution to the total N_2O emissions.

The uncertainties in an N_2O emission inventory compiled using the "bottom-up" approach have been assessed by comparison with the results from Prinn et al. (1990) using the "top-down" approach (Bouwman et al., 1995). Furthermore this inventory was used as an a priori emission estimate to prescribe a 3-D CTM (Bouwman and Taylor, 1996) with the resulting modelled atmospheric concentrations being compared with those measured by ALE/GAGE and CMDL. Bouwman and Talyor (1996) concluded that more long-term observational sites are required, particularly in the continental regions, to better constrain the emissions.

Several regional emission estimates have been made for N_2O using the ratios of observed concentrations. Prather (1985, 1988) determined European emissions relative to CFC-11. Biraud et al. (2000) and Schmidt et al. (2001) have used ratios with ^{222}Radon from Mace Head and from Heidelberg, Schauinsland and Izania, respectively, to determine European emissions of N_2O. By specifically tailored data selection Schmidt et al. (2001) were able identify the influence of a local point source and determine regional and continental scale emission rates. Wilson et al. (1997) have also used ratios with ^{222}Radon to determine emissions from south-eastern Australia, whilst Bakwin et al. (1997) have used ratios with carbon tetrachloride to determine emissions from the south-eastern United States. European emissions of N_2O have also been determined using the trajectory climatology and dispersion model techniques (Derwent et al., 1998a; Ryall et al., 2001) (see section 5.3).

5.3. Emissions of halocarbons

Emissions of halocarbons as determined from their measured concentrations have been used to evaluate industrial emission estimates, determine emissions for gases for which there are no industrial estimates, determine compliance (or non-compliance) to international agreements and to evaluate regional emission estimates. Some case studies are presented below as examples of these different types of application.

For many of the anthropogenic ozone depleting substances estimates have been made of their emissions based on reported industrial production and predicted release functions (i.e. the rate at which they are released to the atmosphere) (e.g. McCulloch et al., 1994). To evaluate these estimates, they have been input into CTMs and the predicted model concentrations compared to those observed. Similarly the CTMs have been used to determine the emissions required to reproduce the observed concentrations so that they can be compared to those estimated by industry.

Figure 7 shows the concentrations of HCFC-142b measured at Cape Grim (Oram et al., 1995, plus updates from D.E. Oram, pers. comm., 2002) along with various global emission estimates. The University of East Anglia (UEA) Modelled Emissions are those which when used in a 2-D global model give concentrations at the latitude of Cape Grim (UEA Modelled Concentrations) that are very similar to those observed. The AFEAS (Alternative Fluorocarbon Environmental Acceptability Study (http:/www.afeas.org/)) emissions estimates are those generated using a "bottom -up" approach based on production data reported by the major chemical companies and knowledge of the rate at which the compound is released to the atmosphere.

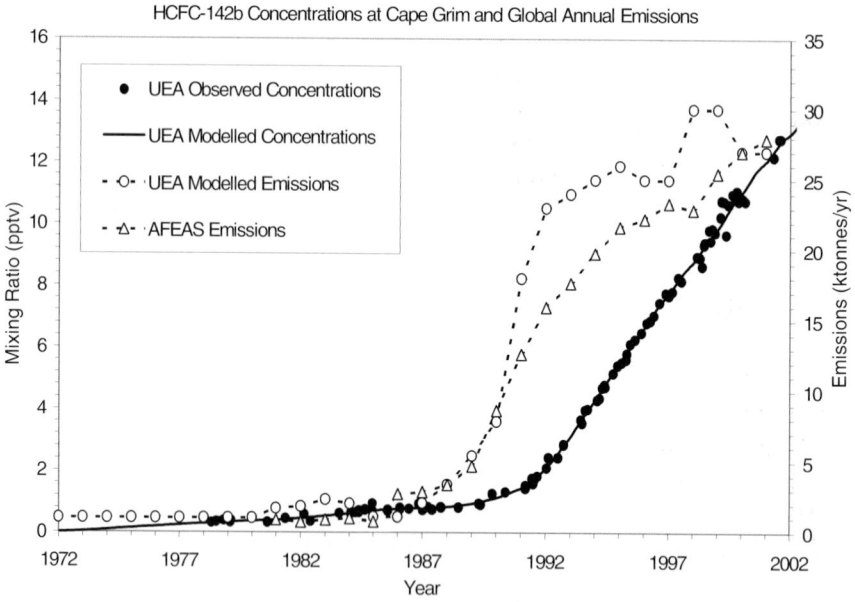

Figure 7. Concentrations of HCFC-142b measured (solid dots) and modelled (solid line) at Cape Grim (Oram *et al.*, 1995, plus updates) along with annual emissions estimates (dashed lines) derived from measurements (open dots) and industry (AFEAS: triangles).

Back in 1995 emissions estimated by industry were much less than those derived from observations (Oram et al., 1995), but subsequently the AFEAS estimates have been updated using improved release functions (AFEAS Emissions) and these emission estimates compare well with those derived from the atmospheric observations. Note, that the AFEAS estimates, as expected, are slightly lower than those derived from the observations, since they do not represent total global values, with some additional production in countries not responding to the AFEAS survey.

A similar situation occurred for the global emissions of CFC-11 except that industry estimates were higher than those derived from the observed concentrations (Cunnold et al., 1997). McCulloch et al. (2001) have recently revised the industry estimates to reflect a change in the usage/release pattern in recent years, which results in better agreement with the emissions derived from the measurement network.

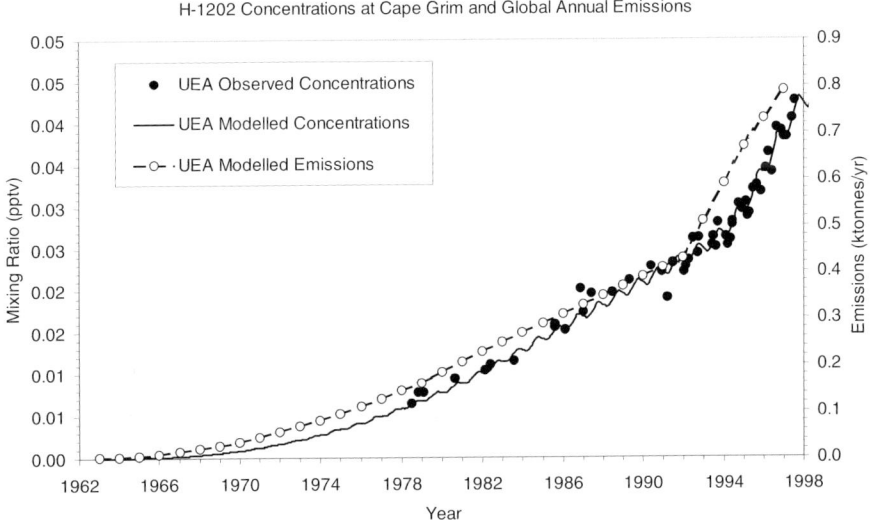

Figure 8. Concentrations of Halon-1202 measured (solid dots) and modelled (solid line) at Cape Grim along with measurement derived annual emissions estimates (open dots and dashed line) (Fraser *et al.*, 1999). (© American Geophysical Union).

For some gases there are no industrial emission estimates. An example of this is Halon-1202 (H-1202) and it has only been through the measurement of its concentration in the atmosphere and subsequent modelling using a CTM that its emissions have been estimated (Fraser et al., 1999) (Figure 8).

Many trace gases, like the halons and CFCs, are thought to be purely anthropogenic in origin and this has been confirmed for several of these compounds by measurement in firn air. Butler et al. (1999) has shown that the concentrations of CFC-11, CFC-12, CFC-113, Halon-1211 and Halon-1301, for example, essentially reached zero at the bottom of the firn air depth profiles. This indicates that the gases were not present in the atmosphere during the early twentieth century and confirms their exclusive anthropogenic origin. Other gases, such as dibromomethane, bromochloromethane and bromoform have been shown to have little or no trend in Antarctic firn air, indicating that, in the southern hemisphere, they are entirely of natural origin (Sturges et al., 2001). Methyl bromide, on the other hand, has been shown to have both existed in the early twentieth century and to have increased significantly over that century consistent with it having both natural and anthropogenic sources (Butler et al., 1999; Sturges et al., 2001). Reeves (2003) used a 2-D global model to examine this historical trend and the constraints it puts on the various source strengths and lifetime of methyl bromide. This has illustrated that not only is the current understanding of the present day atmospheric budget of methyl bromide incomplete, but so too is our understanding of the budget of methyl bromide prior to major industrial emissions and anthropogenic changes to it. It implies that either the estimate of the sink strength is too large or that there is an underestimate of both the "non-industrial" sources and the "anthropogenically influenced" sources.

The production and consumption of a number of ozone depleting substances are now controlled under the Montreal Protocol. Observed concentrations have been used to determine emissions of such compounds to establish whether there is compliance (or non-compliance) with this agreement. Figure 9 compares annual emissions from industry (full black lines based upon data from McCulloch et al., (2003) with emissions derived using the 12-box model (Cunnold et al., 1997).

Determination of Emissions from Observations

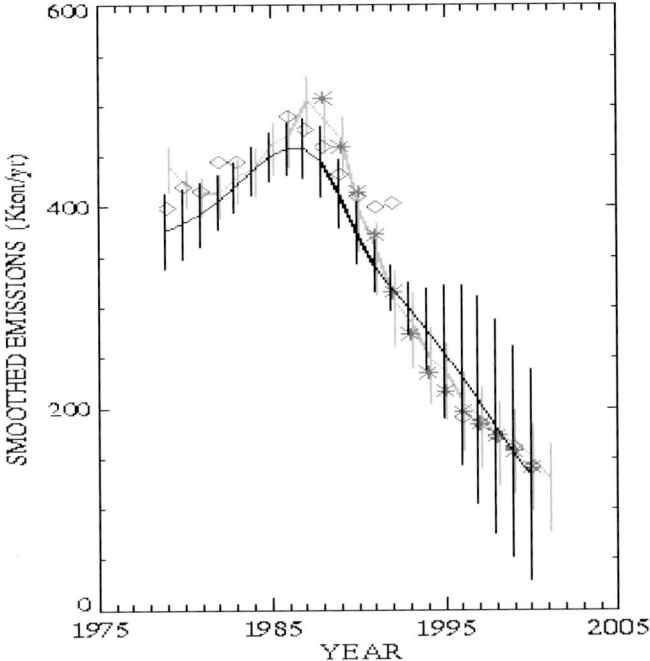

Figure 9 (see plate 24). Industry estimates of the annual releases of CFC-12 (full black lines) (McCulloch et al., 2003). The green lines indicate release estimates based on atmospheric measurements by ALE/GAGE/AGAGE using lifetimes of 100 years with the effect of lifetime uncertainties, associated with an ensemble of model calculations reported in WMO (1999), indicated by the green error bars. The red diamonds show emission estimates derived using the same AGAGE model applied to CMDL flask sample measurements and the asterisks are based on CMDL RITS measurements. All the emission estimates have been smoothed over three year intervals in order to reduce the effects of measurement imprecisions.

Those based on the AGAGE measurements (Prinn et al., 2000) are indicated in green, whilst those from the CMDL (Montzka et al., 1999) flask sample measurements are given as red diamonds and CMDL in situ measurements as red asterisks. All emission estimates derived from atmospheric measurements have been smoothed over three year intervals in order to reduce the effects of measurement imprecision. The emissions inferred from measured atmospheric trends and their comparison with the industrial estimates are consistent with compliance with the fully amended and adjusted Montreal Protocol.

With international legislation governing emissions of ozone depleting gases it is becoming increasingly important to quantify emissions on a regional (or national scale). Using a climatological analysis of trajectories and an a priori emission distribution based on population density European emissions have been derived from the concentrations measured at Mace Head, Ireland (Simmonds et al., 1996; Derwent et al., 1998a).

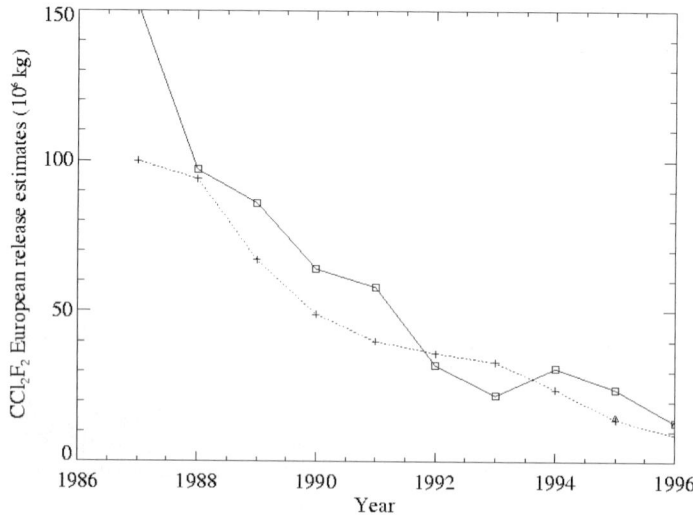

Figure 10. European annual emissions of CFC-12 estimated from industry data (McCulloch and Midgley, 1997) (dotted lines with pluses) compared against estimates derived by Derwent et al. (1998a) from measurements at Mace Head, Ireland (full line with squares). also shown are estimates from the measurements for 1995 and 1996 (triangles) using the NAME model (Ryall et al., 2001).

The calculations showed declines in European emissions of CFC-11 and CFC-12, CFC-113, methyl chloroform and carbon tetrachloride from 1987 to 1994. These declines were consistent with those calculated from industry production and sales figures by McCulloch and Midgley (1997) except that the industry derived figures for 1992-1994 for CFC-11 indicated larger emissions. McCulloch et al. (2001) revised the industry estimates to reflect a change in the usage/release pattern in recent years resulting in better agreement. The results for CFC-12 are illustrated in Figure 10. The uncertainty in the annual emission estimates for individual years is stated to be 25% (Simmonds et al., 1996).

Derwent et al. (1998b) and Ryall et al. (1998) have used a Lagrangian, mesoscale, model (horizontal resolution 50 km) to determine European emissions of many ozone depleting and radiatively active gases based on measured concentrations at Mace Head. Particles are released continuously from over Europe with source strengths initially determined from national emission estimates by McCulloch et al. (1994) and McCulloch and Midgley (1997) combined with distributions within countries based on population density. Then from the slope of a scatter plot of the pollution estimates from measurements versus those from modelling, a factor by which to uniformly multiply the source strength distribution is determined. This approach results in excellent agreement in the timing of pollution events but with considerable scatter in representing the magnitude of the pollution events (Figure 11, left panel).

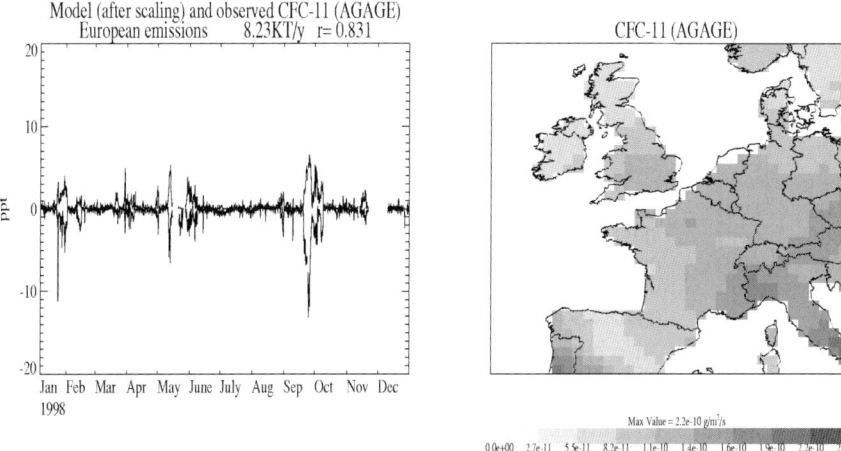

Figure 11. Left panel: CFC-11 concentrations at Mace Head, Ireland in 1997 predicted using the NAME dispersion model compared against observations. The observations are plotted as negative values. Right panel: Annual mean distribution of the emissions over Europe predicted from the Mace Head measurements (Ryall et al., 2001). (Reproduced with permission from Elsevier).

Ryall et al. (2001) have attempted to extend their Lagrangian modelling to the estimation of the annual mean spatial pattern of the European source distribution. They indicate that this is an overdetermined problem, which can give some negative source regions. By focussing on the larger pollution events, they are however able to obtain a robust estimate of the emission pattern (e.g. Figure 11, right panel). The accuracy of the resulting source distribution is unknown and as Ryall et al. (2001) point out, the accuracy and spatial resolution of the source distribution is expected to decrease with distance from Mace Head.

In addition to the application of long range transport models, ratios of measured trace gas concentrations can be used to determine regional emissions. Prather (1985, 1988) examined the correlation between the concentrations of several halocarbons with CFC-11 to determine their European emission strengths. Concurrent measurement of the trace gas concentrations with those of ^{222}Radon, a radioactive short half-life noble gas emitted by soils, has been used by Biraud et al. (2000) to determine European emissions of CFC-11 of 1.8-2.5 kg km^{-2} yr^{-1} for the year 1996. This estimate of European CFC emissions compares closely with those of Derwent et al. (1998a) and Ryall et al. (1998) using the Lagrangian dispersion model method when scaled up with a European surface area of the order of 1.45 x 10^7 km^2.

Estimates of halocarbon emissions for North America have been made by Bakwin et al. (1997) and Hurst et al. (1998) using the simultaneous hi-frequency measurements of carbon tetrachloride and a range of halocarbons made on a 610-m tall television tower in North Carolina. The North American halocarbon source strengths for CFC-11: 12.9 thousand tonnes yr^{-1}, CFC-12: 49 thousand tonnes yr^{-1}, CFC-113: 3.9 thousand tonnes yr^{-1}, methyl chloroform: 47.9 thousand tonnes yr^{-1}, carbon tetrachloride: 2.2 thousand tonnes yr^{-1} are reported by Bakwin et al. (1997) and these estimates are significantly lower than emission inventory estimates of Midgley and McCulloch (1995). Downward trends are reported for North American emissions of the major man-made halocarbons (Hurst et al., 1998)

Similar calculations, but with less sophisticated modelling, have been made by Dunse (1997) for Cape Grim, Tasmania. This site is relatively close to Melbourne and pollution events are predominantly related to air coming from the greater Melbourne area.

5.4 Emissions of CO and NO$_X$.

Parrish et al. (2002) have successfully examined concentration ratios of CO and total oxidized nitrogen species (NO$_Y$) to evaluate CO and NO$_x$ emissions from on-road vehicles in the United States. After emission to the atmosphere, NO_X is oxidized to nitrates within a few hours to days. NO_Y comprises these product species plus unreacted NO_X and therefore, under the conditions considered here, represents the total of the originally emitted NO_X. Figure 12a shows ambient concentrations of CO and the NO_Y at the top of a 110 m building in the centre of Nashville, Tennessee. At this height emissions from individual sources over a relatively large region of the surrounding urban area are well mixed. The slope of the linear fit to the measured CO and NO_Y levels gives the CO to NO_X emission ratio. To discern a significant temporal trend, measurements were made during short (4 to 6 weeks) periods at the same site, time of day, and season (summer) during three years over a six-year period. For each year the large r^2 value indicates that CO and NO_Y are well fitted by the derived linear relationship. Each succeeding year revealed a discernable decrease in the vehicle emission ratio, and the six-year period was long enough to define the statistically significant trend shown in Figure 12b. The focus on ratios of ambient concentrations, rather than the concentrations themselves, avoids effects from varying meteorological conditions, changing traffic density or driving patterns, etc.

Figure 12b compares the summer Nashville measurements with late autumn to early spring measurements made in Boulder, Colorado, a summer determination reported for 1987 in the Los Angeles area (Harley et al., 1997; Fujita et al., 1992)], and the ratio from the national emission inventory estimated by the U.S. EPA (USEPA, 2000a). The Los Angeles result is in excellent agreement with the extrapolation of the temporal trend derived for Nashville. The Boulder ratios are significantly larger than the L.A.-Nashville trend throughout the period of measurements, consistent with the colder seasons and higher altitude of the Boulder data. The Nashville-Los Angeles and Boulder trends suggest that the CO to NO_X emission ratio from vehicles in the U.S. has decreased by 7 to 9% per year, corresponding to a total decrease from 1987 to 1999 of a factor of 2.4 to 3.1. These decreases are a factor of 2 to 3 larger than the annual average decrease of 3.4% for 1988-1998 included in the national emission inventory (USEPA, 2000a). Comparison of the ambient and inventory ratios show that the U.S. EPA inventory seriously underestimates the CO to NO_X ratio in vehicle exhaust before 1990, but overestimates that ratio by nearly a factor of 2 by the end of the decade. It is important to note that the U.S. EPA inventory is detailed down to the county level, and that the inventory for both Boulder and Nashville are similar to the national average.

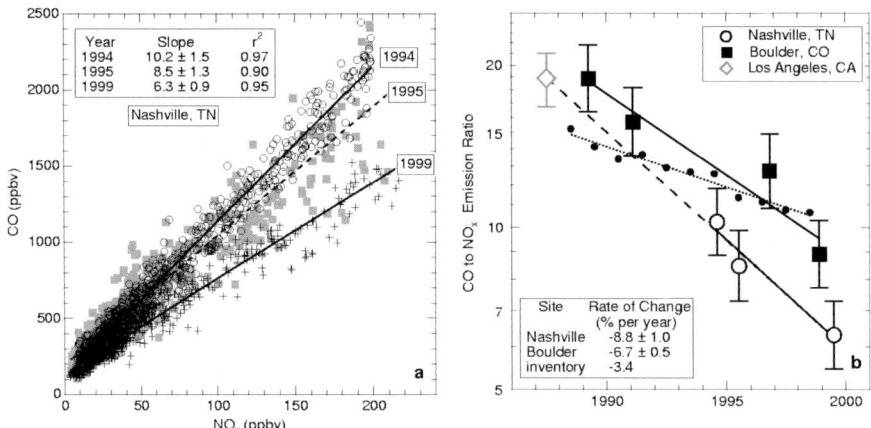

Figure 12. a) Coincident 5 minute average measurements of CO and NO_X in Nashville from three different years (From Parrish et al (2002)). The lines show linear regression fits to each year's data, and the slopes with estimated 95% confidence limits and corresponding squared correlation coefficients are annotated in the figure. b) Temporal trend of vehicular exhaust emission ratios of CO to NO_X on a logarithmic scale (adapted from Parrish et al (2002)). The large symbols with error bars give the slopes for Nashville, TN (from a), Boulder, CO and the Los Angeles, CA area. The error bars indicate the estimated 95% confidence limits. The solid lines give weighted linear fits to the log-transformed data, and the dashed line indicates the extrapolation of the derived Nashville trend back to the year of the Los Angeles measurements. The slopes of these fits with estimated 95% confidence limits and corresponding squared correlation coefficients are annotated in the figure. The small circles give the ratio from national U.S. emission inventory for on-road vehicles (USEPA, 2000a), and the dotted line gives the linear regression fit to the log-transformed inventory ratios. (© American Geophysical Union).

Decreasing CO and/or increasing NO_x vehicular emissions must cause the observed decrease in the CO to NO_x emission ratio. The U.S. EPA reports ambient CO measurements (USEPA, 2000b) from over 350 predominately urban and suburban monitoring sites nationwide, which are primarily influenced by vehicular emissions. These levels have decreased over the last decade. The average annual national decrease is 5.2 ± 0.8% with statistically identical values found for Nashville (4.8 ± 2.4%) and Boulder (6.3 ± 1.5%), indicating that the trends found in these two urban areas are nationally representative. Further, these decreases in CO ambient levels suggest that decreasing CO emissions account for most of the observed 7 to 9% annual decrease in the CO to NO_x emission ratio. However NO_x

emissions from vehicles must have increased by 2 to 3% annually to fully account for the observed decrease in the emission ratio.

The Global Emissions Inventory Activity (GEIA) and Emission Database for Global Atmospheric Research (EDGAR) emission inventories (Olivier et al., 1999; see chapter 1) estimate annual U.S. on-road vehicle emissions as 65 Tg CO in 1990. The rate of decrease of 5.2% per year infers that these emissions have decreased to 38 Tg CO in 2000. The reduction of 27 Tg CO emitted annually corresponds to a 6% reduction in annual global fuel-use CO emissions (480 Tg CO in 1990) or a 3% reduction in total global anthropogenic emissions (970 Tg CO in 1990, which includes all landuse and agricultural waste burning). This reduction in U.S. emissions and the decreasing trend in global background CO levels that has been reported (Novelli et al., 1998, Khalil and Rasmussen, 1994) have been cited as the reason for the decrease of approximately 5 ppbv/year from 1989 to 1997 in CO levels in the U.S. Mid-Atlantic region [Hallock-Waters et al., 1999].

In summary, the results of Parrish et al. (2002) show that, although the U.S. EPA underestimated the CO emissions from on-road vehicles at the beginning of the decade of the 1990's, emission control efforts since that time have been more successful than assumed in the inventory. By the end of the decade vehicle CO emissions had decreased to below U.S. EPA estimates. These decreases are significant on a global scale.

6. SUMMARY

In summary, the measurements provide important constraints on the global burdens, hemispheric and regional variations, and temporal trends of many atmospheric trace compounds. The uncertainties associated with these constraints cover a wide range, from the global burden and trend, with relatively small uncertainties, to speculative scenarios of changes in specific sources that can then be tested with models.

Emissions as determined from this "top-down" approach have been used to evaluate "bottom-up" emission estimates and to identify sources of disagreement that require further study. The "top-down" approach has been used where no data with which to build a "bottom-up" emission estimate are available. It has also provided constraints on regional emissions and on different emission types (e.g. natural versus anthropogenic) and has been valuable in the assessment of compliance with international agreements on emissions.

Improvements in using measurements to constrain the budgets will come by expansion of the current networks, with more sites, in particular downwind of major source regions, and higher frequency measurements, and with improvements to models.

7. REFERENCES

Bakwin, P. S., D. F. Hurst, P. P. Tans, and J. W. Elkins, Anthropogenic sources of halocarbons, sulfur hexafluoride, carbon monoxide, and methane in the south eastern United States, J. Geophys. Res., 102, 15,915-15,925, 1997.

Beardsmore, D.J., G.I. Pearman, P.J. Fraser, and J.G. O'Toole, The CSIRO (Australia) atmospheric carbon monoxide monitoring program: The first 6 years of Data, CSIRO Division of atmospheric physics technical paper No. 6, 115 pp, 1978.

Biraud, S., P. Ciais, M. Ramonet, P. G. Simmonds, V. Kazan, P. Monfray, S. O'Doherty, T. G. Spain, and S. G. Jennings, European greenhouse gas emissions estimated from continuous atmospheric measurements and radon 222 at Mace Head, Ireland, J. Geophys. Res., 105, 1351-1366, 2000.

Brasseur, G.P., J.J. Orlando, G.S. Tyndall, Atmospheric Chemistry and Global Change. Oxford University Press, New York, 654 pp, 1999.

Bouwman, A.F., K.W. Van der Hoek, and J.G.J. Olivier, Uncertainties in the global source distribution of nitrous oxide, J. Geophys. Res., 100, 2785-2800, 1995.

Bouwman, A.F., and J.A. Taylor, Testing high-resolution nitrous emission estimates against observations using an atmospheric transport model, Global Biogeochem. Cycles, 10, 307-318, 1996.

Brunke, E.-G., H.E. Scheel, and W. Seiler, Trends of tropospheric CO, N2O and CH4 as observed at Cape Point, South Africa, Atmos. Environ., 24A, 585-595, 1990.

Burrows, J.P., Current and future passive remote sensing techniques used to determine atmospheric constituents. In Approaches to Scaling of Trace Gas Fluxes in Ecosystems. Ed: A.F. Bouwman, Elsevier, Amsterdam, pp362, 1999.

Butler, J.H., M. Battle, M.L. Bender, S.A. Montzka, A.D. Clarke, E.S. Saltzman, C.M. Sucher, J.P. Severinghaus, and J.W. Elkins, A record of atmospheric halocarbons during the twentieth century from polar firn air, *Nature, 399*, 749-755, 1999.

Conway, T.J., P. Tans, L.S. Waterman, K.W. Thoning, K.A. Masarie, and R.H. Gammon, Atmospheric carbon dioxide measurements in the remote global troposphere, 1981-1984, Tellus, 40B, 81-115, 1988.

Cunnold, D. M., P. J. Fraser, R. F. Weiss, R. G. Prinn, P. G. Simmonds, B. R. Miller, F. N. Alyea, and A. J. Crawford, Global trends and annual releases of CCl_3F and CCl_2F_2 estimated from ALE/GAGE and other measurements form July 1978 to June 1991, J. Geophys. Res., 99, 1107-1126, 1994.

Cunnold, D. M., R. F. Weiss, R. G. Prinn, D. E. Hartley, P. G. Simmonds, P. J. Fraser, B. R. Miller, F. N. Alyea, and L. Porter, GAGE/AGAGE measurements indicating reductions in global emissions of CCl_3F and CCl_2F_2 in 1992-1994, J. Geophys. Res., 102, 1259-1269, 1997.

Cunnold, D. M., L. P. Steele, P. J. Fraser, P. G. Simmonds, R. G. Prinn, R. F. Weiss, L. W. Porter, S. O'Doherty, R. L. Langenfelds, P. B. Krummel, H. J. Wang, L. Emmons, X. X. Tie, and E. J. Dlugokencky, In situ measurements of atmospheric methane at GAGE/AGAGE sites during 1985 to 2000 and resulting source inferences, J. Geophys. Res., 107, D14, 10.1029/2001JD001226, 2002.

Derwent, R. G., P. G. Simmonds, S. O'Doherty, and D. B. Ryall, The impact of the Montreal Protocol on halocarbon concentrations in Northern Hemisphere baseline and European air masses at Mace Head, Ireland from 1987-1996, Atmos. Environ., 32, 3689-3702, 1998a.

Derwent, R. G., P. G. Simmonds, S. O'Doherty, P. Ciao, and D. B. Ryall, European source strengths and northern hemisphere baseline concentrations of radiatively active trace gases at Mace Head, Ireland, Atmos. Environ., 32, 3703-3715, 1998b.

Dlugokencky, E.J., E.G. Dutton, P.C. Novelli, P.P. Tans, K.A. Masarie, K.O. Lantz, and S. Madronich, Changes in CH_4 and CO growth rates after the eruption of Mt. Pinatubo and their link with changes in tropical tropospheric UV flux, Geophys. Res. Lett., 23, 2761-2764, 1996.

Dlugokencky, E.J., K.A. Masarie, P.M. Lang, and P.P. Tans, Continuing decline in the growth rate of the atmospheric methane burden, Nature, 393, 447-450, 1998.

Dlugokencky, E. J., B. P. Walker, K. A. Masarie, P. M. Lang, and E. S. Kasischke, Measurements of an anomalous global methane increase during 1998, Geophys. Res. Lett., 28, 499-502, 2001.

Dunse, B. L., Estimating trace gas emissions from Melbourne by analysis of routine measurements of atmospheric composition at the Cape Grim baseline air pollution station in Tasmania, M.S. Thesis, Department of Environ. Management, Victoria University of Technology, St. Albans, Victoria, Australia, March 1997.

Elkins, J. W., J. H. Butler, T. M. Thompson, S. A. Montzka, R. C. Myers, J. M. Lobert, S. A. Yoon, P. R. Wamsley, F. L. Moore, J. M. Gilligan, D. F. Hurst, A. D. Clarke, T. H. Swanson, C. M. Volk, L. T. Lock, L. S. Geller, G. S. Dutton, R. M. Dunn, M. F. Dicorleto, T. J. Baring, and A. H. Hayden, 5 Nitrous oxide and halocompounds, in Climate Monitoring and

Diagnostics Laboratory Summary Report #23, 1994-1995, U.S. Department of Commerce, Boulder, Colorado, 1996.

Francey, R.J., et al., Global Atmospheric Sampling Laboratory (GASLAB): Supporting and extending the Cape Grim trace gas programs, Baseline Atmospheric Program Australia 1993, edited by: R.J. Francey, A.L. Dick, and N. Derek, Bureau of Meteorology and CSIRO Division of Atmospheric Research, Melbourne, 1996.

Fraser, P.J., P. Hyson, R.A. Ramussen, A.J. Crawford, and M.A.K. Khalil, Methane, carbon monoxide and methylchloroform in the Southern Hemisphere, J. Atmos. Chem., 4, 1-42, 1986.

Fraser, P. J., D. M. Cunnold, F. N. Alyea, R. F. Weiss, R. G. Prinn, P. G. Simmonds, B. R. Miller, and R. L. Langenfelds, Lifetime and emission estimates of 1,1,2-trichlorotrifluoroethane (CFC-113) from daily global background observations, June 1982 to June 1994, J. Geophys. Res., 101, 12,585-12,599, 1996.

Fraser P.J., D.E. Oram, C.E. Reeves, S.A. Penkett and A. McCulloch, Tropospheric Halon Trends in the Southern Hemisphere and Global Halon Emissions, *J. Geophys. Res.*, *104*, 15985-15999, 1999.

Fujita, E.M., B.E. Croes, C.L. Bennett, D.R. Lawson, F.W. Lurmann, and H.H. Main, Comparison of emission inventory and ambient concentration ratios of CO, NMOG, and NO_x in California's South Coast air basin, J. Air Waste Manage. Assoc., 42 , 264-276, 1992.

Fung, I., J. John, J. Lerner, E. Matthews, M. Prather, L.P. Steele, and P.J. Fraser, Three-dimensional model synthesis of the global methane cycle, J. Geophys. Res., 96, 13,033-13,065, 1991.

Golombek, A., and R. G. Prinn, A global three-dimensional model of the circulation and chemistry of CCl_3F, CCl_2F_2, CH_3CCl_3, CCl_4, and N_2O, J. Geophys. Res., 91, 3985-4001, 1986.

Hall, B. D., J. W. Elkins, J. H. Butler, S. A. Montzka, T. M. Thompson, L. Del Negro, G. S. Dutton, D. F. Hurst, D. B. King, E. S. Kline, L. Lock, D. MacTaggart, D. Mondeel, F. L. Moore, J. D. Nance, E. A. Ray, and P. A. Romashkin, 5 Halocarbons and other atmospheric trace species, in Climate Monitoring and Diagnostics Laboratory Summary Report #25, 1998-1999, U.S. Department of Commerce, Boulder, Colorado, 2001.

Hallock-Waters, K.A., B.G. Doddridge, R.R. Dickerson, S. Spitzer, and J.D. Ray, Carbon monoxide in the U.S. Mid-Atlantic troposphere: evidence for a decreasing trend, Geophys. Res. Lett., 26, 2861-2864, 1999.

Hansen, J. D. Johnson, A. Lacis, S. Lebedeff, P. Lee, D. Rind, and G. Russell, Climate impact of increasing carbon dioxide, Science, 213, 957-966, 1981.

Harley, R.A., R.F. Sawyer, and J.B. Milford, Updated photochemical modeling for California's South Coast air basin: Comparison of chemical mechanisms and motor vehicle emission inventories, Environ. Sci. Technol., 31, 2829-2839, 1997.

Houweling, S., F. Dentener, J. Lelieveld, B. Walter, and E. Dlugokencky, The modeling of tropospheric methane: How well can point measurements be reproduced by a global model?, J. Geophys. Res., 105, 8981-9002, 2000.

Hurst, D. F., P. S. Bakwin, and J. W. Elkins, Recent trends in the variability of halogentated trace gases over the United States, J. Geophys. Res., 103, 25,299-25,306, 1998.

Khalil, M.A.K., and R.A. Rasmussen, The global sources of nitrous oxide, *J. Geophys. Res.*, 97, 14651-14660, 1992.

Khalil, M.A.K., and R.A. Rasmussen, Global decrease in atmospheric carbon monoxide concentration, Nature, 370, 639-641, 1994.

Kidder, S.Q., and T.H. Vonder Haar, Satellite Meteorology: An Introduction. Academic Press, San Diego, pp 466, 1995.

King, M.D., Y.J. Kaufman, D. Tanre, and T. Nakajima, Remote sensing of tropospheric aerosols from space: Past, present, and future. Bull. Am.. Met. Soc., 80, 2229-2259, 1999.

Krol, M.C., J. Lelieveld, D.E. Oram, G.A. Sturrock, S.A. Penkett, C.A.M. Brenninkmeijer, V. Gros†, J. Williams, and H.A. Scheeren, Continuing emissions of methyl chloroform from Europe, Nature, 421, 131-135, 2003.

Langenfelds, R., P. Fraser, R. Francey, P. Steele, L. Porter, and C. Allison, The Cape Grim Air Archive: The first seventeen years 1978-1995, in Baseline 94-95, edited by R. Francey, A. Dick, and N. Derek, pp. 53-70, Bur. of Meteorol., Commonw. Sci. and Ind. Res. Org., Melbourne, Victoria, Australia, 1996.

Langenfelds, R.L., R.J. Francey, B.C. Pak, L.P. Steele, J. Lloyd, C.M. Trudinger, and C.E. Allison, Interannual growth rate variations in CO2 and it's ^{13}C, H2, CH4, and CO between 1992 and 1999 linked to biomass burning, Global Biogeochem. Cycles, 16, 1-21, 2002

Lovelock, J., R. Maggs, and R. Wade, Halogenated hydrocarbons in and over the Atlantic, Nature, 241, 194-196, 1973.

Masarie et al, NOAA/CSIRO flask air intercomparison experiment: A strategy for directly assessing consistency among atmospheric measurements made by independent laboratories, J. Geophys. Res., 106, 20,445-20,464, 2001.

McCulloch, A., P. Midgley, and D. Fisher, Distribution of emissions of chlorofluorocarbons (CFCs) 11, 12, 113, 114, and 115 among reporting and non-reporting countries in 1986, Atmos. Environ., 28, 2567-2582, 1994.

McCulloch, A. and P. M. Midgley, Estimated historic emissions of fluorocarbons from the European Union, Atmos. Environ., 32, 1571-1580, 1997.

McCulloch, A., P. Ashford, and P. M. Midgley, Historic emissions of trifluorochloromethane (CFC-11) based on a market survey, Atmos. Environ., 35, 4387-4397, 2001.

McCulloch, A., P.M. Midgley and P. Ashford, Releases of refrigerant gases (CFC-12, HCFC-22 and HFC-134a) to the atmosphere, Atmos. Environ., 37, 889-902, 2003.

Midgley, P., and A. McCulloch, The production and global distribution of emissions to the atmosphere of 1,1,1-trichloroethane, Atmos. Environ., 29, 1601-1608, 1995.

Montzka, S. A., J. H. Butler, J. W. Elkins, T. M. Thompson, A. D. Clarke, and L. T. Lock, Present and future trends in the atmospheric burden of ozone-depleting halogens, Nature, 398, 690-694, 1999.

Novelli, P.C., K.A. Masarie, P. Tans and P.M. Lang, Recent changes in atmospheric carbon monoxide, Science, 263, 1994.

Novelli, P.C., K.A. Masarie, and P.M. Lang, Distributions and recent changes of carbon monoxide in the lower troposphere, *J. Geophys. Res.*, *103*, 19,015-19,033, 1998.

Novelli, P.C., K.A. Masarie, P.M. Lang, B.D. Hall, R.C. Myers, and J.W. Elkins, Re-analysis of tropospheric CO trends: Effects of the 1997-1998 wild fires, J. Geophys. Res., in press, 2003.

O'Doherty, S., D. Cunnold, G. A. Sturrock, D. Ryall, R. G. Derwent, H. J. Wang, P. Simmonds, P. J. Fraser, R. F. Weiss, P. Salameh, B. R. Miller, and R. G. Prinn, In-situ chloroform measurements at AGAGE atmospheric research stations from 1994-1998, J. Geophys. Res., 106, 20,429-20,444, 2001.

Olivier, J.G.J., J.P.J. Bloos, J.J.M. Berdowski, A.J.H. Visschedijk, and A.F. Bouwman, A 1990 global emission inventory of anthropogenic sources of carbon monoxide on 1°x1° developed in the framework of EDGAR/GEIA, Chemosphere: Global Change Science, 1 (1-17), 1999.

Oram, D. E., C. E. Reeves, S. A. Penkett, and P. Fraser, Measurements of HCFC-142b and HCFC-141b in the Cape Grim air archive: 1978-1993, Geophys. Res. Lett., 22, 2741-2744, 1995.

Oram, D.E., W.T., Sturges, S.A. Penkett, A. McCulloch, and P.J. Fraser, Growth of fluoroform (CHF3, HFC-23) in the background atmosphere, Geophys. Res. Lett., 25, 35-38, 1998.

Parrish, D.D., M. Trainer, D. Hereid, E.J. Williams, K.J. Olszyna, R.A. Harley, J.F. Meagher, and F.C. Fehsenfeld, Decadal change in carbon monoxide to nitrogen oxide ratio in the U.S. vehicular emissions, J. Geophys. Res., 107, 4140, doi:10.129/2001/2001JD000720, 2002.

Prather, M.J., Continental sources of halocarbons and nitrous oxide, Nature, 317, 221-225, 1985.

Prather, M.J., European sources of halocarbons and nitrous oxide: Update 1986, J. Atmos. Chem, 6, 375-406, 1988.

Prinn, R., D. Cunnold, R. Rasmussen, P. Simmonds, F. Alyea, A. Crawford, P. Fraser, and R. Rosen, Atmospheric emissions and trends of nitrous oxide deduced from 10 years of ALE-GAGE data, J. Geophys. Res., 95, 18,369-18,385, 1990.

Prinn, R., R. F. Weiss, P. J. Fraser, P. G. Simmonds, D. M. Cunnold, F. N. Alyea, S. O'Doherty, P. Salameh, B. R. Miller, J. Huang, R. H. J. Wang, D. E. Hartley, C. Harth, L. P. Steele, G. Sturrock, P. M. Midgley, and A. McCulloch, A history of chemically and radiatively important gases in air deduced from ALE/GAGE/AGAGE, J. Geophys. Res, 105, 17,751-17,792, 2000.

Prinn, R. G., J. Huang, R. F. Weiss, D. M. Cunnold, P. J. Fraser, P. G. Simmonds, A. McCulloch, C. Harth, P. Salameh, S. O'Doherty, R. H. J. Wang, L. Porter, and B. R. Miller, Evidence for substantial variations of atmospheric hydroxyl radicals in the last two decades, Science, 292, 1882-1888, 2001.

Ramanathan, V., R.J. Cicerone, H.G. Singh, and J.T. Kiehl, Trace gas trends and their potential role in climate change, J. Geophys. Res., 90, 5547-5566, 1985.

Reeves, C.E., Atmospheric budget implications of the temporal and spatial trends in methyl bromide concentrations, J. Geophys. Res., 108 (D11), 10.1029/2002JD002943, 11 June 2003.

Ryall, D. B., R. H. Maryon, R. G. Derwent, and P. G. Simmonds, Modelling long-range transport of CFCs to Mace Head, Ireland, Q. J. R. Meteorol. Soc., 124, 417-446, 1998.

Ryall, D. B., R. G. Derwent, A. J. Manning, P. G. Simmonds, and S. O'Doherty, Estimating source regions of European emissions of trace gases from observations at Mace Head, Atmos. Environ., 35, 2507-2523, 2001.

Schmidt, M., H. Glatzel-Matthier, H. Sartorius, D. E. Worthy, and I. Levin, Western Euorpean N_2O emissions: A top-down approach based on atmospheric observations, J. Geophys. Res., 106, 5507-5516, 2001.

Simmonds, P. G., R. G. Derwent, A. McCulloch, S. O'Doherty, and A. Gaudry, Long-term trends in concentrations of halocarbons and radiatively active trace gases in Atlantic and European air masses monitored at Mace Head, Ireland from 1987-1994, Atmos. Environ., 30, 4041-4063, 1996.

Simmonds, P. G., S. O'Doherty, J. Huang, R. G. Prinn, R. G. Derwent, D. Ryall, G. Nickless, and D. M. Cunnold, Calculated trends and the atmospheric abundance of 1,1,12-tetrafluoroethane, 1,1-dichloro-1-fluoroethane, and 1-chloro-1,1-difluoroethane using automated in situ gas chromatography-mass spectrometry measurements recorded at Mace Head, Ireland, from October 1994 to March 1997, J. Geophys. Res., 103, 16,029-16,038, 1998.

Spivakovsky, C. M., J. A. Logan, S. A. Montzka, Y. J. Balkanski, M. Foreman-Fowler, D. B. A. Jones, L. W. Horowitz, A. C. Fusco, C. A. M. Brenninkmeijer, M. J. Prather, S. C. Wofsy, and M. B. McElroy, Three-dimensional climatological distribution of tropospheric OH: Update and evaluation, J. Geophys. Res., 105, 8931-8980, 2000.

Steele, L.P., P.J. Fraser, R.A. Rasmussen, M.A.K. Khalil, T.J. Conway, A.J. Crawford, R.H. Gammon, K.A. Masarie, and K.W. Thoning, The global distribution of methane in the troposphere, J. Atmos. Chem., 5, 125-171, 1987.

Stephens, G.L., Remote Sensing of the Lower Atmosphere. Oxford University Press. New York, pp 523, 1994.

Sturges, W.T., H.P. McIntyre, S.A. Penkett, J. Chappellaz, J.-M. Barnola, R. Mulvaney, E. Atlas, and V. Stroud, Methyl bromide and other brominated methanes and methyl iodide in polar firn air, *J. Geophys. Res.*, *106*, 1595-1606, 2001.

Tans, P. P., P. S. Bakwin, L. Bruhwiler, T. J. Conway, E. J. Dlugokencky, D. W. Guenther, D. R. Kitzis, P. M. Lang, K. A. Masarie, J. B. Miller, P. C. Novelli, K. W. Thoning, M. Trudeau, B. H. Vaughn, J. W. C. White, and C. Zhao, 2. Carbon cycle, in Climate Monitoring and Diagnostics Laboratory Summary Report #25, 1998-1999, U.S. Department of Commerce, Boulder, Colorado, 2001.

USEPA, National Air Pollutant Emission Trends, 1900-1998, U.S. Environmental Protection Agency, 2000a.

USEPA, National Air Quality and Emission Trends Report, 1998, U.S. Environmental Protection Agency, 2000b.

Volk, C. M., J. W. Elkins, D. W. Fahey, G. S. Dutton, J. M. Gilligan, M. Loewenstein, J. R. Podolske, K. R. Chan, and M. R. Gunson, Evaluation of source gas lifetimes from stratospheric observations, J. Geophys. Res., 102, 25543-25564, 1997.

Volz, A., D.H. Ehalt, and R.G. Derwent, Seasonal and latitudinal variation of ^{14}CO and the tropospheric concentration of OH radicals, *J. Geophys. Res.*, *86*, 5163-5171, 1981.

Wang, W.C., Y.L. Yung, A.A. Lacis, T. Mo, J.E. Hansen, Greenhouse effects due to man-made perturbation of trace gases, Science, 194, 685-689, 1976.

Wilson, S.R., A.L. Dick, P.J. Fraser, and S. Whittlestone, Nitrous oxide flux estimates for South-Eastern Australia, J. Atmos. Chem., 26, 169-188, 1997.

WMO, Scientific Assessment of Ozone Depletion: 1998, Global Ozone Research and Monitoring Project-Report no. 44, World Meteorological Organization, Geneva, Switzerland, 732 pp., 1999.

WMO, The strategic plan of the Global Atmospheric Watch, no. 113, WMO, Geneva, 83 pp, 1997.

WMO, Global Atmospheric Watch Measurements Guide, no. 143, WMO, Geneva, 79 pp, 2001.

Data Assimilation and Inverse Methods

Richard Ménard, Sandrine Édouard, Sander Houweling, Gabrielle Pétron, Claire Granier and Claire Reeves

1. INTRODUCTION

In the previous chapter we reviewed the principal methods of observations of atmospheric chemical constituents and showed how they can be used in relatively simple models, often 2-dimensional, to provide constraints on emission estimates. We will now discuss how these observations can be used in conjunction with more complex 3-dimensional chemical transport models to yield useful knowledge about surface emissions and the chemical state of the atmosphere by employing methods based on estimation theory, called inverse method and data assimilation. Although there is a rich history of application of this theory in other fields, such as meteorology, seismology, and remote sensing, it is only recently that data assimilation and inverse techniques have been developed to address atmospheric constituent problems.

The objective of chemical data assimilation is to estimate the chemical state of the atmosphere given prescribed sources, whereas the objective of inverse methods is to estimate the chemical sources or to improve emission inventories. These methods have remarkable similarities, for instance they share the same mathematical apparatus of estimation theory, but their application is quite different. Data assimilation uses a short-term chemical forecast, typically of 6 hours or less, and combines this information with observations to get the best estimate of the chemical state of the atmosphere in space and evolving in time. Data assimilation is particularly useful to optimally fill in the gaps in the observation network, to maximize the use of information content of satellite observations which often provide only column amounts, and to infer information about unobserved species important, for example fast reacting compounds. In contrast, inverse methods use much longer integration times, typically of a month, and were developed primarily to infer surface emissions of long-lived gas species using surface observations. Inverse modelling tools will have important implications to monitor the application of international treaties on greenhouse gas emissions. However, inverse methods have important limitations. Long time integrations of models representing chemical tracers distributions, which are used in order to avoid estimating the spatial structure of the initial condition, result in a loss of spatial information due to diffusion and mixing, so that only the large-scale, slowly time-varying sources can be estimated.

In this chapter we will outline the basic formulation of both data assimilation and inverse methods, and draw the attention to their similarities and differences. We will then proceed in extending these methods to a simultaneous state-source estimation scheme. The following section gives an overview on chemical transport, with emphasis on the numerical properties that transport schemes should have for chemical modelling. In section 3, we discuss the issue of data assimilation and inverse modelling by drawing into the analogy that those two methods have. Section 4 provides examples of application of data assimilation, inverse modelling, and dual state-source estimation. We conclude with a few remarks on future challenges.

2. CHEMISTRY-TRANSPORT MODELS

Atmospheric chemical transport models play a key role in providing a link between the sources and sinks of a compound with its distribution at a specific location and time. The transport is basically derived from the mass conservation law also called continuity equation. Chemical species can be lost by dry and wet deposition, gained by surface emissions, and chemical reactions can result in either gain or loss of the mass of chemical species. Except for exceptional and simple case scenarios where analytical solutions can be found, the solution of these chemistry-transport models is carried out numerically. This means that space and time are discretized into grid points and time steps, and that the solution is carried out at each time step for each grid point. The models are usually three-dimensional; they are quite complex and computationally expensive to run. Typical models have grid point spacing of a few tens (regional models) to a few hundred kilometres (global models), and time-step sizes of a few tens of minutes to an hour. The real transport is, however, not entirely accounted for by the resolved transport of the model, so that unresolved transport processes usually modelled by diffusion type parameterizations need to be included in the model in order to obtain accurate transport predictions. The best example of this is the turbulent transport in the planetary boundary layer. Convection is also not explicitly computed with typical spatial and temporal resolution of chemical-transport models, and so needs to be accounted by modelling of its average effect. These subgrid scale effects introduce significant uncertainties in transport predictions, as well as numerical discretization errors in the resolved part of the model, that will accumulate over time and create large uncertainties. A thorough discussion of all these aspects of chemical-transport modelling is beyond the scope of this book; for further reading we suggest chapter 12 of Brasseur et al. (1999) for an introduction and Jacobson (2000) for a thorough discussion. Here, we will discuss the numerical properties of the different transport schemes used to solve the continuity equation.

Despite intensive research over the last 30 years, no universal numerical scheme has been developed to date which satisfies all the required numerical properties of transport of mass by the continuity equation. Ideally, a numerical method should meet a subset of the following properties: accuracy, stability, transportivity (requires that any perturbation is advected downwind), locality (the solution of the advection equation at a given point is not significantly influenced by the field far from that point), conservation, monotonicity or shape-preserving (provides no overshoot or undershoot near regions of strong gradients), efficiency (limited computer time and storage requirement) and low numerical diffusion. Depending on the applications, the subsets of required numerical properties may be different.

Very sharp gradients may occur in the vicinity of sources and sinks regions or during the day-night crossing period. The numerical modelling of these features requires monotonic schemes to avoid spurious oscillations in these regions and negative values which may lead to instabilities. Numerical noise might be acceptable when advection is the only concern, but may lead to large errors as a result of nonlinear interactions between chemical tracers. In addition to being definite-positive, chemistry-transport algorithms have to satisfy the essential mass-conservation property. Minor non-conservation, probably unimportant in meteorological simulations, can be amplified in chemistry simulations solving 2nd or higher order chemical reactions. A simpler way to ensure that total mass is conserved is to use the flux-form, or conservative form (the constituent continuity equation is written using the constituent flux), rather than the advected form when solving the chemical species continuity equations.

Several numerical algorithms meet these properties using appropriate adjustments and specific limiters. However, the CPU (Central Processing Unit) time and memory requirements are usually specific to each numerical scheme and are determinant elements for the choice of a specific transport scheme to be implemented in atmospheric-chemistry models. Typically, full-comprehensive chemistry models simulate 50 to 100 species which react with each other in hundreds of nonlinear chemical reactions. The efficiency of the model then strongly depends on the storage capacity and the CPU time requirements. Therefore, additional numerical cost using accurate but expensive transport algorithms could be a real issue when processes other than large-scale advection such as vertical mixing by convection or small-scale inhomogeneities advection due to nonlinear reactions, are concerned.

Actually, three families of advection schemes can be grouped in: Lagrangian, semi-Lagrangian and Eulerian schemes. The main difference between those families is the observer situation: in the Lagrangian and semi-Lagrangian methods, the observer is f attached on air parcels displaced along trajectories whereas it is fixed at the spatial grid points in the Eulerian approach. The transport algorithms are then totally different and do not satisfy the same numerical properties.

In the past, spectral and pseudo-spectral methods have also been largely used because of their attractive advantage of accurately estimating the spatial derivatives (Gottlieb and Orszag, 1977; Machenauer 1979). However, the truncated series expansion of the transport equations tend to produce spurious oscillations in the vicinity of sharp gradients, which is a major drawback for transport-chemistry modelling, and therefore these methods have been progressively abandoned. In semi-Lagrangian formulation, the solution at prescribed grid points is derived at each time step on the basis of a Lagrangian backward calculation. An interpolation technique is used to estimate the required value residing between grid points (Ritchie 1987, McDonald 1986). These methods have received considerable attention in the 1980's in general circulation modelling, because of the use of much larger time steps than Eulerian schemes. The advantages of these methods are their accuracy, monotonicity and shape-preserving properties, which explain their frequent utilization to simulate the transport of water vapour (Rasch et al., 1990) or stratospheric aerosols (Boville, 1991). The success of semi-Lagrangian techniques is nevertheless greatly dependent on the interpolation scheme used to estimate the velocity at the grid points (Staniforth and Côté, 1991). Linear interpolation, while simple and monotonic, introduces strong numerical diffusion. Higher order methods can avoid this numerical problem, but their implementation is complex and expensive. In addition, these methods are not intrinsically conservative, a major drawback in chemistry modelling, which can not be totally corrected by the introduction of a "fixer" applied after the advection calculation to ensure exact mass-conservation (Williamson & Rasch, 1989; Rasch, 1995) which further increases significantly the numerical cost of the model and degrades the overall accuracy. Some limiters are also necessary to achieve positivity. Eluszkiewicz et al. (2000) have compared several transport schemes in the Geophysical Fluid Dynamics Laboratory SKYHI GCM and shown that semi-Lagrangian models are much more expensive (by about 30%) than centred-differences and finite-volumes schemes, the incremental cost to add tracers being nevertheless lower than the other schemes. The non-conservation and complexity of the adjusted semi-Lagrangian transport schemes tend to have led to these techniques being progressively abandoned in chemistry applications in favour of Eulerian approach, despite the CFL stability criteria limitation.

An extensive class of Eulerian methods is available and new schemes, based on classic algorithms, are still being developed and proposed to satisfy the required numerical properties of chemistry-transport models. Elementary one-order upstream schemes, introduced in the 1960's, 1970's (see Godunov, 1959, Van Leer, 1973) are intrinsically positive, monotonic, very cheap and conservative in contrast to semi-Lagrangian transport techniques, but are very diffusive and dispersive. Such nonlinear schemes are far more complex than the traditional linear schemes but offer strong stability, low numerical diffusion, preserve the monotonicity and maintain steep gradients of the advected tracers. These numerical properties are very important, especially for tracer species with sources and sinks near the surface. These properties must be combined with the mass-conservation and positive-definite properties to be attractive for chemical-transport modelling.

In the late 1980's, Bott introduced a mass-conservative, positive-definite, 2nd-order algorithm that offers a good compromise between accuracy and numerical cost. It has been developed from the Tremback et al. (1987) scheme, which estimates tracer advection fluxes and then applies a nonlinear renormalization step to achieve its positive definiteness. Such numerical schemes, called "flux type schemes", are very attractive and recommended for models where resolving sharp vertical gradients of atmospheric species is important. Li & Chang (1996) presented a 4th-order version of Bott's advection scheme which exhibits low numerical diffusion and phase speed errors. This scheme is shown to be mass-conservative, positive-definite with very competitive performances in terms of accuracy and numerical diffusion with semi-Lagrangian and other higher-order Eulerian methods, such as the advection scheme proposed by Smolarkiewicz (1983), or the 2nd-order moments method designated by Prather (1986). The latter algorithm represents the tracer concentration inside a grid box by 2nd-order polynomials and its spatial distribution by 2nd-order moments. Often taken as a reference, its accuracy is significantly increased by advecting the higher-order moments. This method is then characterized by small numerical diffusion, but is computationally expensive in terms of CPU and storage requirements, since at each grid point 10 variables (in the 3D-case) need to be stored, which is a real issue in chemistry modelling, when a large number of species are advected in a 3D global comprehensive chemistry atmospheric model. In addition, Prather's scheme requires limits to completely avoid spurious numerical oscillations, is not inherently positive-definite and is very complicated to code. Finally, even on a coarse grid, Prather's scheme is not able to correctly treat high-order moments in strongly nonlinear processes, such as chemical reactions. Rather than going to a multi-moment scheme, it could be preferable to compute advection or other processes as chemistry or turbulent mixing on a finer grid, using a cheaper numerical algorithm. Van Leer's scheme (1979) was first introduced in the meteorological community by Allen et al. (1991), who abandoned Prather's scheme because of the storage problem, and has then been used at NASA Goddard Space Flight Center for various atmospheric studies. Van Leer's basic method uses a polynomial expression to represent the constituent within each box; and then a monotonic upstream-centred conservative scheme is used to advect the piecewise continuous function. Various schemes were then proposed by Van Leer (1977,1979) based on different estimations of the slope calculated from the tracer mixing ratio extrapolated at the mesh boundary. The Van Leer scheme called slopes scheme, developed by Russel and Lerner (1981), gives better results than Van Leer in terms of accuracy (Lin & Rood, 1996) without any disadvantage other than a small increase of CPU time, which is a minor problem compared to the storage requirement problem for chemistry applications. Although Prather's scheme is more accurate than Van Leer schemes, a factor of 2 in terms of CPU time was found between Prather and the slopes schemes and a factor of 4 between Prather and Van Leer (Hourdin et al., 1999). Recently, Petersen et al. (1998) have presented 4 new advection schemes for use in global chemistry models. Their motivation came from the need of chemistry modellers to have efficient accurate and simple transport algorithms to transport more species with increased spatial resolution. Focusing on both memory requirement and cost in terms of CPU time, they propose 3rd-order flux-schemes using Van Leer flux limiting (1979) to ensure monotonicity. The main difference between the schemes is the interface interpolation of the mixing ratio from the cell centre values. The schemes are mass-conservative, free from over- and undershoot (essential properties for chemistry modelling), positive-definite, easy to implement, and have the attractive advantage of being very efficient in terms of memory and CPU time requirements, since they need 4 to 10 times less memory than the slopes and Prather schemes respectively.

Numerical considerations may also be taken into account in the discussion between off-line and on-line treatment of transport in atmospheric chemistry models. The on-line mode configuration means that the chemistry-transport model is run interactively with a general circulation model or a weather forecast model (Feichter et al., 1992) in contrast with the off-line mode, where the tracer is advected by wind fields originating from the output of an atmospheric circulation model running independently (Jacob, 1987). When numerical weather prediction models are concerned, off-line models are currently driven by analyses (Rasch, 1997). Both treatments have advantages and drawbacks, and the debate is still open as to which method is preferable, depending on the needs of each atmospheric-transport model.

Using an off-line model to advect two tracers within a full hydrological cycle, Rasch et al. (1997) obtained a gain by a factor of 12 overt the cost of a full GCM calculation. Off-line models are then generally less expensive than the on-line models, saving in computer time and memory. However, a potential conflict may exist in the off-line configuration between mass-conservation, uniform mixing ratio conservation, numerical accuracy and fluid mass consistency. Such conflict does not exist in on-line models, where a consistent set of equations is used with possible interactions between tracers and dynamics or thermodynamics. In addition, the sampling of winds every few hours in trajectory calculations can lead to spurious vertical transport caused by the aliasing of vertical velocities. Rasch et al. (1997) have shown that sampling intervals on the order of 6 hours or less is adequate to drive off-line models, which is not the case at 24 hours intervals where numerical errors appear. Further, using 6 hours sampling intervals in global models, it is possible to treat in the off-line mode many problems as accurately as in the on-line mode.

Computational efficiency can also achieved with appropriate splitting schemes. The use of an adequate operator splitting to solve combined advective-diffusion–chemistry equations is crucial in chemistry-transport modelling. This technique allows each physical process to be treated separately using appropriate numerical schemes. The symmetric Strang operator splitting (Strang, 1968; Hundsdorfer and Spee, 1995; Petersen et al., 1998) is commonly used in many chemistry-transport models. Dimensional splitting schemes are also developed in finite-volume type schemes, where 1-D subprocesses are solved with explicit numerical schemes in each spatial direction. The order of the spatial discretizations determines the global order of the transport scheme. Hundsdorfer obtained satisfactory results using 2nd – order splitting discretizations.

In conclusion, the recent algorithms developed to transport chemical species in global 3-D atmospheric models are not fundamentally different from the numerical schemes developed in the 1980's and 1990's, but rather tend to be more efficient in order to include increasingly complex chemistry systems on finer resolution grids. The actual trend seems to be in favour high-order flux Eulerian schemes such as 4th-order Bott's schemes or finite-volume schemes with Van Leer flux limiting, intrinsically mass-conservative, positive-definite, monotonic, stable, shape-preserving and overall efficient algorithms, hoping for the increasing power of massive supercomputers, the newer architecture.

3. DATA ASSIMILATION AND INVERSE MODELLING METHODS

Data assimilation and synthetic inversion of the sources share the same mathematical apparatus of estimation theory and variational methods. Data assimilation refers to the integration of observational data into numerical models in order to estimate the initial conditions, whereas in synthetic inversion problems, the observational data is used to determine the sources using an atmospheric transport model. Chemical data assimilation and synthetic inversion of the sources are thus complementary problems, and are in fact interconnected. Not only that both methods use a chemical transport model to relate the unknown with the observations, but we have now clear indications that having the wrong sources in a chemical data assimilation will always result in a biased estimate of the chemical state, no matter how many and how perfect the observations may be (Dee and da Silva, 1998), and that

Data assimilation and inverse methods 485

having a wrong initial condition in a synthetic inversion of the sources creates a bias in the source estimate (Gilliland and Abitt, 2001). In this section we will draw a parallel between the two methods by emphasizing their analogous formulation, and discuss the simultaneous state/source estimation problem.

3.1. Basic principles

At the root of both methods is the transport of a chemical tracer, which can be described by the linear advection diffusion equation,

$$\frac{\partial C}{\partial t} + L(x,t)C = S(x,t) , \quad \text{(Eq.1)}$$

where C is the concentration expressed as a mixing ratio. L contains the advection and diffusion operators,

$$L(x,t)(\bullet) = V \cdot \nabla(\bullet) - \nu \nabla^2(\bullet) , \quad \text{(Eq.2)}$$

and can also include deposition and linearized chemistry, S is the source/sink term, x is the spatial coordinate, and t the temporal coordinate. Advection is replaced by a flux form if partial density or number density is used as the measure of concentration. Equation (1) has a formal solution of the form

$$C(x,t) = \Phi(t,0) C_0(x) + \int_0^t \Phi(t,\tau) S(x,\tau) d\tau \quad \text{(Eq.3)}$$

where $\Phi(t,\tau)$ is the fundamental solution operator (or the state transition matrix for discretized models). The fundamental solution operator $\Phi(t,\tau)$ is basically the Green function (Enting, 2002) from time τ to time t. Observations are generally carried out at a discrete set of point locations $\{x_1, x_2, ..., x_p\}$. The sampling of the function $C(x,t)$ at those points is what is observed. The forward problem consists in finding the simulated concentrations at the observation locations, given the knowledge of the model inputs $C_0(x)$ (initial concentration), $S(x,t)$ (sources distribution). The inverse problem, on the contrary, consists in estimating the model input based on observed concentrations. Often, the inverse problem is ill-conditioned in the sense that observations alone are insufficient to determine a unique solution of the model input at the resolution of the model. Ill-conditioning may also manifest in an inverse solution, which is very sensitive to observation errors.

Tracer data assimilation and integral tracer inversion are actually both "inverse methods", but their objectives and applications are different. The objective of tracer data assimilation is to estimate the initial (or instantaneous) concentration field (e.g. Ménard, 2000). The chemical source and sinks are prescribed or sometimes ignored as in the stratospheric problems.

In essence, tracer data assimilation aim at finding the "inverse" of the homogeneous equation,

$$C(x,t) = \Phi(t,0) \, C_0(x) + \text{prescribed source effects} \quad \text{(Eq.4)}$$

i.e. $C_0(x)$. In contrast, the objective of integral tracer inversion is to estimate the source/sink field, or to find the "inverse" of the inhomogeneous equation,

$$C(x,t) = \int_0^t \Phi(t,\tau) \, S(x,\tau) \, d\tau + \text{constant} \quad \text{(Eq.5)}$$

i.e. $S(x,t)$ (e.g., Enting, 2000). In general, however, both input fields $C_0(x)$, $S(x,t)$ are either unknown or can be considered as uncertain. Depending on the application and how the method is applied, one or the other method can be used and yield a reasonable estimate. There are several factors that must be taken into account in order to choose which inverse problem to treat. Perhaps the most relevant one is the model time integration period, and other parameters such as the diffusion time scale, the chemical life-time, the initial error variance, the source error variance, and the initial-source error covariance must also be considered (e.g., see Enting, 2002). Although there is still ongoing research in this area, some general statements can be made.

When the chemical transport model is integrated over a long time period such that diffusive effects dominate the homogeneous solution, then $\Phi(t,0)C_0(x)$ is nearly a constant term (or is spatially homogeneous). The problem is then appropriate for integral tracer inversion. Also when the chemical transport model is integrated over a short time scale, comparable to the advection time scale, the solution is then sensitive to the initial conditions, and if the initial state error variance is larger than the initial-source error covariance, then the problem can be formulated as a tracer data assimilation problem. A summary of the distinctive features between tracer data assimilation and integral tracer inversion is given in table 1.

Table 1. Basic features of data assimilation and inverse methods

	Data assimilation	Inverse methods
Problem	Solves for the concentration using the homogeneous solution $$C(x,t) = \Phi(t,0) C_0(x) +$$ source term prescribed	Solves for the fluxes using the inhomogeneous solution $$C(x,t) = \int_0^t \Phi(t,\tau) S(x,\tau) \, d\tau + \text{constant}$$
Time integration	t is smaller or comparable to the advection time scale, so that transport accounts for most of the changes in observed tracer data.	t is large enough so that the homogenous solution becomes basically a uniform concentration field
Ill-conditioning	Usually observations alone do not contain enough information to provide complete coverage of the concentration field. Prior model integrations are weighted with observation residuals to control the estimate.	Diffusion and mixing over long integration times results in a loss of spatial information about sources/sink. Spatial modelling of flux variations (i.e. using pre-specified functions of space and time) is used to constrain the solution.

3.2 Duality of state and source estimation

In order to cast the duality of tracer data assimilation and integral tracer inversion in mathematical terms, let us now write the chemical transport model in a form suitable for inverse methodologies. Except in rare cases, the solution of the chemical transport model can only be calculated by using numerical methods. Let $\mathbf{C}_n$ be the concentration vector at time t_n (a vector of concentrations at the different grid points). The solution of the transport with no source/sink term is of the form

$$\mathbf{C}_n = \mathbf{M}_{n-1}\mathbf{C}_{n-1} \qquad (\text{Eq.6})$$

where the model matrix $\mathbf{M}_{n-1}$ is the discrete time-space linear equivalent of the fundamental solution operator from t_{n-1} to t_n,

$$\Phi(t_{n-1}, t_n) \leftrightarrow \mathbf{M}_{n-1}. \qquad (\text{Eq.7})$$

With the source/sink term, the discrete solution takes the following from

$$\mathbf{C}_n = \mathbf{A}(n,0)\,\mathbf{C}_0 + \Delta t \sum_{k=0}^{n-1} \mathbf{A}(n, k+1)\mathbf{S}_k, \qquad (\text{Eq.8})$$

where

$$\mathbf{A}(n,k) = \begin{cases} \mathbf{M}_{n-1}\mathbf{M}_{n-2}\cdots\mathbf{M}_k & \text{for } n > k \\ \mathbf{I} & \text{for } n = k \end{cases}. \qquad (\text{Eq.9})$$

where Δt is the time step and S_k the source at time k. To simplify, we may assume that the source/sink term is constant in time. In this case, the solution of equation (8) takes the form

$$\mathbf{C}_n = \mathbf{A}(n,0)\,\mathbf{C}_0 + \mathbf{B}(n,0)\,\mathbf{S}, \qquad (\text{Eq.10})$$

where

$$\mathbf{B}(n,0) = \Delta t \sum_{k=0}^{n-1} \mathbf{A}(n, k+1). \qquad (\text{Eq.11})$$

We should note that the concentration at time t_n is linear in $\mathbf{C}_0$ and in the source/sink term $\mathbf{S}$. In data assimilation we usually assume S to be known, and we use the observations to estimate the initial condition $\mathbf{C}_0$. Traditionally in synthetic inversion of sources, we assume the initial condition $\mathbf{C}_0$ to be known, and we use the observations to estimate the source/sink term $\mathbf{S}$.

The problem of estimating an unknown (either the initial condition $\mathbf{C}_0$ or the source $\mathbf{S}$) from observations is generally formulated in a weighted least squares sense. Suppose we have a set of measurements $\{\mathbf{Y}_n, \mathbf{Y}_{n-1}, \ldots, \mathbf{Y}_0\}$

Data assimilation and inverse methods

from time t_0 to time t_n. $\mathbf{Y}_n$ is a vector of observed values, if there is more than one measurement at time t_n. With a discrete point model, the observation location will generally lie in between model grid points. In order to estimate the model's unknown quantities, it is necessary to compute the model equivalent value at the observation location. One way is to interpolate the model grid values at the observation location, and this is denoted by an interpolation operator $\mathbf{H}$. If the true model values were given at the model grid points, and denoted by $\mathbf{C}_n^t$, a perfect observation or true observation equivalent, $\mathbf{Y}_n^t$, would be given /by

$$\mathbf{Y}_n^t = \mathbf{H}_n \mathbf{C}_n^t \qquad \text{(Eq.12)}$$

$\mathbf{H}_n$ is also called the observation operator, as it maps the model space to the observation space. In reality, the observations are noisy, so that the observed values are now given by

$$\mathbf{Y}_n = \mathbf{H}_n \mathbf{C}_n^t + \varepsilon^o \qquad \text{(Eq.13)}$$

where ε^0 is the observation error, and we will assume that we know its statistics.

The time-discrete state equation (10) shows how the concentration at time t_n depends on the initial conditions and the source, while the measurement equation (13) indicates how the observed quantity is related to the concentration. In the following subsections we will derive the estimation schemes for: 1) initial condition or state estimate only, 2) source estimation only, and 3) simultaneous source and state estimation.

3.2.1. State estimation

Let us assume that the concentration uncertainties arise only from uncertainties in the initial conditions. Using a weighted least square method, we weight the observation residuals, $Y_n - \mathbf{H}_n\mathbf{C}_n$, by the inverse of the observation error covariance matrix, $\mathbf{R}_n = \langle \varepsilon_n^o (\varepsilon_n^o)^T \rangle$. The following cost or merit function

$$J = (\mathbf{Y}_n - \mathbf{H}_n \mathbf{C}_n)^T \mathbf{R}_n^{-1} (\mathbf{Y}_n - \mathbf{H}_n \mathbf{C}_n), \qquad \text{(Eq.14)}$$

is then minimized with respect to C_0. The explicit dependence on C_0 is given by substituting (10) into (14),

$$J(C_0) = \left(Y_n - H_n A(n,0) C_0 - H_n B(n,0) S\right)^T R_n^{-1} \left(Y_n - H_n A(n,0) C_0 - H_n B(n,0) S\right).$$
(Eq.15)

To obtain the minimum, consider a perturbation δC_0 of the initial condition C_0. The difference in cost function can be written as,

$$J(C_0 + \delta C_0) - J(C_0) = -(\delta C_0)^T A^T(n,0) H_n^T R_n^{-1} \left(Y_n - H_n A(n,0) C_0 - H_n B(n,0) S\right)$$
$$- \left(Y_n - H_n A(n,0) C_0 - H_n B(n,0) S\right)^T R_n^{-1} H_n A(n,0) (\delta C_0)$$
$$+ O(\delta C_0)^2$$
(Eq.16)

The minimum is obtained when the right hand side (r.h.s.) of Eq. 16 is identical to zero for any perturbation, δC_0. Neglecting the high-order terms in δC_0, and noting that the first and second term on the r.h.s. are the transpose of one another, we get

$$A^T(n,0) H_n^T R_n^{-1} \left(Y_n - H_n A(n,0) \hat{C}_0 - H_n B(n,0) S\right) = 0$$
(Eq.17)

as the condition for the minimum. In (17) the caret (^) indicates the estimated value or value which minimizes the cost function. Rewriting (17) we get,

$$\hat{C}_0 = \left(A^T H_n^T R_n^{-1} H_n A\right)^{-1} A^T H_n^T R_n^{-1} \left(Y_n - H_n B S\right),$$
(Eq.18)

where we have dropped the time indices with the A and B matrices. Such an estimate is possible only when the matrix $A^T H^T R^{-1} H A$ is invertible or of full rank, which will occur only if the number of observations exceeds that of the dimension of the state. In almost all cases of atmospheric interest, we generally have fewer observations than the dimension of the atmospheric model state. The system is then underdetermined, and thus (18) cannot be used.

A solution to this problem is to fit observations over a whole time period rather than at a single time step, so that many more observations are used to determine the same number of unknowns. The method then consists of finding $\mathbf{C}_0$, such that the following cost function is minimized

$$J(\mathbf{C}_0) = \sum_{n=0}^{N} (\mathbf{Y}_n - \mathbf{H}_n \mathbf{C}_n)^T \mathbf{R}_n^{-1} (\mathbf{Y}_n - \mathbf{H}_n \mathbf{C}_n). \qquad \text{(Eq.19)}$$

Following a similar procedure as above, the solution must satisfy the condition

$$\sum_{n=0}^{N} \mathbf{A}^T(n,0) \mathbf{H}_n^T \mathbf{R}_n^{-1} \left(\mathbf{Y}_n - \mathbf{H}_n \mathbf{A}(n,0) \hat{\mathbf{C}}_0 - \mathbf{H}_n \mathbf{B}(n,0) \mathbf{S} \right) = 0 \qquad \text{(Eq.20)}$$

and in principle, similarly to the case above, a solution can be obtained by inverting a matrix $\Sigma \mathbf{A}^T \mathbf{H}^T \mathbf{R}^{-1} \mathbf{H} \mathbf{A}$. This approach is, however, cumbersome and it involves inverting a large matrix and thus is never used in practice. A popular and considerably much more efficient method consists in searching for the minimum of the cost function (19) using a steepest descent method. These descent methods usually require the knowledge of the gradient of the cost function, and proceed iteratively until the minimum is reached. To understand how this algorithm works, we should note first that for an arbitrary $\mathbf{C}_0$ the left-hand-side of equation (20) is actually the gradient of the cost function (and is identical to zero at the minimum). The gradient of the cost function can be easily computed by running the model forward in time from t_0 to t_N and followed by a backward integration of the adjoint model from t_N to t_0 and using residuals $\mathbf{Y}_n - \mathbf{H}_n \mathbf{C}_n$ (obtained from the forward integration) as forcing terms in the backward integration. This technique is known as the adjoint method or the four-dimensional variational method (4-D Var). The adjoint model is defined as follows. If we were to define a single time-step prediction as the result of a model matrix multiplying a vector of concentration, the adjoint model would be equivalent to the transpose of that model matrix. We should also note that

$$\mathbf{A}^T(n,0) = (\mathbf{M}_{n-1} \cdots \mathbf{M}_0)^T = \mathbf{M}_0^T \cdots \mathbf{M}_{n-1}^T \qquad \text{(Eq.21)}$$

and so when $\mathbf{A}^T(n,0)$ is applied to a vector, this is equivalent to applying a sequence of the adjoint model to the same vector, going backwards in time from t_N to t_0. With these remarks, it is then easy to show that the gradient of the cost function, or the left-hand-side of equation (20), is then obtained by stepping backwards in time with the following equation,

$$\mathbf{p}_n = \mathbf{M}_n^T \mathbf{p}_{n+1} + \mathbf{H}_n^T \mathbf{R}_n^{-1}(\mathbf{Y}_n - \mathbf{H}_n \mathbf{C}_n) \qquad \text{(Eq.22)}$$

using $\mathbf{p}_{N+1} = 0$ initial condition at time t_{N+1}. The solution $\mathbf{p}_0$ at time t_0 is the gradient of the cost function, see Giering (2000) or Wang et al. (2001) for further details. In practice and even in the case where the total number of observation exceeds the dimension of the state vector, the solution may still be sensitive to errors in the observational data. This problem arises especially with sparse observations, where small errors in the data are amplified in the solution.

To better constrain the inverse problem, an a priori term can be added in the cost function. Considering a cost function over a single time, the new term on the cost function (15) is:

$$J(\mathbf{C}_0) = (\mathbf{Y}_n - \mathbf{H}_n \mathbf{A}(n,0)\mathbf{C}_0 - \mathbf{H}_n \mathbf{B}(n,0)\mathbf{S})^T \mathbf{R}_n^{-1}(\mathbf{Y}_n - \mathbf{H}_n \mathbf{A}(n,0)\mathbf{C}_0 - \mathbf{H}_n \mathbf{B}(n,0)\mathbf{S}) + (\mathbf{C}_0 - \mathbf{C}_0^p)^T \mathbf{P}_0^{-1}(\mathbf{C}_0 - \mathbf{C}_0^p) \qquad \text{(Eq.23)}$$

where $\mathbf{C}_0^p$ is an a priori on the initial concentration and $\mathbf{P}_0$ is the error covariance (matrix) of the a priori, assumed to be full rank. The a priori can be a climatology or a first guess. The derivation of the estimate follows the minimization procedure used in (16)-(18) above, and we get in this case,

$$\hat{\mathbf{C}}_0 = \mathbf{C}_0^p + \mathbf{K}_n (\mathbf{Y}_n - \mathbf{H}_n \mathbf{A} \mathbf{C}_0^p - \mathbf{H}_n \mathbf{B}\mathbf{S}), \qquad \text{(Eq.24)}$$

where

$$\mathbf{K}_n = \mathbf{P}_0 (\mathbf{H}_n \mathbf{A})^T (\mathbf{H}_n \mathbf{A} \mathbf{P}_0 \mathbf{A}^T \mathbf{H}_n^T + \mathbf{R}_n)^{-1}, \qquad \text{(Eq.25)}$$

is called the Kalman gain matrix (in reference to the Kalman filter). The derivation above used the following identity

$$(\mathbf{F}^T \mathbf{R}^{-1} \mathbf{F} + \mathbf{P}^{-1})^{-1} = \mathbf{P} - \mathbf{P}\mathbf{F}^T (\mathbf{F}\mathbf{P}\mathbf{F}^T + \mathbf{R})^{-1} \mathbf{F}\mathbf{P}, \qquad \text{(Eq.26)}$$

which is valid for any matrix **F** as long as **P** and **R** are full rank covariance matrices. The matrix to invert in (25) is of size p x p where p is the number of observations (at a given time t_n). For observations exceeding 1,000 in number, the matrix inversion in (25) becomes rapidly computationally expensive and sensitive to truncation error, and alternative methods should be used. One way is to minimize the cost function by using an unconstrained minimization routine (e.g. descent methods); the resulting algorithm is called 3-D Var. By extending the cost function to include a summation over time for the observation misfit, we obtain

$$J(\mathbf{C}_0) = \sum_{n=0}^{N} \left(\mathbf{Y}_n - \mathbf{H}_n \mathbf{C_n}\right)^T \mathbf{R}_n^{-1}\left(\mathbf{Y}_n - \mathbf{H}_n \mathbf{C_n}\right) + \left(\mathbf{C}_0 - \mathbf{C}_0^p\right)^T \mathbf{P}_0^{-1}\left(\mathbf{C}_0 - \mathbf{C}_0^p\right),$$

(Eq.27)

which is the basis for the 4-D Var algorithm.

An important quantity that appears in (25) is the chemical model prediction error that is represented by the action **A** of the model on the a priori error covariance $\mathbf{P}_0$. The ratio of observational error **R** with the model error interpolated at the observation location $\mathbf{HAP}_0\mathbf{A}^T\mathbf{H}^T$ determines the weight given to the observations. The spatial structure of the chemical prediction error also determines the spatial extent of the impact of an observation. To see this more clearly, set p=1 and in (25) the "matrix" to invert turns out to be a scalar. The whole spatial structure of the impact of a single observation then depends solely on $\mathbf{P}0(\mathbf{HA})^T$. As an example, Lyster et al. (1997), Ménard et al. (2000), and Ménard and Chang (2000) have developed a two-dimensional assimilation scheme for long-lived species on isentropic surfaces, and have applied it to the assimilation of stratospheric observations of Upper Atmosphere Research Satellite. Figure 1 shows the spatial structure of the impact of single methane observations from the Halogen Occultation Experiment (HALOE) on board UARS at three locations during the month of September 1992. We observe strong anisotropy as a result of shear flow that dominates the southern mid-latitudes.

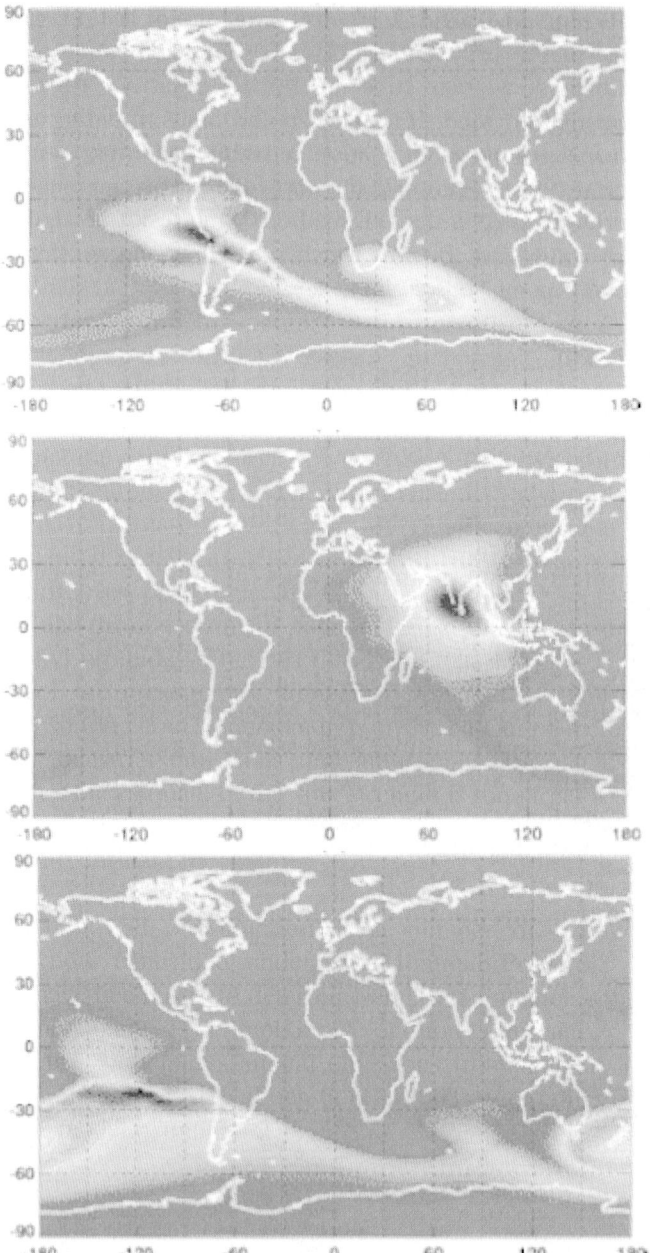

Figure 1 (see plate 25). Spatial error correlation structure at three HALOE data points (Reproduced from Ménard and Chang, 2000)

In general, with three-dimensional flows the propagation of the initial error covariance is computationally prohibitive and approximations are sought. One approximation that was developed by Cohn (1993), tested in a two-dimensional assimilation system by Ménard and Chang (2000) and implemented in three-dimensional assimilation systems by Khattatov et al. (2000), consists in propagating only the error variance. This scheme, known as the sequential scheme, has been implemented in a number of research and operational centres throughout the world and has been applied to many long-lived tracer and aerosols in both troposphere and stratospheric applications. Figure 2 shows a typical result of the sequential scheme for the assimilation of ozone observations from the Microwave Limb Sounder (MLS) on UARS. Note the valleys (relative minima) in the error variance field that are the results of fresh new observations, impacting positively on the uncertainty of the state estimated field.

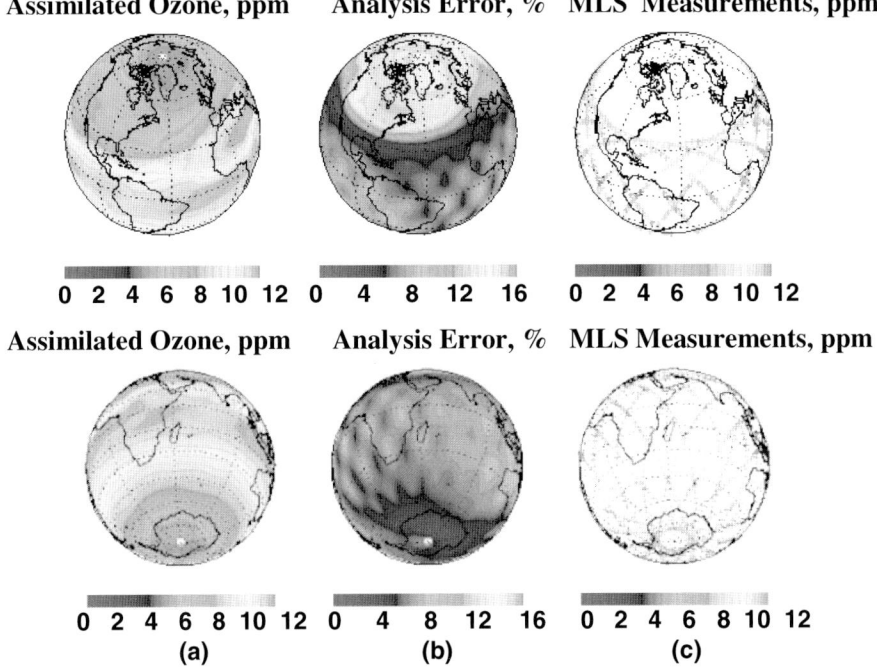

Figure 2 (see plate 26). Result of assimilation with the sequential method. Panel (a) analysed ozone, Panel (b) analysis error in percent, and Panel (c) ozone observations (Reproduced from Khattatov et al., 2000), (© American Geophysical Union).

3.2.2. Source estimation

Now let us assume that the concentration uncertainties arise only (and in practice primarily) from uncertainties in the source/sink term **S**. The algebra to derive the estimate is analogous to that used in the previous subsection. Minimizing the cost function (15), but now with respect to **S**, yield a source estimate

$$\hat{\mathbf{S}} = \left(\mathbf{B}^T \mathbf{H}_n^T \mathbf{R}_n^{-1} \mathbf{H}_n \mathbf{B}\right)^{-1} \mathbf{B}^T \mathbf{H}_n^T \mathbf{R}_n^{-1}\left(\mathbf{Y}_n - \mathbf{H}_n \mathbf{A} \mathbf{C}_0\right). \quad \text{(Eq.28)}$$

As with the state estimation problem, an a priori source estimate can be included in the cost function. Following a similar derivation to (24), the sources estimate with prior, can be obtained as

$$\hat{\mathbf{S}} = \mathbf{S}^p + \mathbf{G}_n \left(\mathbf{Y}_n - \mathbf{H}_n \mathbf{A} \mathbf{C}_0 - \mathbf{H}_n \mathbf{B} \mathbf{S}^p\right), \quad \text{(Eq.29)}$$

where

$$\mathbf{G}_n = \mathbf{P}_S (\mathbf{H}_n \mathbf{B})^T (\mathbf{H}_n \mathbf{B} \mathbf{P}_S \mathbf{B}^T \mathbf{H}_n^T + \mathbf{R}_n)^{-1}, \quad \text{(Eq.30)}$$

is the gain matrix and $\mathbf{P}_S$ is the covariance (matrix) of the uncertainty of the source/sink prior $\mathbf{S}^p$.

We note that the source/sink estimate $\hat{\mathbf{S}}$ (28)(29) has a contribution from the initial concentration $\mathbf{C}_0$, but it is assumed that the initial error variance is much smaller than the a-priori source error variance. In classical synthetic inversion techniques, the contribution from the initial condition can be taken as a constant or spatially uniform field, provided that the time integration of the model allows for sufficient mixing of the initial condition.

3.2.3. Simultaneous state and source estimation

We have seen that either state-only or source-only estimation is incomplete and can introduce biases in the estimates. Let us now consider that both the source/sink a priori and the initial condition have uncertainties. The error covariance matrices are respectively $\mathbf{P}_S$ and $\mathbf{P}_0$. The cost function to minimize should then have the form,

$$J(\mathbf{C}_0, \mathbf{S}) = \left(\mathbf{Y}_n - \mathbf{H}_n \mathbf{A}(n,0)\mathbf{C}_0 - \mathbf{H}_n \mathbf{B}(n,0)\mathbf{S}\right)^T \mathbf{R}_n^{-1}\left(\mathbf{Y}_n - \mathbf{H}_n \mathbf{A}(n,0)\mathbf{C}_0 - \mathbf{H}_n \mathbf{B}(n,0)\mathbf{S}\right)$$
$$+ \left(\mathbf{S} - \mathbf{S}^p\right)^T \mathbf{P}_s^{-1}\left(\mathbf{S} - \mathbf{S}^p\right)^T + \left(\mathbf{C}_0 - \mathbf{C}_0^p\right)^T \mathbf{P}_0^{-1}\left(\mathbf{C}_0 - \mathbf{C}_0^p\right)$$

(Eq.31)

Then, from

$$\frac{\partial J}{\partial \mathbf{C}_0} = 0 \Rightarrow (\mathbf{H}_n \mathbf{A})^T \mathbf{R}_n^{-1}\left(\mathbf{Y}_n - \mathbf{H}_n \mathbf{A}\hat{\mathbf{C}}_0 - \mathbf{H}_n \mathbf{B}\hat{\mathbf{S}}\right) = \mathbf{P}_0^{-1}(\hat{\mathbf{C}}_0 - \mathbf{C}_0^p)$$

(Eq.32)

and from,

$$\frac{\partial J}{\partial \mathbf{S}} = 0 \Rightarrow (\mathbf{H}_n \mathbf{B})^T \mathbf{R}_n^{-1}\left(\mathbf{Y}_n - \mathbf{H}_n \mathbf{A}\hat{\mathbf{C}}_0 - \mathbf{H}_n \mathbf{B}\hat{\mathbf{S}}\right) = \mathbf{P}_s^{-1}(\hat{\mathbf{S}} - \mathbf{S}^p).$$

(Eq.33)

the solution is obtained by solving this system of equation, or by a direct minimization of the cost function using a descent method. The solution of (32) and (33) gives

$$\begin{aligned}\hat{\mathbf{C}}_0 &= \mathbf{C}_0^p + \mathbf{K}_n\left(\mathbf{Y}_n - \mathbf{H}_n \mathbf{A}\mathbf{C}_0^p - \mathbf{H}_n \mathbf{B}\hat{\mathbf{S}}\right) \\ \hat{\mathbf{S}} &= \mathbf{S}^p + \mathbf{G}_n\left(\mathbf{Y}_n - \mathbf{H}_n \mathbf{A}\hat{\mathbf{C}}_0 - \mathbf{H}_n \mathbf{B}\mathbf{S}^p\right)\end{aligned}.$$

(Eq.34)

where

$$\mathbf{K}_n = \mathbf{P}_0 (\mathbf{H}_n \mathbf{A})^T \left[(\mathbf{H}_n \mathbf{A})\mathbf{P}_0 (\mathbf{H}_n \mathbf{A})^T + \mathbf{R}_n\right]^{-1}.$$

(Eq.35)

and

$$\mathbf{G}_n = \mathbf{P}_s (\mathbf{H}_n \mathbf{B})^T \left[(\mathbf{H}_n \mathbf{B})\mathbf{P}_s (\mathbf{H}_n \mathbf{B})^T + \mathbf{R}_n\right]^{-1}.$$

(Eq.36)

However, this system of equations is coupled and thus not easy to use. Fortunately, an uncoupled system of equations can be obtained, as in Dee and da Silva (1998). After considerable algebra it can be shown that the estimates can be computed from

$$\hat{\mathbf{C}}_0 = \mathbf{C}_0^p + \mathbf{K}_n\left(\mathbf{Y}_n - \mathbf{H}_n\mathbf{A}\mathbf{C}_0^p - \mathbf{H}_n\mathbf{B}\hat{\mathbf{S}}\right)$$
$$\hat{\mathbf{S}} = \mathbf{S}^p + \mathbf{L}_n\left(\mathbf{Y}_n - \mathbf{H}_n\mathbf{A}\mathbf{C}_0^p - \mathbf{H}_n\mathbf{B}\mathbf{S}^p\right)$$
(Eq.37)

where **K** is the Kalman gain (36) above, and

$$\mathbf{L}_n = \mathbf{P}_s(\mathbf{H}_n\mathbf{B})^T\left[(\mathbf{H}_n\mathbf{A})^T\mathbf{P}_0(\mathbf{H}_n\mathbf{A}) + (\mathbf{H}_n\mathbf{B})^T\mathbf{P}_s(\mathbf{H}_n\mathbf{B}) + \mathbf{R}_n\right]^{-1}$$
(Eq.38)

The system (37) can now be solved, first by computing the source estimate $\hat{\mathbf{S}}$, then by using the source estimate in the state estimate equation to compute $\mathbf{C}_0$.

3.3. Example of assimilation and inverse modelling of chemical species

3.3.1. Assimilation of satellite data

Carbon monoxide CO is a key component of the troposphere. It is the principal sink of the hydroxyl radical OH in the free troposphere, and thus controls indirectly the lifetime of many other species, such as methane (CH_4) (Isaksen and Hov, 1987). CO global average lifetime is 2 months; its reaction with OH represents 85-90% of its sink. Approximately 10% of tropospheric CO is lost at the surface through biological processes in the soil, and a small amount of CO is transported to the stratosphere. In the presence of nitrogen oxides, NO_x (>10-15 pptv), and sunlight, CO is a precursor of tropospheric ozone (O_3). The indirect greenhouse effect of anthropogenic CO is believed to be larger than N_2O radiative forcing (Daniel and Solomon, 1998).

Carbon monoxide observations from Measurements Of Pollution In The Troposphere (MOPITT) on board the Earth Observing System (EOS) Terra satellite, have been assimilated with the Canadian 3-D Var system (Gauthier et al., 1999). The assimilation system uses a three-dimensional a priori that is given by the Global Environmental Model (GEM) developed at the Meteorological Service of Canada (MSC) for numerical weather prediction (Coté et al., 1998), where online chemistry has been added. The model GEM can work with global uniform resolution, variable resolution geometry, or in a limited area configuration. In online model both the meteorological and chemical variables are solved simultaneously as part of the same numerical

Data assimilation and inverse methods 499

code. Here a simplified tropospheric chemistry for CO has been developed with chemical loss due to the OH and chemical source by the reaction of CH_4 with OH. In these experiments, both OH and CH_4 are prescribed. A global value for CH_4 mixing ratio is given and a zonal time-mean value for OH, obtained from a comprehensive chemical transport model simulation, has been used. Prescribed sources of CO at 1°×1° from GEIA depicted below for the month of October were used as mass flux at the surface (Müller, 1992).

The 3-D Var system uses a relative error covariance formulation where the error variance is proportional to the state. The error correlation is a separable horizontal/vertical model, and horizontally homogeneous. The horizontal correlation length scale, the observation error, and forecast error are usually determined by using the maximum likelihood method for covariance parameter estimation (see Ménard, 2000). The vertical correlation structure is obtained from lagged-forecasts, also known as the NMC method (Parrish and Derber, 1992).

The results presented here were obtained using a non-official and early version of the MOPITT retrieval of CO total column amount. The assimilation conducted with these pre-released observations is nevertheless useful for establishing a proof of concept, and illustrates some of the usefulness of data assimilation. An assimilation of six weeks has been conducted, starting September 26, 2000.

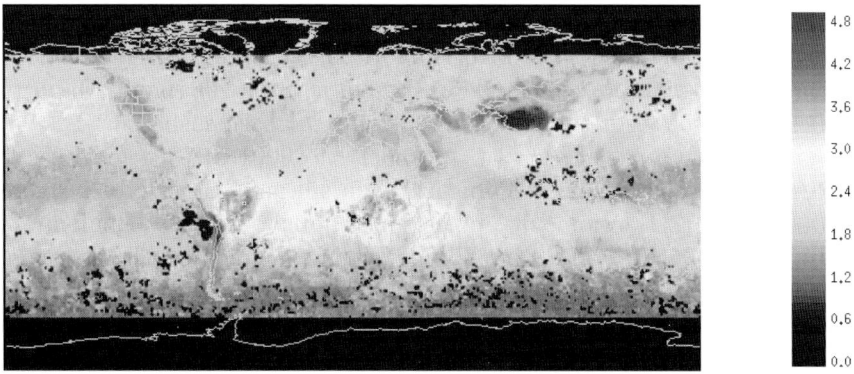

Figure 3 (see plate 27). MOPITT CO observations in molecules $cm^{-2} \times 10^{+18}$. (© American Geophysical Union).

Figure 3 depicts the CO total column observations aggregated for a period of two weeks, in order to have a global coverage. We observe that significant CO in biomass burning areas in the tropics, and out of southeastern Asia is transported across the Pacific Ocean. A closer examination with the emission inventory (not shown) indicates that the geographical location of the tropical biomass burning is significantly offset.

The result of assimilation is presented in Figure 4. To compare the assimilation results with the Figure 3, the analyses (or state estimates) were averaged in time over the same two week period, i.e. from October 12, 2000 to October 27, 2000, see Figure 4. We observe in this comparison a general agreement between observations and analyses as one would expect. It should be noted, however, that instantaneous fields, contrary to the average presented in Figure 4, show much more details in flow structure.

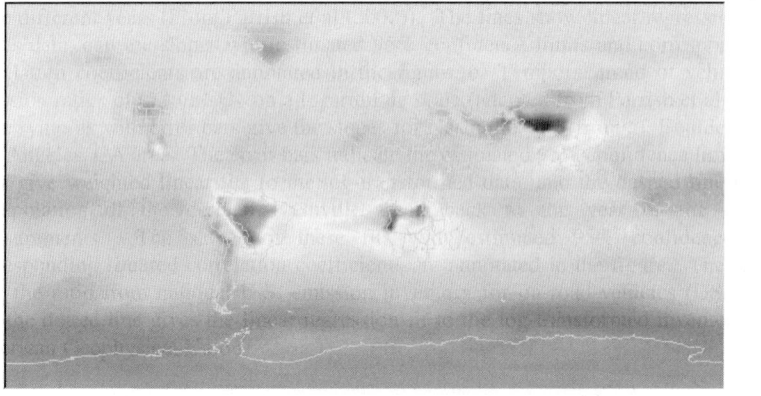

Figure 4 (see plate 28). Assimilation of CO. Time average of total column amount depicted. (© American Geophysical Union).

3.3.2. Inverse modelling

CH_4 inversion

So far, inverse modelling studies of CH_4 mainly focused on multi-annual mean fluxes, by filtering out any inter-annual variations from the measured time series (Hein, 1997; Houweling, 1999). The measurements come from global networks of remote monitoring stations (mainly NOAA/CMDL, see chapter 11). In addition, measurements of the isotopic composition of CH_4

($^{12}C/^{13}C$) are used to discriminate between different methane producing and consuming processes (Hein, 1997; Bermaschi, 2001). In broad lines, these studies show that this type of inversion primarily constrains the global budget of CH_4, and the latitudinal distribution of sources and sinks. This can be explained by the configuration of the measurements networks and characteristics of the atmospheric circulation. The results indicate a relatively low ratio of the Northern versus Southern Hemispheric net source of CH_4 that is difficult to reconcile with the existing emission inventories. Tropical natural wetlands, however, are an important and uncertain term that introduces a large degree of freedom. Natural wetland emissions that are in the upper part of the reported range of estimates (90-250 $TgCH_4$/yr) could explain the observed north-south concentration gradient of CH_4 (Bergamaschi, 2001). The required ratio of emissions from tropical and temperate/boreal latitudes, however, seems implausible. Besides this, the observed increase of atmospheric methane concentrations since pre-industrial times points to lower natural wetland emissions (Houweling, 2000). We may need a regional rather than a global scale approach to resolve this issue.

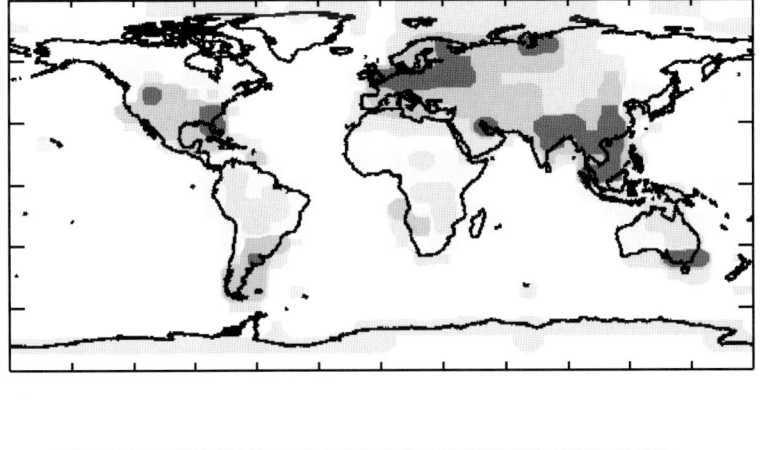

Figure 5 (see plate 29). A comparison of inversion derived estimates of climatological CH_4 sources and sinks (a posteriori minus a priori on a 8x10 degree grid)

Regional scale inversions require stations in the vicinity of the (continental) sources, measuring at relatively high sampling frequencies (Houweling, 2000). Some recent studies (Vermeulen, 1999; Ryall, 2001) indicate that emission estimates for Europe could already be improved by a relatively sparse network of continuously sampling measurement towers. However, the quality of the applied atmospheric transport model becomes a critical factor, in particularly its ability to reproduce the boundary layer height (Vermeulen, *1999*). The use of measurements of diagnostic tracers such as ^{222}Rn and ^{212}Pb has been proposed to eliminate such model errors (Biraud, 2002).

Inverse modelling can also contribute to our understanding of the mechanisms behind the CH_4 growth rate anomalies that were observed during the nineties, for example, the growth rate decrease following the Mt. Pinatubo eruption (June 15, 1991), and the increase during the exceptionally strong 1997-1998 El-Nino. Recently Dentener et al. (*2002*) estimated that the amplitudes of the anomalies during the 1980's and early 1990's correspond to source anomalies up to 15 $TgCH_4$/yr. Preliminary results by Houweling et al. (in preparation, 2003) for the period 1990-2000 suggest that the major source anomalies were about a factor of two larger (see Figure 6).

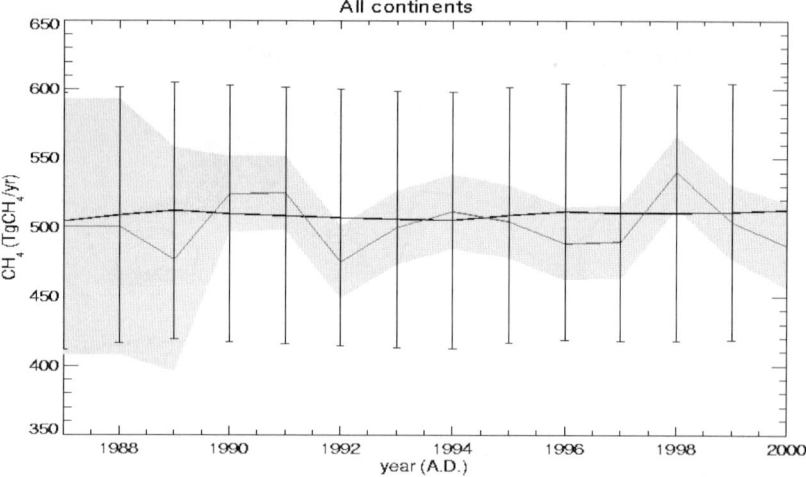

Figure 6. Inverse modelling derived annual CH_4 sources and sinks during the 1990's. Black and error bar, a priori and 1 sigma uncertainty; red and grey shading, a posteriori and 1 sigma uncertainty. Note that the first three years include inversion spin-up.

The interpretation of these results is complicated by the fact that inversions cannot distinguish between surface source anomalies and atmospheric sink anomalies since the measurements are all taken at the boundary between them (the surface). Without a priori information, a range of combinations of sources and sinks form valid solutions to the inverse problem. Unfortunately, the a priori term cannot add a significant constraint, since the inter-annual variation of the hydroxyl radical and therefore the CH_4 sink are poorly understood.

A promising new direction is the inverse modelling of CH_4 sources and sinks on the basis of satellite measurements. At present, three satellite instruments are in orbit that measure vertical columns of CH_4: MOPITT (Gauthier et al., 1999), AIRS (Aumann and Pagano, 1994) and SCIAMACHY (Bovensmann, 1999). The first two measure thermal infrared radiation emitted by the Earth, while the latter measures near infrared solar radiation that is reflected by the Earth's surface. These techniques are similar in that they generate orders of magnitude more data then surface networks of monitoring stations, but unfortunately, of a much lower precision. Synthetic studies indicate that the increased number of measurements may potentially overcompensate the increased measurement uncertainty (Houweling, in preparation, 2003). The measurements, however, have yet to prove how realistic the assumed uncertainty estimates really are. One should realise that remote sensing of longer-lived greenhouse gases is still at an early stage of development, and more sophisticated instruments have the potential to make this approach much more powerful in the near future.

CO inversion

The processes leading to the emission of CO are fairly well established. CO is a byproduct of fossil fuel use and incomplete biomass combustion. The incomplete oxidation of hydrocarbons also produces substantial amounts of CO (Kanakidou and Crutzen, 1999; Granier et al., 2000). Uncertainties attached to CO sources are however still high. Several studies in the past years have presented global budgets of CO obtained using either bottom-up approach (extrapolation of measurements to build inventories: Olivier et al., 1996) or forward modelling (Logan et al., 1981; Hauglustaine et al., 1998; Kanakidou et al., 1999; Holloway et al., 2000). Because of its relatively short lifetime compared to the interhemispheric transport in the atmosphere, and the spatial and time variability of its sources and sink, CO distribution in the troposphere shows strong gradients (Novelli et al., 1994). The seasonality of CO distribution has been related to the seasonality of its sources and sinks but also to the seasonality of the transport in the atmosphere. Supposing the transport is quite well known, the variations in CO mixing ratio can be used to better estimate the intensity and timing of its sources and sinks.

Since the early 1980's, measurements of CO mixing ratio have been performed regularly at different locations (See Chapter 11). In the last two decades, important efforts have been made to combine the information coming from CO observations together with the state-of-the-art numerical representations of the atmospheric chemistry and transport. This top-down approach is also called source estimation and its general theory is described in section 13.2.

The optimization of CO emissions using the inverse modelling technique is based 1) on a simulation of the tropospheric chemistry and transport, which numerically links the emissions inventory to the CO distribution at each model grid point, 2) on measurements of the CO distribution in the real atmosphere, and 3) on the knowledge of *a priori* values and uncertainties for the emissions and observations errors. This section gives a synthesis of recent studies (indicated in Table 2) on the optimization of CO emissions at the global scale.

Table 2. Previous studies of the optimization of CO sources using inverse modelling techniques

STUDY	MODEL	OBSERVATIONS	TYPE OF INVERSION
Manning et al., 1997	2D GFDL	[CO] and $\delta^{13}C$ at Baring Head monthly median values (June 1989-June 1995)	Time independent inversion (TII) of global sources Climatological study
Bergamaschi et al., 2000a	3D TM2	[CO] at 31 CMDL stations monthly means 1993-1995	TII of global sources, fixed methane oxidation source
Bergamaschi et al., 2000b	3D TM2	Same as above + $\delta^{13}C$ and $\delta^{18}O$ of CO at 5 stations	TII of global sources
Kasibhatla et al., 2002	3D GFDL	[CO] at 38 CMDL stations monthly means 1994	TII of global sources
Pétron et al., 2002	3D IMAGES	[CO] at 39 CMDL stations monthly means 1990-1996	Time dependent inversion (TDI) of monthly regional direct sources and global annual secondary source. Climatological study

To prevent the problem from being under-determined, the number of variables to be estimated is limited: the sources are aggregated in space and time. The first inverse studies of CO emissions focused on optimizing global and annual emissions of various CO sources. Pétron et al. (2002) present the first time-dependent inversion of CO emissions. In Manning et al. (1997) and Bergamaschi et al. (2000b), the isotopic signatures of the various sources are also being optimized. In all studies, *a priori* estimates for the variables are used, otherwise the inverse problem would be ill-constrained and there would be multiple solutions. Gaussian statistics are assigned to both the emissions and the observations. The covariances matrices are diagonal. Pétron et al. (2002) showed that the problem is weakly non-linear (because of the CO + OH chemistry). The iteration of the inversion did not change the results significantly. Even though the OH field is tested or calibrated to match recent estimates of methyl chloroform lifetime in the troposphere, the major sink of CO has not been optimized yet.

The emission estimates resulting from the first 4 studies can be compared as the emission categories used for the inversion are similar. The diagram in Figure 7 shows the *a priori* and *a posteriori* scenarios for (from left to right) Manning et al. (1997), Bergamaschi et al. (2000ab) (*a posteriori* using only [CO], *a posteriori* using [CO] and $\delta^{13}C$) and Kasibhatla et al. (2002). Manning et al. estimates are the lowest except for the source due to terpene oxidation. The total in situ oxidation source (968-821) is closer to Hauglustaine et al. (1998) 881 TgCO/yr. Yet, it is much lower than in the other studies: 1420 in Bergamaschi et al. (2000b), 1376 in Kasibhatla (2002) and 1536 in Pétron et al. (2002). IPCC 2001 (Houghton et al., 2000) recommends 1230 TgCO/yr for year 2000. In Manning et al., most of the constraint on the emissions comes from a remote site in the Southern Hemisphere. This explains that the fossil-fuel/biofuel emissions do not change much as they are not well-sampled at the station. On the contrary, the emissions due to biomass burning increase by 131 TgCO/yr. The 3D studies, which use more stations, should be able to give a better picture of CO global budget. The inversion results still depend however on the variables chosen: global or regional sources. Indeed, the large increase in Asian emissions is derived in both Kasibhatla et al. and Pétron et al.. Yet, there is no such increase in Bergamaschi et al. estimates as the optimization is made on the global scale. Bergamaschi et al. (2000a) showed that the selection of stations does have some impact on the results. Biases for example can be introduced when using stations close to local sources. As CMDL filters local pollution events, similar discrimination should be used with the model outputs. This filtering process will be more accurate as modellers now use assimilated winds to drive the transport in their models.

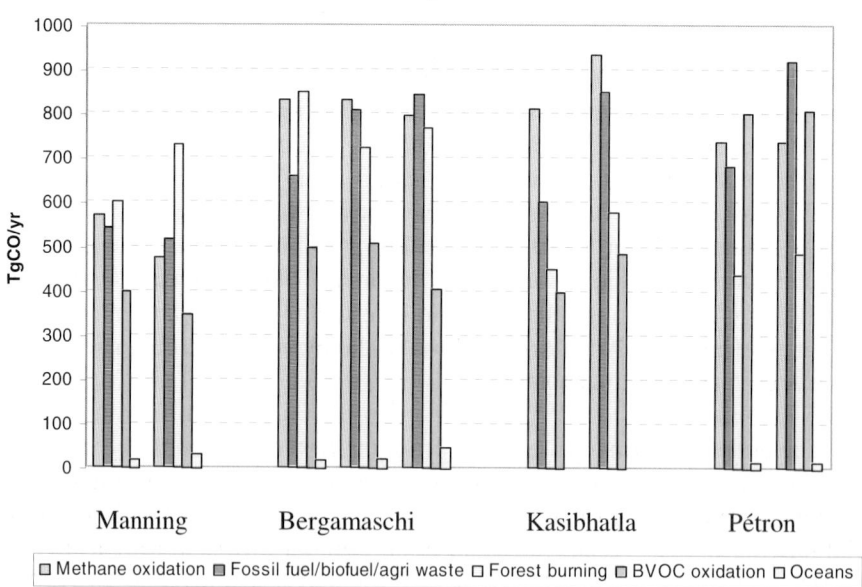

Figure 7. CO global emissions from a priori (left) and a posteriori (right) studies of Manning et al. (1997), Bergamaschi et al. (2000ab) (*a posteriori* using only [CO], *a posteriori* using [CO] and $\delta^{13}C$), Kasibhatla et al. (2002) and Pétron et al. (2002).

The emission categories considered in Pétron et al. are: technological activities, forest and savannah burning, agricultural waste and biofuel use, vegetation/soil emissions, oceanic emissions. These sources are optimized on a monthly and regional basis while the chemical production of CO is kept fixed. The TDI changes the seasonality of biomass burning in Southern Africa and Southern America, the *a priori* of which were taken from Hao and Liu (1994). This result agrees with conclusion from other studies based on remote sensing measurements (Dywer and Gregoire, 1998). Especially for TII of non- averaged observations, it is critical to use emission distributions (in time and space) which are already derived for the target time frame. This is even more important for emissions with high inter-annual variability. In the same way, the impact of the terpene emission distribution (GEIA, Guenther et al., 1995 versus Hough, 1991) used in the inversion by Bergamaschi et al. was quite large. Indeed, the GEIA inventory led to inconsistencies (negative source). As the inversions with isotopic data showed, a source becomes better constrained (methane oxidation with $\delta^{13}C$ and anthropogenic emissions with $\delta^{18}O$) if its isotopic signature is well known.

To reduce the uncertainties on the emissions, Pétron et al. have performed an inversion with a reduced number of variables. Total monthly fluxes over the regions are optimized together with the global and annual chemical production of CO. The *a priori* and *a posteriori* repartitions of CO direct emissions are shown in Table 3.

Table 3. A *priori* and *a posteriori* repartitions of CO direct emissions on a regional basis

Tg CO/year REGION	A priori annual emission flux	A priori uncertainty	A posteriori annual emission flux	A posteriori uncertainty
Europe	162.7	81.3	173.2	23.2
Asia	395.2	197.6	719.5	96.2
Northern Africa	197.4	98.7	226.3	96.4
Southern Africa	98.5	49.2	75.5	46.4
Oceania	31.1	15.5	12.1	12.3
Northern America	192.7	96.3	190.2	32.2
Southern America	202.1	101.0	114.7	90.9
Oceans	16.4	4.0	16.2	2.9
Global surface emissions	1296		1528	
Global chemical production	1373	686.5	1536	41
Total source	2669		3064	

The studies described here have shown that the optimization of CO surface emissions has proved to significantly improve the agreement between modeled and observed CO mixing ratios at surface monitoring stations. The most recent inversions based on *in situ* data tend to converge towards the upper end of the IPCC (Houghton et al., 1995) range for CO total source : 1800-2700 TgCO/yr. Further work is needed however to also constrain CO sink. The use of methylchloroform lifetime may prove to not be an enough accurate proxy of OH field, especially at mid and high latitudes. In the near future, more detailed improvements of surface emissions are expected from the combination of satellite observations (fires, tracer distribution) together with higher resolution models driven by assimilated meteorology.

3.4. Future challenges

The remote sensing of atmospheric chemistry and aerosols is rapidly evolving, from passive remote sensing to active lidar from space that will be launched in a few years, to geostationary platforms that will observe continuously the Earth disk by the end of this decade. There is an urgent need for an integrated approach to best use all the available data in a dynamically and chemically consistent way. Data assimilation is such a technique that uses an atmospheric model to make the links between observations and provide the consistency. In just a few years, chemical data assimilation schemes have been developed from a distinct numerical weather prediction heritage to nearly operational chemical assimilation schemes. Yet, several outstanding problems still remain in several areas of chemical data assimilation. For instance, the correlation between long-lived species is a property that needs to be included in chemical data assimilation schemes. Better and computationally more efficient methods to assure chemical balance are still needed, in order to make a better use of observations of reacting species. Another example is the issue of coupling between meteorology with chemistry. We know, for instance, that tropospheric ozone is greatly influenced by temperatures, in a coupled model the assimilation of ozone will thus create a correction on temperature – is this desirable? But perhaps the most important issues are the need to eliminate the a priori in passive remote soundings for assimilation purposes, and the need for high resolution source estimates. Passive remote sensing instruments do not offer the necessary vertical resolution to be correctly accounted by transport, and they use a priori in their retrieval to obtain a vertical profile (also for column amounts). The a priori may differ considerably from the instantaneous model value, and differs between different instruments, thus introducing additional errors hard to account for and that can jeopardized the effective use of satellite observations in data assimilation. Despite the considerable improvement in modelling the sources, their strength on a finite grid cell, their time and spatial variation still have large modelling uncertainties.

The assimilation and inverse modelling with chemically active species is also very important, since emission estimates of chemically active species are needed to better understand atmospheric chemistry, but also for air quality prediction. Several countries such as the USA, Japan, Canada, and European countries are actively involved in the development of real time air quality prediction that includes chemically actives gases and aerosols. Over the USA and Canada, for instance, nearly 1300 stations take surface ozone measurements each hour and distribute it through the US EPA in near real time. Likewise, there are at the time of this writing about 125 stations that provide fine particle measurements in near real time, and this network is still increasing. Research in data assimilation and inverse modelling indicates that advanced data assimilation techniques such as 4-D Var and the simultaneous state-source estimation are needed to address this issue. One of the very first assimilation experiments of chemically active species was performed with Lagrangian models. Fisher and Lary (1995) showed with the assimilation of UARS data using 4-D Var, that information about unobserved species can be obtained as a result of assimilation. Using the same Lagrangian formulation and similar hypotheses, the assimilation problem was solved using a Kalman filter approach by Khattatov et al. (1999). It was found in particular, that the cross-species error covariance exhibits a diurnal variation and clustering such that the system can be determined by a few linear combinations of the species. With the development of more powerful and distributed memory computers, the development of assimilation schemes that include chemistry and transport processes are well underway. Segers (2002) has developed a reduced rank Kalman filter and in a series of papers (Elbern, 1997, Elbern et al., 1997, Elbern and Schmidt, 1999), Elbern has developed a 4-D Var tropospheric assimilation scheme on distributed memory computers, and was able to show the impact of surface observations in a regional air quality model. Sensitivity to initial condition, sources, and reaction rates were also studied by Menut et al. (2000ab). Source estimation with complex photochemical models were also performed by van Loon et al. (2000) using a reduced rank Kalman filter, and by Elbern et al. (2000) with 4D Var, and with Gilliland and Abitt (2001) using a simpler scheme.

4. REFERENCES

Allen, D., A.R. Douglass, R.B. Rood, and D.P Guthrie, Application of a monotonic upstream-biased transport scheme to three-dimensional constituent transport calculations, Mon. Weather Rev., 119, 2456-2464, 1991.

Bergamaschi P., Hein R., Brenninkmeijer C.A.M. and Crutzen P.J., Inverse modeling of the global CO cycle: 1. Inversion of CO mixing ratios, J. Geophys. Res., 105, 1909-1927, 2000a.

Bergamaschi P., Hein R., Brenninkmeijer C.A.M. and Crutzen P.J., Inverse modeling of the global CO cycle: 2. Inversion of $^{13}C/^{12}C$ and $^{18}O/^{16}O$ isotope ratios, J. Geophys. Res., 105, 1929-1945, 2000b.

Biraud, S., and P. Ciais, M. Ramonet, P. Simmonds, V. Kazan, P. Monfray, S. O'Doherty, and G. Spain, Quantification of carbon dioxide, methane, nitrous oxide and chloroform emissions over Ireland from atmospheric observations at Mace Head. Tellus, Ser. B, 54, 41-60., 2002.

Brasseur, G.P., J.J. Orlando, G.S. Tyndall, Atmospheric Chemistry and Global Change, Oxford University Press, New York, 654 pp, 1999.

Bovensman, H., J.-P. Burrows, M. Buchwitz, J. Frerick, S. Noel, and V.V. Rozanov, SCIAMACHY: Mission objectives and measurement modes. J. Atmos. Sci., 56, 127-150. 1999.

Boville, B.A., and Holton, J.R., and Mote, P.W., Simulation of the Pinatubo aerosol cloud in general circulation model, Geophys. Res. Lett., 18, 2281-2284, 1991.

Burrows, J.P., Current and future passive remote sensing techniques used to determine atmospheric constituents, In Approaches to Scaling of Trace Gas Fluxes in Ecosystems. Ed: A.F. Bouwman, Elsevier, Amsterdam, pp362, 1999.

Carlson, T.N., and J.M. Prospero, The large-scale movement of Saharan air outbreaks over the northern equatorial Atlantic, J. Appl. Meteorol., 11, 283-29, 1972.

Cohn, S.E., Dynamics of short-term univariate forecast error covariances, Mon. Wea. Rev., 121, 3123-3148, 1993.

Coté, J., S. Gravel, A. Méthot, A. Patoine, M. Roch, and A. Staniforth, The operational CMC-MRB Global Environmental Multiscale (GEM) model. Part I: Design considerations and formulation, Mon. Wea. Rev, 126, 1373-1395, 1998.

Daley, R, Atmospheric Data Analysis. Cambridge University Press, New York, pp 457, 1991.

Daniel J.S. ans Solomon S., On the climate forcing of carbon monoxide, J. Geophys. Res., 103, 13,249-13,260, 1998.

Dee, D.P., and A.M. daSilva, Data assimilation in the presence of forecast bias, Q. J. R. Meteorol. Soc., 124, 269-295, 1998.

Dywer E. and Gregoire J.M., A global analysis of vegetation fires using satellites images – spatial and temporal dynamics, Ambio, 27, 175-181, 1998.

Elbern, H., Parallelization and load balancing of a comprehensive atmospheric chemistry transport model, Atmos. Environ., 31, 3561-3575, 1997.

Elbern, H., H. Schmidt, and A. Ebel, Variational data assimilation for tropospheric chemistry modeling, J. Geophys. Res., 102, 15,967-15,985, 1997.

Elbern, H., and H. Schmidt, A four-dimensional variational chemistry data assimilation scheme for Eulerian chemistry transport model, J. Geophys. Res., 104, 18,583-18,598, 1999.

Elbern, H., H. Schmidt, O. Talagrand, and A. Ebel, 4D-Variational data assimilation with an adjoint air quality model for emission analysis, Environ. Mod. Softw., 15, 539-548, 2000.

Eluszkiewicz, J, R.S. Hemler, J.D. Mahlman, L. Bruhwiler, and L.L. Tucaks, Sensitivity of Age-of-Air Calculations to the choice of the advection scheme, J. Atmos. Sci., 57, 3185-3201, 2000.

Enting, I.G., Green's function methods of tracer inversion. In "Inverse methods in Global Biogeochemical Cycles", Eds. Kasibhatla et al., American Geophysical Union Press, Washington DC, 19-31, 2000.

Enting, I.G., Inverse Problems in Atmospheric Constituent Transport, Cambridge University Press, New York, pp 408, 2002.

Feichter, J.E., E. Roeckner, U. Schlese, and M. Windelband, Tracer transport in the Hamburg climate model. In Air Pollution Modeling and Its Application VIII, H. van Dop, Ed., Plenum, pp. 497-506, 1992.

Ferrare, R.A., R.S. Fraser, and Y.J. Kaufman, Satellite remote sensing of large-scale air pollution: Measurements of forest fire smoke, J. Geophys. Res., 95, 9911-9925, 1990.

Fisher, M., and D.J. Lary, Lagrangian four-dimensional variational data assimilation of chemical species, Q. J. R. Meteorol. Soc., 121, 1681-1704, 1995.

Gauthier, P., C. Charette, L. Fillion, P. Koclas, and S. Laroche, Implementation of a 3D variational data assimilation system at the Canadian Meteorological Centre. Part I: The global analysis, Atmosphere Ocean, 37, 103-156, 1999.

Gelb, A., (Ed.), Applied Optimal Estimation. The MIT Press, Cambridge, pp 374, 1989.

Giering, R., Tangent linear and adjoint biogeochemical models. In "Inverse Methods in Global Biogeochemical Cycles", eds. Kasibhatla et al., Geophysical Monograph 114, American Geophysical Union Press, 33-48, 2000.

Gilliland, A., and P.J. Abitt, A sensitivity study of the discrete Kalman filter (DKF) to initial condition discrepancies, J. Geophys. Res., 106, 17,939-17,952, 2001.

Granier C., Pétron G., Müller J.-F., Brasseur G., The impact of natural and anthropogenic hydrocarbons on the tropospheric budget of carbon monoxide, Atm. Env., 34, 5255-5270, 2000.

Godunov, S.K., Finite difference method for numerical computation of discontinuous solutions of the equations of fluid dynamics, Mat. Sb., 47, 271, 1959.

Gottlieb, D. and Orszag, S.A., Numerical Analysis of Spectral Methods: Theory and Applications, Society for Industrial and Applied Mathematics, Philadelphia, Pa, pp 172, 1997.

Guenther A., Hewitt C.N., Erickson D.J., Fall R., Geron C., Graedel T., Harley P., Klinger L., Lerdau M., McKay W., Pierce T., Scholes B., Steinbrecher R., Tallamraju R., Taylor J.,

Zimmermann P., A global model of natural volatile organic compound emissions, J. Geophys. Res., 100, 8873-8892, 1995.

Hao W.M. and Liu M.H., Spatial and temporal distribution of tropical biomass burning, Global Biogeochem. Cycles, 8, 495-503, 1994.

Hauglustaine D.A., Brasseur G.P., Walters S., Rash P.J., Müller J.-F., Emmons L.K. and Carroll M.A., MOZART, a global chemical transport model for ozone and related chemical tracers: 2. Model results and evaluation, J. Geophys. Res., 103, 28,291-28,335, 1998.

Hein, R., and M. Heimann, Determination of global scale emissions of atmospheric methane uswing an inverse modeling method, in Ham, v. J. (eds), Non-CO2 greenhouse gases, Kluwer, The Netherlands, 271-281, 1994.

Holloway, T., H. Levy, and P. Kasibhatla, Global distribution of carbon monoxide, J. Geophys. Res., 105, 12123-12147, 2000.

Houghton J. T., Ding Y., Griggs D.J., Noguer N., van der Linden P.J., Dai X., Maskell K., and Johnson C. A., Climate Change 2001: The Scientific Basis, Cambridge university Press, 2000.

Houghton J.T., Meira Filho L.G., Bruce J., Hoesung Lee, Callander B.A., Haites E., Harris N., Maskell, K., Climate Change 1994: Radiative forcong of Climate Change and An Evaluation of the IPCC IS92 Emission Scenarios, Cambridge university Press, 1995.

Hourdin, F., and Armengaud, A., The use of finite-volume methods for atmospheric advection of trace species. Part I: Test of various formulations in a general circulation model, Mon. Wea. Rev., 127, 822-837, 1999.

Houweling, S., F. Dentener, J. Lelieveld, T. Kaminski, and M. Heimann, Inverse modeling of methane sources and sinks using the adjoint of a global transport model, J. Geophys. Res., 104, 26137-26160, 1999.

Hundsdorfer, W., and Spee, E.J., An Efficient Horizontal Advection Scheme for the Modeling of Global Transport of Constituents, Mon. Wea. Rev., 123, 3554-3564, 1995.

Isaksen I.S.A. and Hov O., Calculations of trends in the tropospheric concentrations of O_3, CO, CH_4 and NOx, Tellus, 39B, 271-283, 1987.

Jacob, D.J., M.J. Prather, S.C. Wofsy, and M.B. McElroy, Atmospheric distribution of 85Kr simulated with a general circulation model, J. Geophys. Res., 92, 6614-6626, 1987.

Jacobson, M.Z., Fundamentals of Atmospheric Modeling. Cambridge University Press, New York, 656 pp, 1999.

Kanakidou M. and Crutzen P.J., The photochemical source of carbon monoxide: Importance, uncertainties and feedbacks, Chemosphere Global Change Science 1, 91-109, 1999.

Kanakidou M., Dentener F.J., Brasseur G.P., Berntsen T.K., Collins W.J., Hauglustaine D.A., Houweling S., Isaksen I.S.A., Krol M., Lawrence M.G., Müller J.-F., Poisson N., Roelofs G.J.,

Wang Y., Wauben W.M.F., 3D global simulations of tropospheric CO distributions – results of the GIM/IGAC intercomparison 1997 exercice, Chemosphere Global Change Science, 1, 263-282, 1999.

Kasibhatla P., Arellano A., Logan J., Palmer P., Novelli P., Top-down estimate of a large source of atmospheric carbon monoxide associated with fuel combustion in Asia, Geophys. Res.Let., in press, 2002.

Khattatov, B.V., J.C. Gille, L.V. Lyjak, G.P. Brasseur, V.L. Dvortsov, A.E. Roche, and J. Waters, Assimilation of photochemically active species and a case analysis of UARS data, J. Geophys. Res., 104, 18,715-18,737, 1999.

Khattatov, B.V., J.-F. Lamarque, L.V. Lyjak, R. Ménard, P. Levelt, X. Tie, G.P. Brasseur, and J.C. Gille, Assimilation of satellite observations on long-lived chemical species in global chemistry transport models, J. Geophys. Res., 105, 29,135-29,144, 2000.

Kidder, S.Q., and T.H. Vonder Haar, Satellite Meteorology: An Introduction. Academic Press, San Diego, pp 466, 1995.

King, M.D., Y.J. Kaufman, D. Tanre, and T. Nakajima, Remote sensing of tropospheric aerosols from space: Past, present, and future. Bull. Am.. Met. Soc., 80, 2229-2259, 1999.

Li, Y., and Chang, J.S., A Mass-Conservative, Positive-Definite, and Efficient Eulerian Advection Scheme in Spherical Geometry and on a Nonuniform Grid System, J. Applied Met., 35, 1897-1913, 1996.

Lin, S.J., and Rood, R.B., Multi-dimensional flux form semi-Lagrangian transport schemes, Mon. Wea. Rev., 124, 2046-2070, 1996.

Logan J.A., Prather M.J., Wofsy S.C., and M.B. McElroy, Tropospheric chemistry: a global perspective, J. Geophys. Res., 86, 7210-7254, 1981.

Lyster, P.M., S.E. Cohn, R. Ménard, L.-P. Chang, S.-J. Lin, and R.G. Olsen, Parallel implementation of a Kalman filter for constituent data assimilation, Mon. Wea. Rev., 125, 1674-1686, 1997.

Machenauer, B., The spectral method. In Numerical Methods Used in Atmospheric Models, vol 2, GARP Publ Ser, 17, World Meteorological Organization, Geneva, 124-277, 1979.

Manning M.R., Brenninkmeijer C.A.M. and Allan W., The atmospheric carbon monoxide budget of the southern hemisphere : Implications of $^{13}C/^{12}C$ measurements, J. Geophys. Res., 102D, 10,673-10,682, 1997.

McDonald, A., A semi-Lagrangian and semi-implicit two time-level integration scheme, Mon. Wea. Rev., 114, 824-830, 1986.

Ménard, R., S.E. Cohn, L.-P. Chang, and P.M. Lyster, Assimilation of stratospheric chemical tracer observations using a Kalman filter: Part I: Formulation, Mon. Wea. Rev., 128, 2654-2671, 2000.

Ménard, R., and L.-P. Chang, Assimilation of stratospheric chemical tracer observations using a Kalman filter: Part II: Chi-squared-validated results and analysis of variance and correlation dynamics, Mon. Wea. Rev., 128, 2672-2686, 2000.

Ménard, R, Tracer assimilation. In "Inverse methods in Global Biogeochemical Cycles", Eds. Kasibhatla et al., American Geophysical Union Press, Washington DC, 67-79, 2000.

Menut, L., R. Vautard, C. Flamant, C. Abonnel, M. Beekmann, P. Chazette, P.H. Flamant, D. Gombert, D. Guedalia, D. Kley, M.P. Lefebvre, B. Lossec, D. Martin, G. Mégie, M. Sicard, P. Perros, and G. Toupance, Measurement and modeling of atmospheric pollution over the Paris area: An overview of the ESQUIF project, Annales Geophysicae, 18, 1467-1481, 2000a.

Menut, L., R. Vautard, M. Beekmann and C. Honoré, Sensitivity of photochemical pollution usig the adjoint of a simplified chemistry-transport model, J. Geophys. Res., 105, 15,379-15,402, 2000b.

Müller, J.F., Geographical distribution and seasonal variation of surface emissions and deposition velocities of atmospheric trace gases, J. Geophys. Res., 97, 3787-3804, 1992.

Novelli P.C., Masarie K., Tans P., and Lang P., Recent changes in atmospheric carbon monoxide, Science, 263, 1587-1590, 1994.

Parrish, D., and J. Derber, The national meteorological center's spectral statistical-interpolation analysis system, Mon. Wea. Rev., 120, 1747-1763, 1992.

Petersen, A.C., E.J. Spee, H. van Dopp, and W. Hundsdorfer, An evaluation and intercomparison of four new advection schemes for use in global chemistry models, J. Geophys. Res., 103, 19,253-19,269, 1998.

Pétron G., Granier C., Khattatov K., Lamarque J.-F., Yudin V., Müller J.-F., Gille J., Inverse modeling of carbon monoxide surface emissions using CMDL network observations, J. Geophys. Res., J. Geophys. Res., 107, D24, 4761, doi:10.1029/2001JD001305, 2002.

Prather, M.J., Numerical advection by conservation of second-order moments, J. Geophys. Res., 91, 6671-6681, 1986.

Rasch, P.J., and Williamson, D.L., Computational aspects of moisture transport in global models of the atmosphere, Q. J. R. Meteorol. Soc., 116, 1071-1090, 1990.

Rasch, P.J., and Boville, B.A., and Brasseur, G.P., A three-dimensional general circulation model with coupled chemistry for the middle atmosphere, J. Geophys. Res., 100, 9041-9071, 1995.

Rasch, P.J., and Mahowald, N.M., and Eaton, B.E., Representations of transport, convection, and the hydrologic cycle in chemical transport models: Implications for the modeling of short-lived and soluble species, J. Geophys. Res., 102, 28,127-28,138, 1997.

Ritchie, H., Semi-Lagrangian advection on a Gaussian grid, Mon. Wea. Rev., 115, 608-619, 1987.

Russel, G.L., and Lerner, J.A., A new finite-differencing scheme for the tracer transport equation, J. Applied Met., 20, 1483-1498, 1981.

Segers, A., Data Assimilation in Atmospheric Chemistry Models using Kalman Filtering. Ph.D. Thesis. University of Delft. [Available from Delft University Press, P.O. Box 98, 2600, MG Delft, The Netherlands], pp 212, 2002.

Smolarkiewicz, P.K., A simple positive definite advection scheme with small implicit diffusion, Mon. Wea. Rev., 111, 479-486, 1983.

Staniforth, A., and Cote, J., Semi-Lagrangian integration schemes for atmospheric models - A review, Mon. Wea. Rev., 119, 2206-2223, 1991.

Strang, G., On the construction and comparison of difference schemes, SIAM J. Numer. Anal., 5, 506-517, 1968.

Stephens, G.L., Remote Sensing of the Lower Atmosphere. Oxford University Press. New York, pp 523, 1994.

Torres, O., P.K. Bhartia, J.R. Herman, Z. Ahmad, and J. Gleason, Derivation of aerosol properties from satellite measurements of backscattered ultraviolet radiation: Theoretical basis. J. Geophys. Res., 103, 17,099-17,110, 1998.

Tremback, C., J, J. Powell, W.R. Cotton, and R. Pielke, The forward-in-time advection transport algorithm. Extensions to higher orders, Mon. Wea. Rev., 115, 540-555, 1987.

Van Leer, B., Towards the ultimate conservative difference schemes, I, The quest of monotonicity, Lect. Notes Phys., 18, 164-168, 1973.

Van Leer, B., Towards the ultimate conservative difference schemes, III, Upstream-centered finite-difference schemes for ideal incompressible flow, J. Comput. Phys., 23, 263-275, 1977.

Van Leer, B., Towards the ultimate conservative difference schemes, V, A second-order sequel to Godunov's method, J. Comput. Phys., 32, 101-136, 1979.

Van Loon, M., P.J.H. Builtjes, and A.J. Segers, Data assimilation of ozone in the atmospheric transport chemistry model LOTOS, Environ. Mod. Softw., 15, 603-609, 2000.

Vermeulen, A.T., R. Eisma, A. Hensen, and J. Slanina, Transport model calculations of NW-European methane emissions, Env. Sci. and Policy, 2, 315-324, 2000.

Wang, K.Y., D.J. Lary, D.E. Shallcross, S.M. Hall, and J.A. Pyle, A review on the use of adjoint method in four-dimensional atmospheric-chemistry data assimilation. Q. J. R. Meteorol. Soc., 127, 2181-2204, 2001.

Williamson, D.L., and Rasch, P.J., Two dimensional semi-Lagrangian transport with shape-preserving interpolation. Mon. Wea. Rev., 117, 102-129, 1989.

List of acronyms

Organizations

CCGG: CMDL Carbon Cycle Greenhouse Gases
CGEIC: Canadian Global Interpretation Center
CLRTAP: Convention on Long Range Transboundary Air Pollution
CMDL: Climate Monitoring and Diagnostics Laboratory
CNES: French National Center for Space Studies
CNRS: French National Center for Scientific Research
CSIRO: Commonwealth Scientific and Industrial Research Organisation
ECMWF: European Centre for Medium-Range Weather Forecasting
EEA: European Environment Agency
EMEP: Co-operative Programme for Monitoring and Evaluation of the Long-range Transmission of Air Pollutants in Europe
EPA: U.S. Environmental Protection Agency
ESA: European Space Agency
FAO: Food and Agriculture Organization (United Nations)
FRSGC: Frontier Research System for Global Change
FSU: Former Soviet Union
GEIA: Global Emissions Inventory Activity
GMCC: NOAA Geophysical Monitoring for Climate Change laboratory
HATS: CMDL Halocarbon and Other Atmospheric Trace Species
IEA: International Energy Agency
IGAC: International Global Atmospheric Chemistry Project
IGBP: International Geosphere-Biosphere Program
IIASA: International Institute for Applied Systems Analysis
INPE: Brazilian National Institute for Space Research
INSTAAR: Institute for Arctic and Alpine Research
IPCC: Intergovernmental Panel on Climate Change
NASA: US National Aeronautics and Space Administration
NASDA: Japanese Space Agency
NCEP: US National Center for Environmental Prediction
NOAA: US National Oceanic and Atmospheric Administration
OECD: Organization for Economic Cooperation and Development
RIVM: National Institute for Public Health and the Environment, The Netherlands
SEPA: State Environmental Protection Administration, China
SU: Soviet Union
TNO: Netherlands Organization for Applied Scientific Research

UEA: University of East Anglia
UN: United Nations
UNDP: United Nations Development Program
UNECE: United Nations Economic Commission for Europe
UNFCCC: United Nations Framework Convention on Climate Change
USDA: United States Department of Agriculture
USEPA: United States Environmental Protection Agency
USGS: United States Geological Survey
WWF: World Wildlife Fund

Satellites or satellite experiments

ADEOS: Advanced Earth Orbiting Satellite
AIRS: Atmospheric Infrared Sounder
ATSR: Along Track Scanning Radiometer sensor
AVHRR: Advanced Very High Resolution Radiometer
BIRD: Bi-spectral infrared detection
CALIPSO-CENA: Cloud Aerosol Lidar and Infrared Pathfinder Satellite Observation
ENVISAT: ENVIronmental SATellite
EOS : Earth Observing System
ERS: European Remote-sensing Satellite
GOES: Geostationary Operational Environmental Satellite
GOME: Global Ozone Monitoring Experiment
IMG: Interferometric Monitor for Greenhouse Gases
ISCCP: International Satellite Cloud Climatology Project
MIPAS: Michelson Interferometer for Passive Atmospheric Sounding
MISR: Multi-Angle Imaging SpectroRadiometer
MLS: Mesospheric Limb Sounder
MODIS: Moderate Resolution Imaging Spectroradiometer
MOPITT: Measurement of Pollution in the Troposphere
OMI: Ozone Monitoring Instrument
PARASOL: Polarization and Anisotropy of Reflectances for Atmospheric Sciences Coupled with Observations from a Lidar
POES: Polar Orbiting Environmental Satellite
POLDER: Polarization and Directionality of the Earth's Reflectance
SAGE: Stratospheric Aerosol and Gas Experiment
SAM: Stratospheric Aerosol Measurement
SCIAMACHY: SCanning Imaging Absorption spectroMeter for

List of Acronyms

Atmospheric CHartographY
SEAWIFS: SEA viewing Wide FIeld-of-view Sensor
SPOT: Système pour l'Observation de la Terre
TES: Tropospheric Emission Spectrometer
TOMS: Total Ozone Mapping Spectrometer
TOVS: TIROS Operational Vertical Sounder
UARS: Upper Atmosphere Research Satellite

Other

2-D: two-dimensional
3-D: three-dimensional
AFEAS: Alternative Fluorocarbon Environmental Acceptability Study
AGA: ALE/GAGE/AGAGE
AGAGE: Advanced GAGE network
AI: Aerosol Index
ALE: Atmospheric Lifetime Experiment
AMVER: Automated Mutual Assistance Vessel Rescue System
AOD: Aerosol Optical Depth
ATHAM: Active Tracer High resolution Atmospheric Model
BIBEX: Biomass burning Experiment
CAA: Clean Air Act, USA
CAT: Coastal Aerosol Transport model
CCN: Cloud Condensation Nuclei
CEC: Cation Exchange Capacity
CEM: Continuous Emissions Monitoring
Chl: Chlorophyll
COADS: Comprehensive Atmosphere Ocean Database
CORINAIR: CORINE AIR emissions inventory
CORINE: Coordination d'Information Environnementale
COSPEC: Correlation Spectrometer
CPU: Central Processing Unit
CRF : Canopy Reduction Factor
CTM: Chemistry-Transport Model
DNDC: Denitrification-decomposition model
DOC: Dissolved Organic Carbon
DOE-CDIAC: Department of Energy-Carbon Dioxide Information Analysis Center

DOT: Dust Optical Thickness
DVI: Dust Veil Index
ECD: Electron Capture Detection
EDGAR: Emission Database for Global Atmospheric Research
EF: Emission Factor
EUROTRAC: European Experiment on Transport and Transformation of Environmentally Relevant Constituents
EXPRESSO: EXPeriment for Regional Sources and Sinks of Oxidants
FID: Flame Ionization
FIS: Fast Isoprene Sensor
FPAR: Fractional Photosynthetically Active Radiation
FTIR: Fourier Transform Infrared Spectroscopy
FWI: Fire Weather Index
GAGE: Global Atmospheric Gases Experiment
GC: Gas chromatography
GCM: General Circulation Model
GENEMIS: GENeration and Evaluation of EMISsion data
GIS: Geographical Information System
HAP: Hazardous Air Pollutant
HYDE: History Database of the Global Environment
IAE: Indirect Aerosol Effect
IDDI: Infrared Difference Dust Index
ISCCP: International Satellite Cloud Climatology Project
IVI: Ice core Volcanic Index
KIE: Kinetic Isotope effect
LAI: Leaf Area Index
LBA: Large-scale Biosphere Atmosphere Experiment
LFSM: Landscape-fire-succession model
NAPAP: U.S. National Acid Precipitation Assessment Program
NDIR: Non-Dispersive Infrared Analysis
NEI: US National Emission Inventory
NTI: National Toxics Inventory
nM: $nmol\ l^{-1}$
NPP: Net Primary Production
PAR: Photosynthetically Active Radiation
PBL: Planetary Boundary Layer
PFT: Plant Functional Types
PID: PhotoIonization Detector
PM: Particulate Matter

List of Acronyms

POP: Persistent Organic Pollutants
PTR: Proton Transfer Reaction
QBO: Quasi Biennial Oscillation
RAINS: Regional Air Pollution Information and Simulation Project
RGD: Reduction Gas Detectors
SNAP: Selected Nomenclature for Air Pollution (UNECE/CORINAIR)
SOM: Soil Organic Matter
TCD: Thermal Conductivity Detection
TDLS: Tunable Diode Laser Spectroscopy
Tgdm: Teragrams of dry matter
THC: Total hydrocarbon
TPM: Total Particulate Matter
TSP: Total Suspended Particulates
UV: ultra violet
VAI: Volcanic Aerosol Index
VEI: Volcanic Explosivity Index
VSI: Volcanic SO_2 index
WFPS: Water Filled Pore Space

List of chemical species

BC: Black carbon
BVOC: Biogenic Volatile Organic Compound
$CBrClF_2$: Halon-1211
$CBrF3$: Halon-1301
CCl_4: carbon tetrachloride
CH3Br: methyl bromide
CH3Cl: methyl chloride
CH3CFCl2: HCFC-141b
CH3CF2Cl: HCFC-142b
CH_4: methane
C_2H_4: ethylene (ethene)
C_2H_6: ethane
C_3H_6: propylene
C_3H_8: propane
C_5H_8: isoprene
$C_{10}H_{16}$: terpenes (e.g., α-pinene)
CFCl3: CFC-11
CFCs: chlorofluorocarbons
CF2Cl2: CFC-12
CF2ClCFCl2: CFC-113
CF3CH2F: HFC-134a
$CHCl_3$: chloroform
CH_3CCl_3: methylchloroform
CHCl3: chloroform
CHF2Cl: HCFC-22
CO: carbon monoxide
CO_2: carbon dioxide
DMS: dimethylsulfide (CH_3SCH_3)
DMSO: dimethylsulfoxide (CH_3SOCH_3)
DMSP: dimethylsulfoniopropionate
H2: hydrogen
HCHO: formaldehyde
HCFC: hydrochlorofluorocarbons
HFC: Hydrofluorocarbons
HO_2: hydroperoxy radical
MBO: 2-methyl-3-buten-2-ol
N_2O: nitrous oxide
NH_3: ammonia

NO: nitrogen monoxide (nitric oxide)
NO_2: nitrogen dioxide
NO_x: Nitrogen oxides (NO + NO_2)
NOy: odd nitrogen
O_3 : ozone
OC: Organic carbon
OCS: carbonyl sulphide
OH: hydroxyl radical
PAN: PeroxyAcetyl Nitrate (peroxyacetic nitric anhydride)
Pb: lead
PFC: Perfluorocarbons
$PM_{2.5}$: Particulate Matter with diameter < 2.5 μm
PM_{10}: Particulate Matter with diameter < 10 μm
PPN: peroxypropionic nitric anhydride
SF_6: sulphur hexafluoride
VOC: Volatile Organic Compound

Index

activity data 22, 30, 39, 42, 52
activity rate 18, 24, 59
aerosol index 249, 251, 256, 445
aerosol optical depth 91, 262, 333
agricultural fires 72, 76, 87, 94, 99, 104
ammonia 37, 171, 180, 182, 189
advective transport 6, 324, 451, 479, 481, 485
air-sea exchange 305, 321
ash 270, 273, 283
atmospheric chemistry 14, 75, 123, 146, 158, 172, 213, 279, 334, 361, 386, 428
atmospheric models 6, 8, 14, 27, 39, 115, 131, 305, 448, 478, 483
biofuel 35, 51, 73, 79, 87, 94, 99
biogenic aerosols 116, 153, 156
biomass burning 9, 11, 31, 48, 71, 76, 94, 175, 250, 253, 283, 365, 374, 381, 387, 393, 500
biomass density 77, 86, 100, 104
bottom-up approach 18, 28, 52, 76, 80, 102, 458, 469, 503
burnt biomass 76, 80, 86, 95
canopy models 207
carbon monoxide isotope 404
carbonyls 149
charcoal 72, 81, 88, 99
chemodenitrification 171 189 195 204
chlorophyll 116, 137, 239, 306, 309, 312, 317, 327
cloud condensation nuclei 2, 153, 333
CMDL network 102, 430, 435, 437, 440, 452, 454, 458, 463
CO isotopes 384
convective transport 4, 6, 101, 324, 479
cosmic rays 385
data assimilation 477, 484, 487, 508
deforestation 35, 74, 175
denitrification 8, 171, 177, 187, 192, 195, 204, 214, 398, 404
dry deposition 4, 7, 171, 175, 191, 199, 207, 218, 251, 278
ecoregions 121, 124, 145
EDGAR inventory 22, 29, 34, 41, 56, 75, 87, 324, 340, 347, 469
EMEP inventory 19, 37, 41, 376, 379
emission factor 9, 18, 22, 30, 42, 53, 56, 72, 80, 86, 99, 101, 116, 133, 142, 148, 151, 243, 249
emission model 14, 100, 116, 120, 129, 146, 211, 219
emission ratio 81, 467
EPA inventory 43, 468
eruption column 276, 285

fire pixels 94, 103
flaming phase 81, 101, 367
foliar density 127
FPAR 131
friction velocity 241, 243, 248
greenhouse effect 2, 284
halocarbons 13, 436, 440, 459, 466
ice cores 13, 274, 279, 288, 458
indirect aerosol effect 333
injection height 73, 88, 90, 101
inverse methods 102, 427, 458, 477, 484, 498, 500, 598
isoprene 117, 122, 129, 132, 136, 141, 147
kinetic isotope effect 368, 387, 401
land use 26, 122, 158, 171, 193, 212, 219
leaf area index 127, 158, 217
methane isotope 362
mineralization 182, 184, 194, 205
monoterpenes 8, 117, 122, 129, 138, 147, 154
N-fixation 182, 183, 204
nitrification 8, 171, 177, 182, 185, 192, 195, 204, 214, 397, 404, 410
nitrous oxide isotopes 396
NSS-sulfate 325, 326, 327
PAR 129, 133, 137, 143
phenology 128, 133
plant functional types 118, 121
radiative forcing 71, 78, 283, 287, 333, 498
RAINS inventory 22, 48
roughness height 240, 241, 262
satellite observations 27, 75, 87, 94, 97, 104, 121, 131, 239, 244, 260, 273, 290, 307, 320, 328, 477, 507
secondary organic aerosols 116, 153, 155
shipping 22, 27, 31
shrubs 86, 127, 144
smoldering phase 81, 85, 101, 367
soil erosion 246, 255
source sector 20, 24
sulphate aerosols 9, 272, 279, 282, 287, 305, 445, 457
top-down approach 18, 52, 102, 173, 411, 427, 451, 459, 469, 504
tree inventories 124, 143
volatilization 174, 182, 189, 193, 199, 215
volcanic indices 290
volcanism 283, 289
whitecap cover 334, 340, 344

Color Plates

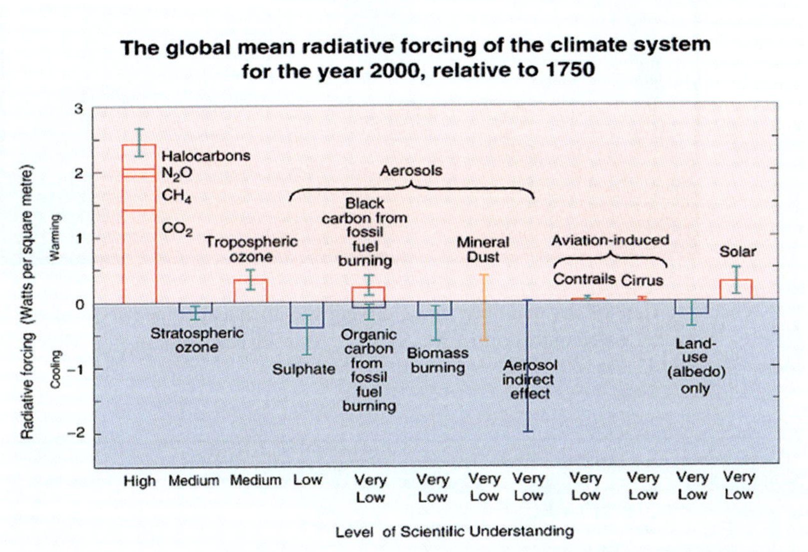

Plate 1: Global, annual mean radiative forcing (Wm^{-2}) for the period from pre-industrial (1750) to present (IPCC, 2001).

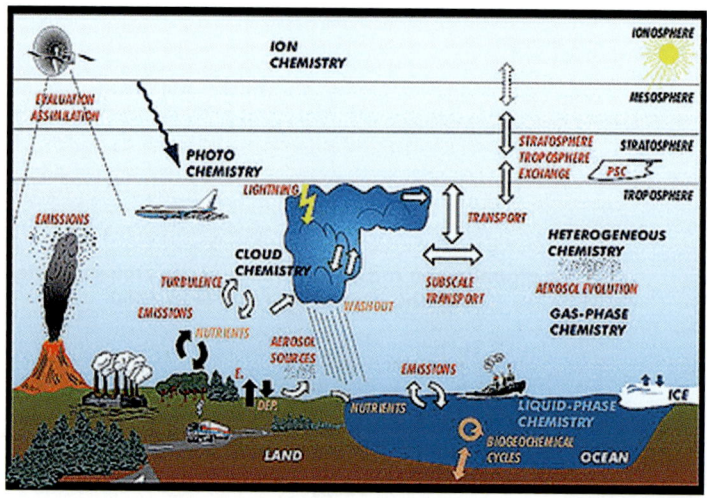

Plate 2: schematic view of the main processes driving the chemical composition of the lower atmosphere. (Personal communication of M. Schultz, Max Planck Institute for Meteorology, Hamburg, Germany.)

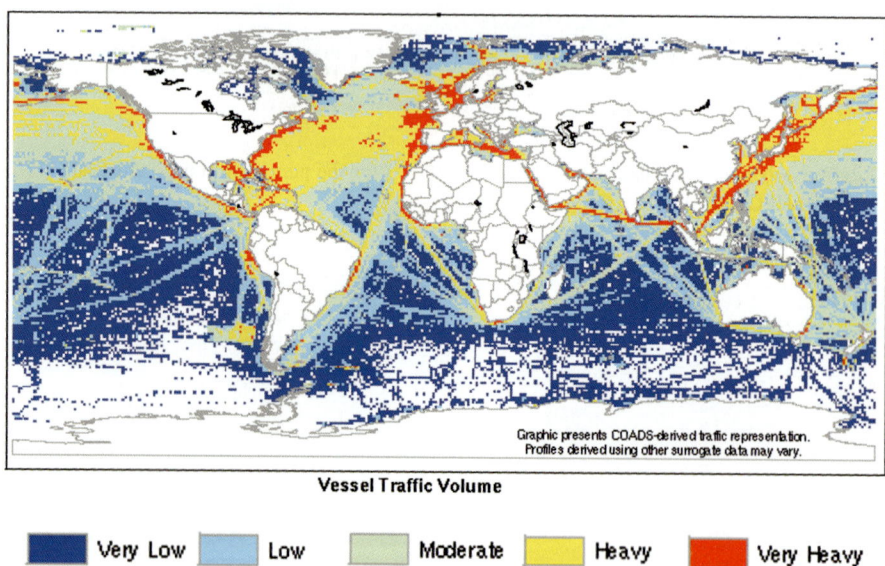

Plate 3. International ship traffic for 1990 (adapted from Corbett et al., 1999 and Skjølsvik et al. 2000)

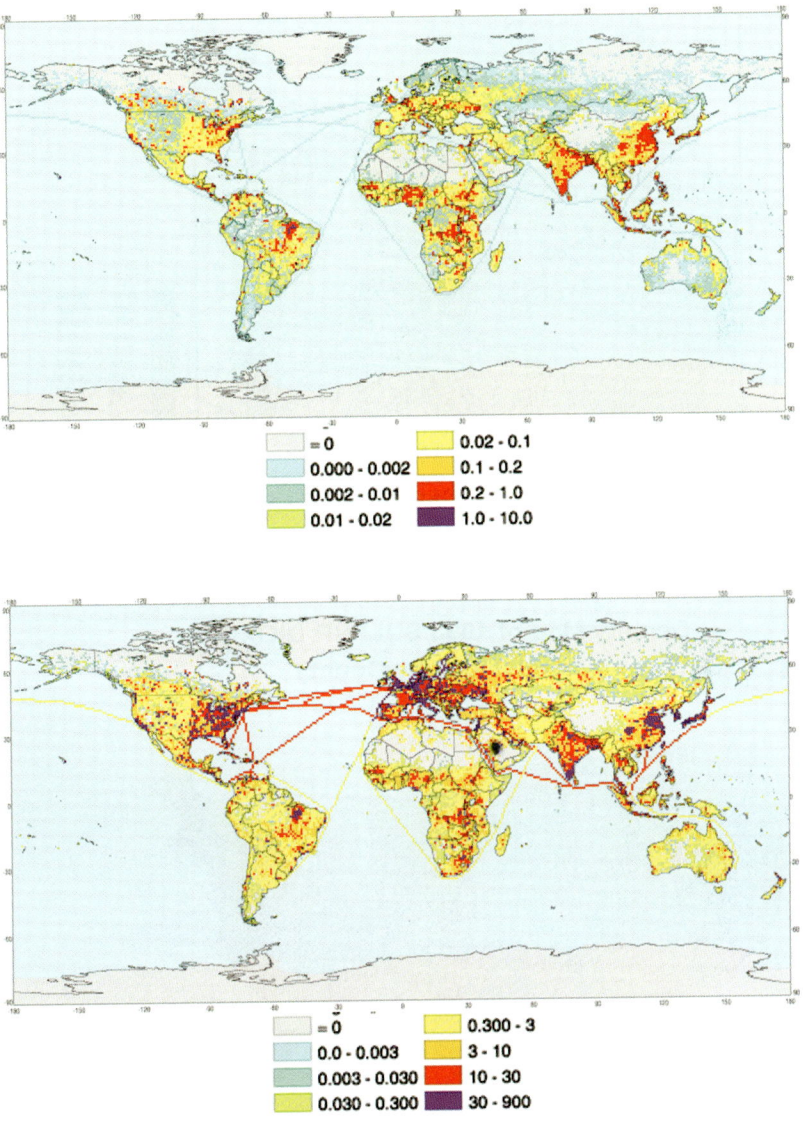

Plate 4: Global distribution of NO_x (top) and CO (bottom) anthropogenic emissions in 1995. Source: EDGAR 3.2

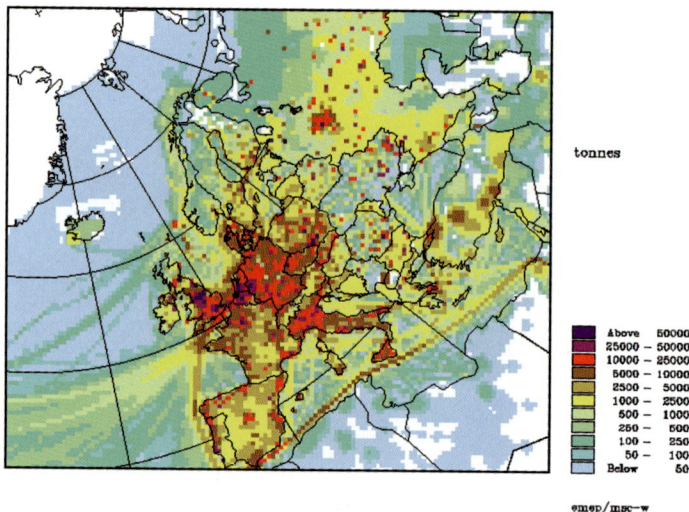

Plate 5. European emissions of NO_x in 1995 at 50 km grid resolution (Mg as NO_2)

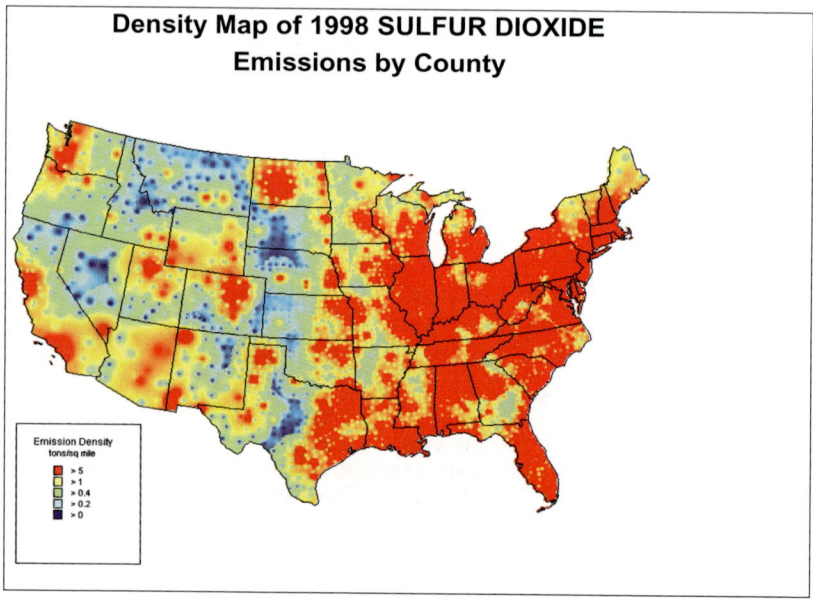

Plate 6. Emissions of SO_2 in the United Stated in 1998 (short tons/mile^2). From EPA web site http://www.epa.gov/ttn/chief/trends.

Color Plates

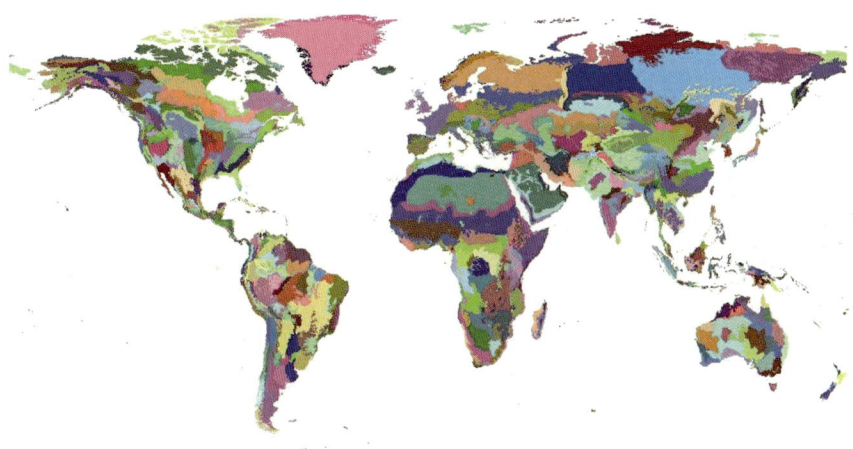

Plate 7. The global distribution of ecoregions as assigned by the World Wildlife Fund ecoregion scheme. Each color represents a different ecoregion (over 850 ecoregions are assigned to the global land area) (Based on Olson et al., 2001). For more information, visit http://www.worldwildlife.org/ecoregions.

Eastern U.S.	Western U.S.	Eastern Canada	Western Canada
Pinus taeda	*Pseudotsuga menziesii*	***Populus tremuloides***	***Picea spp***
Acer rubrum	***Pinus ponderosa***	***Picea spp***	***Populus tremuloides***
Quercus alba	*Juniperus osteosperma*	*Abies spp*	*Pinus banksiana*
Liquidambar styraciflua	***Pinus contorta***	*Pinus banksiana*	*Abies spp*
Acer saccharum	*Tsuga heterophylla*	*Thuja occidentalis*	*Tsuga spp*
Quercus rubra	*Abies concolor*		
Pinus elliottii	***Picea engelmannii***	Northern Mexico	Southern Mexico
Liriodendron tulipifera	*Abies grandis*	*Pinus durangensis*	***Quercus resinoa***
Populus tremuloides	*Pinus edulis*	*Pinus arizonica*	*Pinus oocarpa*
Quercus virginiana	*Abies lasiocarpa*	***Quercus* spp**	*Acacia* spp

Plate 8. Dominant tree species by region in North America. Bold indicates high (> 10 μg C g^{-1} h^{-1}) VOC emission rates at standard (leaf temperature of 30°C and PAR of 1000 μmol m^{-2} s^{-1}) conditions. Red indicates species adapted to warm sunny climates, green indicates temperate adapted species, and blue denotes species found in cool, or montane, climates.

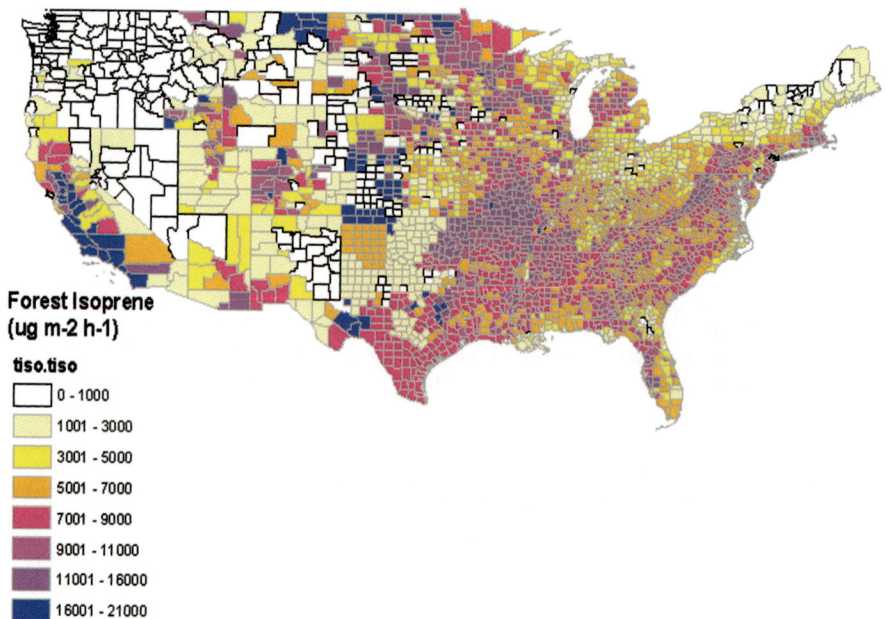

Plate 9. County-level isoprene emission rates ($\mu g\ m^{-2}\ h^{-1}$) for midday summertime conditions (above-canopy 30°C and 1500 $\mu mol\ m^{-2}\ s^{-1}$). The values have been averaged over the total forest land area (not over total land area) for each county in the United States (based on Guenther et al., 2000).

Color Plates

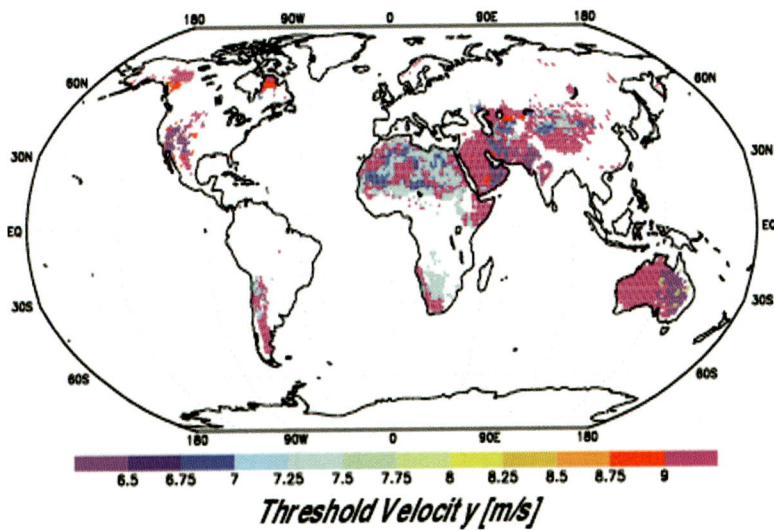

Plate 10. Global map of threshold velocities over arid and semi-arid areas. Soil types were used to infer threshold velocities representative of the FAO soil types. The global FAO map was then used to affect the threshold velocities to soil types for other areas over the world.

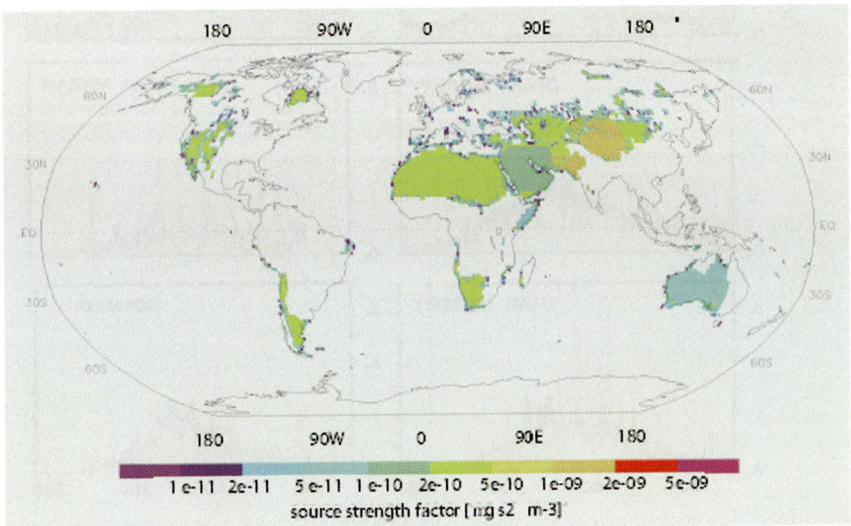

Plate 11. Global distribution of source strength factors (kg m^{-5} s^2) (C' in equation 3) derived by calibrating the model derived optical depth to the optical depth corrected from the TOMS aerosol index. The comparison was made for 1990 and these coefficients are weighted by the fraction of erodible land of each model grid box.

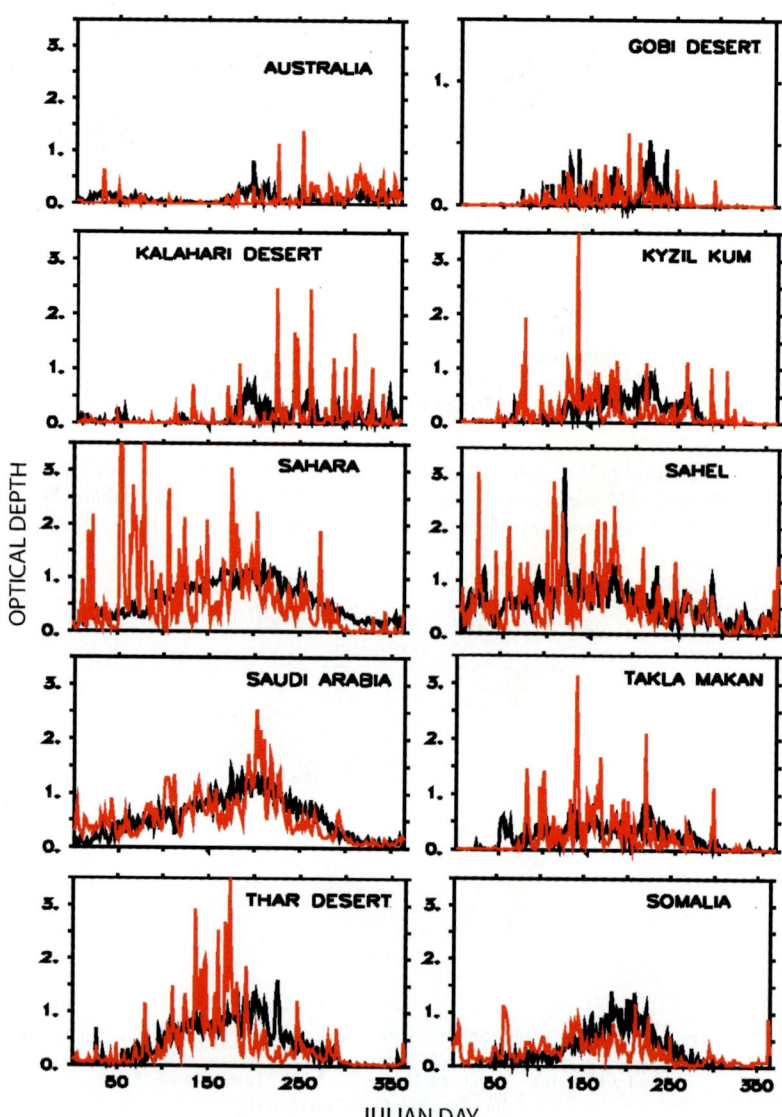

Plate 12. Time series of optical depth over the areas listed in Table 3 for 10 of the 12 deserts. The black line represents TOMS-derived optical depth corrected to account for wavelength and the altitude dependence of TOMS retrieved signal. The red line is the modelled optical depth. No cloud screening was applied on the model results.

Color Plates

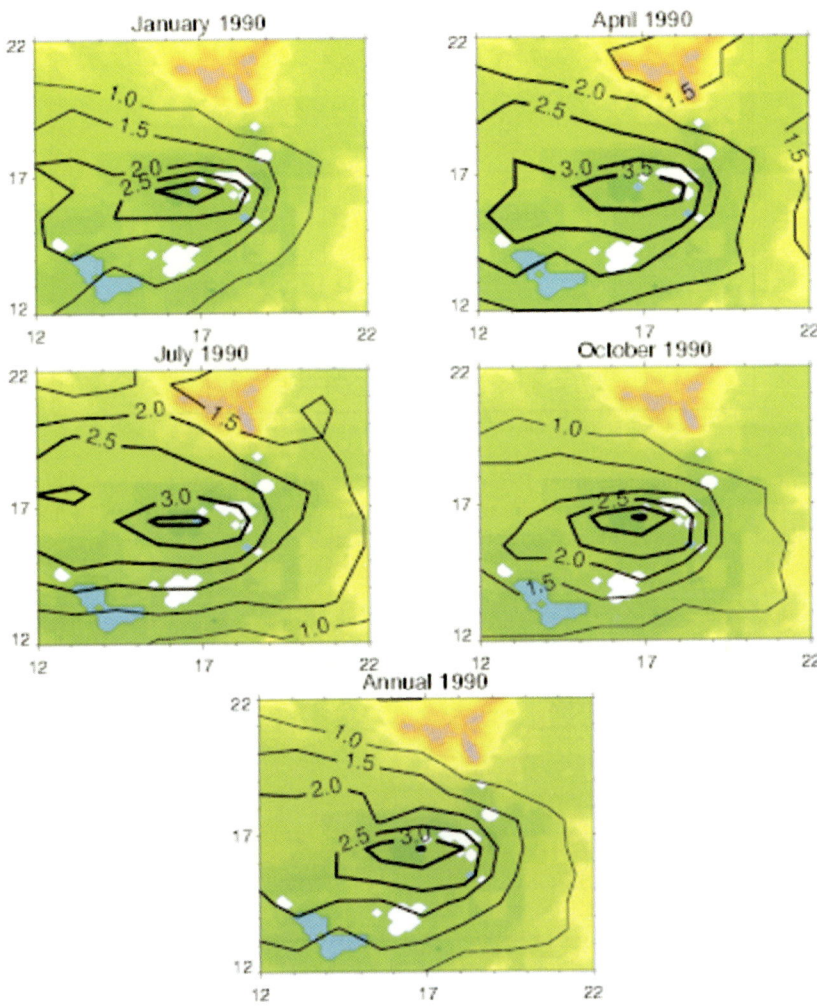

Plate 13. TOMS Aerosol distribution for 1990 (bold lines) and the orography of the Bodele Depression which lies south of the Tibesti and west of the Ennedi mountains. The blue area are lakes and the white spots are dry lakes or salt pans.

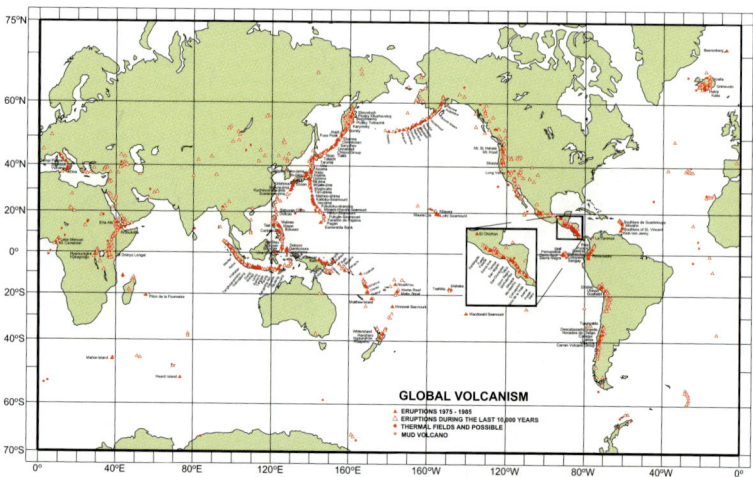

Plate 14: Active volcanoes from 1975-1985 (solid triangles) and sites with volcanic activity during the last 10,000 years (open triangles). Taken from Graf et al. (1997), based on McClelland et al. (1989), adapted by permission of Prentice Hall.

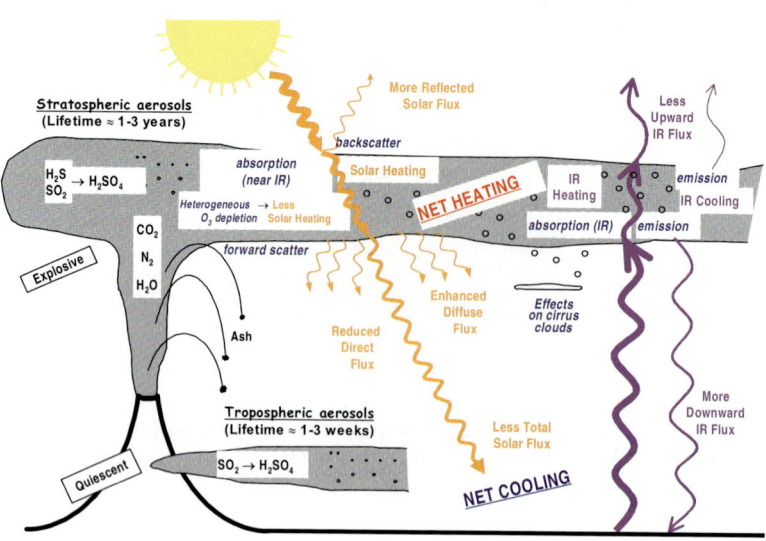

Plate 15: Schematic diagram of volcanic inputs to the atmosphere and their effects.

Reproduced with permission from Alan Robock (2000).

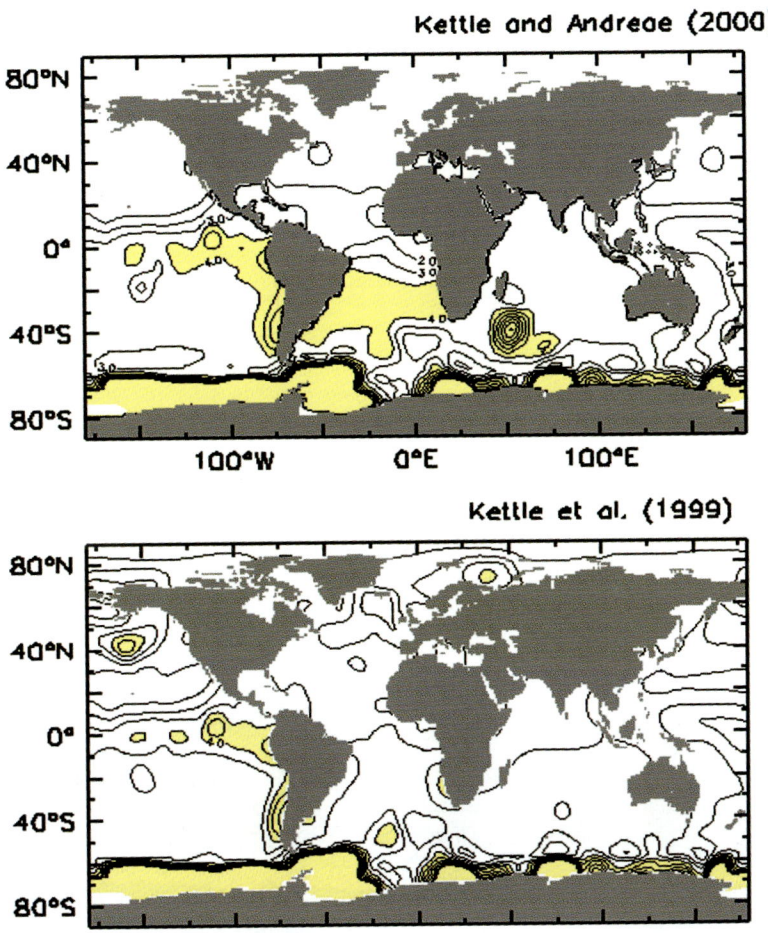

Plate 16. Comparison of predicted oceanic DMS concentrations from the work of Kettle and Andreae, 2000 (upper panel) and Kettle et al., 1999 (lower panel) for the month of December. Shaded areas denote regions where DMS is higher than 4 nM. Isolines are every 1 from 1 to 10 nM.

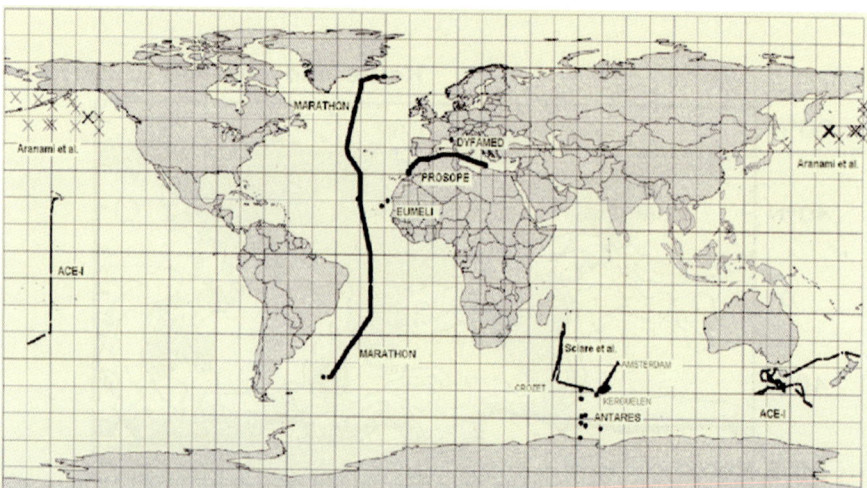

Plate 17. Geographical coordinates of seawater samples used to establish the parameterizations of equations 1-3 (thick dots and lines: Mediterranean Sea (PROSOPE and DYFAMED projects), Atlantic Ocean (EUMELI and MARATHON projects) and Indian Sector of the Austral Ocean (ANTARES project)). Observations used to evaluate predicted sea surface DMSP and DMS levels (crosses and thin dots) are taken from the North Pacific Ocean (Aranami et al., 2001), the Central and South Pacific Ocean (cruise ACE-I, Bates et al., 1998), and the Indian Sector of the Austral Ocean (Sciare et al., 1999).

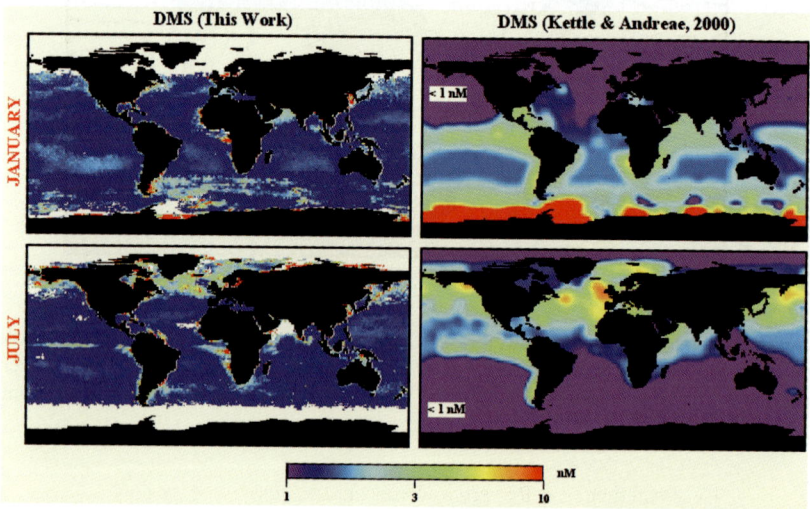

Plate 18. Fields of sea surface DMS concentration (nM) for January (upper panels) and July (lower panels). This work (left column) and Kettle and Andreae (2000) (right column).

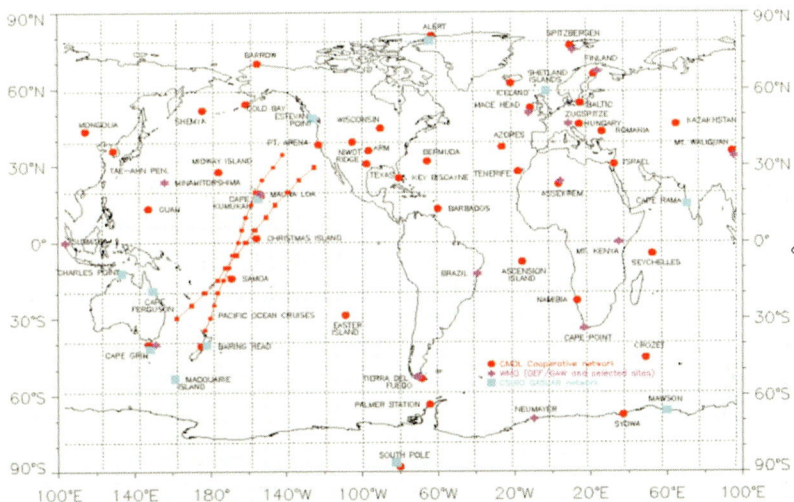

Plate 19. The distribution of 'global monitoring' sites in 2003. Large circles indicate sites that comprise the NOAA/CMDL Carbon Cycle Group Cooperative Air Sampling Network (the small connected symbols show the shipboard sampling tracts). Squares show the CSIRO sampling sites, and stars indicate sites in the WMO GAW. Several locations are used by two or more networks.

Plate 20: Species observed by current and planned satellite experiments

Sensor	TOMS	GOME	IMG	MOPITT	SAGE III	SCIA-MACHY	MIPAS	AIRS	TES	MLS	HRDLS	OMI
Launch		04/95	10/96	12/99	12/01	02/02	02/02	05/02	01/04	01/04	01/04	01/04
O₃					>6 km	T	>8 km	P	P	>8 km	>7 km	T
OClO												
CO				P		T			P	>8 km		
CO₂			P			T						
CH₄						T	>8 km				>7 km	
NO									>5 km			
NO₂						T			>5 km			
N₂O						T	>8 km				>7 km	
H₂O		P				T	>8 km	P	P	>5 km	>7 km	
HNO₃							>8 km		P			
HCOH												
HCN										>8 km		
BrO												
SO₂												
CFC-11											>7 km	
CFC-12											>7 km	

Near-real time   P : vertical profile information
Research mode T : tropospheric column by combining nadir and limb soundings
Under special conditions only. Profile information from limb sounders (e.g. > 8 km)

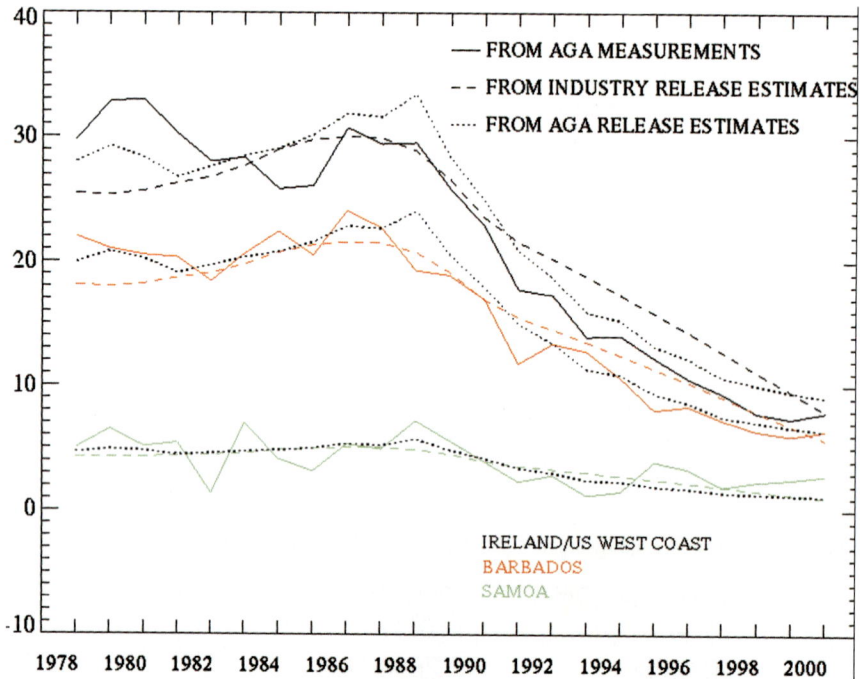

Plate 21. ALE/GAGE/AGAGE measured differences between annual mean CFC-12 mole fractions in Ireland, Barbados and Samoa (full lines, top to bottom) and those in Tasmania between 1978 and 2000. These are compared against the differences calculated with the 12-box 2-D model (Cunnold et al., 1994) using the emissions estimated by McCulloch et al. (2003) (dashed lines), emissions estimated from the AGA measurements (dotted line), and a lifetime of 100 years.

Color Plates 541

Plate 22. Smoothed, zonally-averaged representation of the global distribution of CH_4 in the boundary layer for 1992-2001. Grid spacing is 10° latitude by 1 week. It is constructed from curves fitted to 43 sites in the CMDL cooperative air sampling network.

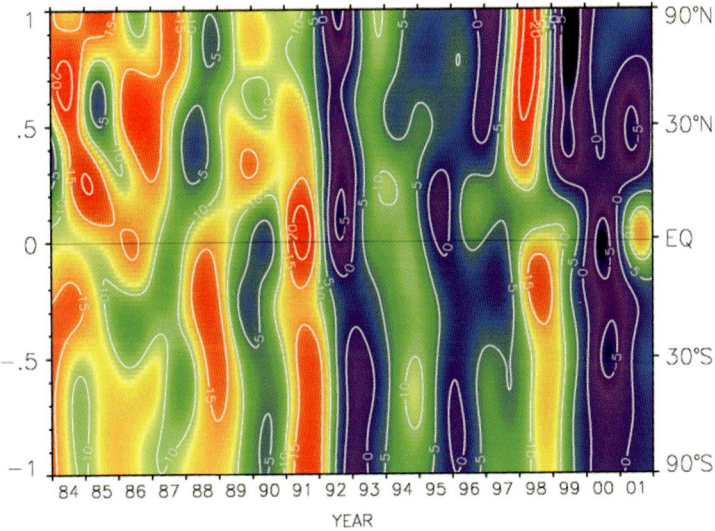

Plate 23. Contour plot showing spatial and temporal variations in the growth rate of atmospheric CH_4 for 1984-2001. Cool colors (blue, violet, and black) represent regions and periods of low growth rate; warm colors (yellow, orange, and red) identify high growth rate.

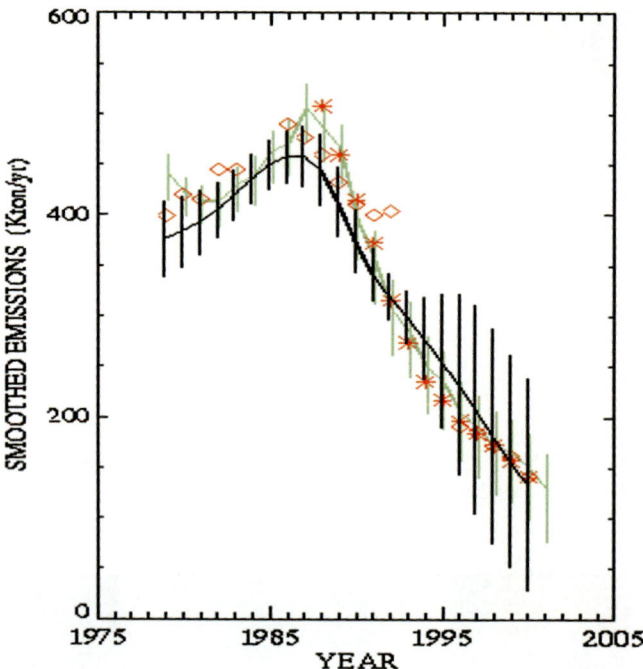

Plate 24. Industry estimates of the annual releases of CFC-12 (full black lines) (McCulloch et al., 2003). The green lines indicate release estimates based on atmospheric measurements by ALE/GAGE/AGAGE using lifetimes of 100 years with the effect of lifetime uncertainties, associated with an ensemble of model calculations reported in WMO (1999), indicated by the green error bars. The red diamonds show emission estimates derived using the same AGAGE model applied to CMDL flask sample measurements and the asterisks are based on CMDL RITS measurements. All the emission estimates have been smoothed over three year intervals in order to reduce the effects of measurement imprecisions.

Color Plates 543

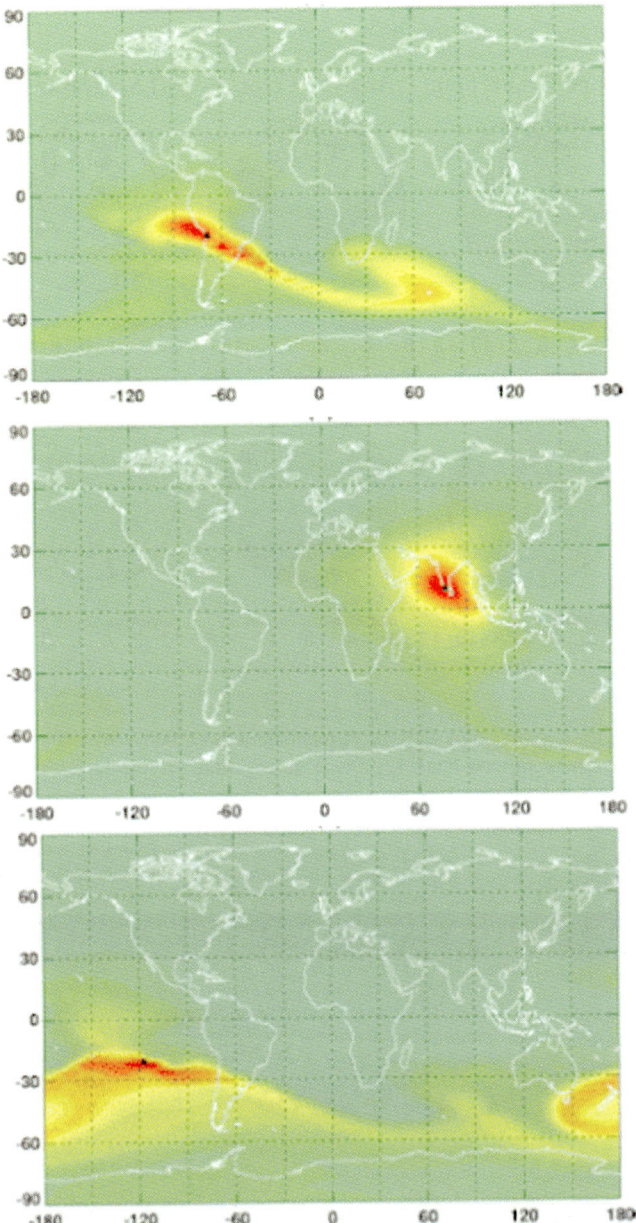

Plate 25. Spatial error correlation structure at three HALOE data points (Reproduced from Ménard and Chang, 2000).

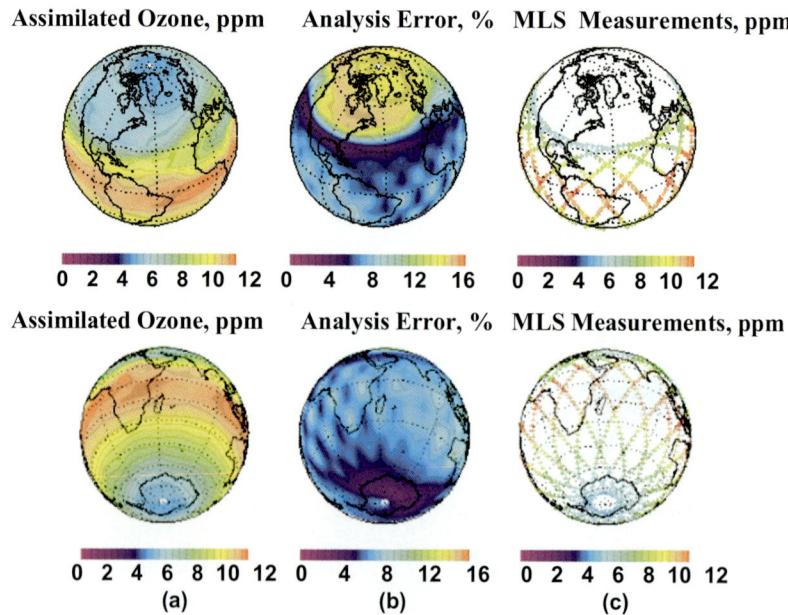

Plate 26. Result of assimilation with the sequential method. Panel (a) analysed ozone, Panel (b) analysis error in percent, and Panel (c) ozone observations (Reproduced from Khattatov et al., 2000)

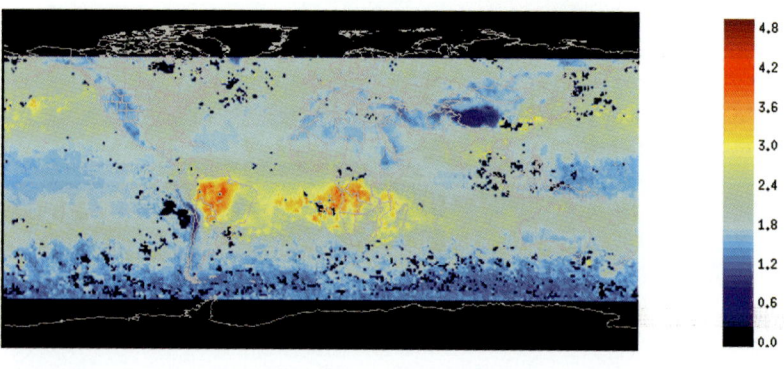

Plate 27. MOPITT CO observations in molecules cm$^{-2} \times 10^{+18}$

Color Plates

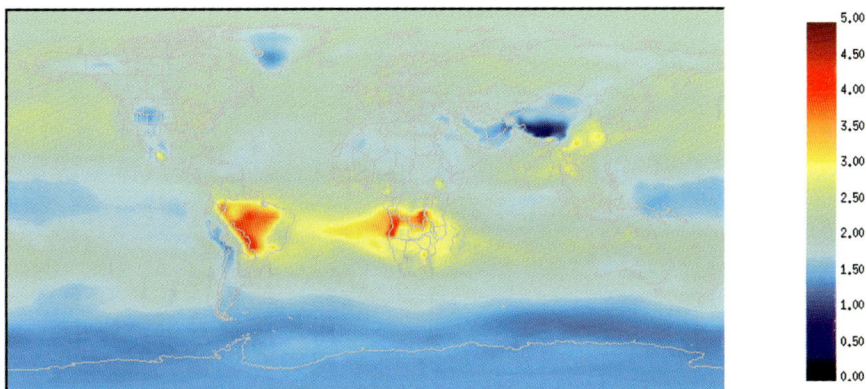

Plate 28. Assimilation of CO. Time average of total column amount depicted.

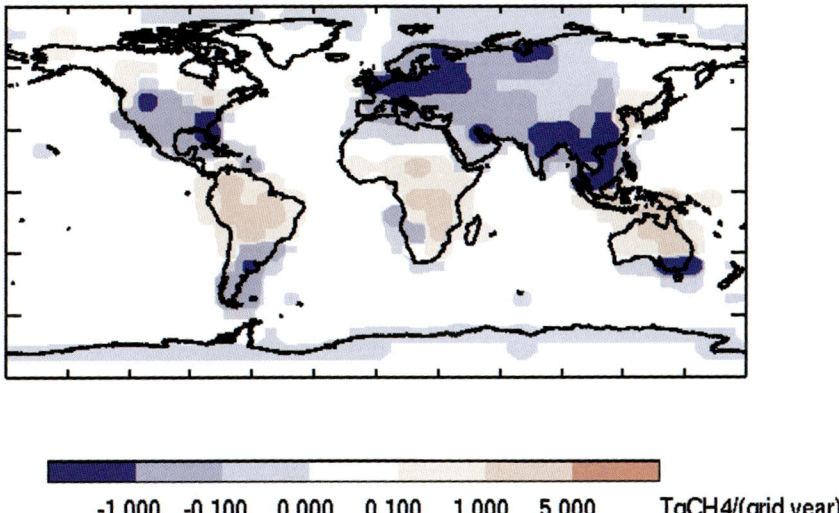

Plate 29. A comparison of inversion derived estimates of climatological CH_4 sources and sinks (a posteriori minus a priori on a 8x10 degree grid)

Copyright Acknowledgments

We gratefully acknowledge the following copyright holders, who have kindly provided permission to reproduce the figures indicated. Sources of figures are given in the *References* section of each chapter.

Chapter 1: Fig.1, from IPCC (2001), by permission of Intergovernmental Panel on Climate Change.
Chapter 2: Fig.5, reproduced with permission of Norwegian Meteorological Institute/EMEP/MSC-W
Chapter 10, Fig. 6 and Chapter 11, Fig. 11: reproduced with permission from Elsevier; Fig.7: reproduced with permission from John Wiley & Sons Limited.
Copyright American Geophysical Union: Chapter 3, Fig. 6; Chapter 4: Figs. 3 and 4; Chapter 5: Figs. 1, 4 and 5; Chapter 6, Fig.2; Chapter 7, Fig. 2; Chapter 8: Figs. 1 and 4; Chapter 10: Fig. 3; Chapter 11, Figs. 8 and 12; Chapter 12: Figs. 2, 3 and 4.